REVIEWS IN MINERALOGY & GEOCHEMISTRY

VOLUME 51   2002

PLASTIC DEFORMATION
OF MINERALS AND ROCKS

Editors:

Shun-ichiro Karato
      Yale University, New Haven, Connecticut

Hans-Rudolf Wenk
      University of California, Berkeley, California

COVER:

Upper left: Results of a computer simulation of dislocation motion in olivine (courtesy of J. Durinck, University of Lille, France).

Upper right: A digital orientation mapping of experimentally deformed olivine constructed from electron back-scattered diffraction patterns.

Lower left: A TEM micrograph of wadsleyite deformed under mantle transition zone conditions showing evidence of dissociation (courtesy of Patrick Cordier).

Lower right: An optical micrograph (polarized light) of naturally deformed peridotite from the Ivrea zone (Italy) showing evidence of grain-size reduction along grain-boundaries due to dynamic recrystallization.

Series Editor: **Paul H. Ribbe**
Virginia Polytechnic Institute and State University
Blacksburg, Virginia

MINERALOGICAL SOCIETY of AMERICA
Washington, DC

COPYRIGHT 2002

MINERALOGICAL SOCIETY OF AMERICA

The appearance of the code at the bottom of the first page of each chapter in this volume indicates the copyright owner's consent that copies of the article can be made for personal use or internal use or for the personal use or internal use of specific clients, provided the original publication is cited. The consent is given on the condition, however, that the copier pay the stated per-copy fee through the Copyright Clearance Center, Inc. for copying beyond that permitted by Sections 107 or 108 of the U.S. Copyright Law. This consent does not extend to other types of copying for general distribution, for advertising or promotional purposes, for creating new collective works, or for resale. For permission to reprint entire articles in these cases and the like, consult the Administrator of the Mineralogical Society of America as to the royalty due to the Society.

REVIEWS IN MINERALOGY AND GEOCHEMISTRY

(Formerly: REVIEWS IN MINERALOGY)

ISSN 1529-6466

Volume 51

Plastic Deformation and Deformation Microstructure in Earth Materials

ISBN 0-93995063-4

Additional copies of this volume as well as others in this series may be obtained at moderate cost from:

THE MINERALOGICAL SOCIETY OF AMERICA
1015 EIGHTEENTH STREET, NW, SUITE 601
WASHINGTON, DC 20036 U.S.A.

REVIEWS IN MINERALOGY AND GEOCHEMISTRY
VOLUME 51

Plastic Deformation of Minerals and Rocks

FOREWORD

Volume 51 was prepared for a short course on Plastic Deformation of Minerals and Rocks, sponsored by the Mineralogical Society of America (MSA) in Berkeley, California, December 4 and 5, 2002. Shun-ichiro Karato (Yale University) and Hans-Rudolf Wenk (University of California–Berkeley) organized both the course and this volume, and the National Science Foundation supported the short course with a generous grant through Yale University.

Visit MSA's web site, http://minsocam.org , for any errata that may eventually be posted for this volume and for a listing of other volumes in the *Reviews* series. Books may be ordered over the internet.

Paul H. Ribbe
Virginia Tech
Blacksburg, Virginia

PREFACE

This volume highlights some of the frontiers in the study of plastic deformation of minerals and rocks. The research into the plastic properties of minerals and rocks had a major peak in late 1960s to early 1970s, largely stimulated by research in the laboratory of D.T. Griggs and his students and associates. It is the same time when the theory of plate tectonics was established and provided a first quantitative theoretical framework for understanding geological processes. The theory of plate tectonics stimulated the study of deformation properties of Earth materials, both in the brittle and the ductile regimes. Many of the foundations of plastic deformation of minerals and rocks were established during this period. Also, new experimental techniques were developed, including deformation apparatus for high-pressure and high-temperature conditions, electron micros-copy study of defects in minerals, and the X-ray technique of deformation fabric analysis. The field benefited greatly from materials science concepts of deformation that were introduced, including the models of point defects and their interaction with dislocations. A summary of progress is given by the volume "Flow and Fracture of Rocks: The Griggs Volume," published in 1972 by the American Geophysical Union.

Since then, the scope of Earth sciences has greatly expanded. Geodynamics became concerned with the Earth's deep interior where seismologists discovered heterogeneities and anisotropy at all scales that were previously thought to be typical of the crust and the upper mantle. Investigations of the solar system documented new mineral phases and rocks far beyond the Earth. Both domains have received a lot of attention from mineralogists (e.g., summarized in MSA's *Reviews in Mineralogy*, Volume 36, "Planetary Materials," and Volume 37, "Ultra-High Pressure Mineralogy"). Most attention was directed towards crystal chemistry and phase relations, yet an understanding of the deformation behavior is essential for interpreting the dynamic geological processes from geological and geophysical observations. This was largely the reason for a rebirth of the study of rock plasticity, leading to new approaches that include experiments at extreme conditions and modeling of deformation behavior based on physical principles. A wide spectrum of communities emerged that need to use information about mineral plasticity, including mineralogy, petrology, structural geology, seismology, geodynamics and engineering. This was the motivation to organize a workshop, in December 2002 in Emeryville, California, to bridge the very diverse disciplines and facilitate communication. The volume that was written for this workshop should help one to become familiar with a notoriously difficult subject, and the various contributions represent some of the important progress that has been achieved.

The spectrum is broad. High-resolution tomographic images of Earth's interior obtained from seismology need to be interpreted on the bases of materials properties to understand their geodynamic significance. Key issues include the influence of deformation on seismic signatures, such as attenuation and anisotropy, and a new generation of experimental and theoretical studies on rock plasticity has contributed to a better understanding. Extensive space exploration has revealed a variety of tectonic styles on planets and their satellites, underlining the uniqueness of the Earth. To understand why plate tectonics is unique to Earth, one needs to understand the physical mechanisms of localization of deformation at various scales and under different physical conditions. Also here important theoretical and experimental studies have been conducted. In both fields, studies on anisotropy and shear localization, large-strain deformation experiments and quantitative modeling are critical, and these have become available only recently. Complicated interplay among chemical reactions (including partial melting) is a key to understand the evolution of Earth.

This book contains two chapters on the developments of new techniques of experimental studies: one is large-strain shear deformation (Chapter 1 by Mackwell and Paterson) and another is deformation experiments under ultrahigh pressures (Chapter 2 by Durham et al.). Both technical developments are the results of years of efforts that are opening up new avenues of research along which rich new results are expected to be obtained. Details of physical and chemical processes of deformation in the crust and the upper mantle are much better understood through the combination of well controlled laboratory experiments with observations on "real" rocks deformed in Earth. Chapter 3 by Tullis and Chapter 4 by Hirth address the issues of deformation of crustal rocks and the upper mantle, respectively. In Chapter 5 Kohlstedt reviews the interplay of partial melting and deformation, an important subject in understanding the chemical evolution of Earth. Cordier presents in Chapter 6 an overview of the new results of ultrahigh pressure deformation of deep mantle minerals and discusses microscopic mechanisms controlling the variation of deformation mechanisms with minerals in the deep mantle. Green and Marone review in Chapter 7 the stability of deformation under deep mantle conditions with special reference to phase transformations and their relationship to the origin of intermediate depth and deep-focus earthquakes. In Chapter 8 Schulson provides a detailed description of fracture mechanisms of ice, including the critical brittle-ductile transition that is relevant not only for glaciology, planetology and engineering, but for structural geology as well. In Chapter 9 Cooper provides a review of experimental and theoretical studies on seismic wave attenuation, which is a critical element in interpreting distribution of seismic wave velocities and attenuation. Chapter 10 by Wenk reviews the relationship between crystal preferred orientation and macroscopic anisotropy, illustrating it with case studies. In Chapter 11 Dawson presents recent progress in poly-crystal plasticity to model the development of anisotropic fabrics both at the microscopic and macroscopic scale. Such studies form the basis for geodynamic interpretation of seismic anisotropy. Finally, in Chapter 12 Montagner and Guillot present a thorough review of seismic anisotropy of the upper mantle covering the vast regions of geodynamic interests, using a global surface wave data set. In Chapter 13 Bercovici and

Karato summarize the theoretical aspects of shear localization. All chapters contain extensive reference lists to guide readers to the more specialized literature.

Obviously this book does not cover all the areas related to plastic deformation of minerals and rocks. Important topics that are not fully covered in this book include mechanisms of semi-brittle deformation and the interplay between microstructure evolution and deformation at different levels, such as dislocation substructures and grain-size evolution ("self-organization"). However, we hope that this volume provides a good introduction for graduate students in Earth science or materials science as well as the researchers in these areas to enter this multidisciplinary field.

We thank all the authors who spent their precious time in making a great contribution to our community. We also are grateful to the Mineralogical Society of America for sponsoring this book and the accompanying workshop and to the National Science Foundation for generous financial support of the entire effort. Paul Ribbe, the series editor, did a superb job in coordinating the book and managing to unify all the different formats into a coherent volume by delicately pressuring authors, without breaking their enthusiasm.

Shun-ichiro Karato
Yale University
New Haven, Connecticut

Hans-Rudolf Wenk
University of California
Berkeley, California

December 2002

Table of Contents

1 New Developments in Deformation Studies: High-Strain Deformation
Stephen J. Mackwell, Mervyn S. Paterson

INTRODUCTION	1
EXPERIMENTAL METHODS FOR LARGE STRAINS	1
Coaxial deformation	2
Rotational deformation	2
LARGE-STRAIN BEHAVIOR IN ROCKS — GENERAL CONSIDERATIONS	4
LARGE-STRAIN BEHAVIOR IN ROCKS — CASE STUDIES	6
Monomineralic crustal rocks	6
Monomineralic mantle rocks	9
Polyphase systems	12
SUMMARY	13
APPENDIX I: Stress Analysis for Diagonally Split-Cylinder Shear Test	14
APPENDIX II: The Torsion Test	15
REFERENCES	17

2 New Developments in Deformation Experiments at High Pressure
William B. Durham, Donald J. Weidner, Shun-ichiro Karato, Yanbin Wang

INTRODUCTION	21
Why is pressure important?	22
Terminology related to strength and deformation	23
A BRIEF HISTORY OF HIGH-PRESSURE APPARATUS	24
To 5 GPa: Cylindrical devices	24
5 GPa and above: Anvil devices	26
MEASUREMENT METHODS AT HIGH PRESSURE	27
Stress measurement	27
Strain rate measurement	32
MODERN TECHNIQUES FOR DEFORMATION AT HIGH PRESSURES	34
Sample assemblies	34
Modifications to the sample assembly in a multianvil press	36
Diamond-anvil cell	37
Deformation-DIA	37
Rotational Drickamer apparatus (RDA)	40
SUMMARY AND PERSPECTIVES	44
ACKNOWLEDGMENTS	45
REFERENCES	46

3 Deformation of Granitic Rocks: Experimental Studies and Natural Examples
Jan Tullis

INTRODUCTION	51
CRYSTAL PLASTICITY PROCESSES	52
MONOMINERALIC AGGREGATES: QUARTZ	55
Water	55
Slip systems	56
Recrystallization mechanisms and microstructures	56
LPOs	61

Flow laws and piezometers 64
Comparisons with naturally deformed quartzites 65
MONOMINERALIC AGGREGATES: FELDSPAR 70
Water 71
Slip systems 71
Recrystallization mechanisms and microstructures 71
LPOs 74
Flow laws and piezometers 74
Comparisons with naturally deformed feldspars 75
POLYPHASE AGGREGATES 78
Quartz-feldspar aggregates 78
Effects of mica 79
Comparisons with naturally deformed granitic rocks 81
APPLICATIONS AND IMPLICATIONS 85
Information from dislocation creep microstructures 85
Flow laws for crustal rocks 86
Assessing the strength of the crust 88
ACKNOWLEDGMENTS 89
REFERENCES 89

4 Laboratory Constraints on the Rheology of the Upper Mantle
Greg Hirth

INTRODUCTION 97
BACKGROUND 97
Brittle deformation and low-temperature plasticity 97
High-temperature creep 99
Brittle-ductile/brittle-plastic transitions 102
INSIGHTS, CAVEATS AND QUESTIONS ABOUT APPLYING LABORATORY
DATA TO CONSTRAIN THE RHEOLOGY OF THE OCEANIC MANTLE 103
Strength of the lithosphere and the depth of oceanic earthquakes 106
Strength of plate boundaries in the viscous regime 109
High-temperature creep and the viscosity of the mantle 112
ACKNOWLEDGMENTS 116
REFERENCES 116

5 Partial Melting and Deformation
David L. Kohlstedt

INTRODUCTION 121
MELT DISTRIBUTIONS IN NON-DEFORMING ROCKS 121
DEFORMATION OF PARTIALLY MOLTEN ROCKS 127
MELT DISTRIBUTIONS IN DEFORMING ROCKS 131
ACKNOWLEDGMENTS 134
REFERENCES 134

6 Dislocations and Slip Systems of Mantle Minerals
Patrick Cordier

INTRODUCTION	137
BASIC CONSIDERATIONS	137
EXPERIMENTAL ADVANCES	142
High-pressure deformation experiments	143
Transmission electron microscopy	144
X-ray diffraction peak broadening	147
PLASTICITY OF MANTLE PHASES	147
SiO_2 system	149
Olivine, wadsleyite, ringwoodite	153
Garnets	167
WHERE DO WE STAND?	171
PERSPECTIVES: FROM ATOMIC TO THE GLOBAL SCALE	173
REFERENCES	174

7 Instability of Deformation
Harry W. Green, II, Chris Marone

INTRODUCTION	181
SHEARING INSTABILITY	182
EXPERIMENTAL HIGH-PRESSURE FAULTING MECHANISMS	185
Dehydration-induced embrittlement	185
Transformation-induced faulting	188
"Brittle" versus "plastic" shear failure	192
Thermal runaway due to shear heating	192
APPLICATION TO EARTHQUAKE MECHANISMS	193
Earthquake distribution with depth	193
Mineral reactions available to trigger earthquakes	194
CONCLUSIONS AND SPECULATIONS	196
REFERENCES	197

8 Brittle Failure of Ice
Erland M. Schulson

INTRODUCTION	201
FAILURE UNDER TENSION	202
Characteristics	202
Failure mechanisms	205
Ductile ice	206
BRITTLE FAILURE UNDER COMPRESSION	207
Overview	207
Brittle compressive failure under multiaxial loading	210
MICROSTRUCTURAL FEATURES AND MECHANISMS OF BRITTLE COMPRESSIVE FAILURE	222
Longitudinal splits, material collapse and wing cracks	222
Shear faults and comb-cracks	225
DUCTILE-TO-BRITTLE TRANSITION UNDER COMPRESSION	234
Definition	234
Transition models	235
ON THE FORMATION OF "LEADS" IN THE ARCTIC SEA ICE COVER	239
ICE AND ROCK	245
ACKNOWLEDGMENTS	246
REFERENCES	246

9   Seismic Wave Attenuation: Energy Dissipation in Viscoelastic Crystalline Solids
Reid F. Cooper

INTRODUCTION	253
LINEAR VISCOELASTICITY: A CHEMICAL KINETICS PERSPECTIVE	253
SPRING AND DASHPOT MODELS OF VISCOELASTICITY	256
ATTENUATION AND THE LINEAR VISCOELASTIC MODELS	259
THE ATTENUATION BAND/HIGH-TEMPERATURE BACKGROUND	262
ISOLATION/CHARACTERIZATION OF A SINGLE PHYSICAL MECHANISM PRODUCING A POWER-LAW ATTENUATION SPECTRUM: THE INTRINSIC TRANSIENT IN DIFFUSIONAL CREEP	266
DATA EXTRAPOLATION AND APPLICABILITY OF EXPERIMENTS TO GEOPHYSICAL CONDITIONS	271
SUBGRAIN ABSORPTION AND THE ATTENUATION BAND	276
IMPACT OF PARTIAL MELTING	278
IMPACTS OF DEFECT CHEMISTRY (INCLUDING WATER) AND THE STRUCTURE(S) OF INTERFACES	283
FINAL COMMENTS	286
ACKNOWLEDGMENTS	286
REFERENCES	287

10   Texture and Anisotropy
Hans-Rudolf Wenk

INTRODUCTION	291
MEASUREMENTS OF TEXTURES	291
Overview	291
X-ray pole figure goniometer	292
Synchrotron X-rays	292
Neutron diffraction	294
Transmission electron microscope (TEM and HVEM)	295
Scanning electron microscope (SEM)	296
Comparison of methods	296
DATA ANALYSIS	297
Orientation distributions	297
From pole figures to ODF	298
Use of whole diffraction spectra	299
Statistical considerations of single orientation measurements	300
From textures to elastic anisotropy	301
POLYCRYSTAL PLASTICITY SIMULATIONS	302
General comments	302
Deformation	303
Recrystallization	306
APPLICATIONS OF POLYCRYSTAL PLASTICITY	307
Introduction	307
Coaxial thinning versus non-coaxial shearing (calcite)	307
Anisotropy in the upper mantle (olivine)	309
Lower mantle	312
Core	314
PROBLEMS AND OPPORTUNITIES	316
CONCLUSIONS	322
REFERENCES	323

11   Modeling Deformation of Polycrystalline Rocks
Paul R. Dawson

MODELING PRELIMINARIES	331
Length scales	331
General comments	332
Small-scale simulations	332
Large-scale simulations	333
SINGLE CRYSTAL CONSTITUTIVE BEHAVIOR	334
Slip systems	334
Crystal kinematics	334
Crystal compliance and stiffness	335
POLYCRYSTAL CONSTITUTIVE EQUATIONS	336
Orientational averages	336
Linking crystal responses to continuum scale motion	337
SMALL-SCALE HYBRID ELEMENT FORMULATION	337
LARGE-SCALE VELOCITY-PRESSURE FORMULATION	339
HALITE TEXTURE EVOLUTION: A SMALL-SCALE APPLICATION	340
Generalities	340
Simulation specifics	342
Simulation results	342
MANTLE CONVECTION — A LARGE-SCALE APPLICATION	343
Generalities	343
Simulation specifics	345
Simulation results	346
SUMMARY	350
ACKNOWLEDGMENTS	350
REFERENCES	350

12   Seismic Anisotropy and Global Geodynamics
Jean-Paul Montagner, Laurent Guillot

INTRODUCTION	353
CAUSES OF SEISMIC ANISOTROPY FROM MICROSCOPIC TO LARGE SCALE	354
Shape Preferred Orientation (S.P.O.)	354
Lattice Preferred Orientation (L.P.O.)	355
EFFECT OF ANISOTROPY ON SEISMIC WAVES	357
Body waves	358
Surface waves	359
Comparison between surface wave anisotropy and SKS splitting data	363
ANISOTROPY IN THE DIFFERENT LAYERS OF THE EARTH AND THEIR GEODYNAMIC APPLICATIONS	365
Reference 1-D Earth models	366
Evidence of anisotropy in the upper 410 km of the mantle	367
Oceanic plates	367
Continents	370
Anisotropy in the transition zone	372
Anisotropy in the D″-layer	372
NUMERICAL MODELING AND BOUNDARY LAYERS	373
ACKNOWLEDGMENTS	377
APPENDIX A. Basic Theory of Wave Propagation in Anisotropic Media	377
APPENDIX B. Tensors and Matrices Manipulations	378
REFERENCES	380

13 Theoretical Analysis of Shear Localization in the Lithosphere
David Bercovici, Shun-ichiro Karato

INTRODUCTION	387
THEORETICAL PRELIMINARIES	388
SHEAR LOCALIZING FEEDBACK MECHANISMS	389
Thermal feedback with decay-loss healing	390
Simple-damage feedback with decay-loss healing	396
Grain-size feedback	399
Thermal and simple-damage feedbacks with diffusive-loss healing	405
OTHER CONSIDERATIONS	411
Two-dimensional examples	411
More sophisticated damage theories	413
SUMMARY AND CONCLUSIONS	413
REFERENCES	418

1 New Developments in Deformation Studies: High-Strain Deformation

Stephen J. Mackwell
Bayerisches Geoinstitut
Universität Bayreuth
95440 Bayreuth, Germany

Mervyn S. Paterson
Research School of Earth Sciences
Australian National University
Canberra, Australia

INTRODUCTION

Plastic deformation of rocks in the Earth is most strikingly illustrated in exposures of shear zones. Very large strains (shear strains of $\gamma = 10$ or more) may be developed in these zones. They appear to be ubiquitous within the continental and oceanic crust and may provide the dominant locus for deformation in these regions and, potentially, also in the lithospheric upper mantle. It is therefore of the greatest interest to obtain information about the deformation properties of rock at large strains

In the past, various approaches have been made to achieving in the laboratory the types of deformation seen in natural rocks. In some cases it has been possible to achieve plastic deformation in single crystals and fine-grained aggregates of minerals at atmospheric pressure at sufficiently high temperatures and a number of such experiments have been carried out (Mackwell et al. 1990, Dimanov et al. 1998). However, it is generally necessary to conduct experiments under conditions of high confining pressure as well as high temperature. The earliest experiments of this nature were those of Adams and Nicholson (1901) and von Kármán (1911), followed by the pioneering work of Bridgman (1949) and Griggs (1936, and many subsequent papers). These studies were carried out using fluid pressure media. Although Otto Mügge carried out some deformation experiments on single crystals of minerals in solid confining media already in the late 19th to early 20th century, the extensive use of solid media for high-pressure experiments was not introduced until the development of the Griggs solid medium apparatus in the 1960's (e.g., Griggs et al. 1966, Griggs 1967).

Since the 1960s, extensive studies have been carried out in both gas-medium and solid-medium apparatus. The apparatus and techniques have been reviewed in several publications (Griggs et al. 1960, Paterson 1964, 1970, 1978, 1990, Tullis and Tullis 1986, Kohlstedt et al. 1995). These studies have concentrated on axisymmetric triaxial deformation (the so-called "triaxial test"). However, the total amount of deformation in such tests is limited by several considerations discussed below. Also, axisymmetric tests do not give information about rotational types of strain history such as occur in shear zones. Therefore attention has turned to shear tests in which both large strains and rotational strain histories can be studied. This paper deals with the various experimental approaches to achieving large deformations.

EXPERIMENTAL METHODS FOR LARGE STRAINS

Experimental studies of the mechanical behavior and textural evolution of rocks to high strains necessitate apparatus that permits large changes in shape or configuration of samples while maintaining stable control for the purpose of measuring the deformation and the deforming forces. There are two types of homogeneous deformation path that can be studied, which can be distinguished as, respectively, *coaxial* and *rotational*. Either type of deformation path may be followed in attempting to reach large deformations. The properties and potentialities of these types of deformation for large strains are as follows:

Coaxial deformation

The principal axes of the strain increments remain parallel to each other during the finite coaxial deformation. This is illustrated in Figure 1a for the case of simple shortening deformation. The "true" measure of the finite strain in this case is the integrated strain based on current length l, that is, $e = -\int dl/l = -\ln l_1/l_0$ where l_1 and l_0 are the final and initial lengths, respectively (shortening strain positive). Practical examples of coaxial deformation tests include the triaxial compression and extension tests.

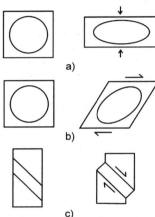

Figure 1. Schematic of the evolution of the principal strain axes of samples deformed in (a) axial compression, (b) simple shear and (c) using the split-cylinder shear configuration.

If the tests are carried out to large strains in compression, the increase in cross-sectional area tends to lead to the specimen overlapping the compression platens unless it is re-machined to smaller diameter at stages during the deformation. Further, the associated decrease in length leads to an unfavorable length-to-diameter ratio, with greater distortion of the stress field due to end-friction effects.

In extension, the uniform strain is limited in magnitude by necking instability of the specimen (Hart 1967). Relatively large strains may develop within the neck (Griggs et al. 1960) but the stresses and strains are now inhomogeneous and a sophisticated analysis is required to deduce stress-strain properties in this case (Bridgman 1952). Rutter (1998) has carried out extension tests to obtain information about the behavior of marble in large coaxial deformation, with natural strains up to about $\gamma = 2.4$.

In either case, there are practical limitations in accommodating the displacements involved in reaching large strains. In compression, the increase in diameter of the specimen may lead to interference with the bore of the furnace in the deformation apparatus. In extension the maximum displacement may be limited by the range of movement in the loading system or by the length of the region of uniform temperature in the furnace.

Rotational deformation

The principal strain axes rotate relative to external coordinates during the finite deformation, as illustrated in Figure 1b for simple shearing deformation. Simple shear is an idealization approximated in greater or less degree in several types of experimental shearing arrangements under high pressure. These include the diagonally split-cylinder test and the torsion test.

In the *diagonally split-cylinder shear test*, a non-deformable cylinder is cut at an acute angle θ, usually 45°, to the cylindrical axis and the deformable specimen inserted as a layer within the cut (Fig. 1c). The cylinder is then loaded axially, giving rise to a shear loading on the deformable layer. This arrangement can be conveniently inserted in a triaxial testing rig with a jacket enclosing the specimen assembly. However, it should be noted that the stress distribution is not that of a simple shear test because there is an additional normal stress acting on the layer. Moreover, in addition to the axial force F applied by the testing machine, there will also be a lateral force P acting on the layer due to constraints brought into play by the lateral component of the shearing displacement. Thus, as shown in the Appendix I, the shear stress τ parallel to the layer will be

$$\tau = \frac{F}{A}\sqrt{1+\left(\frac{P}{F}\right)^2}\cos\theta'\sin\theta \quad \text{where} \quad \theta' = \theta + \frac{\pi}{2} - \arctan\frac{F}{P},$$

which will be accompanied by a normal stress σ_n on the layer of

$$\sigma_n = \frac{F}{A}\sqrt{1+\left(\frac{P}{F}\right)^2}\sin\theta'\sin\theta.$$

The presence of the lateral force component P is often not acknowledged but it may be quite important, especially in situations such as in a solid-medium machine where the loading piston is short and the medium also provides a resistance to lateral displacement. The actual stresses are difficult to calculate in such a case, but an estimate can be made of the lateral constraints in a fluid-medium machine where it corresponds to the bending of the loading pistons (see Appendix I). Thus additional stress components of tens of MPa can arise from lateral constraints. This effect will be most serious for weak specimens. The other important feature of the split-cylinder shear test is the presence of the substantial normal stress component on the layer in addition to the shear stress. The normal stress tends to lead to thinning of the specimen layer and its lateral extrusion.

The advantage of the split-cylinder shear test is that it is often a convenient way of achieving substantial shear strains in thin layers of material, permitting the study of the evolution of microstructures to moderately large strains. However, it must be borne in mind that the stress conditions are not simple to analyse and that the deformation may include substantial flattening strains as well as the shear strains.

The *torsion test* is, in effect, another type of shear test. It has proved to be the most effective way of achieving very large strains. Moreover, the non-coaxial deformation path closely simulates that in geological shear zones. The deformation conditions at any point in a cylindrical torsion specimen correspond to simple shear and are not complicated by the presence of a normal stress component as in the split-cylinder test, provided no axial force is applied to the cylinder. The plane of shearing is normal to the axis of the cylinder. Practical aspects of the torsion test and its application are given in Appendix II.

The analysis of the torsion test has been set out by Paterson and Olgaard (2000). In torsion of a cylinder, the shear strain varies linearly with the radius, being a maximum at the outer surface of the cylinder. If the angle of twist is θ in a cylinder of diameter d and length l, then the maximum shear strain γ at the surface of the cylinder is given by

$$\gamma = \frac{d\theta}{2l}$$

However, the stress distribution is not linear with radius and depends on the nature of the stress-strain relationship. If the form of the stress-strain relationship can be assumed, for example, if it can be expected to be of power-law form, the torque-twist observation can be inverted to obtain the stress-strain relationship. Thus if a power law with stress exponent n is assumed, the shear stress τ at the surface of the specimen is given by

$$\tau = \frac{4M\left(3+\frac{1}{n}\right)}{\pi d^3}$$

where M is the measured torque. Since the stress increases with radius more rapidly than linearly for $n > 1$, and the contribution of each radial element is proportional to the radius, most of the torque derives from the stress in the outer layers of the specimen. Thus the circumstance that there is a radial gradient of stress and strain in the specimen is generally of relatively minor importance, and Paterson and Olgaard (2000) argue that there is little point in attempting to carry out experiments with hollow specimens in order to achieve more nearly homogeneous deformation, especially in view of the experimental difficulties of jacketing specimens inside and outside.

Microstructural observations are most appropriately made on thin sections cut tangentially to the outer surface of the specimen. In this case, the section represents the plane of simple shearing. Texture observations can also be made in such a section.

LARGE-STRAIN BEHAVIOR IN ROCKS — GENERAL CONSIDERATIONS

Considerations of large-strain behavior in rocks call for a broader view of time/strain scales than has been usual for small-strain studies. In particular, new perspectives on "steady state" are needed. In constant strain-rate axial deformation tests, it is common for the stress-strain curve to become more or less flat, that is, strain-independent, after a true strain of between 0.1 and 0.3. The behavior is said to have reached a "steady state", with the implication that the behavior has become independent of strain and that the structure or state of the material is not longer changing significantly. However, observations on specimens at this stage often reveal that a partial degree of recrystallization or some change in the grain size has occurred and that there is the potential for further change. Thus factors that influence the level of the flow stress have become nearly constant over the relatively short strain intervals of order 0.1 but in fact slower or longer-term changes are still in the process of occurring. Torsion tests on rocks so far have revealed that evolutionary changes such as recrystallization tend to require strains of at least $\gamma = 5$ or more to come to completion and that the stress-strain curve also shows non-steady-state trends over this range (Fig. 2). It is also possible that grain growth can proceed over even longer strain intervals.

Since it has become clear from large strain studies that apparent "steady state" behavior after relatively small strains does not generally persist over larger ranges of strain, interest has shifted more to the evolutionary aspects of deformation properties and structural change. The quantitative description of the mechanical behavior is no longer simple, as embodied in the steady-state power law under the assumption of isotropic behavior, and it is difficult to establish simple expressions that describe stress-strain curves such as those now commonly obtained from large-strain torsion tests (e.g., Pieri et al. 2001a, Mackwell and Rubie 2000). Also a treatment that takes into account the developing anisotropy has not yet been elaborated. An alternative approach is to use an expression that is valid over a limited part of the stress-strain curve and determine the variation in the parameters over the full strain range to give a full description. Such an expression is

$$\dot{\gamma} = A\gamma^{-p}\tau^n \exp(-Q/RT) \tag{1a}$$

or

$$\tau = A^{-\frac{1}{n}}\gamma^{\frac{p}{n}}\dot{\gamma}^{\frac{1}{n}} \exp(Q/nRT) \tag{1b}$$

where τ is the shear stress, γ the shear strain, and T the temperature (R is the gas constant). In theoretical continuum mechanics it is sometimes preferred to write the temperature term as T^ν rather than in the Arrhenius form above (e.g., Fressengeas and Molinari 1987); in this case, $\nu = -\dfrac{Q}{nRT_0}$.

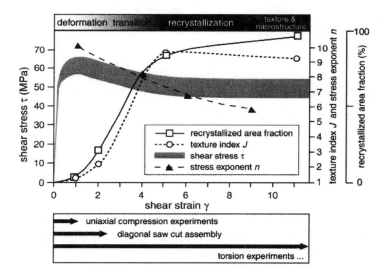

Figure 2. Rheological, textural and microstructural evolution of Carrara marble deformed to high shear strain in torsion. Shear stress τ, stress exponent n, texture index J and area fraction of recrystallized grains are given as a function of shear strain γ. The strain intervals for the various types of experiments are given at the bottom. [Used by permission of the editor of the *Journal of Structural Geology*, from Pieri et al. (2001), Fig. 17, p. 1411.]

The quantities A, p, n and Q are now parameters whose values gradually change or evolve in the course of the deformation but which can be taken as constants over limited ranges of strain. Parameters n and Q give measures, respectively, of the stress sensitivity and temperature sensitivity of the strain rate, and the ratio p/n is a measure of the strain hardening. Thus

$$n = \frac{\partial \ln \dot{\gamma}}{\partial \ln \tau} \quad \text{at constant } \gamma \text{ and } T,$$

$$Q = -R\frac{\partial \ln \dot{\gamma}}{\partial (1/T)} = RT^2 \frac{\partial \ln \dot{\gamma}}{\partial T} \quad \text{at constant } \gamma \text{ and } \tau$$

or

$$Q = nR\frac{\partial \ln \tau}{\partial (1/T)} = -nRT^2 \frac{\partial \ln \tau}{dT} \quad \text{at constant } \gamma \text{ and } \dot{\gamma},$$

and

$$p = n\frac{\partial \ln \tau}{\partial \ln \gamma} \quad \text{at constant } \dot{\gamma} \text{ and } T \text{ (stress-strain test)}$$

or

$$p = -\frac{\partial \ln \dot{\gamma}}{d \ln \gamma} \quad \text{at constant } \tau \text{ and } T \text{ (creep test)}$$

The last expressions give directly a measure of the strain hardening in terms of the slope of the stress-strain curve or creep curve, so

$$\frac{d\tau}{d\gamma} = \frac{p}{n}\frac{\tau}{\gamma} \quad \text{at constant } \dot{\gamma} \text{ and } T \text{ (stress-strain test)}$$

or

$$\frac{d\dot{\gamma}}{d\gamma} = -p\frac{\dot{\gamma}}{\gamma} \quad \text{at constant } \tau \text{ and } T \text{ (creep test).}$$

Parameter A is a related to the absolute level of the stress-strain curve, and it can also evolve with strain. In special cases, A can be expanded to show an explicit variability of the flow behavior with particular factors such as initial grain size or other internal variables, oxygen or water fugacity, etc.

If the parameters in (1) are constant and $p = 0$, we have the "steady state" flow law that is often quoted in geodynamical discussions. It is now evident that the steady state discerned at strains of order 0.2 is commonly ephemeral but there may be an approximation to steady state again at large strains, with $\gamma > 5$ or 10.

LARGE-STRAIN BEHAVIOR IN ROCKS — CASE STUDIES

There have been a number of studies in recent years of the high-strain behavior of rocks using the techniques described in the previous sections. In some of these studies, measurements have been made on the same rock type using complementary techniques. Many of these measurements have been performed relatively recently, notably due to technological developments, such as the utilization of the torsional configuration in the gas-medium apparatus. In the following, we will review the current state of the field in high-strain experimental studies of rocks, focussing on monomineralic crustal and mantle rocks and then briefly summarizing research on two-phase and polyphase systems.

Monomineralic crustal rocks

Perhaps the most studied rock in terms of deformation behavior, Carrara marble has been investigated using a range of techniques, under compressional, extensional, split-cylinder and torsional configurations. The axial compression studies have provided a comprehensive understanding of low-strain deformation mechanisms, notably dislocation creep (e.g., Rutter 1974, Schmid et al. 1980). Under extensional deformation conditions to high strains, Rutter (1998) performed experiments at 973 and 1073 K, utilizing the necking instability to attain maximum shear strains of around $\gamma = 2.4$. He reported significant dynamic recrystallization at high strains. The observed weakening with progressive strain may have resulted from purely geometric constraints due to necking at constant displacement rate. A further study that made use of the split-cylinder configuration (Schmid et al. 1987), which allowed constant strain rates, also reported significant grain-size reduction due to dynamic recrystallization but only a modest reduction in shear strength to a shear strain of around $\gamma = 3$. In neither case was a change in deformation mechanism from dislocation creep to diffusional creep inferred despite the significant grain refinement.

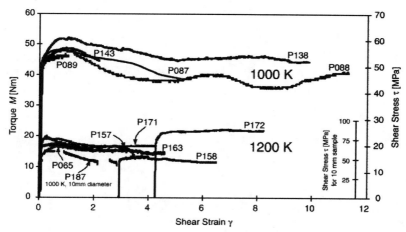

Figure 3. Torque versus shear strain for all constant strain rate tests on Carrara marble at 1000 and 1200 K. The torque is converted into shear stress on the scale to the right assuming 15 mm (inset 10 mm) diameter samples and stress exponent of $n = 10$. [Used by permission of the editor of the *Journal of Structural Geology*, from Pieri et al. (2001), Fig. 6, p. 1399.]

More recently, Pieri et al. (2001a,b) used the torsional configuration to study high-strain deformation of Carrara marble at 1000 - 1200 K to shear strains as high as $\gamma = 11$. After an initial peak at a shear strain of around $\gamma = 1$, the strength decreased to a steady level at a shear strain of around $\gamma = 5$ (Fig. 3). The low-strain microstructures showed undulose extinction, deformation bands and the development of subgrain boundaries within the grains, as well as some bulging of grain boundaries (Fig. 4). A strong shape preferred fabric developed with increasing strain, but was destroyed as the samples became fully recrystallised by a strain of between $\gamma = 5$ and 11, depending on temperature. The grain-size reduction from around 150 to 10 μm occurred predominantly due to nucleation and growth of strain-free grains on grain boundaries. The lattice preferred orientation that developed at low strains, which was consistent with the slip systems identified in axial compression experiments, was progressively replaced by a recrystallization texture.

Experiments on fine-grained Solnhofen limestone show behavior that is quite distinct from that the coarser grained marble. Low-strain axial compression experiments show several deformation fields. At higher-stress, lower-temperature conditions, deformation is dominated by dislocation creep, while at lower stresses behavior is consistent with a mechanism dominated by grain-boundary sliding accommodated by diffusion along grain boundaries (Schmid et al. 1977). Schmid et al. interpreted this latter behavior as "superplastic" creep. They observed no net grain growth (due presumably to the presence of impurity phases) and no associated changes in deformation mechanism during the experiments. Studies of the texture development in such samples (e.g., Rutter et al. 1994) showed strong development of a lattice preferred orientation in the dislocation creep field, but only weak development in the "superplastic" field. Using the split-cylinder configuration, Schmid et al. (1987) extended the work to shear strains of $\gamma \cong 1$, while Casey et al. (1998) used the torsion design to reach shear strains of up

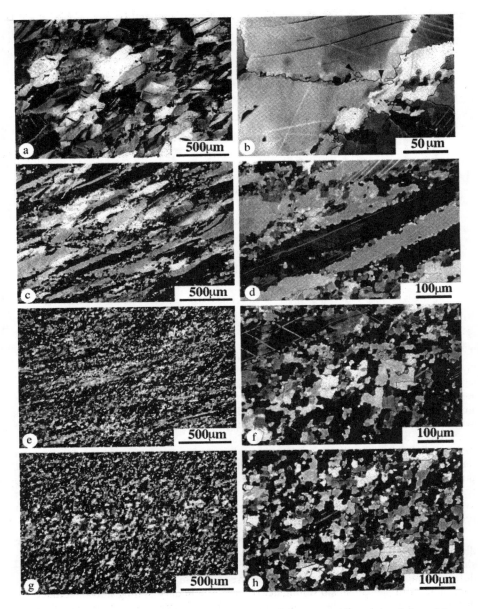

Figure 4. Microphotographs in cross-polarized light of Carrara marble samples deformed in torsion at 1000 K to shear strain of $\gamma = 1$ (a,b), $\gamma = 2$ (c,d), $\gamma = 5$ (e,f) and $\gamma = 11$ (g,h). Overviews at low magnifications (at left) and details at higher magnification (at right). Shear zone boundary is oriented parallel to the long edge of the image and the sense of shear is dextral. The arrows in (b) point to a small relict grain and to bulging of the grain boundary, indicating that the horizontal grain boundary is migrating downwards. [Used by permission of the editor of the *Journal of Structural Geology*, from Pieri et al. (2001), Fig. 8, p. 1402.]

to $\gamma = 12$. Within the region near the transition from dislocation to "superplastic" creep, little further textural or mechanical evolution was observed beyond that seen in the low-strain axial experiments, as illustrated in Figure 5. The presence of a lattice preferred orientation and near-equant grains argues for deformation by both dislocation creep and grain boundary sliding. At high strains, both textures and mechanical behavior appeared to be near steady state.

Another material that has been the focus of many deformation studies is quartzite. Deformation to small strains in compression has been performed on a range of quartzites with large variability in grain size, impurity phase and water contents. Deformation under dry conditions yields highly variable rheologies characterized by mostly brittle mechanisms. Under wet high-pressure conditions, quartzites can deform by fully plastic mechanisms, but the measured rheologies are highly variable in terms of the calculated constitutive parameters (e.g., Kronenberg and Tullis 1984, Luan and Paterson 1992, Gleason and Tullis, 1995). High-strain deformation experiments on untreated (and hence wet) samples of several quartzites (mostly flint) in torsion by Schmocker et al. (2002) reveal deformation by granular flow (a brittle to semi-brittle deformation mechanism), with formation of bands of localized deformation, which appear to form in regions of excess fluid pressure, and development of no significant texture or shape fabric.

Anhydrite is the only material studied to date that shows coupled textural evolution and changes in deformation mechanism (Stretton and Olgaard 1997). As with the marble, recrystallization occurs predominantly by nucleation and growth of strain-free grains on grain boundaries. The high-strain experiments have been performed in torsion. Bulging of grain boundaries begins at relatively low shear strains of around $\gamma = 0.5$ and recrystallization is essentially completed by $\gamma = 3.7$. A parallel and relatively abrupt strength reduction occurs as the deformation mechanism changes from dislocation to diffusional creep at a shear strain of about $\gamma = 1.5$ (Fig. 5). Subsequently samples appear to reach a near-steady-state microstructure and rheology. Deformation textures develop quite rapidly at lower shear strains but weaken with progressive deformation and recrystallization (Heidelbach et al. 2001). After recrystallization, there is no evidence of significant grain growth to a shear strain of $\gamma = 20$, stabilizing deformation in the grain boundary diffusional creep field. The reason for the inhibition of grain growth in the diffusion creep field despite initial rapid growth of recrystallized grains in the dislocation creep field is not clear. In this case, pinning by impurity phases seems unlikely.

For this material progressive recrystallization with increasing strain in torsion may result in a sample with a rim deforming by diffusional creep surrounding an interior deforming by dislocation creep, complicating interpretation of the stress distribution within the sample. However, there is no clear signature of such a complex deformation geometry in the mechanical behavior, which shows a stress exponent of near $n = 1$ at higher strains (Stretton and Olgaard 1997).

Monomineralic mantle rocks

A significant body of both low- and high-strain deformation data has been collected for magnesiowüstite, the second most abundant phase in the Earth's lower mantle. While important as a component of aggregate flow, it is not clear that single-phase flow has a direct relevance to convection in this region of the Earth. Stretton et al. (2001) have performed axial compression experiments on $(Mg_{0.8}Fe_{0.2}O)$ to determine the low-strain mechanical behavior. The material deforms in the dislocation creep regime with a stress exponent of near $n = 4$, with the development of a weak lattice preferred orientation, consistent with the dominant slip systems for rock-salt-structure materials. High-strain experiments have been performed in both torsion (Heidelbach et al. 2002) and split-sphere (Yamazaki and Karato 2002) configurations. The torsion experiments on

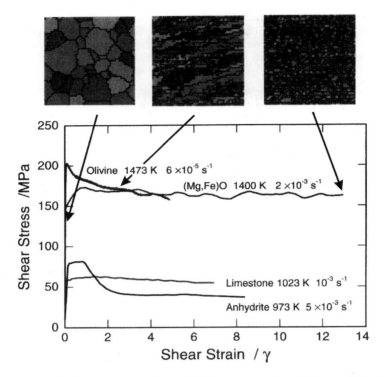

Figure 5. Shear stress versus shear strain for tests on polycrystalline samples of magnesiowüstite (Heidelbach et al. 2002), olivine (Bystricky et al. 2000), Solnhofen limestone (Casey et al. 1998) and anhydrite (Stretton and Olgaard 1997) at the temperatures and shear strain rates shown on the right. The three small images are microstructure maps derived from SEM electron backscatter diffraction investigations of tangential sections of magnesiowüstite samples deformed to shear strains of $\gamma = 0$, 2.3 and 13.

($Mg_{0.8}Fe_{0.2}O$) show deformation by dislocation creep (the low-strain behavior is largely consistent with the axial experiments) with the development of a strong shape preferred fabric and initiation of dynamic recrystallization by progressive subgrain rotation (Fig. 6). Full recrystallization occurs by a shear strain of about $\gamma = 6$ with loss of the shape preferred fabric. The initial lattice preferred orientation, which is consistent with the dislocation slip systems expected for the rock-salt structure, weakens with progressive strain and is overprinted by a recrystallization texture. No change in deformation mechanism has been observed and, after an initial peak stress (in some cases) the strength of the samples remains essentially constant to shear strains as high as $\gamma = 16$ (Fig. 5). While the split-sphere experiments of Yamazaki and Karato (2002) on ($Mg_{0.75}Fe_{0.25}O$) were not able to access the full range of conditions sampled in the torsion experiments (maximum $\gamma = 8$), the overall conclusions are essentially the same.

High-strain deformation studies have also been performed on upper mantle rocks. Of particular note, dry olivine aggregates have been deformed to high strains in torsion (Bystricky et al. 2000) and using the split-cylinder configuration (Zhang and Karato 1995, Zhang et al. 2000). The low-strain behavior in these tests is consistent with the

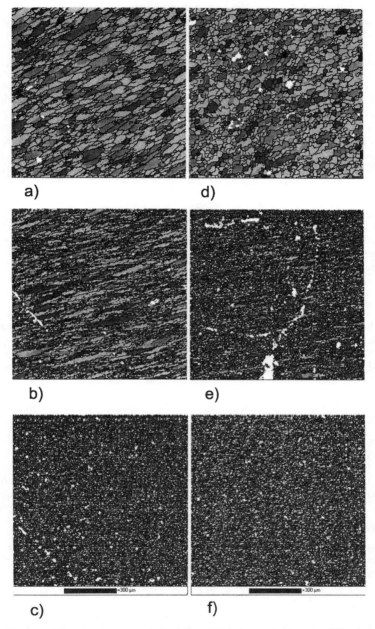

Figure 6. Microstructure maps derived from SEM electron backscatter diffraction investigations of tangential sections of magnesiowüstite samples, with a map size 1×1 mm^2, and step width 5 µm; (a) $\gamma = 1.1$, (b) $\gamma = 3.1$, (c) $\gamma = 15.5$, (d) $\gamma = 2.3$, (e) $\gamma = 5.8$, (f) $\gamma = 13$; (a-c) T = 1300 K, (d-f) T = 1400 K; grain boundaries were drawn in black for misorientation angles $\omega > 10°$. [Used by permission of the authors, from Heidelbach et al. (2002), Fig. 6.]

large body of data on rocks of similar compositions from axial compression tests performed in the dislocation creep regime (e.g., Chopra and Paterson 1984, Karato et al. 1986). At larger strains recrystallization by progressive subgrain rotation is apparent resulting in a net reduction in grain size. No change in deformation mechanism has been observed. Although some bulging of grain boundaries is apparent in the optical thin sections, subgrain rotation seems to be the dominant recrystallization mechanism. As with the marble and magnesiowüstite, a strong shape preferred fabric develops at lower strains but is destroyed by the recrystallization. The lattice preferred orientations measured for the high-strain samples are consistent with the dislocation slip systems observed in low-strain experiments. Interestingly, although both Bystricky et al., and Zhang et al. reported textures with the [100] axis parallel to the shear direction, Zhang et al. noted a (010) plane parallel to the shear plane while Bystricky et al. showed girdles of [010] and [001] normal to the shear direction. Unlike the magnesiowüstite, the aggregate strength decreases steadily with increasing strain to the maximum reported shear strain of around $\gamma = 5$ (Fig. 5).

Recent experiments on olivine aggregates under hydrous conditions by Jung and Karato (2001b) using the split cylinder configuration yield lattice preferred orientations consistent with a change from the dominant a-slip (dominance of [100] Burgers vector dislocations) to c-slip (dominance of [001] dislocations) with increasing water content. While the cause of the changing textures is debated (Kaminski 2002), such a difference in the deformation textures for wet and dry regions would have important ramifications in the interpretation of seismic anisotropy observations for the lithospheric and asthenospheric mantle.

Tullis and Yund (1999) and Stünitz et al. (2001) have performed experiments on plagioclase aggregates under predried and water-added conditions to shear strains of around $\gamma = 2$ using the split-cylinder configuration in the solid-medium apparatus. They observed that the predried plagioclase stayed within the dislocation creep regime but showed some weakening with increasing strain, while the wet plagioclase deformed in the diffusional creep field. Transmission electron microscopy images of the predried samples showed remanent high-dislocation-density grains surrounded by small polygonal recrystallized grains.

A preliminary study of high-strain deformation of clinopyroxenite to relatively high strains by Bystricky and Mackwell (2001) yielded the same general observations as for olivine, with development of an initial shape-preferred fabric, subgrain rotation recrystallization and maintenance of dislocation creep as the dominant deformation mechanism. The observed lattice preferred orientations are also consistent with the dislocation slip systems operative in previous low-strain axial deformation experiments.

Polyphase systems

Obviously the Earth is not comprized of significant regions where the rock is monomineralic and an important challenge for the future is to characterize the deformation behavior of multi-phase systems, both to low and high strains. In particular, we aim to mimic processes occurring in plastic shear zones (regions where deformation occurs by plastic deformation that is relatively localized spatially). Such zones are often the downward continuation of shallower brittle fault systems. Of special interest, we would like to see how the individual minerals contribute to the deformation and whether the deformation leads to phase separation into interlayered structures where deformation is concentrated in the layers comprized of the weaker minerals. Several recent studies have begun to address these questions.

Deformation experiments in torsion have been performed on mixtures of quartz and calcite with variable modal abundances by Rybacki et al. (2002). In these experiments, the quartz essentially acts as a non-deforming component in a plastic calcite matrix and is observed to exert a strong effect on aggregate strength. To a large extent the behavior mimics the high-strain deformation of marble, but strengthened by the presence of a second harder phase. Rybacki et al. also observe a strong temperature effect on the extent of recrystallization and strain softening behavior even to high strains.

Experiments have also been performed to high strains in 2-phase systems where the individual phases are not so different in strength. Barnhoorn et al. (2002) chose a range of modal compositions between pure anhydrite and pure calcite and investigated deformation under conditions where the two phases have similar strengths. For a shear strain of about $\gamma = 1$, the pure anhydrite samples show significant weakening relative to pure calcite, with intermediate behavior for mixed aggregates. With increasing strain, the steady-state stress values for all compositions converge, but deformation becomes localized, with phase separation, seemingly mimicking observations of natural plastic shear zones.

While beyond the scope of the present review, there is some interesting work currently investigating the effect of high shear strains on the texturing of melt within partially molten systems (e.g., Zimmerman et al. 1999, Bruhn et al. 2000, Holtzman et al. 2002), with implications for such diverse processes as core formation and melt segregation at mid-ocean ridges.

SUMMARY

While numerous experiments have been performed to high strains on a variety of materials using the various technologies described in this paper, this is still a relatively new field of study and much of the work reported above is research in progress. In the near future, we will certainly see major advances, particularly using the torsional configuration of the gas-medium apparatus. There are, however, a number of interesting observations that can be made based on the experimental studies to date on high-strain deformation of earth materials:

- In almost all materials to date, dynamic recrystallization resulted in a reduction of grain size, although only in the case of anhydrite did that result in a change in deformation mechanism. Recrystallization occurred by subgrain rotation, by nucleation and growth of new grains on grain boundaries, or a combination of both. Complete recrystallization required shear strains of at least $\gamma = 3$, which is considerably higher than can be attained in axial compression experiments. Thus, textural and mechanical steady state can only be attained during high-strain experiments.
- Strain softening was observed even to relatively high shear strains in a number of cases, presumably reflecting progressive alignment of favorable slip systems or incomplete recrystallization.
- In cases where shear strains higher than $\gamma = 10$ have been attained for high-temperature deformation in the dislocation creep field, initial textures reflecting the active dislocation slip systems evolve towards a recrystallization texture.
- The role of water in terms of textural and mechanical behavior has been relatively little studied to date. However, the results for olivine (Jung and Karato 2001a,b) and plagioclase (Tullis and Yund 1999, Stünitz et al. 2001) suggest that it may play a major role, with important implications for the behavior of the Earth's interior.

APPENDIX I:
STRESS ANALYSIS FOR DIAGONALLY SPLIT-CYLINDER SHEAR TEST

The specimen layer in the diagonally split-cylinder test is subjected to a force F parallel to the cylinder axis, apart from the hydrostatic loading from the confining pressure. If the cylinder cut is at an angle θ to the cylinder axis, then the layer can be regarded as being initially subject to a shear force component $F \cos\theta$ parallel to the layer and a normal force component $F \sin\theta$ normal to the layer. If the cylinder is of cross-sectional area A, the layer is initially subject to a shear stress component $(F/A) \cos\theta \sin\theta$ and a normal stress component $(F/A) \sin^2\theta$. However, this stress state strictly only applies at the initial stage of loading. As soon as some shearing displacement takes place, there is also a lateral component of displacement, normal to the cylinder axis, which will result in a reaction force from the constraints of the testing machine that are intended to ensure primarily axial displacement.

As an illustration, suppose that the split cylinder is loaded by two pistons of diameter D and length L, and that a shearing displacement s has occurred across the specimen layer. This displacement has a lateral component $s \sin\theta$, requiring a lateral displacement $(s/2) \sin\theta$ of the end of each piston. Suppose that the loading pistons are rigidly constrained at their outer ends and that the lateral deflection of the inner ends is accommodated entirely by elastic bending of the pistons. Since the deflection of the end of a cantilever beam is given by $PL^3/3EI$ (Gere and Timoshenko 1984, p. 369), where P is the lateral force, E the Young's modulus, and $I = \pi D^4/64$, the moment of inertia of the beam cross-section, we then have

$$P = \frac{3\pi \sin\theta}{128} \frac{ED^4}{L^3} s$$

For example, if $D = 15$ mm, $E = 200$ GPa, $L = 100$ mm, and $\theta = 45°$, then $P = 527 \left(\frac{s}{mm}\right)$ N, or 527 N if $s = 1$ mm. If the split cylinder were of 10 mm diameter, this lateral force would thus introduce an additional lateral component of normal stress of about 7 MPa when the shearing displacement is 1 mm.

The resultant force R on the specimen layer is therefore

$$R = \sqrt{F^2 + P^2} = F\sqrt{1 + \left(\frac{P}{F}\right)^2}$$

which is inclined at an angle θ' to the specimen layer, where

$$\theta' = \theta + \frac{\pi}{2} - \arctan\frac{F}{P}$$

The resultant shear stress τ parallel to the layer is then

$$\tau = \frac{F}{A}\sqrt{1 + \left(\frac{P}{F}\right)^2} \cos\theta' \sin\theta$$

and the normal stress σ_n acting on the layer is

$$\sigma_n = \frac{F}{A}\sqrt{1 + \left(\frac{P}{F}\right)^2} \sin\theta' \sin\theta.$$

The principal stress components in the plane of the cylinder axis and the normal to the specimen layer, in addition to the hydrostatic stress, are

$$\sigma_{1,2} = \frac{\sigma_n}{2} \pm \sqrt{\left(\frac{\sigma_n}{2}\right)^2 + \tau^2}$$

and their axes are inclined to the specimen layer at angles α and $\pi/2 - \alpha$ such that

$$\cos 2\alpha = \frac{\sigma_n}{2\sqrt{\left(\frac{\sigma_n}{2}\right)^2 + \tau^2}} \quad \text{and} \quad \sin 2\alpha = \frac{\tau}{\sqrt{\left(\frac{\sigma_n}{2}\right)^2 + \tau^2}}$$

(Gere and Timoshenko 1984, p. 288)

APPENDIX II: THE TORSION TEST

A discussion of the torsion test as applied to experimental rock deformation has been given by Paterson and Olgaard (2000). References are given there to applications in materials science and to earlier applications in the testing of rocks, ice, and rock gouges. Notable was the pioneering work of Handin et al. (1960, 1967).

The particular test configuration described by Paterson and Olgaard (2000) is an add-on module for the Paterson gas-medium HPT testing machine (Paterson 1990); see Figure A1. The combination with an axial deformation facility gives considerable flexibility of operation, including the ability to check length changes in specimens during torsion and even to apply axial force in combination with a torque so as to vary all three principal stresses independently (Handin et al. 1967). By using an internal furnace, temperatures of up to 1600 K can be reached at confining pressures of up to 500 MPa. A combined internal load cell permits the measurement of axial force and torque simultaneously, while avoiding uncertainty from friction corrections.

The most convenient way of gripping the specimen is to utilize the friction between the driving piston and the specimen, arising from the confining pressure, to transmit the torque to the specimen. However, for very strong specimens, inadequate frictional force may lead to slipping at this interface. In such cases, it may be necessary to resort to "dog-bone" shaped configurations to increase the available torque. In any case, it is important to vent the inside of the jacketed assembly in order not to reduce the friction by the presence of a pore fluid pressure. Conversely, this consideration markedly limits the scope for combining pore fluid studies with torsion testing.

Jacketing itself may present problems with very large strain tests because the jacket has also to be capable of undergoing the same strains as the specimen without disintegrating. Thus it has been observed that iron (low-carbon steel) is capable of very large strains at moderate temperatures, up to around 1000 K, but at higher temperatures it seems to develop excessive grain growth and to disintegrate by parting at the grain boundaries at strains of $\gamma = 5$ to 10. The selection of appropriate jacket material is part of the preparation for high-strain experiments.

An advantageous aspect of torsion testing is that end effects associated with friction at the contact with the driving pistons are much less than in axial deformation tests where friction gives rise to constraints on the change in diameter. Thus, if need be, very short specimens can be used. Another aspect of torsion testing is that shear localization with respect to the axial distribution of deformation can develop with little constraint if unstable behavior arises in the outer layers of the specimen. However, theoretical considerations (Fressengeas and Molinari 1987) suggest that localization is unlikely to develop in torsion tests with even moderately large strain softening rates if the tests are conducted at constant twist rate, but may be more likely in creep tests at constant torque.

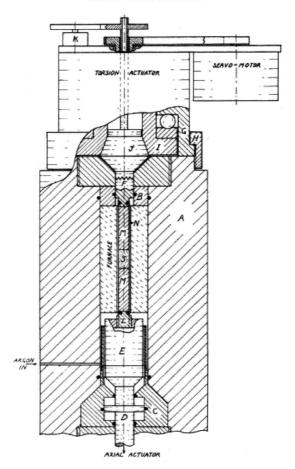

Figure A1. Schematic diagram of high-pressure high-temperature testing machine with torsion actuator and specimen assembly. [Used by permission of the editor of the *Journal of Structural Geology*, from Paterson and Olgaard (2000), Fig. 5, p. 1346.]

The strain rates that can be used in torsion testing are limited by time constraints if large strains are to be achieved. Supposing that it were desired not to exceed about one day (say, 10^5 seconds) for a given experiment, then to achieve a strain of $\gamma = 10$ the strain rate would need to be not less than about 10^{-4} s^{-1}. Thus the trend is to use higher strain rates in large-strain torsion tests than in axial deformation tests. However, the strain rate dependence of the flow stress of the material at any point in its evolution can be tested by strain-rate stepping to lower strain rates. The stress exponent n for a power-law fit can be obtained directly from the torque-twist measurements through

$$n = \frac{d \ln \dot{\theta}}{d \ln M}$$

where $\dot{\theta}$ is the twist rate and M is the torque. Similarly the apparent activation energy Q can be obtained from

$$Q = nR \frac{d \ln M}{d(1/T)}$$

(Paterson and Olgaard 2000).

In order to compare rheological behavior in torsion and axial deformation, some assumption must be made concerning the nature of the constitutive relationship for general stress states. If the von Mises-type relationship for isotropic behavior is assumed under which the flow behavior is the same for a given shear strain energy, then if the flow stress in a conventional triaxial is σ the flow stress τ in torsion for the same "equivalent" strain rate is $\tau = \frac{1}{\sqrt{3}} \sigma$. Alternatively, if the comparison is made at the same nominal strain rate, we have $\tau = \frac{1}{3^{\frac{1+n}{2n}}} \sigma$, varying from $\tau = \frac{1}{3} \sigma$ to $\tau = \frac{1}{\sqrt{3}} \sigma$ as n increases from 1 to infinity (Paterson and Olgaard 2000). Thus, to a rough approximation, the axial deformation flow stress ("differential stress") is generally about double the shear flow stress in torsion.

REFERENCES

Adams FD, Nicolson JT (1901) An experimental investigation into the flow of marble. Phil Trans R Soc A195:363-401

Barnhoorn A, Bystricky M, Kunze K, Burlini L (2002) Localization of deformation in a two-phase rock: torsion experiments on calcite-anhydrite aggregates. EMPG IX J Conf Abstr 7:10

Bridgman PW (1949) The Physics of High Pressure (reprinted with Supplement), G. Bell and Sons, London, 445 p

Bridgman PW (1952) Studies in Large Plastic Flow and Fracture. McGraw Hill, New York, 362 p

Bruhn D, Groebner N, Kohlstedt DL (2000) An interconnected network of core-forming melts produced by shear deformation. Nature 403:883-886

Bystricky M, Kunze K, Burlini L, Burg J-P (1999) Rheology and texture of olivine aggregates deformed to high strains in torsion. EOS Trans AGU 80:F973.

Bystricky M, Kunze K, Burlini L, Burg J-P (2000) High shear strain of olivine aggregates: rheological and seismic consequences. Science 290:1564-1567

Bystricky M, Mackwell S (2001) Creep of dry clinopyroxene aggregates. J Geophys Res 106: 13,443-13,454

Casey M, Kunze K, Olgaard DL (1998) Texture of Solnhofen limestone deformed to high strains in torsion. J Structural Geology 20:255-267

Chopra PN, Paterson MS (1984) The role of water in the deformation of dunite. J Geophys Res 89: 7861-7876

Dimanov A, Dresen G, Wirth R (1998) High-temperature creep of partially molten plagioclase aggregates. J Geophys Res 103:9651-9664

Fressengeas C, Molinari A (1987) Instability and localization of plastic flow in shear at high strain rates. J Mech Phys Solids 35:185-211

Gere JM, Timoshenko SP (1984) Mechanics of Materials. PWS Engineering, Boston, Massachusetts, 768 p

Gleason GC, Tullis J (1995) A flow law for dislocation creep of quartz aggregates determined with the molten salt cell, Tectonophysics 247:1-23

Griggs DT (1936) Deformation of rocks under high confining pressures. J Geol 44:541-577

Griggs DT (1967) Hydrolytic weakening of quartz and other silicates. Geophys J R Astron Soc 14: 19-31

Griggs DT, Blacic JD, Christie JM, McLaren AC, Frank FC (1966) Hydrolytic weakening of quartz crystals. Science 152:674

Griggs DT, Turner FJ, Heard HC (1960) Deformation of rocks at 500° to 800°C. *In* Griggs D, Handin J (eds) Rock Deformation. Geol Soc Am Mem 79:39-104

Handin J, Heard HC, Magouirk JN (1967) Effects of the intermediate principal stress on the failure of limestone, dolomite, and glass at different temperatures and strain rates. J Geophys Res 72: 611-640

Handin J, Higgs DV, O'Brien JK (1960) Torsion of Yule marble under confining pressure. *In* Griggs D, Handin J (eds) Rock Deformation, Geol. Soc. Am Mem 79, p 245-274

Hart EW (1967) The theory of the tensile test. Acta Metall 15:351-355

Heidelbach F, Stretton IC, Kunze K (2001) Texture development in polycrystalline anhydrite experimentally deformed in torsion. Intl J Earth Sciences 90:118-126

Heidelbach F, Stretton IC, Langenhorst F, Mackwell S (2002) Fabric evolution during high shear-strain deformation of magnesiowüstite ($Mg_{0.8}Fe_{0.2}O$). J Geophys Res (in press)

Holtzman BK, Groebner NJ, Zimmerman ME, Ginsberg SB, Kohlstedt DL (2002) Deformation-driven melt segregation in partially molten rock. G^3 (in press)

Jung H, Karato S (2001a) Effect of water on the size of dynamically recrystallized grains in olivine. J Struct Geol 23:1337-1344

Jung H, Karato S (2001b) Water-induced fabric transition in olivine. Science 239:1460-1463

Kaminski E (2002) The influence of water on the development of lattice-preferred orientation in olivine aggregates. Geophys Res Lett 29(12):10.1029/2002GL014710

Karato S, Paterson MS, Fitz Gerald JD (1986) Rheology of synthetic olivine aggregates: influence of grain size and water. J Geophys Res 91:8151-8179

Kohlstedt DL, Evans B, Mackwell SJ (1995) Strength of the lithosphere: constraints imposed by laboratory experiments. J Geophys Res 100:17,587-17,602

Kronenberg AK, Tullis J (1984) Flow strengths of quartz aggregates: Grain size and pressure effects due to hydrolytic weakening. J Geophys Res 89:4281-4297

Luan F, Paterson MS (1992) Preparation and deformation of synthetic aggregates of quartz. J Geophys Res 97:301-320

Mackwell S, Rubie D (2000) Earth under strain. Science 290:1514-1515

Mackwell SJ, Kohlstedt DL, Durham WB (1990) High-resolution creep apparatus. *In* Duba AG, Durham WB, Handin JW, Wang HF (eds) The Brittle-Ductile Transition in Rocks. *The Heard Volume*. American Geophysical Union, Washington, DC, Geophys Monogr 56:235-238

Paterson MS (1964) Triaxial testing of materials at pressures up to 10,000 kg/sq cm (150,000 lb/sq in). J Institution Engin Australia 36:23-29

Paterson MS (1970) A high-pressure, high-temperature apparatus for rock deformation. Intl J Rock Mech Min Sci 7:517-526

Paterson MS (1978) Experimental Rock Deformation—The Brittle Field, Minerals and Rocks. Springer-Verlag, Berlin, 254 p

Paterson MS (1990) Rock deformation experimentation. *In* Duba AG, Durham WB, Handin JW, Wang HF (eds) The Brittle-Ductile Transition in Rocks. *The Heard Volume*. American Geophysical Union, Washington, DC, Geophys Monogr 56:187-194

Paterson MS, Olgaard DL (2000) Rock deformation tests to large shear strains in torsion. J Struct Geol 22:1341-1358

Pieri M, Burlini L, Kunze K, Stretton I, Olgaard DL (2001a) Rheological and microstructural evolution of Carrara marble with high shear strain: results from high temperature torsion experiments. J Struct Geol 23:1393-1413

Pieri M, Kunze K, Burlini L, Stretton I, Olgaard DL, Burg J-P, Wenk H-R (2001b) Texture development of calcite by deformation and dynamic recrystallization at 1000 K during torsion experiments of marble to large strains. Tectonophysics 330:119-140

Rutter EH (1974) The influence of temperature, strain rate and interstitial water in the experimental deformation of calcite rocks. Tectonophysics 22:311-334

Rutter EH (1998) Use of extension testing to investigate the influence of finite strain on the rheological behavior of marble. J Struc Geol 20:243-254

Rutter EH, Casey M, Burlini L (1994) Preferred crystallographic orientation development in during the plastic and superplastic flow of calcite rocks. J Struct Geol 16:1431-1446

Rybacki E, Paterson MS, Wirth R, Dresen G (2002) Rheology of calcite-quartz aggregates deformed to large stain in torsion. J Geophys Res (submitted)

Schmid SM, Boland JN, Paterson MS (1977) Superplastic flow in finegrained limestone. Tectonophysics 43:257-291

Schmid SM, Panozzo R, Bauer S (1987) Simple shear experiments on calcite rocks: rheology and microfabric. J Struct Geol 9:747-778

Schmid SM, Paterson MS, Boland JN (1980) High temperature flow and dynamic recrystallization in Carrara marble. Tectonophys 65:245-280

Schmocker M, Bystricky M, Kunze K, Burlini L, Stünitz H, Burg J-P (2002) Granular flow and Reidel band formation in water-rich quartz aggregates experimentally deformed in torsion. J Geophys Res (in review)

Stretton IC, Heidelbach F, Mackwell S, Langenhorst F (2001) Dislocation creep of magnesiowüstite ($Mg_{0.8}Fe_{0.2}O$). Earth Planet Sci Lett 194:229-240

Stretton I, Olgaard DL (1997) A transition in deformation mechanism through dynamic recrystallization - evidence from high strain, high temperature torsion experiments. EOS, Trans Am Geophys Union 78:F723

Stünitz H, Tullis J (2001) Weakening and strain localization produced by syn-deformational reaction of plagioclase. Intl J Earth Sciences 90:136-148

Tullis J, Yund R (1999) Shear deformation of plagioclase aggregates: effects of composition and water. EOS Trans AGU 80:F1054

Tullis TE, Tullis J (1986) Experimental rock deformation techniques. *In* Hobbs BE, Heard HC (eds) Mineral and Rock Deformation: Laboratory Studies. *The Paterson Volume*. American Geophysical Union, Washington, DC, Geophys Monogr 36:297-324

von Kármán T (1911) Festigkeitsversuche unter allseitigem Druck. Z Verein Deutsch Ing 55: 1749-1757

Yamazaki D, Karato S (2002) Fabric development in (Mg,Fe)O during large strain, shear deformation: Implications for seismic anisotropy in Earth's lower mantle. Phys Earth Planet Inter (in press)

Zhang S, Karato S (1995) Lattice preferred orientation of olivine aggregates deformed in simple shear. Nature 375:774-777

Zhang S, Karato S, Fitz Gerald JD, Faul U, Zhou Y (2000) Simple shear deformation of olivine aggregates. Tectonophysics 316:133-152

Zimmerman ME, Zhang S, Kohlstedt DL, Karato S (1999) Melt distribution in mantle rocks deformed in shear. Geophys Res Lett 26:1505-1508

2 New Developments in Deformation Experiments at High Pressure

William B. Durham
*University of California,
Lawrence Livermore National Laboratory, P.O. Box 808
Livermore, California 94550*

Donald J. Weidner
*Department of Geosciences
State University of New York
Stony Brook, New York 11794*

Shun-ichiro Karato
*Departmentof Geology and Geophysics
319 Kline Geology Laboratory, Yale University
New Haven, Connecticut 06520*

Yanbin Wang
*Center for Adanced Radiation Studies
University of Chicago
Chicago, Illinois 60439*

INTRODUCTION

Although the importance of rheological properties in controlling the dynamics and evolution of the whole mantle of Earth is well-recognized, experimental studies of rheological properties and deformation-induced microstructures have mostly been limited to low-pressure conditions. This is mainly a result of technical limitations in conducting quantitative rheological experiments under high-pressure conditions. A combination of factors is changing this situation. Increased resolution of composition and configuration of Earth's interior has created a greater demand for well-resolved laboratory measurement of the effects of pressure on the behavior of materials. Higher-strength materials have become readily available for containing high-pressure research devices, and new analytical capabilities—in particular very bright synchrotron X-ray sources—are now readily available to high-pressure researchers.

One of the biggest issues in global geodynamics is the style of mantle convection and the nature of chemical differentiation associated with convectional mass transport. Although evidence for deep mantle circulation has recently been found through seismic tomography (e.g., van der Hilst et al. (1997)), complications in convection style have also been noted. They include (1) significant modifications of flow geometry across the mantle transition zone as seen from high resolution tomographic studies (Fukao et al. 1992; Masters et al. 2000; van der Hilst et al. 1991) and (2) complicated patterns of flow in the deep lower mantle (~1500-2500 km), perhaps caused by chemical heterogeneity (Kellogg et al. 1999; van der Hilst and Karason 1999).

These studies indicate that while large-scale circulation involving the whole mantle no doubt occurs, significant deviations from simple flow geometry are also present. Two mineral properties have strong influence on convection: (1) density and (2) viscosity (rheology) contrasts. In the past, the effects of density contrast have been emphasized

(Honda et al. 1993; Kellogg et al. 1999; Tackley et al. 1993), and the influence of rheology has been demonstrated by geodynamic modeling (Davies 1995; Karato et al. 2001).

Rheological properties of the Earth's mantle can be inferred from analyses of geodynamic data such as post-glacial crustal movement and gravity anomalies (Mitrovioca and Forte 1997; Peltier 1998). While such geodynamic inference provides an important data set on rheological structures of the Earth, there are serious limitations for these approaches. First, the resolution of post-glacial rebound data to infer mantle viscosity is limited to ~1200 km. Below that depth, there is little constraint for viscosity from post-glacial rebound (Mitrovioca and Peltier 1991). Furthermore, strain magnitude involved in post-glacial rebound is much smaller than strain magnitude involved in convection, and this raises the issue of effects of transient rheology (Karato 1998). Gravity data (e.g., the geoid) have better sensitivity to radial variation in viscosity in the deep mantle (Hager 1984), but they suffer from non-uniqueness (King 1995; Thoraval and Richards 1997). Therefore, experimental studies on mineral rheology remain a vital component in inferring mantle rheology.

High-pressure rheology may also hold some of the keys for understanding deep earthquakes. Four classes of mechanism have been suggested over the years, including thermal runaway instabilities (Hobbs and Ord 1988; Ogawa 1987), instabilities accompanying recrystallization (Post 1977), dehydration embrittlement (Jiao et al. 2000; Raleigh and Paterson 1965), and instabilities associated with polymorphic phase transformations (see also reviews by Green and Houston (1995) and Kirby et al. (1996)). Again, the main reason uncertainty still exists is the poor knowledge of materials properties at high pressures. Experimental testing of the thermal runaway involved in plastic instability requires high rates of deformation, which are difficult to obtain in small samples; modeling would be a useful tool for this mechanism, but critical parameters for models (rheological properties of deep Earth materials) have been missing so far (Hobbs and Ord 1988; Karato et al. 2001). Well-controlled deformation experiments under deep mantle conditions are needed to provide greater understanding of the shearing instability associated with the phase transformations of olivine to wadsleyite and ringwoodite (Burnley et al. 1991; Green et al. 1990) and the possibility of faulting associated with dehydration of the dense hydrous magnesium silicates (Green 2001).

Why is pressure important?

Pressure is one of the most important variables in defining the ductile rheology of mantle minerals. Where the deformation mechanism is fixed, the effect of pressure on creep strength is given by

$$\sigma(P)/\sigma(0) \propto \exp(PV^*/nRT), \tag{1}$$

where $\sigma(P)$ is the creep strength at pressure P, V^* is activation volume, n is the stress exponent, R is the gas constant, and T is temperature. For olivine, assuming an activation volume of $V^* = 15$ cm^3/mol, $T = 1600$ K and $n = 3$, $\sigma(10\text{ GPa})/\sigma(0) = 43$. This is to be compared with the effect of water, which will change the creep strength of olivine by a factor of ~10 from dry to water-saturated (Hirth and Kohlstedt 1996; Karato and Jung 2002 in press) or partial melting that will affect creep strength by no more than a factor of 2 (Kohlstedt and Chopra 1994). However, the pressure effect is small at the low pressures where studies are made, e.g. $\sigma(0.5\text{ GPa})/\sigma(0) = 1.2$. Therefore, high-pressure studies are necessary in order to determine the flow properties in the deep Earth.

Such large effects of pressure on the flow laws can result in changes of dominant flow mechanisms with depth in the Earth. Karato and Wu (1993) suggest that dislocation flow is overtaken by diffusion flow as the creep mechanism for olivine within the upper mantle. Their reasoning is based on the differences in the pressure dependence, i.e., in the value of V^* in Equation (1), of the two flow laws.

Pressure is an agent for phase transformation within the Earth. These high-pressure phases, with different crystal structures, will also differ in their mechanical properties. Rheology of high-pressure phases needs to be investigated within the stability field of the mineral. Because of equipment limitations, the flow properties of minerals such as ringwoodite or majorite, which set the strength characteristics of the transition zone, remain virtually unknown.

The effects of water can be well characterized only by high-pressure experiments. At low pressures ($P<0.5$ GPa), the creep strength of olivine *decreases* with pressure under water-saturated conditions (Mei and Kohlstedt 2000b; Mei and Kohlstedt 2000a). However, at higher pressures ($P>1$ GPa), the creep strength of olivine *increases* with pressure even at water-saturated conditions (Karato and Jung 2002 in press). Therefore results that can be extrapolated to infer creep strength (viscosity) at deep portions (>30 km) of Earth can be obtained only by quantitative experiments at pressures higher than ~1 GPa.

Terminology related to strength and deformation.

Usage sometimes varies among research disciplines, so let us define a few key terms, taking the language of rock mechanics as a basis (e.g., Jaeger and Cook (1976)). It is assumed that the meaning of the *stress* (σ_{ij}) and *strain* (ε_{ij}) tensors are understood. *Deformation* (used already in the title without definition) refers loosely to any process that results in strain. We take positive values of stress and strain to indicate compression, following geologic convention. *Pressure*, *hydrostatic pressure*, and *mean normal stress* are all equivalently the trace P of the stress tensor, namely,

$$\sigma_{ij} = \sigma'_{ij} + P, \qquad (2)$$

where σ'_{ij} is the *deviatoric* stress tensor, $i, j = 1, 2, 3$, representing orthogonal directions. *Confining pressure* P_c is the hydrostatic stress generated by laboratory devices to simulate geologic overburden. Although P and P_c are different when σ'_{ij} is non-zero, the difference in most geologic settings is small and the distinction is very often ignored. Different terms for the same physical quantity are sometimes used to connote different situations. Strain can be divided into two categories: *elastic*, the recoverable portion, which owes its existence to the state of non-zero σ_{ij}, and *inelastic*, the permanent strain that remains forever after stress has been removed. Classic rock mechanics usage recognizes two categories of deformation that result in inelastic strain: *brittle* and *ductile*. Brittle deformation is associated with discontinuities of displacement and sudden loss of strength (a term to be defined shortly) and is generally not volume conservative. Ductile deformation is associated with finite displacement gradients, retention of strength, and volume conservation at scales above the nm. Note that there are important strain-producing phenomena that do not fit well into either category, in particular those that involve polymorphism (phase change) or chemical reaction. They may be ignored for the purposes of this review. The *strength* of a material (in specific circumstances called the *yield strength*, *ultimate strength*, etc.) is the maximum stress the material can support under the stated environmental conditions. *Flow* and *failure* are closely related terms for the state that exists the moment that applied stress matches strength for ductile and brittle materials, respectively.

The term *plasticity* requires special note. It was originally defined for metals as essentially equivalent to the phenomenon of flow, just defined, but was defined in terms of the yield envelope (Hill 1950) without qualification as to ductile or brittle yield. While metallurgists and most rock rheologists have retained this meaning for the term plasticity (e.g., Evans and Kohlstedt [1995] and Poirier [1985], many in the computational branches of geology have taken plasticity to mean any yield process, whether volume conservative or not. In this review, we follow tradition and use the terms ductile and plastic interchangeably.

Here we focus exclusively on ductile deformation, so we use the terms *rheology* (flow of materials under stress), deformation, and flow interchangeably. The rheology of a material is described by a constitutive law of the type

$$\dot{\varepsilon}_{ij} = Af(\sigma'_{ij}), \quad A = A(P, T, ...) \tag{3}$$

where all relevant state variables are included in the term A. Engineering the containment of very high pressures in the laboratory generally requires high symmetry in the sample assembly. As will be seen below, two basic types of high-symmetry strain environments exist: axisymmetric and rotational. A shorthand form of Equation (3) has evolved for each of these geometries and we will use that shorthand here. For axisymmetric deformation (whose two-dimensional analog is called *pure shear*), the values of the three principal stresses are $\sigma_1 \neq \sigma_2 = \sigma_3$, and we simplify the strain rate tensor to a scalar $\dot{\varepsilon}$, which we call the *strain rate* or *shortening rate*. The deviatoric stress tensor can be characterized by the scalar $\sigma = \sigma_1 - \sigma_3$, which we call the *differential stress* or simply *stress*. For rotational deformation, the strain is approximately *simple shear*, and Equation (3) becomes a scalar relationship between the *shear strain rate*, $\dot{\gamma}$, and the *shear stress*, τ.

Despite the apparent simplicity of this language of strength and deformation, one should always keep two things in mind: (1) the terminology is not universal, and (2) the simple language belies the flow of rocks in nature, which can be counted on to be both inhomogeneous and low symmetry.

A BRIEF HISTORY OF HIGH-PRESSURE APPARATUS

To 5 GPa: Cylindrical devices

Confining pressure in rock mechanics testing serves several purposes: (1) simulating geologic overburden, (2) activating mechanisms of flow and fracture that depend on both normal and shear components of stress, and (3) suppressing brittle deformation where one is interested in gathering information on purely ductile phenomena. The scientific needs for doing measurements under pressure are illustrated in detail in several other contributions to this volume. These needs have driven the development of experimental pressure systems since the early days of rock mechanics.

The earliest experimental machines for deforming rocks under elevated confining pressure were some version of a piston pushing on a cylindrical rock held fast in a cylindrical container. A solid or fluid medium within the container, or even walls of the container itself, provided confining pressure to the rock sample. The reader may find an early history of such machines in Tullis and Tullis (1986), for example. The direction of apparatus development after 1900 was influenced primarily by P.W. Bridgman, who made such rapid advances in the design of hydrostatic equipment that the only reasonable path for rock mechanics to follow was to adapt deformation to Bridgman's vessels. Bridgman, who conceived the dynamic piston seal still commonly used, was involved in that adaptation (Bridgman 1952c), although he was no longer alone in his efforts. Continuing to this day, virtually every new deformation rig is some adaptation of a hydrostatic design. The backbone of the history of deformation machines is mostly a history of pressure vessels.

The technological breakthrough behind modern gas and liquid pressure vessels was the Bridgman unsupported area packing (Bridgman 1914), a geometry that intensified the vessel's own pressure at the point of seal and made it possible to "reach without leak any pressure allowed by the mechanical strength of the walls of the containing vessels" (Bridgman 1952b). Bridgman regularly worked at pressures in excess of 1 GPa and occasionally reached 2 GPa with his fluid vessels (Bridgman 1935b). The application of the Bridgman-seal vessel to rock deformation led to the creation of what is commonly called in rock mechanics the "triaxial" apparatus (which is somewhat inappropriate, because the stress environment in the fluid pressure medium can be no lower than axisymmetric). Griggs, who learned the techniques of high pressure from Bridgman and shared his "gift for gadgets" (Rubey 1972), initiated this application by developing a version of the unsupported area packing around a solid moving piston (Fig. 1), thereby allowing sample shortening to be equated (with appropriate corrections for elasticity) to piston displacement (Griggs 1936). Griggs (1936) added a second innovation to the design, a pressure-compensating double piston that automatically kept pressure constant even as the sample was shortened (Fig. 1). Further development of the triaxial rig was aided by advances such as the controlled-clearance (Newhall) packing to reduce piston friction (Handin 1953; Turner et al. 1956); internal heating to achieve higher sample temperatures (Griggs et al. 1960); and the internal force gauge, which removed piston friction from the measurement of differential force on the sample (Heard and Carter 1968; Paterson 1970). This list is by no means exhaustive. A notable recent development is the proliferation in many of the world's most renown rock mechanics laboratories of one particular triaxial apparatus, the Paterson rig, which is a 0.3-GPa ready-to-use system complete with actuation and internal force gauging for both axial and high-displacement torsional deformation (Paterson 1990; Paterson and Olgaard 2000). As of this writing, there are ten such vessels in use in North America and Europe.

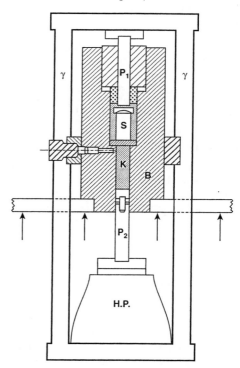

Practical matters related to sealing a high-pressure fluid have generally limited triaxial work to about $P = 1$ GPa, with most triaxial machines, in fact, having maximum design pressures of 0.3-0.5 GPa (Paterson 1978). At higher pressures, the solid pressure medium piston-cylinder apparatus is used. The solid-medium press used most widely for geological applications is the Griggs apparatus (Griggs 1967), adapted from the hydrostatic solid-medium piston-cylinder apparatus (Boyd and England 1960), and subsequently improved to provide a stress resolution of a few MPa (Borch and Green 1989; Green and Borch 1989; Tingle et al. 1993). Since a solid medium has a finite shear strength, measurements are limited to materials

Figure 1. The Griggs triaxial deformation apparatus, including the yoked pistons to hold pressure constant during deformation. As piston P_1 moves against the sample, piston P_2 automatically withdraws from the vessel at the same rate, holding the pressurized volume constant. After Griggs (1936), Fig. 2; reproduced with permission.

that are stronger than the medium, so very low strain rates and temperatures near the sample melting point must be avoided, but pressures of several GPa are now easily accessible. The Griggs apparatus has been regularly used to 1.5-1.8 GPa, and to a maximum pressure of 4 GPa.

A recent development is an apparatus built by Getting (Getting 1998; Getting and Spetzler 1993) that combines aspects of the solid-medium and fluid-triaxial deformation machines. Getting's is a piston-cylinder device that features precision packings that are sufficiently tight to contain solid argon (an exceedingly weak solid) but sufficiently reproducible in their frictional behavior that much of the uncertainty in stress introduced by piston friction can be accounted for by calibration. Getting (1998) claims pressure uncertainty at $P = 3$ GPa at the extraordinary level of a few MPa. If this resolution applies also to the sum of $\sigma_1 + P$, as expected, it may mean that rock strengths can be measured with a resolution that is comparable to that of many triaxial gas rigs that operate at far lower pressures.

The practical pressure limit for all cylindrical machines is around 5 GPa. While the pressure capacity of a hollow cylinder (or any shape, for that matter) is a function of wall thickness and therefore virtually unlimited (Eremets 1996), the same is not true for the piston. As one approaches $P = 5$ GPa, the stress required to drive the piston into the pressure cylinder (which must be 5 GPa plus the stress to overcome packing friction plus the strength of the sample) causes pistons constructed of even the strongest of steels to creep. Pistons will swell and seize in a short amount of time. Note that this 5-GPa limit exists for hydrostatic piston-cylinder vessels as well. Bridgman reached higher pressures in hydrostatic vessels by confining the entire piston and cylinder in a larger vessel, thus raising σ_3 on the piston to the pressure in the outer vessel and allowing the piston to apply correspondingly more stress to the inner vessel without exceeding its strength limit. His vessel-inside-vessel multi-staged devices have been used to reach pressures of at least 10 GPa (e.g., Bridgman (1942; 1948)). Because of the complexity of those vessels, however, they are not well suited to being instrumented for deformation studies.

5 GPa and above: Anvil devices

A much simpler method for achieving an elevated σ_3 is to give the pistons a tapered shape. (The change from rectangular to trapezoidal section also merits a name change: from piston to *anvil*.) Now a component of the force applied in the σ_1 direction at the larger end of the piston/anvil is directed in the σ_3 direction at the smaller end (platen face), making $\sigma_3 > 0$ at the platen face. The drawback of this solution is, of course, that one cannot expect to contain the pressure medium over large displacements in a conventional cylindrical pressure vessel if the sides of the (moving) anvil are tapered.

Sealing against a moving tapered surface can be achieved for small displacements through use of a crushable gasket material; so if one is satisfied with anvil displacements that are less than gasket thickness, one can reach high values of σ_1 at the platen faces without exceeding the failure criterion of the anvils. The first application of this principle is generally credited to Bridgman's (Bridgman 1952a) opposed-anvil device, which is capable of 10 GPa, although Bridgman used the same principle many years earlier in confining samples in a 5-GPa shear apparatus (Bridgman 1935a). It is also the underlying principle of the Drickamer cell (see below). Note that the original purpose of anvil displacement in these designs was pressure generation rather than sample deformation, although as we will see below, the possibility of the latter was not ignored.

A slight variation, which allows for larger volumes at a given pressure, is to combine the function of pressure container and pressure generator by clustering several anvils in a quasi-spherically symmetric pattern. These are called multianvil devices. Four-anvil tetrahedral (Hall 1958; Houck and Hutton 1963; Lloyd et al. 1959), six-anvil cubic (Carter et al. 1964; Houck and Hutton 1963; Osugi et al. 1964), and eight-anvil octahedral (Kawai and Endo 1970; Walker et al. 1990) have been constructed. The eight-anvil devices are multistage, with the eight anvils of the final stage contained within the working volume of an intermediate cubic stage. They are often referred to, therefore, as 6/8 devices. Hydraulic actuation of the several anvils was sometimes independent or in opposed pairs (Carter et al. 1964; Hall 1958), but better symmetry of anvil displacement was achieved, along with higher pressures, when the anvils were driven by wedge-shaped guide blocks that then required only one hydraulic actuator (Lloyd et al. 1959). There are at least two versions of the wedge-type cubic multianvil device: a split-sphere device (Houck and Hutton 1963) and the DIA (Osugi et al. 1964; Shimomura et al. 1985); the latter will be describe below in detail.

Wedged guide block machines achieve high pressures more reliably than machines with multiple actuators, and for that reason they have proliferated, especially the DIA and 6/8 devices (e.g., see review by Onodera (1987)). Note, however, that their design function is exclusively hydrostatic pressure generation. Advances in high-pressure rheology required that the quasi-isotropy within the pressurized volume be broken; those advances are the main subjects of this review.

Another reason for the recent proliferation of multianvil devices is a serendipitous advantage they have over pressure cylinders: optical and/or X-ray transparency along a line of sight leading directly to and from the sample (Fig. 2). In the case of diamond (or sapphire) anvil cells, the anvils themselves are transparent. For devices where anvils are opaque, lines of sight are always available in the narrow gaps between anvils, and the gaskets compressed in those gaps can be made of X-ray-transparent weak material (e.g., pyrophyllite). The analytical potential of high-energy X-rays and their recent availability at very bright synchrotron sources has been the motivation for development of a new generation of deformation machines that have the pressure capacity of anvil devices and the lowered symmetry of deformation machines.

MEASUREMENT METHODS AT HIGH PRESSURE

Stress measurement

Measurement of deviatoric stress is key to quantifying the rheological properties of a material. Stress provides information about strength relative to the laboratory time scale. Stress, measured in conjunction with ductile strain, can yield flow laws within the context of the state of the sample. The crucial ingredient to facilitate such studies is the development of a stress meter, or piezometer.

In experiments operating below 5 GPa, the deforming force is applied directly to the sample and various strategies are designed to measure the magnitude of this deforming force. Stress is then given as the ratio of force to sample area. Typically, a load cell placed in series with the deforming piston defines this force along the deforming column. Combinations of friction between the force gauge and the sample and forces that support the piston, such as in the multianvil system, continually degrade the ability to relate a measured force to sample stress. These problems begin to interfere with the resulting accuracy at 0.3 GPa and completely overwhelm the measurements by 5 GPa.

A revolutionary technique for stress and strain measurements under high-pressure (and temperature) conditions *in situ* has now become feasible using high-energy X-rays generated by a synchrotron radiation facility. These new tools are just now being explored and their limitations defined. The exceptional quality of these tools rests in the fact that they are directly monitoring the sample. Stresses are measured in the sample.

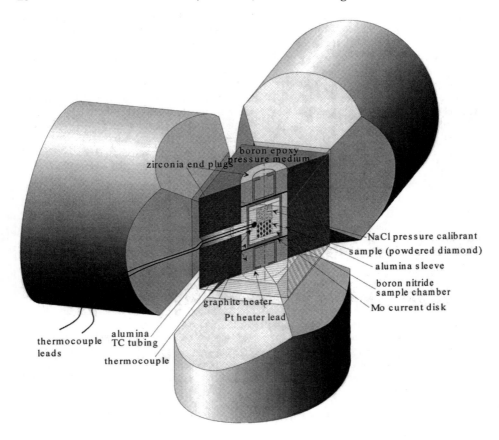

Figure 2. Sample assembly in the DIA cubic apparatus, shown with three of the six anvils removed and with the sample assembly itself, i.e., the cube, shown in cross section. Analytical synchrotron X-rays enter and exit along a horizontal line through the sample and normal to the section as shown here. The X-rays pass through the vertical gaps between anvils. Drawing by M. Vaughan.

Different positions in the sample can be isolated to test for stress uniformity. Measurements of stress and strain can be time resolved with a precision of about one minute. Strain is also obtained by images of the sample. Distribution of strain with position and time can be defined. All of this is done with a current accuracy of a few tens of MPa in differential stress and 10^{-4} in strain.

Strategies for measuring stress are found through use of the X-ray diffraction or pressure-sensitive probes (such as ruby fluorescence), and are being made available by exploiting powerful synchrotron X-ray sources. The basic principle of X-ray piezometers is that lattice strain is sensitive to elastic strains and not to plastic strains. Changes in distances between atoms are expected to balance stress fields that are present. These stress fields may be created by dislocations or external forces, but departures from the lattice spacing that the material experiences at zero pressure and stress are sensibly related to the stress field within the sample. Furthermore, the coupling between the strain and stress is expressed by the elastic moduli. There are two styles of X-ray piezometers, one based on stress gradient and the other based on strain anisotropy.

Stress gradient. Where the inertia term can be neglected, variation in the stress field is restricted to obey the mechanical equilibrium relations:

$$\sum_{j=1}^{3}\frac{\partial \sigma_{ij}}{\partial x_j}=0$$

Three equations, one for each value of i, represent the vector force balance equations. The first of these equations is:

$$\frac{\partial \sigma_{11}}{\partial x_1}+\frac{\partial \sigma_{12}}{\partial x_2}+\frac{\partial \sigma_{13}}{\partial x_3}=0$$

Thus, if there is a variation of the normal stress, σ_{11}, with x_1, there must also be a variation of the shear stresses, σ_{12}, in the x_2 direction and/or a variation of σ_{13} in the x_3 direction. The peak magnitude of the shear stress is related to the variation of the normal stress (or pressure) and the length scales of two types of stress. This has been exploited in the diamond anvil cell geometry by measuring the radial gradient of pressure to deduce the axial (along the axis of the diamonds) variation of shear stress. This is illustrated in Figure 3. The shear stress, τ, vanishes along the center plane of the sample and along the central axis of the cell where $\partial P/\partial r$ is zero. The maximum shear stress is at the sample-diamond interface and is given by

$$\tau_{max}=(\partial P/\partial r)_{max}\, d/2$$

where d is the thickness of the sample. This technique was developed by Sung et al. (1977) and was further used by Meade and Jeanloz (1988a,b; 1990). These experiments relied on ruby dust distributed across the sample as a pressure marker for delineating the

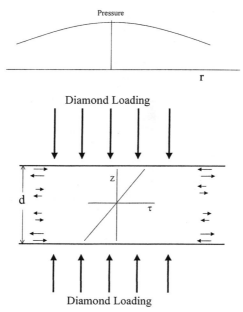

Figure 3. Illustration of stress field in a diamond-anvil cell. The radial distribution of pressure, illustrated at the top, is coupled to the shear stress distribution along the axial direction. Force couples indicate the sense of the shear stress distribution within the cell.

pressure gradient. As Wu and Bassett (1993) illustrate, the pressure marker can also be an X-ray diffraction standard such as gold. The objective is to measure the radial pressure profile to define the shear-stress distribution. Often, the peak shear stress is taken as the yield point of the solid.

Stress gradients coupled with deviatoric stress affect the diffraction peak shape as described in detail by Weidner(1998) and Weidner et al. (1998). In this case, heterogeneities in the stress field broaden the diffraction peaks. Since the length scale of stress variation is the same for normal and shear stress, then the magnitude of the heterogeneity for all stresses is similar. Weidner et al. (1998) model the relationship between the observed strain (that creates the broadened diffraction peak) and the stress magnitude and conclude that Young's modulus is the appropriate coupling elastic modulus, or

$$\sigma_{differential} = E\ \varepsilon_{differential},$$

where the differential strain is deduced as the "strain broadening" contribution to the peak profile. Since for the peak-broadening measurement, the full-width, half-height of the diffraction peak is used to define the strain broadening and represents roughly two-thirds of the scattering diffraction planes, then most of the sample must be experiencing the inferred shear stress. Deviatoric stress can originate from dislocations, from heterogeneities in material or properties, or any number of sources. Peak-broadening analysis is often used in commercial applications to determine stresses induced by processing. For example, internal stress generated by welding is often analyzed by this method. In high-pressure studies, the stress is most commonly generated by compressing powdered samples (e.g., Chen et al. 1998). As a rheological tool, this approach resembles an indentation experiment where the indenter is the same material as the sample. Weidner et al. (1994) studied the high-temperature and -pressure strength of diamond with this method. Chen et al. (1998) studied plastic deformation of olivine, wadsleyite, and ringwoodite. Flow can also be studied with this piezometer, with the constraint that this becomes a relaxation experiment (Weidner et al. 2001).

Strain anisotropy. A uniaxial stress will introduce elastic strains in the sample. The strain parallel to the axis of compression will generally be larger than the strain perpendicular. The material's elastic moduli quantitatively define the relationship between these strains and the imposed stresses. X-rays sample the distance between lattice planes (called *d*-spacing) whose normals are parallel to the diffraction vector, which is the bisector between the incident X-ray and the detector. The lattice spacing is insensitive to strain history (plastic strain) but reflects the elastic strain field. Figure 4 illustrates a Debye ring that would be observed from a powder sample in a stress field using a monochromatic X-ray beam and an area detector. The lattice spacing is related to the distance from the center of the image to the Debye ring through Bragg's law. For the angle Ψ of zero, the lattice spacing reflects that measured parallel to the x_1 direction. For Ψ of ninety degrees, the X-rays are sampling grains aligned parallel to the x_2 direction. The strains deduced from these measurements compared with ambient conditions are related to the stresses by

$$\varepsilon_{ij} = \sum_{k=1}^{3}\sum_{l=1}^{3} s_{ijkl}\sigma_{kl},$$

where **s** is the elastic compliance tensor.

In the actual case, more than one set of diffraction planes will produce Debye rings similar to the one illustrated here. Each will be approximately circular in shape, but distorted by the stress field. The amount of distortion may be different if the elastic

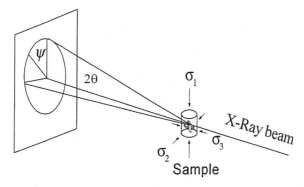

Figure 4. X-ray diffraction line on a two-dimensional detector from a sample under stress. The distance from the center to the ring defines the scattering angle, 2θ, and hence the lattice spacing through Bragg's law. The presence of the stress field distorts the Debye ring from its normally circular shape. The usual stress field is axisymmetric, with $\sigma_2 = \sigma_3$. The diffraction vector at $\psi = 0$ lies at an angle θ from the direction of σ_1. θ is typically only 3 to 4 degrees.

compliance tensor is not isotropic. To evaluate this in detail, we need to be careful to account for the crystal orientation when defining the compliance tensor. Thus, in a non-hydrostatic stress field, crystallographic planes with different orientations change their spacing by different amounts. For cubic crystals, the relationship between the Ψ dependence of strain and the differential stress, $\Delta\sigma$, is given by:

$$\varepsilon_{hkl}(\Psi) = \Delta\sigma /3 \{1 - 3\cos^2\Psi\} \{(S_{11} - S_{12})[1 - \Gamma(hkl)] + S_{44} \Gamma(hkl)/2\}$$

Each diffraction peak corresponds to a set of lattice planes defined by the Miller index, hkl. In an anisotropic crystal, the elastic moduli will depend on the particular set of planes that are considered. The function, $\Gamma(hkl)$, varies from 0 to 1 and accounts for this variation as it multiplies the different elastic compliance moduli, S_{ij}. Thus, for a particular diffraction plane, hkl, the right hand bracket contains a constant value. Then ε varies in a fixed manner with the angle, Ψ.

The above expression is based on the assumption that each grain experiences the same stress field, or the Reuss condition. A similar relation can be derived by assuming that each grain satisfies the Voigt condition of iso-strain. In this case the equation is altered by replacing the elastic shear compliances, S_{44} and $(S_{11} - S_{12})$ by their Voigt equivalent. This relation for cubic symmetry is generalized by Singh et al. (1998) for all crystal symmetries.

This phenomenon gives rise to a couple of strategies for a piezometer. Stress has a manifestation even for a single value of Ψ. For an elastically anisotropic material, the strain for a given crystallographic plane will depend on the orientation of crystal with respect to the applied stress and the stress magnitude. Therefore by measuring the strain for several diffraction peaks corresponding to different orientations, one can determine the stress magnitude using the known elastic constants of the crystal. Such a measurement can now be done in a multianvil high-pressure apparatus using a synchrotron X-ray source (Weidner et al. 1992). The accuracy relies on knowing the elastic anisotropy at the pressure/temperature conditions of the experiment.

A more robust strategy comes from mapping out the d spacing as a function of Ψ. The values of the elastic moduli are still required, but the details of anisotropy are less critical to yielding precise measurements. With a precision of 10^{-4} for the lattice spacing and a typical elastic modulus of 200 GPa, it should be possible to resolve differential stresses of ~20 MPa. Singh has reduced the elasticity equations to a closed-form solution for several crystal symmetries and has given a concise summary (Singh et al. 1998).

This methodology was first applied in a diamond-anvil cell by Kinsland and Bassett (1977) by passing the X-ray beam through a Be gasket, perpendicular to the axis of the diamonds. The multianvil system has not been accessible by this technique because the anvils themselves cast a shadow and limit the range of Ψ where the diffracted signal can be observed. However, we have recently experimented with cubic BN anvils. These are nearly as hard as diamond but are about an order of magnitude less expensive. Furthermore, they are transparent to X-rays. Thus, by using these anvils it is now possible to observe diffraction spectra in any plane relative to the incident beam.

Here we have illustrated this tool with monochromatic X-rays and angle dispersive measurements. Energy dispersive methods are equally valid for these measurements and are sometimes advantageous. Spectra need to be collected for different values of Ψ through collimators that fix the two-theta value. Weidner's group has designed and used a conical slit system for this purpose as illustrated in Figure 5. Two concentric cones, whose angle is the desired diffraction angle for white, energy dispersive diffraction, create the slit itself. Solid-state detectors with energy discrimination are placed at specific values of Ψ behind the slit system. Each detector is calibrated independently. Lattice spacings are used as illustrated above to define the differential stress.

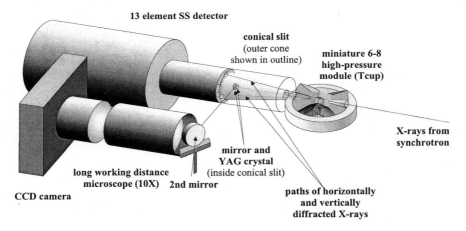

Figure 5. Conical slit assembly in use at National Synchrotron Light Source. The cone defines the two-theta angle and is constructed by two concentric cones with this angle. The X-rays that pass between the two cones originate from the sample at that angle as all other X-rays are blocked by the cones. Multiple energy discriminating detectors, located behind the slit assembly, record the X-ray signals. A YAG crystal located inside the slit assembly fluoresces in the X-ray beam, emitting visible light where an X-ray photon passes. The visible light is magnified by a microscope assembly and recorded by a digital camera.

Strain rate measurement

Synchrotron X-rays can also be used to measure sample length. Direct images of the sample can be obtained using an incident beam whose dimensions are larger than the sample. Platinum or gold foils above and below the sample are easily viewed. The X-ray image is obtained by projecting the X-ray onto a fluorescent screen that is viewed with a CCD camera though a magnifying system. A typical image is shown in Figure 6. The horizontal lines in this image are gold discs placed at the ends of the sample. The black edges are the shadows of the opaque tungsten carbide anvils. Cubic boron nitride anvils are transparent and can be used to view the entire sample because they do not cast such a shadow. Figure 5 illustrates the position of the microscope for these measurements. The fluorescent screen is located on-axis inside the conical slit system. We determine that, by comparing two images, strains of 10^{-4} can be measured. For images taken 100 seconds apart, this allows resolution of strain rates of $10^{-6}\,\text{s}^{-1}$.

Figure 6. X-ray shadowgraph of MgO sample during compression. The horizontal dark lines are images of gold foil that bound the sample whose length is ~1 mm. The horizontal dimension is defined by the opening between the high-pressure anvils. The shadow across the top is caused by a platinum foil wrapped around the end-plug.

The X-ray image yields the total strain of the sample, which is the sum of the elastic and plastic strain or:

$$\varepsilon_{ob} = \varepsilon_{el} + \varepsilon_{pl}$$

In an experiment both the elastic and plastic components may vary with time. Thus, to deduce the plastic strain rate, it is necessary to account for the variations of all strain components. The elastic strain, however, is reflected in the diffraction pattern and can be represented by a unit cell length, a. This cell-edge length is obtained from analysis of all of the diffraction peaks recorded with a diffraction vector parallel to the sample axis. In elastically isotropic samples, there is no ambiguity in defining this value. Anisotropic samples in a differential stress field will exhibit lattice spacings that reflect the aggregate properties. In this case, an average cell dimension based on several diffraction peaks should be obtained. We define the Kung ratio, R_k, as:

$$R_k = (d/d_0) / (a/a_0)$$

where d is the length of the sample measured with an image and a is a unit cell length measured from diffraction; the zero subscript refers to a reference value such as ambient conditions. This ratio, which is normalized to one, represents the length of the sample in the units of the cell dimension. This number is proportional to the number of unit cells that make up the length. The Kung ratio remains constant if all deformations are elastic. Thermal expansion, pressure, and stress will not change the Kung ratio if the sample responds elastically. Plastic strain changes the Kung ratio. In this case the number of unit cells that define the length changes. Thus, the Kung ratio can be used as the proxy for plastic strain, or:

$$\varepsilon_{pl} = [R_k(2) - R_k(1)] / R_k(1)$$

where (1) and (2) represent two states of the material.

Strain is not always uniform in the sample. Visual tools as described here provide opportunity to map the strain field as a function of space and time during the experiment. Vaughan et al. (2000) demonstrate that one can embed strain markers in a sample during preparation and view the position of these markers during the experiment. Figure 7, from this work, shows a grid made from a TEM gold sample grid, in the sample at different P, T conditions. The sample was a sintered olivine cylinder. The cylinder was sliced into two halves, the grid inserted, and the sample halves placed back together. The sample assembly included hard end-plugs in a DIA apparatus, so as to produce shortening during compression. The grid size is initially about 70 microns. With these images, it is possible to map the strain as a function of position in the sample and as a function of time by comparing successive images.

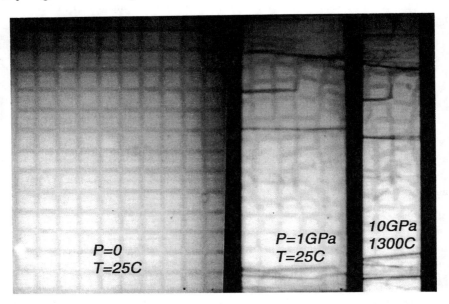

Figure 7. A TEM gold grid embedded on the central plane of a cylindrical sample. Images at different conditions illustrate different amounts of strain. The initial pressurization distorted the grid because surfaces of sample assembly parts were not perfectly flat. Subsequent strain can be mapped by following specific strain markers.

MODERN TECHNIQUES FOR DEFORMATION AT HIGH PRESSURES

In current technology, deformation at $P > 10$ GPa is the realm of anvil devices. We describe in detail four methods currently in use or in active development: (1) sample assembly modification to hydrostatic multianvil devices, (2) sample assembly modification to the DAC, (3) the D-DIA, and (4) the RDA.

Sample assemblies

Sample assemblies used in these machines share several common features of those used in hydrostatic devices, serving common requirements such as high pressure, high temperature, and control of chemical activity in the sample. With the exception of the specialized environment of the DAC, these common features include a pressure medium, such as soft fired pyrophyllite, ZrO_2, $MgO/5\%Cr_2O_3$, castable ceramic or epoxy; a furnace, for example, metal foil, $LaCrO_3$, or graphite; and a soft medium surrounding the sample itself, for example, BN, MgO, NaCl, or glass (the latter two occasionally as liquids). Descriptions of such assemblies can be found in, e.g., Kawai and Endo (1970), Liebermann and Wang (1992), Walker(1991), and Weidner et al. (1992). For control of chemical environment the sample can be encapsulated along with buffering compounds (see e.g., Rubie et al. 1993).

Non-hydrostaticity is introduced by introducing hard platens that impart shear or normal stresses to the sample. Several of the methods for achieving this are compared in Figure 8. They provide axisymmetric deformation in the multianvil devices (Figs. 8a-8c), simple shear in the multianvil (Fig. 8d), axisymmetric deformation in the DAC (Fig. 8e), and rotational (approximately simple shear) deformation in the RDA (Fig. 8f). We compare these assemblies briefly here and in more detail below.

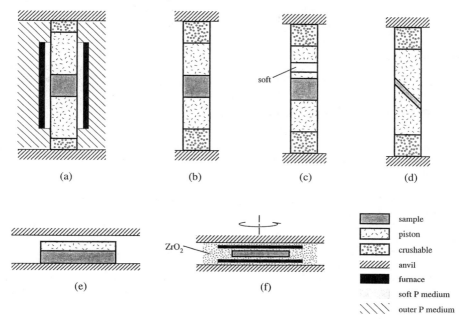

Figure 8. Starting ($P = 0$) configurations of sample columns used for deformation experiments at very high pressures. For simplicity, details such as sample encapsulation, thermocouples, and furnace electrodes are not shown. (a) Stress relaxation. The sample is loaded elastically during pressurization. The furnace and outer and inner pressure media shown here also apply to (b)-(d). (b) Constant stress/strain rate tests. The porosity within the crushable ends of the column takes up most of the column shortening during pressurization. (c) Specialized version of (b), with a fluid in the column assuring strictly hydrostatic loading of the sample during pressurization. (d) Simple shear, high strain. (e) Column for deformation in the diamond-anvil cell. Use of a separate piston is optional, i.e., the diamond anvils can also act as pistons. In some applications, a ruby piston can also serve as a differential stress gage. (f) Assembly for the rotational Drickamer apparatus, also shown in Figure 13.

The adaptations to the DIA and 6/8 discussed in the next section use a column such as shown in Figure 8a, in which the source of deviatoric stress in the sample is the closure of the anvils during initial pressurization. At the scale of Figure 8a, the shortening of the column (elastic plus inelastic) is approximately 20% of the column length (Durham and Rubie 1998), so the "crushable" material at the ends of the column, present to protect thermocouple wires and furnace electrodes, has a porosity somewhat less than 20% of the column length. To avoid deforming the sample during pressurization in devices capable of deforming at fixed pressure (currently only the D-DIA) one uses a column such as in Figure 8b, where the length of crushable material is sufficient to absorb all anvil displacement during pressurization When a strictly hydrostatic environment is required during pressurization, such as when the sample is a fragile single crystal, it is also possible to create a liquid cell around the sample before the pressurizing load is applied (Fig. 8c) (Durham, unpublished results).

Figure 8d shows the method for converting axial displacement to high-strain, simple-shear deformation in the multianvil cell (Karato and Rubie 1997). As discussed below, this design was first introduced as a method for imposing larger strains in the stress-relaxation mode of the hydrostatic multianvil device. However, it can also be used in the D-DIA, presumably with slightly more crushable material at the ends, to carry out constant stress or constant strain-rate tests to very high simple-shear strains.

The very simple deformation assembly for the DAC (Fig. 8e) reflects the greatly reduced available volume and the fact that laser heating can be used in place of a resistance heater to achieve local heating of the sample. Finally, the assembly for the RDA (Fig. 8f) is the rotational analog to the axisymmetric columns in Figures 8a-8c. Because rotational deformation and pressurization are entirely decoupled in the RDA, there is no need for crushable material in the cell.

Modifications to the sample assembly in a multianvil press

Fujimura et al. (1981) developed one of the first techniques for converting the environment of the 6/8 device from hydrostatic to non-hydrostatic, thus making the 6/8 device a sort of deformation apparatus. By making the elastic character and/or mechanical strength of a sample assembly anisotropic, one can create a deviatoric stress in the assembly (e.g., Fig. 8a). This deviatoric stress can be relaxed by plastic flow upon heating. This technique has been used in numerous studies (Bussod et al. 1993; Durham and Rubie 1998; Fujimura 1989; Green et al. 1990; Karato et al. 1998; Karato and Rubie 1997; Weidner et al. 2001; Weidner et al. 1998). One typically achieves the anisotropic strain by embedding a rigid sample column in the compliant pressure medium. As the assembly is pressurized at room temperature (where most relevant geologic materials are very strong), a large deviatoric stress is generated on the sample. When temperature is increased, the sample softens and begins to flow, and the stress gradually relaxes. The analysis by Karato and Rubie (1997) (see also Durham and Rubie [1998]) showed that the mode of deformation in most cases is this "stress relaxation," i.e., the magnitude of deviatoric stress changes significantly within a single experiment.

This technique has several limitations. First, pressurization and plastic deformation are not completely separated; therefore, deformation during pressurization likely occurs unless special care is taken in sample assembly to absorb the initial-stage shape change of sample assembly (Durham and Rubie 1998; Karato and Rubie 1997). Second, deformation by this method is "stress relaxation," and the interpretation is complicated because of the change in stress that could cause the change in deformation mechanisms. Third, the amount of inelastic displacement is limited to the amount that can be imposed elastically during pressurization, usually well under 100 μm. By using a thin sample sandwiched between two pistons cut at 45°, Karato and Rubie (1997) were able to convert this very small displacement to relatively large strains ($\gamma \approx$ 1-2) (Fig. 8d), yet the maximum strain is not large enough to obtain steady-state fabrics. Quasi-constant displacement rate tests can also be made with multianvil apparatus through the continuous movement of the hydraulic ram (Bussod et al. 1993; Green et al. 1990), although one must realize that pressure is increasing steadily and significantly in such tests.

The multianvil deformation technique has been applied to ~25 GPa, ~2000 K with a sample dimension of ~1-2 mm diameter and ~2-4 mm long, or a thin (~0.2 mm) disk sample with a similar diameter (Chen et al. 1998; Cordier and Rubie 2001). Sacrificing pressure for sample size, Green et al. (1990) deformed 3-mm diameter x 6-mm length samples in a 6/8 device. The major advantage of this technique as compared with a diamond-anvil cell technique is that rheological properties can be investigated under more homogeneous high temperature and pressure conditions with a better-controlled chemical environment. Also, because relatively large samples can be used, microstructural evolution during deformation and its effects on rheology can be investigated (Green et al. 1990, 1992; Karato 1998a; Karato et al. 1998).

The limited plastic deformation is a serious shortcoming of the method. The displacement available is generally not sufficient to probe important rheological questions about lattice preferred orientation and strength in the steady state. Second, because the stress magnitude changes during an experiment, dominant mechanisms of deformation may change in a single run, making interpretation of mechanical data and microstructures difficult. This last point, however, may turn out to be an advantage for the study of rheology at a very small strain-rate (small stress) that is relevant to Earth.

Diamond-anvil cell

The highest pressures obtainable in the laboratory are in the diamond-anvil cell. This device loads the sample uniaxially and, thus, naturally produces a deviatoric stress environment. The first examinations of differential stresses in the diamond cell were motivated by the need to produce a hydrostatic environment. Use of the diamond cell as a deformation device was pioneered by Kinsland and Bassett (1977) and Sung et al. (1977). The former used X-ray transparent gaskets and passed the X-ray beam through the gaskets, perpendicular to the diamond-cell axis. Stress was measured from the shape of the Debye rings as discussed above. The measurements provided the first estimates of the room temperature strength of MgO at high pressures. Sung et al. (1977) developed the methodology for using pressure gradients to deduce shear stress. In this case, a sample is sandwiched between two single crystals of diamond, and radial distribution of pressure is determined by measuring the shift of fluorescence lines of ruby crystals located at various points in the sample space. Then, assuming that the pressure gradient is supported by the sample strength, one can estimate the sample's strength from the equation for force balance. This technique has been used extensively by Meade and Jeanloz (1988a,b; 1990) at room temperature to the pressure of ~40 GPa.

Chai and Brown (1996) developed a diamond-cell piezometer based on the splitting of the ruby fluorescence lines. They find that the R_2 line reflects the average stress, that the splitting between the R_1 and R_2 lines is sensitive to the differential stress, and that the character of the sensitivity depends on whether the load is applied parallel to the *c*-axis or the *a*-axis. This phenomenon suggests experiments that can be done with a diamond-anvil cell at room temperatures using single crystal ruby as the pressure/stress calibrant.

The major advantage of diamond cell studies is the high pressures that can be used under which a sample can be plastically deformed Wenk et al. (2000) recently deformed Fe at ~220 GPa). However, there are major limitations with this approach: homogeneous heating is difficult in a diamond-anvil cell, and almost all previous results were obtained at room temperature. In addition, the strain rate is not well constrained even though deformation is likely to be time-dependent in these tests. Furthermore, the sample space is so small that some important effects such as grain-size sensitivity of strength are difficult to measure.

Poirier et al. (1981) and Sotin and Poirier (1990) report on a sapphire cell that enables larger sample volumes than a diamond cell and is more versatile for higher temperature studies. They calibrate the stress field from the characteristics of the loading system and observe the movement of strain markers in the sample. This enables them to define strain rate along with the stress determinations.

Deformation-DIA

Samples of MgO and polycrystalline tantalum were deformed in the deformation-DIA (D-DIA) at the National Synchrotron Light Source (NSLS) in February 2002. This marked the first materials test ever above $P = 10$ GPa in which virtually all relevant rheological independent variables were under full and independent operator control. We provide here a detailed description of this new apparatus.

The D-DIA (Wang et al. 2002) is a modification of the cubic anvil device known as the DIA (e.g., Osugi et al. (1964) and Shimomura et al. (1985)) a single-stage, wedged guide block machine. The hydrostatic DIA usually operates to ~15 GPa and ~2000 K. The D-DIA modification gives independent displacement control to one pair of opposing anvils, reducing the cubic symmetry of the DIA to tetragonal, thus allowing high-strain deformation experiments to 15 GPa.

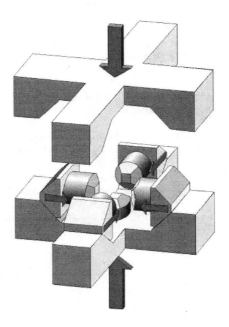

Figure 9. The original hydrostatic DIA apparatus. A single hydraulic ram (large arrows) drives 2 wedge-type guide blocks toward one another. The 6 anvils (truncated pyramids on the guide blocks and wedge blocks; only 5 are visible here) are thus driven toward one another, symmetrically compressing a cubic sample assembly (not shown) at the very center.

The original DIA consists of symmetric upper and lower guide blocks, four wedge-shaped thrust blocks, and six anvils, as indicated in Figure 9. Four of the anvils are mounted on the inside faces of the thrust blocks and the other two are mounted on the inside central faces of the guide blocks, the square fronts of the anvils thus defining a cubic volume at the center of the apparatus. The operation of the DIA can be visualized by recognizing that the eight inner inclined surfaces of the guide blocks define a virtual regular octahedron whose dimension changes with the separation distance between the guide blocks. The six anvils are aligned with the six apices of this virtual octahedron, and parts are machined such that the (fixed) distance from anvil face to associated apex of the virtual octahedron is precisely the same for each anvil. As the guide blocks close or open, all displacements of the anvils are symmetric about the center. The force of a single hydraulic actuator (or ram) applied along the vertical axis in Figure 9 closes the guide blocks and, thus, compresses the cell hydro-statically. Ram forces of 100 T are typically required with various truncation sizes (typically 3 to 6 mm) to reach maximum pressures near 20 GPa (see e.g., Shimomura et al. (1992) and Utsumi et al. (1992)). By the cubic symmetry of the DIA assemby, this force is divided into three equal portions along each of the three orthogonal directions defined by opposing pairs of anvils.

Independent control of one anvil pair in D-DIA is provided by two additional hydraulic actuators—called differential rams—within each guide block (Fig. 10). These rams may be thought of not as parts that are added, but rather as a central portion of each guide block that has been cut free so that it can move vertically. A small hole is introduced into the side of each guide block to allow access for hydraulic fluid to each ram, and seals for the pressurized plenum below each differential ram are provided. The conical shape of the differential rams is rather unimportant; it leaves more steel on the guide blocks and makes them slightly stronger. Note that in the D-DIA modification, the guide blocks not only support the forces confining the thrust blocks, they now have also become thick-wall pressure containers for the hydraulic fluid driving the differential rams. For this reason, the guide blocks of the D-DIA are complete discs rather than crossed steel members as in the original DIA (Fig. 9).

Figure 10. Computer-generated 3/4 section through the D-DIA. Comparing to Figure 9, the guide blocks are now full cylindrical shapes, but the four wedge blocks and six anvils are identical to those in Figure 9. The two differential rams, the large conical pieces within the top and bottom guide blocks, move independently of the guide blocks themselves, and can thus generate a deviatoric state of stress in the assembly at the center (not shown). In actual operation, as the differential rams displace inward toward the center, the guide blocks move apart to allow the four wedge blocks to displace outward. The displacement rates are independently controllable and are usually chosen such that the net volume of the sample assembly, and therefore the confining pressure, remains constant.

The differential rams have a (pressurized) diameter of 89 mm and are therefore capable of generating a force of 125 T at a hydraulic pressure of 0.2 GPa, sufficient to overcome the force of the main ram (up to 33% of the main ram maximum of 300 T) and apply differential force even at the highest confining pressure. Finite element analysis of the D-DIA guide blocks show that the maximum octahedral stress occurs on the inner bore in a "typical" operation where the differential load is 67% of the main ram load (presuming some force is needed to overcome packing friction and sample strength) (Wang et al. 2002). The stress at the base of the inclined surface of the guide block, the next highest point of stress concentration in the guide block, is roughly 90% of the stress in the bore. Note that in the original DIA with very large notches cut in the guide blocks, the stress concentration at the base of the inclined surfaces is twice that of the D-DIA under a similar main ram load.

The key point of independent control of the differential rams is that deformation can be imposed without the necessity of increasing confining pressure (which has been the long-standing limitation of deformation studies in 6/8 multianvil and diamond-anvil assemblies). In normal operation, the D-DIA sample is brought to run conditions of P and T with differential rams fully withdrawn in the same manner as the hydrostatic DIA. Advancing the differential rams then introduces a non-cubic shape change to the assembly. Furthermore, by simultaneously draining hydraulic fluid from the main ram at an appropriate rate, the four side anvils retract, and the total force of the main ram as well as the volume of the sample cell can be held constant. In the synchrotron X-ray beamline) pressure itself can be monitored and, accordingly, serves as the process variable. The operation at NSLS in February 2002, showed that this procedure gave very satisfactory control of pressure over very large displacements of the differential rams (Fig. 11).

The differential rams are driven by high-precision syringe pumps, and their velocities are controllable from approximately 10^{-7} to 10^{-2} mm/s. When both rams are driven symmetrically, this translates to a strain rate on a typical 1-mm-length, 1-mm^3-volume sample of $2 \times 10^{-2} \geq \dot{\varepsilon} \geq 2 \times 10^{-7}$ s^{-1}. These rates are much faster than most relevant rates in geology, but as with all experimental rock mechanics work—for which these rates are typical—it is the human time scale and not the geologic time scale that governs the duration of an experiment. As with most experimental studies of rock deformation, appropriate scaling analysis is critical to apply these results to Earth. Differential ram displacements of >1 mm are possible, so strains in pure shear compression can approach 1. Deformation in simple shear, using 45°-cut pistons (see "Sample Assemblies," above) is also possible, allowing for much higher strains, at some cost to sample volume. Finally, the D-DIA is capable of extensional, as well compressional, deformation, because the sense of motion of the six anvils can be reversed.

Rotational Drickamer apparatus (RDA)

Large strain deformation experiments can be conducted in the torsion mode, allowing a detailed study of deformation microstructures as well as rheology. Paterson and Olgaard (2000) described such an apparatus in which a torsion actuator is attached to a gas-medium high-pressure apparatus. This apparatus contains an internal load-cell, making possible a precise measurement of stress. However, the maximum pressure of operation is limited to ~0.5 GPa. Deformation experiments can be conducted at higher pressures by twisting a thin sample between anvils. Bridgman (1935a; 1937) was the first to apply the technique to 5 GPa with an apparatus that consisted of two fixed anvils bearing on a flat, rotating anvil. The design was later modified by Griggs et al. (1960) to remove the intermediate anvil and rotate one of the anvils instead. Similar attempts were made in the late 1960s at room temperatures and at pressures to 7 GPa (Abey and Stromberg 1969; Riecker and Seifert 1964) in which a sample was sheared, and the strength was determined by the measurements of torque on an anvil needed to deform the sample.

Yamazaki and Karato (2001) have improved these techniques by modifying the Drickamer-type high-pressure apparatus (Fig. 12). This apparatus allows large rotational shear deformation of a sample at higher pressures and temperatures. A Drickamer apparatus consists of a pair of anvils in a cylinder with a gasket material between the anvils and a thin disk of sample squeezed between the anvils to reach high pressure

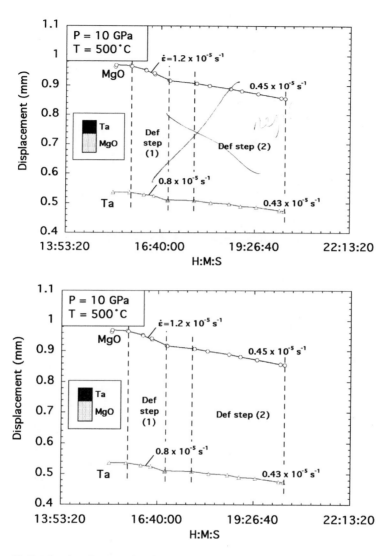

Figure 11. Results plotted vs. run time from a D-DIA experiment on a stacked sample of tantalum (starting length 1 mm) and MgO (starting length 1.5 mm), both polycrystalline. (a) Five traces showing, from top to bottom: measured hydraulic oil pressure in the top and bottom differential rams (scale on right-hand axis); combined centerward displacement of the two rams as measured by displacement transducers outside the sample assembly; shortening of the MgO portion of the sample; and shortening of the Ta component (scale for last three on left-hand axis). Sample length vs. time was measured directly using X-radiography. (b) Detail of the sample shortening, also indicating the strain rates (displacement rates normalized by sample length). The run was conducted by pumping hydraulic fluid to the differential rams at two different constant rates (hence deformation steps (1) and (2)), while draining fluid from the main ram in order to keep pressure at a constant level of 10±1 GPa. Note that shortening rates of the samples were also approximately constant. The steady increase in differential ram pressure during both deformation steps is the result of steadily increasing friction on the gaskets squeezed between the differential and side anvils.

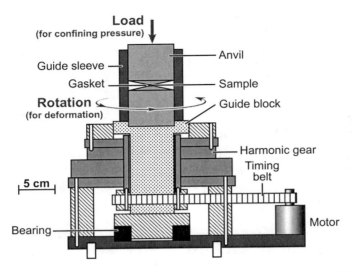

Figure 12. Sketch of a rotational Drickamer apparatus (Yamazaki and Karato 2001).

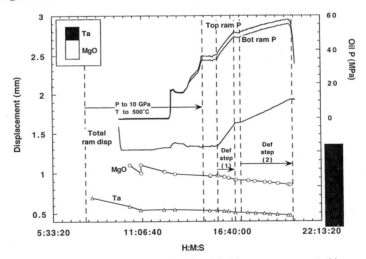

Figure 13. Sample assembly for a rotational Drickamer apparatus. A thin disk of sample is sandwiched between two anvils. A sample is cut into two pieces and W5%Re and W25%Re foils are inserted vertically; these act both as strain markers and as a thermocouple. The change in shape of these foils can be measured after an experiment or during an experiment through X-ray absorption. A small cylindrical W3%Re foil at the center acts as an electrode for the heater and as one component of the thermocouple with a W25%Re foil.

(Fig. 13). The gasket provides an extra support for the anvils and, as a result, this apparatus yields significantly higher pressures than does a Bridgman apparatus without a gasket (Perez-Albuerne et al. 1964; Prins 1984). Because much of the force is supported by the gasket and the sample itself, the cylinder does not support a large load. Consequently, it is possible to make holes in the cylinder to provide a path for X-rays; the thermocouple leads can be taken through these holes. The X-ray path feature is important for high-resolution stress and temperature measurements, both of which are critical for rheological studies. The Drickamer apparatus has recently been used for the *in situ* measurements of static properties using a synchrotron X-ray facility under conditions up to ~35 GPa and ~2000 K (Funamori and Yagi 1993).

In the Rotational Drickamer Apparatus (RDA) modification, one of the anvils is fixed with the frame, whereas another is attached to a rotational actuator. A gearbox with an ac-servo motor provides controlled rotation of one of the anvils. A sample is first pressurized (and heated), and then the rotational actuator is started. The sample is twisted between the two anvils, and the geometry of deformation is rotational shear. The strain (and strain rate) increases from zero to the maximum value, which is determined by the motor rotation speed and sample thickness. With the current motor and gear combination, a shear strain rate from ~0 to ~10^{-3} s^{-1} can be realized. This design has three advantages. First, the motion of the anvil for deformation is *orthogonal* to the motion of the anvil for pressurization. Therefore, pressurization and deformation can be clearly separated. Second, unlike axisymmetric deformation, the shape of the sample does not change appreciably during rotational deformation, allowing very large strain deformation and making possible studies of microstructural evolution. Third, because of the radial gradient in strain (strain rate), microstructural evolution and/or rheology at different strain-rates can be investigated in a single run.

The diameter of the anvil tip in the current RDA is 4 mm (sample diameter is ~1-2 mm, thickness ~0.2-0.4 mm). With a load of ~30 T, a pressure of ~15 GPa can be generated using tungsten carbide anvils. The maximum pressure is determined primarily by the strength of anvil materials (and the diameter of sample space) and is expected to be more than ~30 GPa when a harder material such as sintered diamond or cubic-BN is used. The rotational Drickamer apparatus constructed at Yale University has been tested at P = 12 GPa, T = 1473 K. A polycrystalline sample of MgO was deformed homogeneously to the maximum shear strain of ~3. Two methods can be employed to estimate the stress magnitude. First, X-ray diffraction techniques can be used for deformation experiments with the RDA in a synchrotron radiation facility in the manner discussed previously. At this writing, such experiments are planned but have not yet been carried out. To adopt the diffraction technique for rotation geometry, a conical shape window is made in a cylinder to collect diffracted X-rays for a range of angles (Fig. 14b). One complication with this apparatus is the radial variation of stress. This is inevitable in the torsion test (e.g., Paterson and Olgaard 2000); Fig. 14a). In measuring the stress magnitude by X-ray diffraction, one must consider the spatial resolution. Important factors that control the spatial resolution are the cross-sectional area of the X-ray beam and the 2θ value used for diffraction (with energy dispersive mode). A smaller beam size and a larger 2θ are preferred for a better spatial resolution of stress measurements. However, a smaller beam size results in a weaker signal. Also the 2θ values are limited by the energy of X-ray and the *d*-spacings as well as the geometry of the apparatus (we will use ~10-50 × 10-50 μm^2 beam [slit] size and $2\theta \approx 8$-$10°$). To allow for control of 2θ, the apparatus is set on a mobile stage (with a remote control through a stepping motor) whose angle and position with respect to the beam line can be adjusted to optimize the conditions. The resolution of the stress measurement is also dependent on the detector. We anticipate that a spatial resolution of 100-200 μm and the stress resolution better than 0.1 GPa will be obtained by this technique. The use of a hollow specimen could help solve the problem of stress heterogeneity (Fig. 14d).

We note that this technique not only allows us to estimate the stress magnitude, but also provides information as to the grain-scale stress (strain) distribution in a polycrystalline material (either homogeneous stress or homogeneous strain (Funamori et

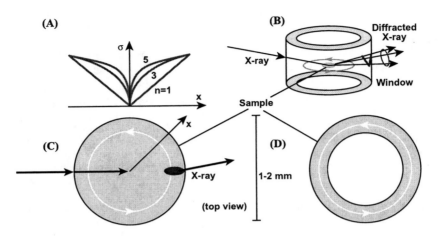

Figure 14. (A) Stress distribution in a sample in a torsion test (n is stress exponent, $\gamma = A\tau^n$). (B) X-ray diffraction geometry. Stress magnitude can be estimated from the difference in d-spacing collected from two different diffracted beams through two windows that are cut at 45° with respect to the vertical axis. (C) A top view of a sample and the X-ray path. In order to determine the stress, X-ray diffraction must occur in a region of a sample where the shear direction is normal to the X-ray beam. (D) The geometry of a hollow sample, where stress heterogeneity is less.

al. 1994). Such information is useful in understanding the physics of deformation of polycrystals (Kocks 1970).

Another method is to use dislocation density as a stress indicator. This technique can be used when the deformation mechanism is dislocation creep and when stress has been held at a constant value for at least a few percent strain. The method does not work well if stress is inhomogeneous. We have recently used this technique for olivine and are able to estimate the stress magnitude with an uncertainty of ~10-15% (Jung and Karato 2001; Karato and Jung 2002 in press). The reliability of such measurements for high-pressure phases is not as high as that for olivine because the calibration curve for dislocation density vs. stress relationship is not available. However, the relation between dislocation density and stress is nearly universal,

$$\rho = \alpha b^{-2}(\sigma/\mu)^2$$

(ρ: dislocation density, α: a constant of order unity, b: the length of the Burgers vector, σ: stress, μ: shear modulus (e.g., Kohlstedt and Weathers 1980) We consider that such measurements should provide at least semi-quantitative data on the stress magnitude.

We will perform both single-layer deformation experiments and two-layer deformation experiments. A single-layer deformation experiment is simpler to interpret, but the big advantage of a two-layer design is that one can determine the relative strength of two samples very accurately. Our experience with this apparatus showed that we can deform our samples to strains up to at least $\gamma \approx 10$ at ~15 GPa. Such a large strain is essential for the study of deformation fabric.

SUMMARY AND PERSPECTIVES

We have summarized some of the recent developments in quantitative characterization of stress and strain and the controlled generation of deviatoric stresses under high pressure (and temperature). These developments will allow us to explore rheological behavior of Earth materials under much deeper conditions than heretofore possible. Such studies will, for the first time, provide the critical data sets necessary for understanding whole mantle dynamics through modeling and/or through the analyses of observations such as seismic anisotropy. Further technical developments need to be made to make these techniques robust. First, stress measurements using different techniques must be benchmarked. Three techniques are currently used to measure stress under high pressure: (1) Measurement of force using an external load-cell with a low-friction sample design (Tingle et al. 1993), (2) synchrotron X-ray measurement of orientation dependence of d spacing (Singh et al. 1998), and (3) inference from dislocation densities (Karato and Jung 2002 in press). Below ~4 GPa, all of these techniques can be employed to determine the strength of a standard materials such as olivine. A comparison of results of these measurements will provide a measure of reliability of each technique. Second, a number of issues still remain regarding the sample assembly. They include a proper control of thermodynamic environment, particularly water fugacity (or water content) and oxygen fugacity. Third, in almost all of these techniques under high pressure, it is almost inevitable to have deformation during pressurization. An experimental procedure (e.g., annealing; a very low strength medium surrounding the sample) must be established to minimize the effects of initial stage deformation.

It must also be emphasized that the sample preparation and characterization are as critical as careful mechanical tests. Samples with negligible porosity must be used and grain size and water content must be measured both before and after each experiment (control of other thermodynamic variables such as oxygen fugacity is also important; see Rubie et al. (1993)). In almost all laboratory deformation experiments, conditions are often close to the boundary between dislocation and diffusion creep (Karato et al. 1998), and, therefore, precise measurements of grain size are needed to interpret the data. Also, water content must be measured both before and after each experiment. Water is known to have significant effects on rheology, but its content in samples is difficult to control during high-pressure experiments.

Using the results of mechanical tests and sample characterization, dominant microscopic mechanisms of deformation must be identified. A meaningful comparison of *strength* of different materials (or for the same material for different conditions) can only be made when the deformation mechanism is the same, and the comparison must be made based on an appropriate *scaling*. For instance, if the dominant mechanism of deformation is diffusion creep, then the data must be compared by normalizing with respect to the grain size.

With the use of these new techniques (combined with careful characterization of samples and the analyses of deformation mechanisms), we will be able to cast first light on the rheology of the more than 90% of Earth's mantle that has not been accessible by quantitative studies. Results of such studies will be critical to our better understanding of dynamics and evolution of this planet.

ACKNOWLEDGMENTS

We are most grateful to H.W. Green, II, for reviewing the manuscript.

Funding was provided by the National Science Foundation under award EAR-0135551. Work by WBD performed under the auspices of the U.S. Department of Energy by the Lawrence Livermore National Laboratory under contract W-7405-ENG-48. Additional portions of this work were performed at GeoSoilEnviroCARS (GSECARS), Sector 13, Advanced Photon Source at Argonne National Laboratory. GSECARS is supported by the National Science Foundation–Earth Sciences, Department of Energy–Geosciences, W.M. Keck Foundation, and the U.S. Department of Agriculture.

Use of the Advanced Photon Source was supported by the U.S. Department of Energy, Basic Energy Sciences, Office of Energy Research, under Contract No. W-31-109-Eng-38.

REFERENCES

Abey AE, Stromberg HD (1969) 70 kilobar shear apparatus. Rev Sci Instrum 40:557
Borch RS, Green HW II (1989) Deformation of peridotite at high pressure in a new molten cell: Comparison of traditional and homologous temperature treatments. Phys Earth Planet Inter 55: 269-276
Boyd FR, England JL (1960) Apparatus for phase equilibrium measurements at pressures up to 50 kilobars and temperatures up to 1750°C. J Geophys Res 65:741-748
Bridgman PW (1914) The technique of high pressure experimenting. Proc Am Acad Arts Sci 49:627-643
Bridgman PW (1935a) Effects of high shearing stress combined with high hydrostatic pressure. Phys Rev 48:825-847
Bridgman PW (1935b) Polymorphism, principally of the elements, up to 50,000 kg/cm^2. Phys Rev 48: 893-906
Bridgman PW (1937) Shearing phenomena at high pressures, particularly in inorganic compounds. Proc Am Acad Arts Sci 72:45-136
Bridgman PW (1942) Pressure-volume relations for seventeen elements to 100,000 kg/cm^2. Proc Am Acad Arts Sci 74:425-440
Bridgman PW (1948) The compression of 39 substances to 100,000 kg/cm^2. Proc Am Acad Arts Sci 76: 55-70
Bridgman PW (1952a) The resistance of 72 elements, alloys, and compounds to 100,000 kg/cm^2. Proc Am Acad Arts Sci 81:65-NN
Bridgman PW (1952b) The Physics of High Pressure. G. Bell and Sons, London
Bridgman PW (1952c) Studies in Large Plastic Flow and Fracture with Special Emphasis on the Effects of Hydrostatic Pressure. McGraw-Hill, New York
Burnley PC, Green HW II, Prior D (1991) Faulting associated with the olivine to spinel transformation in Mg_2GeO_4 and its implications for deep-focus earthquakes. J Geophys Res 96:425-443
Bussod GY, Katsura T, Rubie DC (1993) The large volume multi-anvil press as a high P-T deformation apparatus. Pure Appl Geophys 141:579-599
Carter N.L, Christie J.M, Griggs D. T (1964) Experimental deformation and recrystallization of quartz. J Geol 72:687-733
Chai M, Brown JM (1996) Effects of static non-hydrostatic stress on the R lines of ruby single crystals. Geophys Res Lett 23:3539-3542
Chen J, Inoue T, Weidner DJ, Wu Y, Vaughan MT (1998) Strength and water weakening of mantle minerals, olivine, wadsleyite and ringwoodite. Geophys Res Lett 25:575-578
Cordier P, Rubie DC (2001) Plastic deformation of minerals under extreme pressure using a multi-anvil apparatus. Mater Sci Eng A309-310:38-43
Davies GF (1995) Penetration of plates and plumes through the mantle transition zone. Earth Planet Sci Lett 133, 507-516
Durham WB, Rubie DC (1998) Can the multianvil apparatus really be used for high-pressure deformation experiments? In Properties of Earth and Planetary Materials at High Pressure and Temperature, Geophysical Monograph 101. M Manghnani, Y Yagi (eds) p 63-70. American Geophysical Union
Eremets M. I (1996) High Pressure Experimental Methods. Oxford University Press
Fujimura A (1989) Preferred orientation of mantle minerals. In Rheology of Solids and of the Earth. S-i Karato, M Toriumi (eds) p 263-283. Oxford University Press
Fujimura A, Endo S, Kato M, Kumazawa M (1981) Preferred orientation of $\beta-Mn_2GeO_4$. In Programme and Abstracts. Japan Seismological Society, p 185
Fukao Y, Obayashi M, Inoue H, Nenbai M (1992) Subducting albs stagnant in the mantle transition zone. J Geophys Res 97:4809-4822
Funamori N, Yagi T (1993) High pressure and high temperature in situ X-ray observation of $MgSiO_3$ perovskite under lower mantle conditions. Geophys Res Lett 20:387-390
Funamori N, Yagi T, Uchida T (1994) Deviatoric stress measurement under uniaxial compression by a powder X-ray diffraction method. J Appl Phys 75:4327-4331
Getting IC (1998) New determination of the bismuth I-II equilibrium pressure: A proposed modification to the practical pressure scale. Metrologia 35:119-132
Getting IC, Spetzler HA (1993) Gas-charged piston-cylinder apparatus for pressures to 4 GPa. In High-Pressure Science and Technology. SC Schmidt, JW Shaner, GA Samara, M Ross (eds) p 1581-1584. American Institute of Physics

Green HW II (2001) Earthquakes at depth and their enabling mineral reactions. Eleventh Annual V.M. Goldschmidt Conf (abstr)
Green HW II, Borch RS (1989) A new molten salt cell for precision stress measurement at high pressure. European J Mineral 1:213-219
Green HW II, Houston H (1995) The mechanics of deep earthquakes. Ann Rev Earth Planet Sci 23:169-213
Green HW II, Scholz CH, Tingle TN, Young TE, Koczynski TA (1992) Acoustic emissions produced by anticrack faulting during the olivine-spinel transformation. Geophys Res Lett 19:789-792
Green HW II, Young TE, Walker D, Scholz CH (1990) Anticrack-associated faulting at very high pressure in natural olivine. Nature 348:720-722
Griggs DT (1936) Deformation of rocks under high confining pressure. J Geol 44:541-577
Griggs DT (1967) Hydrolytic weakening of quartz and other silicates. Geophys J R Astron Soc 14:19-31
Griggs DT, Turner FJ, Heard HC (1960) Deformation of rocks at 500° to 800°C. *In* Rock Deformation, Geol Soc Am Memoir 79. DT Griggs, JW Handin (eds) p 39-104. Geol Soc Am, Boulder, Colorado
Hager BH (1984) Subducted slabs and the geoid: constraints on mantle rheology and flow. J Geophys Res 89:6003-6015
Hall HT (1958) Some high-pressure, high-temperature apparatus design considerations: Equipment for use at 100,000 atmospheres and 3000°C. Rev Sci Instrum 29:267-275
Handin JW (1953) An application of high pressure in geophysics: experimental rock deformation. Am Soc Mech Eng Trans 75:315-325
Heard HC, Carter NL (1968) Experimentally induced "natural" intergranular flow in quartz and quartzite. Am J Sci 266:1-42
Hill R (1950) The Mathematical Theory of Plasticity. Oxford University Press
Hirth G, Kohlstedt DL (1996) Water in the oceanic upper mantle; implications for rheology, melt extraction and the evolution of the lithosphere. Earth Planet Sci Lett 144:93-108
Hobbs BE, Ord A (1988) Plastic instabilities: implications for the origin of intermediate and deep focus earthquakes. J Geophys Res 89:10521-10540
Honda S, Yuen DA, Balachandar S, Reuteler D (1993) Three-dimensional instabilities of mantle convection with multiple phase transitions. Science 259:1308-1311
Houck JC, Hutton UO (1963) Correlation of factors influencing the pressures generated in multi-anvil devices. *In* High-Pressure Measurement. AA Giardini, EC Lloyd (eds). Butterworths, London
Jaeger JC, Cook NGW (1976) Fundamentals of Rock Mechanics, 2nd edition. Chapman and Hall
Jiao W, Siliver PG, Fei Y, Prewitt CT (2000) Do deep earthquakes occur on preexisting weak zones? An examination of the Tonga subduction zone. J Geophys Res 105:28125-28138
Jung H, Karato S (2001) Effect of water on the size of dynamically recrystallized grains in olivine. J Struct Geol 23:1337-1344
Karato S-i (1998) Micro-physics of post-glacial rebound. *In* Ice Age Dynamics. P Wu (ed) p 351-364. TTP (Trans Tech Publications), Zürich
Karato S-i (1998a) Effects of pressure on plastic deformation of polycrystalline solids: some geological applications. *In* High Pressure Research in Materials Sciences. PY Yu (ed) p 3-14. Materials Research Society, Warrendale, Pennsylvania
Karato S-i, Dupas-Bruzek C, Rubie DC (1998) Plastic deformation of silicate spinel under the transition zone conditions of the Earth. Nature 395:266-269
Karato S-i, Jung H (2002) Effects of pressure on high-temperature dislocation creep in olivine polycrystals. Phil Mag (in press)
Karato S-i, Riedel MR, Yuen DA (2001) Rheological structure and deformation of subducted slabs in the mantle transition zone: implications for mantle circulation and deep earthquakes. Phys Earth Planet Inter 127:83-108
Karato S-i, Rubie DC (1997) Toward experimental study of plastic deformation under deep mantle conditions: a new multianvil sample assembly for deformation studies under high pressures and temperatures. J Geophys Res 102:20111-20122
Karato S-i, Wu P (1993) Rheology of the upper mantle; a synthesis. Science 260:771-778
Kawai N, Endo S (1970) The generation of ultrahigh hydrostatic pressure by a split sphere apparatus. Rev Sci Instrum 41:1178-1181
Kellogg LH, Hager BH, van der Hilst RD (1999) Compositional stratification in the deep mantle. Science 283:1881-1884
King SD (1995) Radial models of mantle viscosity: results from a generic algorithm. Geophys J Intl 122:725-734
Kinsland GL, Bassett W (1977) Strength of MgO and NaCl polycrystals to confining pressures of 250 kbar at 25 C. J Appl Phys 48:978-985
Kirby SH, Stein S, Okal EA, Rubie DC (1996) Metastable mantle phase transformations and deep earthquakes in subducting oceanic lithosphere. Rev Geophys 34:261-306

Kocks UF (1970) The relation between polycrystal deformation and single-crystal deformation. Metall Trans 1:1121-1143

Kohlstedt DL, Chopra PN (1994) Influence of basaltic melt on the creep of polycrystalline olivine under hydrous conditions. *In* Magmatic Systems. Intl Geophysics Series, Vol. 57. MP Ryan (ed) p 37-53

Kohlstedt DL, Weathers MS (1980) Deformation-induced microstructures, paleopiezometers, and differential stresses in deeply eroded fault zones. J Geophys Res 85:6269-6285

Liebermann RC, Wang Y (1992) Characterization of sample environment in a uniaxial split-sphere apparatus. *In* High-Pressure Research: Application to Earth and Planetary Sciences. Y Syono, MH Manghnani (eds) p 19-31. American Geophysical Union, Washington, DC

Lloyd EC, Hutton UO, Johnson DP (1959) Compact multi-anvil wedge-type high pressure apparatus. J Res Nat Bur Stand 59:63C

Masters G, Laske G, Bolton H, Dziewonski AM (2000) The relative behavior of shear velocity, bulk sound speed, and compressional velocity in the mantle: Implications for chemical and thermal structure. *In* Earth's Deep Interior. S Karato, AM Forte, RC Liebermann, G Masters, L Stixrude (eds) p 63-87. American Geophysical Union, Washington, DC

Meade C, Jeanloz R (1988a) The yield strength of B1 sand B2 phases of NaCl. J Geophys Res 93: 3270-3274

Meade C, Jeanloz R (1988b) The yield strength of MgO to 40 GPa. J Geophys Res 93:3261-3269

Meade C, Jeanloz R (1990) The strength of mantle silicates at high pressures and room temperature: implications for the viscosity of the mantle. Nature 348:533-535

Mei S, Kohlstedt DL (2000a) Influence of water on plastic deformation of olivine aggregates, 1. Diffusion creep regime. J Geophys Res 105:21457-21469

Mei S, Kohlstedt DL (2000b) Influence of water on plastic deformation of olivine aggregates, 2. Dislocation creep regime. J Geophys Res 105:21471-21481

Mitrovioca JX, Forte AM (1997) Radial profile of mantle viscosity: results from the joint inversion of convection and postglacial rebound observables. J Geophys Res 102:2751-2769

Mitrovioca JX, Peltier WR (1991) A complete formalism for the inversion of postglacial rebound data: resolving power analysis. Geophys. J Intl 104:267-288

Ogawa M (1987) Shear instability in a viscoelastic material as the cause for deep earthquakes. J Geophys. Res 92:13801-13810

Onodera A (1987) Octahedral-anvil high-pressure devices. High Temp-High Press 19:579-609

Osugi J, Shimizu K, Inoue T, Yasunami K (1964) A compact cubic anvil high pressure apparatus. Rev Phys Chem Japan. 34:1-6

Paterson M. S (1970) A high-pressure, high-temperature apparatus for rock deformation. Intl J Rock Mech Min Sci 7:517-526

Paterson MS (1978) Experimental Rock Deformation: The Brittle Field. Springer-Verlag, Berlin

Paterson MS (1990) Rock deformation experimentation. *In* The Brittle-Ductile Transition in Rocks: The Heard Volume. AG Duba, WB Durham, JW Handin, HF Wang (eds) p 187-194. American Geophysical Union, Washington, DC

Paterson MS, Olgaard DL (2000) Rock deformation tests to large shear strains in torsion. J Struct Geol 22:1341-1358

Peltier WR (1998) Postglacial variation in the level of the sea: implications for climate dynamics and solid-Earth geophysics. Rev Geophys 36:603-689

Perez-Albuerne EA, Forgsgren KF, Drickamer HG (1964) Apparatus for X-ray measurements at very high pressure. Rev .Sci Instrum 35, 29-33

Poirier J-P, Sotin C, Peyronneau J (1981) Viscosity of high-pressure ice VI and evolution and dynamics of Ganymede. Nature 292:225-227

Post RL (1977) High-temperature creep of Mt. Burnett dunite. Tectonophysics 42:75-110

Prins JF (1984) A semiempirical description of pressure generation between Bridgman anvils. High Temp – High Press 16:657-664

Raleigh CB, Paterson MS (1965) Experimental deformation of serpentinite and its tectonic implications. J Geophys Res 70:3965-3985

Riecker RE, Seifert KE (1964) Olivine shear strength at high pressure and room temperature. Geol Soc Am Bull 75:571-574

Rubey WW (1972) Foreword. *In* Flow and Fracture of Rocks. Geophys Monogr 16:*ix-x*. HC Heard, IY Borg, NL Carter, CB Raleigh (eds) p. American Geophysical Union, Washington, DC

Rubie DC, Karato S-i, Yan H, O'Neill HSC (1993) Low differential stress and controlled chemical environment in multianvil high-pressure experiments. Phys Chem Minerals 20:315-322

Shimomura O, Utsumi W, Taniguchi T, Kikegawa T, Nagashima T (1992) A new high pressure and high temperature apparatus with sintered diamond anvils for synchrotron radiation use. *In* High-pressure Research: Application to Earth and Planetary Sciences. Y Syono, MH Manghnani (eds) p 3-11. Terra Scientific Publishing, Tokyo

Shimomura O, Yamaoka S, Yagi T, Wakatsuki M, Tsuji K, Kawamura H, Hamaya N, Fukuoga O, Aoki K, Akimoto S (1985) Multi-anvil type X-ray system for synchrotron radiation. *In* Solid State Physics Under Pressure. S Minomura (ed), p 351-356. Terra Scientific Publishing, Tokyo

Singh AK, Balasingh C, Mao H-k, Hemley R, Shu J (1998) Analysis of lattice strains measured under nonhydrostatic pressure. J Appl Phys 83:7567-7578

Sotin C, Poirier J-P (1990) The sapphire anvil cell as a creep apparatus. *In* The Brittle-Ductile Transition in Rocks: The Heard Volume, Vol. Geophys. Monogr. No. 56. AG Duba, WB Durham, JW Handin, HF Wang (eds) p 219-223. American Geophysical Union, Washington, DC

Sung CM, Goetze C, Mao HK (1977) Pressure distribution in the diamond anvil press and the shear strength of fayalite. Rev Sci Instrum 48:1386-1391

Tackley PJ, Stevenson DJ, Glatzmaier GA, Schubert G (1993) Effects of endothermic phase transition at 670 km depth in a spherical model of mantle convection in the Earth's mantle. Nature 361:699-704

Thoraval C, Richards MA (1997) The geoid constraint in global geodynamics: viscosity structure, mantle heterogeneity models and boundary conditions. Geophys J Intl 131:1-8

Tingle TN, Green HW II, Young TE, Koczynski TA (1993) Improvements to Griggs-type apparatus for mechanical testing at high pressures and temperatures. Pure Appl Geophys 141:523-543

Tullis TE, Tullis J (1986) Experimental rock deformation techniques. *In* Mineral and Rock Deformation: Laboratory Studies: The Paterson Volume. BE Hobbs, HC Heard (eds) p 297-324. American Geophysical Union, Washington, DC

Turner FJ, Griggs DT, Clark RH, Dixon RH (1956) Deformation of Yule marble. Part VII: Development of oriented fabrics at 300°-500°C. Geol Soc Am Bull 67:1259-1294

Utsumi W, Yagi T, Leinenweber K, Shimomura O, Taniguchi T (1992) High-pressure and high-temperature generation using sintered diamond anvils. *In* High-pressure Research: Application to Earth and Planetary Sciences. Y Syono, MH Manghnani (eds), p 37-42. American Geophysical Union, Washington, DC

van der Hilst RD, Engdahl R, Spakman W, Nolet G (1991) Tomographic imaging of subducted lithosphere below northwest Pacific island arcs. Nature 353:37-43

van der Hilst RD, Karason H (1999) Compositional heterogeneity in the bottom 1000 kilometers of Earth's mantle: toward a hybrid convection model. Science 283:1885-1888

van der Hilst RD, Widiyantoro RDS, Engdahl ER (1997) Evidence for deep mantle circulation from global tomography. Nature 386:578-584

Vaughan M, Chen J, Li L, Weidner DJ, Li B (2000) Use of X-ray imaging techniques at high pressure and temperature for strain measurements. *In* Science and Technology of High Pressure. Proceedings of AIRAPT-17. MH Manghnani, WJ Nellis, MF Nicol (eds) p 1097-1098. Universities Press

Walker D (1991) Lubrication, gasketing, and precision in multianvil experiments. Am Mineral 76: 1092-1100

Walker D, Carpenter MA, Hitch CM (1990) Some simplifications to multianvil devices for high-pressure experiments. Am Mineral 75:1020-1028

Wang Y, Durham W, Getting I (2002) D-DIA: A new apparatus for high-temperature deformation at pressures up to 15 GPa. Rev Sci Instrum (submitted)

Weidner DJ (1998) Rheological studies at high pressure. Rev Mineral 39:493-524

Weidner DJ, Chen J, Xu Y, Wu Y, Vaughan MT, Li L (2001) Subduction zone rheology. Phys Earth Planet Inter 127:67-81

Weidner DJ, Vaughan MT, Ko J, Wang Y, Liu X, Yeganeh-Haeri A, Pacalo RE, Zhao Y (1992) Characterization of stress, pressure, and temperature in SAM85, A DIA type high pressure apparatus. *In* High-Pressure Research: Application to Earth and Planetary Sciences. Geophys Monogr 67:13-17. Y Syono, MH Manghnani (eds) Terra Scientific Publishing Company, Tokyo, and American Geophysical Union, Washington, DC

Weidner DJ, Wang Y, Chen G, Ando J, Vaughan MT (1998) Rheology measurements at high pressure and temperature. *In* Properties of Earth and Planetary Materials at High Pressure and Temperature. Geophys Monogr 101:473-482. M Manghnani, Y Yagi (eds) American Geophysical Union, Washington, DC

Weidner DJ, Wang Y, Vaughan MT (1994) Strength of diamond. Science 266:419-422

Wenk H-R, Matthies S, Hemley RJ, Mao HK, Shu J (2000) The plastic deformation of iron at pressure of the Earth's inner core. Nature 405:1044-1047

Wu TC, Bassett WA (1993) Deviatoric stress in a diamond anvil cell using synchrotron radiation with two diffraction geometries. Pure Appl Geophys 141:509-519

Yamazaki D, Karato S (2001) High pressure rotational deformation apparatus to 15 GPa. Rev Sci Instrum 72:4207-4211

3 Deformation of Granitic Rocks: Experimental Studies and Natural Examples

Jan Tullis

Department of Geological Sciences
Brown University
Providence, Rhode Island 02912

INTRODUCTION

Crustal deformation is inherently complex because the crust is inhomogeneous and anisotropic on a wide variety of scales. Advances in our understanding have come from a combination of field observations, laboratory experiments and theoretical models. What necessarily links all these investigations is a focus on process. Crystal plasticity (dislocation creep) is one of several grain-scale deformation processes or mechanisms; each process is dominant over a certain range of conditions in the crust, has a particular form of the flow law, and produces a characteristic set of microstructures. Characterizing the deformation mechanisms for different minerals, over a range of conditions, allows one to link the mechanical behavior or flow law associated with that process to the microstructures it produces. The information on microstructures can be used to help interpret the thermomechanical history of naturally deformed rocks, and flow laws can be used to create more realistic models predicting crustal behavior in different tectonic settings. Iterations among field, theoretical, and experimental approaches are critical for making progress in our understanding.

The deformation process of crystal plasticity is the main focus of this volume. Crystal plastic deformation of the crust is in many ways more complex than that of the mantle, because most crustal rocks are polyphase aggregates, and the strengths of common crustal silicates (such as quartz, feldspar, and mica) are very different at a given set of conditions. Yield strength envelopes for the crust, as shown schematically in Figure 1, are commonly plotted using the dislocation creep flow law for quartz, which inherently assumes that quartz is the controlling phase and undergoes steady state flow. But is the flow law for dislocation creep of quartz a good approximation for the deformation of a

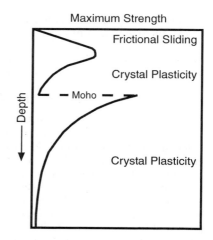

Figure 1. Highly schematic depiction of maximum rock strength as a function of depth for the continental lithosphere. The transition from frictional sliding to crystal plasticity is blunted to indicate the operation of other mechanisms such as cataclastic flow and pressure solution. There may be multiple brittle-viscous transitions, depending on compositional layering in the crust, and the depth of the peak strength will depend on the local geotherm, rock and fluid composition, and previous deformation history.

granite, let alone a crustal 'plum pudding' of regions and layers of different lithologies? The crust also tends to be anisotropic on various scales, due to sedimentary layering, metamorphic foliation, igneous intrusions, lattice preferred orientations (LPOs), and ductile shear zones with reduced grain size. How does this anisotropy evolve with time (strain) and how does it influence crustal strength and strain partitioning? The chemical environment is also important; trace amounts of water profoundly affect the dislocation creep strength of various minerals. Is it possible that 'dry' granulite grade rocks might be stronger than the colder but wetter rocks above them, creating a weak layer in the middle lower crust?

In this contribution I will focus on crystal plasticity in granitic rocks of the continental crust, emphasizing some of the progress that has been made in the recent past from experimental work, with a particular focus on microstructures, and indicating the similarities and differences that are observed in nature. I will begin with brief summaries of our understanding of crystal plasticity in monomineralic aggregates of quartz and of feldspar, then briefly illustrate the similarities and differences in deformation behavior of these phases in polyphase aggregates.

In order to induce crystal plasticity in silicates at experimental strain rates about 5 to 8 orders of magnitude faster than natural rates, it is necessary to trade temperature for time. Early attempts to induce ductile deformation in quartz were unsuccessful; although quartz was obviously weak even at low metamorphic grade in nature, it was strong and brittle at the conditions achievable in the early generations of deformation apparatus, and some workers postulated it must 'flow' by penetrative fracturing and rehealing. It was Griggs' development of apparatus capable of operation at higher confining pressures (to inhibit cracking and, as was later realized, to provide high water fugacity) and temperatures (to increase diffusion rates), plus his and Blacic's discovery of the importance of trace amounts of water, that finally allowed crystal plasticity to be induced experimentally in single crystals (Griggs and Blacic 1964) and 'mylonites' to be produced from quartzites (e.g., Carter et al. 1964; Tullis et al. 1973).

How can we be sure that the crystal plasticity induced in experimental samples is the same process that occurs in nature? Primarily by verifying that the deformation microstructures (grain size, shape, internal defects and LPO) are the same, but also by extrapolating the experimental flow laws to natural conditions and seeing if the predicted stresses appear reasonable. Experimental results on crystal plasticity have proved very useful in understanding natural crustal deformation, but field observations have also shown the importance of other processes, such as a switch to grain-size sensitive creep induced by syntectonic reactions. These applications and interpretations are discussed in the final section, together with an assessment of our current understanding and remaining questions concerning the strength of the crust.

CRYSTAL PLASTICITY PROCESSES

For the grain-scale deformation mechanism of crystal plasticity, strain (change of shape) occurs by propagation of line defects or dislocations through crystals (for a more detailed review see Poirier 1985 or Green 1992). These dislocations glide on specific crystal planes, generally those within which bonds are stronger and across which bonds are fewer and/or weaker, and in specific crystal directions, generally those with shorter unit cell parameters. The combination of a slip plane and direction is termed a slip system and may be specified as basal <a> or (001) [110], for example. Glide of dislocations occurs by expansion of dislocation loops, with edge dislocation components (extra half planes) at the leading and trailing portions propagating in the shear direction, and screw components on the sides propagating normal to the shear direction (Fig. 2a). This incremental glide accomplishes 'card deck shear', with an individual offset equal to the Burgers vector, normally one unit cell length. In contrast to olivine, individual dislocations in quartz and feldspar cannot be observed optically, but arrays of geometrically necessary dislocations produce optical microstructures such as undulatory extinction (Fig. 2b) and low angle tilt or subgrain boundaries (Fig. 2c).

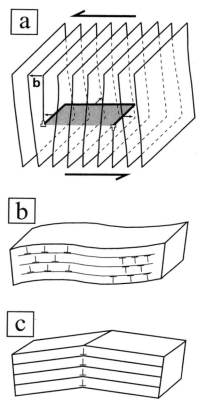

Figure 2. Schematic illustrations of dislocations. (a) Idealization of half of a dislocation loop consisting of pure edge and screw components (heavy line). Light solid and dashed lines represent crystal lattice planes normal to the glide plane of the dislocation; **b** indicates the Burgers vector; T symbols are edge dislocations (extra half planes of atoms). An applied shear stress (heavy arrows) causes expansion of the loop (small arrows), increasing the slipped area (stippled). [Used by permission of Cambridge University Press. From Poirier (1985) *Creep of Crystals*, Fig. 2.4, p. 52.] (b) Schematic representation of undulatory extinction, with distributed edge dislocations of both signs 'wedging' the crystal lattice. (c) Idealized low angle (tilt) subgrain boundary, consisting of an array of edge dislocations. Both glide and climb are required to produce such a boundary.

The stress required for dislocation glide is much lower than that required for fracture because only a few bonds are broken at a time. Experiments on oriented single crystals can be used to determine the critical resolved shear stress (CRSS) for slip on particular crystallographic planes and directions; these CRSS values may change with deformation conditions such as temperature and water fugacity. Most silicates are anisotropic so that one slip plane is significantly easier than others; thus even in pure monophase aggregates, grains in some orientations will be relatively 'hard' and others will be 'soft'.

Dislocation creep refers to steady state flow, which requires a recovery process in addition to the strain-producing glide process. Glide alone tends to result in work hardening, from the tangling of dislocations gliding on intersecting planes or from obstacles such as impurities on the glide plane. Continued strain at constant stress requires an accommodation or recovery process that continually reduces the dislocation density by annihilation and/or rearranges the dislocations into low energy configurations. There are several different recovery processes. At relatively high temperatures where volume diffusion is rapid, climb of edge dislocations allows them to maintain low energy configurations (e.g., Fig. 2c), and to get past obstacles and continue gliding at constant stress. At these conditions the creep process may be termed climb-accommodated dislocation creep. Cross slip of screw dislocations also may contribute to recovery. A different recovery process occurs at low temperatures where volume diffusion is slow and climb is limited. At such conditions, single crystals work harden up to fracture, but in aggregates the grain boundaries allow bulging recrystallization which can serve to soften or recover the structure. This process is driven by high dislocation density contrasts across local grain boundary segments, and results in the formation of very small and initially dislocation-free recrystallized grains (e.g., Bailey and Hirsch 1962). The recrystallized grains achieve a dynamic steady state of continual work hardening and recovery; this process may be termed recrystallization-accommodated dislocation creep (Hirth and Tullis 1992). At all conditions, dynamic recrystallization may relax grain boundary stress and strain mismatches and thus reduce the number of active slip systems from the five that are required for a completely homogeneous strain.

The operation of dislocation creep produces characteristic microstructures, on the optical to TEM scale, which can potentially provide a variety of useful information about the thermomechanical history of naturally deformed rocks. First, there are several different processes of dynamic recrystallization, dominant at different temperatures (flow stresses), which produce distinct microstructures (e.g., Urai et al. 1986; Hirth and Tullis 1992). Thus optical characterization of the size, shape and internal strain energy of porphyroclasts and recrystallized grains can provide useful information about the deformation conditions. Second, transmission electron microscope (TEM) characterization of the type, density and arrangement of dislocations can be used to provide information about dominant slip systems from natural samples, although details of the dislocation microstructure may have changed following the deformation of interest. Third, materials research shows a relation between the magnitude of the steady-state flow stress and the free dislocation density, the subgrain size, and the recrystallized grain size (Takeuchi and Argon 1976). Once experimental calibrations are done for different recrystallization mechanisms in different minerals, such piezometers may allow constraints to be put on the magnitude of the flow stress from different depths and tectonic settings. Fourth, dislocation creep produces LPOs, because slip on specific planes tends to rotate the crystal with respect to the instantaneous stretching axes of bulk flow. Different patterns, reflecting different dominant slip systems as well as different strain geometries, can be used to indicate different deformation conditions.

Dislocation creep is a thermally activated process, and the most commonly used flow law is a power law relation between strain rate $\dot{\varepsilon}$ and differential stress $\sigma_1 - \sigma_3$

$$\dot{\varepsilon} = A(\sigma_1-\sigma_3)^n \exp(-Q/RT) \tag{1}$$

where A is a material parameter, the stress exponent n is typically 3 to 5, Q is the activation energy, R is the gas constant, and T is absolute temperature. For dislocation creep there is no dependence on grain size; thus the strength of materials deforming by this process can be significantly different from that for materials deforming by grain size sensitive creep processes (grain boundary sliding and diffusion creep). A chemical activity or fugacity term typically needs to be included in the flow law to account for the effect of the chemical environment, e.g., $\dot{\varepsilon} \propto (f_{H2O})^m$ (Kohlstedt et al. 1995). Flow law determinations are usually made using gas apparatus (operated at ≤300 MPa), due to their superior stress sensitivity and capability to operate in constant stress mode, although the development of the molten salt cell (MSC) for high pressure Griggs-type apparatus (Green and Borch 1989) has greatly improved the quality of mechanical data achievable at higher pressures (500 MPa to 5 GPa). It is important to note that a flow law can only be extrapolated outside of experimental conditions if it has been determined for a single process which has achieved steady state flow. Flow law determinations are typically done using a series of small strain steps (1-5%) and small total strains (10-20%); in such cases it is important to verify that microstructural re-equilibration has occurred. Obviously a major question concerning extrapolation of experimental flow laws is whether steady state flow by the same (single) process has occurred in the natural area of interest.

The following sections present brief summaries of the processes, microstructures and mechanical behavior or flow laws (where available) for dislocation creep in pure quartz aggregates and pure feldspar aggregates, primarily based on laboratory experiments but including comparisons with naturally deformed examples. These sections are followed by accounts of some of the complications observed to occur in the experimental deformation of polyphase aggregates of quartz, feldspar and mica, and a comparison with granitic rocks naturally deformed over a wide range of conditions. The last section addresses our current understanding and remaining questions concerning the role of crystal plasticity in crustal deformation.

MONOMINERALIC AGGREGATES: QUARTZ

After olivine, quartz is probably the silicate mineral whose crystal plastic behavior has been best characterized. There are extensive data on the processes and microstructures of dislocation creep in quartzites which have come from experiments in solid media (Griggs-type) apparatus, but fewer good mechanical and flow law data because quartzites have not been successfully deformed in a gas apparatus, due to the apparent requirement for water fugacities that can only be achieved at pressures $>\sim 1000$ MPa. The first sections below briefly summarize results from experimental studies; they are followed by a section of comparisons with naturally deformed quartzites.

Water

A discussion of crystal plasticity of quartz must begin with a consideration of the role of water. The importance of trace amounts of water was first demonstrated for quartz by Griggs and Blacic (1964) and Griggs (1967) and termed 'hydrolytic weakening'. The state of our knowledge about the form and effects of water on quartz properties and deformation has been reviewed by Paterson (1989) and Kronenberg (1994); a few major points and recent results are summarized here.

At experimental conditions of 900°C, 1.5 GPa and 10^{-6}/s, clear quartz crystals are extremely strong and experience brittle failure, whereas milky quartz crystals and natural quartz aggregates such as quartzites and novaculites are weak and undergo steady state dislocation creep. Infrared analyses show that natural milky quartz crystals as well as the grains of quartzites contain a high average concentration of water (~2000-10,000 ppm or $H/10^6 Si$), largely in the form of freezable fluid inclusions associated with deformation microstructures such as healed cracks and dislocations (e.g., Kronenberg et al. 1990), whereas clear quartz crystals have a very low water content (~30 ppm). The OH concentration in hydrothermally grown synthetic quartz crystals, as well as in natural amethyst and citrine, can be large but most is in a 'gel-form' and is non-freezable. Although 'wet' synthetic crystals are weak (at least transiently) even at low pressure, it is not clear that they have the same properties or undergo the same deformation processes as the quartz in natural crustal aggregates.

There is still debate about the water-related defect that causes weakening; Griggs (1967) postulated that hydrolyzed Si-O bonds reduce the Peierls stress for glide, whereas Hobbs (1984) postulated that water-related defects act as dopants which influence the concentration of charged defects. Intragranular water certainly facilitates the creep recovery processes; annealing experiments on pre-deformed quartzites show that the rates of both dislocation climb and grain boundary migration are enhanced by water and pressure (Tullis and Yund 1989). Climb depends on volume diffusion, and whereas oxygen volume diffusion in quartz is almost immeasurably slow in a dry CO_2 environment (Sharp et al. 1991), in wet environments its rate increases linearly with water fugacity and the diffusing species appears to be molecular water (Farver and Yund 1991). Whatever the water-related defect may be, the climb-accommodated dislocation creep strength of quartzite depends linearly on water fugacity (Kohlstedt et al. 1995; Post et al. 1996), and failure to take this into account explains some of the previous difficulties in matching experimental flow law extrapolations to reasonable crustal strengths.

The water content of natural quartzites (~0.1 to 0.2 wt %) is greatly in excess of equilibrium, and yet at experimental conditions where dislocation creep occurs, the creep strength can be lowered by adding ~0.2 wt % water to the sample environment (e.g., Kronenberg and Tullis 1984), although greater amounts cause fracturing and dissolution/reprecipitation (e.g., den Brok and Spiers 1991; Post and Tullis 1998). The added water does not result in measurably higher intragranular water contents (den Brok et al. 1994), suggesting that the water residing in fluid inclusions does not act as a free fluid phase to control the water fugacity (e.g., Cordier and Doukhan 1989). Removing the water originally present in quartzites by vacuum drying greatly strengthens them (e.g., Kronenberg and Tullis 1984), but the water can be reintroduced at a rate consistent with 'wet' oxygen diffusivity into non-equilibrium sites such as fluid inclusions, rapidly restoring the low strength of the original 'as-is' aggregate (Post and Tullis 1998). These results suggest that over the range of crustal conditions where quartz deforms by crystal plasticity, it will rapidly readjust to changing water fugacity conditions.

Slip systems

In most crustal environments quartz will be in the low (α) form, with trigonal symmetry, although at high temperatures it may be in the high (β) form, with hexagonal symmetry. In either case slip requires strong Si-O bonds to be broken, and the strength of dry crystals approaches the theoretical shear strength (e.g., Kronenberg et al. 1986). There are few data on the CRSS of different slip systems in quartz crystals which contain water with the same form and concentration as that in natural quartzites, although there have been several studies using synthetic crystals at very low pressures (Baeta and Ashbee 1969). A combination of these experimental data plus inferences from LPOs developed in experimentally and naturally deformed quartzites indicates that basal <a> slip is the easiest system at lower temperatures, prism <a> and rhomb <a> become easier with increasing temperature, and prism [c] becomes operative only at very high temperatures.

Recrystallization mechanisms and microstructures

In quartz the recovery process for dislocation creep and the resulting mechanical behavior change with flow stress, as does the mechanism of dynamic recrystallization. Three different recrystallization mechanisms have been identified in experiments on quartzites (Hirth and Tullis 1992), resulting in different creep 'regimes' characterized by distinct microstructures (Fig. 3). The original documentation of the different regimes was based on samples deformed at constant strain rate to moderately low strains in axial compression, with extensive correlation of TEM and optical microstructures with the mechanical behavior (Fig. 4). The same material was used in MSC experiments to determine the flow law for the higher temperature regimes (Gleason and Tullis 1995) and to provide a preliminary test of published recrystallized grain size piezometers (Gleason and Tullis 1993). Recent shear experiments to much higher strains ($\gamma \sim 7$), also in each of the three regimes, have allowed detailed characterization of the LPOs including the effect of dynamic recrystallization (Heilbronner and Tullis 2002; Tullis and Heilbronner 2002). These experimental dislocation creep regimes are described below, in order of increasing temperature (decreasing flow stress).

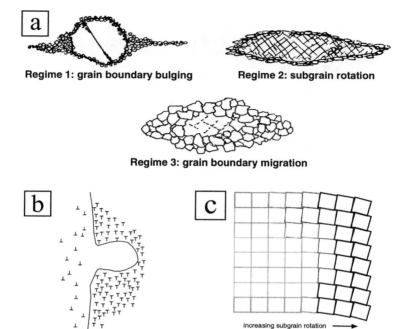

Figure 3. (a) Schematic illustrations of the optical microstructures developed in quartz aggregates deformed in the three dislocation creep regimes, characterized by distinct mechanisms of dynamic recrystallization. Note that the size of the recrystallized grains increases from regime 1 to 2 to 3, as flow stress decreases with increasing temperature and/or decreasing strain rate. Subequant porphyroclasts and finely recrystallized tails are developed in regime 1; homogeneously flattened original grains are only developed in regime 2; complete recrystallization occurs at low strain in regime 3. (b) Schematic illustration of grain boundary bulging recrystallization, driven by local differences in dislocation density across portions of original grain boundaries. (c) Schematic illustration of subgrain rotation recrystallization. Progressive glide and climb of dislocations into low-angle boundaries (lighter lines) results in increasing misorientation of subgrains, until at about 10° the boundaries have the properties of 'high angle' grain boundaries (darker lines).

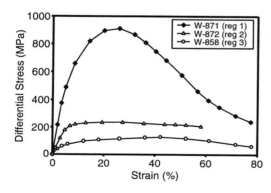

Figure 4. Representative curves of differential stress vs. axial shortening strain for samples of Black Hills quartzite (original grain size ~100 μm) deformed at a confining pressure of 1.5 GPa in dislocation creep regime 1 (bulging recrystallization), regime 2 (sub-grain rotation recrystallization) and regime 3 (migration recrystallization). W871: 850°C, 1.5×10^{-5}/s, as-is; W872: 900°C, 1.5×10^{-5}/s, 0.17 wt % water added; W858: 900°C, 1.5×10^{-6}/s, 0.17 wt % water added.

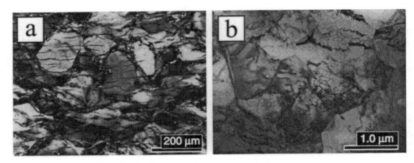

Figure 5. Microstructures developed in quartz aggregates experimentally deformed in axial compression in dislocation creep regime 1. (a) Optical micrograph (crossed polars) of sample shortened 58% (parallel to short edge of photo) at 700°C, 10^{-6}/s; note inhomogeneously deformed grains. The parallel cracks formed at low temperature during unloading. (b) TEM micrograph showing recrystallized grains from sample deformed at 850°C, 10^{-5}/s, 45% strain; note variable dislocation densities and strain-induced grain boundary bulging.

Regime 1: Bulging recrystallization. In the lowest temperature portion of the dislocation creep regime, volume diffusion rates are slow and dislocation climb cannot keep pace with the imposed strain rate; extreme work hardening occurs, with the development of very high densities of straight dislocation segments as well as cells and tangles. Locally along grain boundaries the dislocation density contrasts are sufficient to drive grain boundary bulging (e.g., Bailey and Hirsch 1962) from the lower into the higher density areas. These bulges pinch off to form extremely small (~1-2 μm) recrystallized grains, which are initially dislocation-free and thus much weaker than the work-hardened original grains (Fig. 5b). In this regime (termed regime 1 by Hirth and Tullis 1992), deformation of quartzites with a grain size significantly larger than the final recrystallized grain size is characterized by an initial stress peak, resulting from work hardening of the original grains, followed by strain weakening, resulting from progressive recrystallization (Fig. 4). Mechanical steady state requires substantial recrystallization, and can only be achieved in low strain axial compression experiments by using aggregates such as novaculite with an initial grain size (d ~ 5 μm) closer to the final recrystallized grain size. A component of grain boundary sliding in the extremely fine recrystallized grains cannot be ruled out, although the recrystallized regions develop a strong LPO and TEM shows that some grains are dislocation-free while others have very high densities, consistent with a dynamic steady state.

The optical microstructures characteristic of low to moderate strain samples include relatively equant porphyroclasts with deformation bands and strong undulatory extinction, and anastomosing zones of extremely fine (often not resolvable) recrystallized grains along their boundaries (Fig. 5a). Strain is strongly partitioned into the recrystallized grains once an interconnected grain boundary layer has formed, and further recrystallization progressively reduces the size of the porphyroclasts and results in further weakening. As the porphyroclasts experience decreasing stress and strain rate, they may undergo limited internal recovery. Sheared samples develop an S-C' fabric, with lensoid porphyroclasts defining the foliation and zones of recrystallized grains defining extensional shear bands.

Regime 2: Subgrain rotation recrystallization. At somewhat higher temperatures and/or slower strain rates (lower stress), dislocation climb becomes easy enough to accommodate the imposed strain rate; thus in this regime of dislocation creep (termed regime 2 by Hirth and Tullis 1992) both the mechanical behavior and the recrystallization process are quite different. The free dislocation density remains lower due to climb-assisted annihilation, and geometrically necessary dislocations are arranged in low-energy walls or subgrain boundaries. Gliding dislocations do not become tangled, thus there is no work hardening and original grains are fairly homogeneously flattened with increasing strain. Dynamic recrystallization occurs dominantly by progressive misorientation of subgrain boundaries ('subgrain rotation' of Guillope and Poirier 1979) until they become effectively high angle grain boundaries at a misorientation of ~10° (Fig. 6b). Because the recrystallized grains have about the same density of free dislocations as the subgrains within the original grains, recrystallization in this regime does not result in weakening (Fig. 4). Steady state flow is achieved at low strains regardless of original grain size or degree of recrystallization.

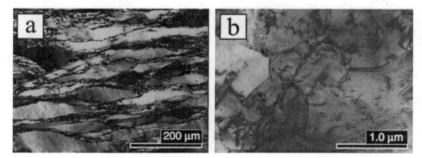

Figure 6. Microstructures developed in quartz aggregates experimentally deformed in axial compression in dislocation creep regime 2. (a) Optical micrograph (crossed polars) of sample shortened 60% (parallel to short edge of photo) at 800°C, 10^{-6}/s; note homogeneously flattened original grains, some with deformation lamellae. (b) TEM micrograph of recrystallized region from same sample; dislocation densities are uniform, and both subgrain and high-angle boundaries are present.

The optical microstructures characteristic of deformation in this regime include relatively homogeneously flattened original grains containing optically visible subgrains, and recrystallized grains with a size (~10 µm) approximately the same as that of the subgrains and significantly larger than that in regime 1 (Fig. 6a). Subgrain rotations tend to be greatest at original grain boundaries, where stresses and strains are somewhat larger, thus producing 'core and mantle' structures (e.g., White 1975) of flattened and polygonized original grains surrounded by a necklace of recrystallized grains. At somewhat lower temperatures original grains may become extremely elongated without being replaced by recrystallized grains, while at slightly higher temperatures a sample may become completely recrystallized. An SC fabric is produced in shear, with slightly inequant recrystallized grains oblique to the shear zone boundary (Dell'Angelo and Tullis 1989), defining a steady state foliation (Means 1981).

Regime 3: Grain boundary migration recrystallization. At still higher temperatures or slower strain rates (lower stress), dislocation climb remains easy but grain boundary mobility is much higher, resulting in a different process of recrystallization and different microstructures. In this regime (regime 3 of Hirth and Tullis 1992) recrystallization occurs chiefly by grain boundary migration; although the driving force is low, due to small differences in dislocation density and subgrain boundary spacing, the grain boundary velocity is high (Fig. 7b). Mechanical (and microstructural) steady state is achieved at very low strain (Fig. 4).

The optical microstructures characteristic of this regime are quite distinct from those produced in the lower temperature regimes. Complete recrystallization occurs at low strain, before significant flattening of porphyroclasts can occur, and the microstructure maintains a dynamic steady state with increasing strain (Fig. 7a). Subgrains rarely accumulate enough misorientation to produce high angle boundaries, but any new grains produced in this way increase their size by boundary migration. The average recrystallized grain size is significantly larger (~40-50 μm) than that in regime 2, and larger than the average co-existing subgrain size (as also observed in halite by Guillope and Poirier 1979).

 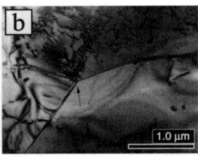

Figure 7. Microstructures developed in quartz aggregates experimentally deformed in axial compression in dislocation creep regime 3. (a) Optical micrograph (crossed polars) of sample shortened 57% (parallel to short edge of photo) at 900°C, 10^{-6}/s with 0.12 wt % water added; the sample is completely recrystallized. (b) TEM of sample deformed at same conditions but to 36% strain; note low dislocation density, and boundary of large recrystallized grain migrating toward subgrain boundary (arrow).

Transitions. It should be emphasized that the transitions between these dislocation creep regimes are not sudden but gradual (Fig. 8a). In regime 1, bulging recrystallization may occur along internal boundaries such as deformation bands or microfractures, in addition to original grain boundaries. In high strain regime 1 samples, after significant strain weakening has occurred, porphyroclasts begin to develop flattened shapes and subgrains due to the decreased stress and strain rate they experience. At low strains in regime 2 recrystallized grains form by grain boundary bulging ('suturing') because subgrains have not accumulated enough misorientation to form high angle boundaries. At higher strain in regime 2 at least an increment of grain boundary migration must occur, because recrystallized grains formed by subgrain rotation retain an equant shape and a steady state oblique foliation angle. Experiments so far have been limited to the lower temperature portion of regime 3, but there is no evidence for any difference in deformation behavior or microstructures in the α– vs. β–quartz fields.

Trace amounts of water affect the operative slip systems as well as the ease of dislocation climb and grain boundary migration, and the regime transitions for water-added quartzites are observed to occur at lower temperatures than those for 'as-is' samples (Hirth and Tullis 1992) (Fig. 8b). Experiments in a MSC and with controlled water fugacity are needed to better define the stresses of the regime transitions and how they depend on water.

A transition from dislocation creep to grain size sensitive flow has been documented in synthetic aggregates deformed at 300 MPa and 1100-1200°C (Brodie and Rutter 2000). Dislocation creep was observed in aggregates with a grain size of 20 μm and 0.6 wt % H_2O, but grain size sensitive creep could only be produced in very dry and

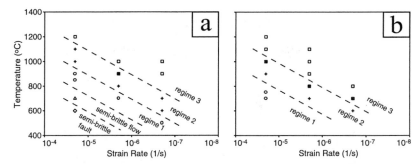

Figure 8. Plots of temperature vs. strain rate showing the location of the dislocation creep regimes for quartz aggregates deformed at a confining pressure of 1.2 GPa and (a) 'as-is' or (b) with 0.17 wt % water added. Note that the boundaries between the regimes are actually gradational. Circles represent regime 1, plus symbols regime 2, solid squares are transitional between regimes 2 and 3, and open squares represent regime 3. For as-is samples, the fields of semi-brittle faulting and semi-brittle flow are included. (Modified from Hirth and Tullis 1992, with additional data from Hirth and Tullis 1994.)

extremely fine-grained (0.4 µm) aggregates. Extrapolation of these data to natural conditions indicates that grain size sensitive creep of aggregates with d ~ 1 µm would require a stress of ~30 MPa at 500°C and thus is unlikely unless grain growth is prevented due to pinning by other phases.

LPOs

LPOs in experimentally deformed quartz aggregates have been measured using several techniques. Optical techniques such as the Universal (U) stage provide information only on the [c] axis orientations, but the computer integrated polarization (CIP) technique (Heilbronner and Pauli 1993, 1994) allows very small grain sizes to be measured and grain orientation maps to be prepared. X-ray techniques yield the complete volume-averaged LPO of all crystal axes for all grains present within an ~100 µm thick section; electron back-scatter diffraction (EBSD) also provides the complete crystal orientation, but on a grain-by-grain basis.

Early X-ray determinations of the complete LPOs for quartzites deformed in axial compression over a range of conditions were generally consistent with U-stage measurements of [c] axes for the same samples. They showed a transition with increasing temperature from a broad maximum of [c] axes parallel to compression to small circle girdles of increasing opening angle, consistent with a shift from dominantly basal slip to increasing contributions from slip on other planes (Tullis et al. 1973). The X-ray measurements showed a big difference in concentration for the positive and negative rhombs, resulting from mechanical Dauphine twinning (Tullis and Tullis 1972).

The high symmetry and relatively low strain of these early experiments, plus the inability to measure grain-by-grain orientations for small grain sizes, did not allow investigation of two major questions about LPO development, namely the effects of shear geometry and dynamic recrystallization. Dell'Angelo and Tullis (1989) made U-stage measurements of porphyroclast LPOs in low shear strain samples, and Gleason et al. (1993) attempted to determine the LPOs of recrystallized grains in axial compression samples for each of the three recrystallization mechanisms identified by Hirth and Tullis (1992), but neither study was fully satisfactory. Recently quartzite samples have been deformed to high strain in general shear in all three dislocation creep regimes (optical microstructures are shown in Fig. 9), and for all samples the [c] axis LPOs have been

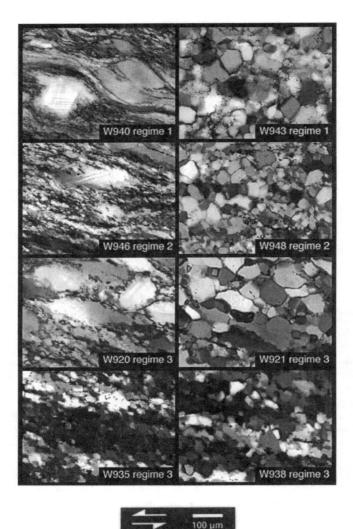

Figure 9. Optical micrographs of samples of Black Hills quartzite deformed in general shear at a confining pressure of 1.5 GPa in dislocation creep regimes 1 (bulging recrystallization), 2 (subgrain rotation recrystallization), and 3 (migration recrystallization), before annealing (left column) and after annealing at the deformation temperature for 4 days (right column). Shear plane is horizontal and shear sense is sinistral. Micrographs taken with circular polarization; light grains = c-axes parallel to plane of thin section and dark grains = c-axes parallel to viewing direction. W940: 850°C, 10^{-5}/s, as-is, $\gamma = 4.3$; W946: 875°C, 10^{-5}/s, 0.17 wt % water added, $\gamma = 7.2$; W920: 900°C, 10^{-5}/s, 0.17 wt % water added, $\gamma = 2.1$; W935: 915°C, 10^{-5}/s, 0.17 wt % water added, $\gamma = 5.6$. [Used by permission of the Geological Society of London. From Heilbronner and Tullis (2002), Fig. 12.]

measured separately for porphyroclasts and recrystallized grains (Heilbronner and Tullis 2002; Tullis and Heilbronner 2002) (Fig. 10). The resulting patterns are very similar to those observed in naturally deformed quartzites; a brief summary is given below.

The [c] axis LPO developed in regime 1 shows only a slight change in pattern with progressive recrystallization. At low strain ($\gamma \sim 2$) the porphyroclasts develop a broad peripheral maximum rotated slightly against the sense of shear, with a small tail toward the center of the pole figure, while the recrystallized grains form two broad maxima ~20° to 30° from the periphery (Heilbronner and Tullis 2002) (Fig. 10). At higher strain ($\gamma \sim 5$) and complete recrystallization, there is a sharper and stronger peripheral maximum that extends inward about 30°, rotated slightly with the sense of shear (Tullis and Heilbronner unpubl.). Although <a> axis orientations have not yet been measured, the approach of Schmid (1994, Fig. 7) suggests that this [c] axis pattern could result from a combination of basal plus rhomb <a> slip.

The [c] axis LPO developed in regime 2 also shows only a slight change in pattern with progressive recrystallization. At low strain ($\gamma \sim 2$) the LPO is dominated by the porphyroclasts, and is a broad peripheral maximum rotated against the sense of shear

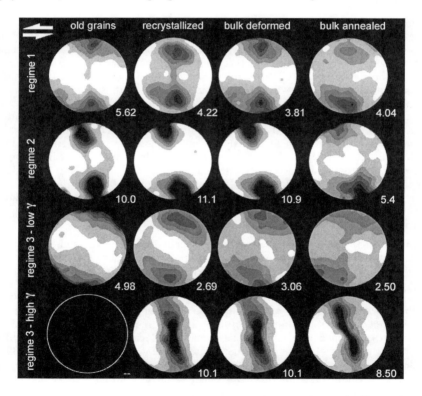

Figure 10. Bulk and partial *c*-axis pole figures (LPOs) of the quartzite samples illustrated in Figure 9, sheared in dislocation creep regimes 1, 2 and 3 (left 3 columns) and after annealing at the deformation temperature for 4 days (right column). Measured area = 0.4 mm² (~80 original grains) except for regime 3 high γ where a combination of 3 areas (~240 original grains) was used. Shear plane is horizontal and shear sense is sinistral; LPO maximum density is indicated to the lower right of each figure. [Used by permission of the Geological Society of London. From Heilbronner and Tullis (2002), Fig. 9.]

(Dell'Angelo and Tullis 1989). The LPO for a sample completely recrystallized at high strain ($\gamma \sim 7$) is a strong peripheral maximum extending inward about 30°, rotated with the sense of shear (Heilbronner and Tullis 2002) (Fig. 10), again consistent with basal plus rhomb <a> slip. Thus in regimes 1 and 2, the evolution of the LPO with increasing shear strain is similar, and there is only a minor difference in pattern for porphyroclasts and recrystallized grains.

It should be noted that Gleason et al. (1993) reported a different role for recrystallization in regime 1 LPO development. In flint samples (containing ~1 wt % water) axially compressed at effectively regime 1-2 conditions, a transient maximum of [c] axes parallel to compression was produced by the growth to impingement of rhomb-bounded porphyroblasts. The porphyroblast growth was apparently facilitated by the special mobility of the rhomb planes in the presence of water, and it favored the 'hard' orientation with no resolved shear stress on base, prism or rhomb <a> systems. After impingement, however, further strain caused deformation and recrystallization of the porphyroblasts, and development of a small-circle girdle of [c] axes very similar to that developed in axially compressed quartzites. Thus the selective growth of 'hard' grains may be an experimental oddity.

The [c] axis LPO developed in quartzites sheared in regime 3 undergoes a profound change at $\gamma \sim 3$-4, associated with complete recrystallization (Heilbronner and Tullis 2002). At lower strains the LPO of porphyroclasts is a broad peripheral maximum rotated against the sense of shear, whereas the recrystallized grains define two maxima about 20-30° from the periphery (Fig. 10). However at $\gamma \sim 4$ and above, with complete recrystallization, the LPO switches to a partial girdle with strong maxima at the intermediate strain axis Y and at 35-40° to Y (Heilbronner and Tullis 2002) (Fig. 10). Although <a> axis orientations have not yet been measured, these [c] axis patterns suggest a switch from dominantly basal plus rhomb <a> slip to dominantly prism <a> and rhomb <a> slip. Although TEM analyses such as those reported by Knipe and Law (1987) for a naturally sheared quartzite have not yet been done, optical orientation analyses suggest that grains originally oriented for basal slip accumulate more internal strain energy compared to grains originally oriented for prism <a> and rhomb <a> slip; the latter grains begin to recrystallize at $\gamma \sim 4$ and the recrystallized grains preferentially grow into and consume grains of other orientations, including those forming the peripheral maximum.

For all three dislocation creep regimes, recent experimental studies show that although static annealing following deformation profoundly changes the grain size and shape, the LPO is little affected (Heilbronner and Tullis 2002) (Fig. 10).

Flow laws and piezometers

Flow law parameters for recrystallization-accommodated (regime 1) dislocation creep have not yet been determined; they will probably require higher stresses and thus confining pressures than are possible in gas apparatus. Because water is known to affect the recovery process of grain boundary migration (e.g., Tullis and Yund 1989), a dependence of the flow law on water fugacity is expected, but it may not have the same dependence as found for climb-accommodated creep in regimes 2-3. A preliminary calibration of the recrystallized grain size piezometer for bulging recrystallization (Bishop, unpubl. data; see Post and Tullis 1999) found a significantly different slope for grain size vs. stress than has been found for subgrain rotation and grain boundary migration recrystallization in other minerals (e.g., Guillope and Poirier 1979; van der Wal et al. 1993).

Early flow laws for quartz (e.g., Parrish et al. 1976; Kronenberg and Tullis 1984;

Jaoul et al. 1984; Koch et al. 1989) cannot be trusted due to errors in the measured stresses. Recent flow laws for climb-accommodated dislocation creep (including regime 2 and part of regime 3) have been determined by Paterson and Luan (1990) and Luan and Paterson (1992) using synthetic aggregates in a gas apparatus at 300 MPa and by Gleason and Tullis (1995) using a fine-grained natural quartzite in a MSC in a Griggs apparatus at 1.5 GPa. Both groups found n = 4 although there was a considerable range in activation energy Q (from 137±34 to 223±56 kJ/mol); however, when adjusted for the difference in water fugacity these flow laws are in remarkably good agreement (Kohlstedt et al. 1995). Constraints from comparisons with natural quartzites suggest that an activation energy of 135±15 kJ/mol and a water fugacity exponent of one are most reasonable (Hirth et al. 2001). Also, Gleason and Tullis (1995) used 'as-is' quartzite samples in their flow law determination at 1.5 GPa, and later analysis indicates that the water fugacity was likely somewhat less than that given by the pressure (Hirth et al. 2001).

Thus far there has not been a calibration of the recrystallized grain size piezometer for subgrain rotation or for grain boundary migration recrystallization in quartz, using molten salt assemblies. Previously determined recrystallized grain size piezometer relations (e.g., Mercier et al. 1977; Christie et al. 1980; Koch 1983) did not take account of different recrystallization mechanisms and used sample assemblies which introduced serious errors in the measured stresses (Gleason and Tullis 1993). In new MSC piezometer calibrations it would be desirable to test for the independent effects of temperature (e.g., de Bresser et al. 2001) and water fugacity.

Comparisons with naturally deformed quartzites

The dislocation creep microstructures described by Hirth and Tullis (1992) have been observed in naturally deformed quartz aggregates, in the same sequence with increasing temperature, although the absolute temperatures of the regime transitions differ for different strain rates (Fig. 11). In addition, quartzites naturally deformed at high

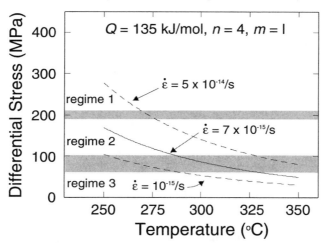

Figure 11. Plot of temperature vs. differential stress, illustrating the range of deformation conditions predicted for best fitting quartzite flow law parameters. Estimated stresses at the transitions between the dislocation creep regimes in the Ruby Gap duplex are shown by the shaded regions. Differential stresses were calculated using a stress exponent (n) of 4 and a water fugacity exponent (m) of 1. An activation energy (Q) of 135 kJ/mol enables the entire range of differential stresses to be achieved within the estimated deformation conditions of temperature and strain rate. [Used by permission of the editor of *International Journal of Earth Sciences*. From Hirth et al. (2001), Fig. 6b, p. 85.]

temperatures show microstructures not yet achieved experimentally. For example, within the regime where grain boundary migration recrystallization is dominant, at lower temperatures migrating boundaries are pinned by impurity phases such as micas, as observed experimentally, whereas at higher temperatures they are able to engulf the second phase particles (e.g., Drury and Urai 1990). In addition, at upper amphibolite to granulite grade, depending on water availability, one observes 'chessboard' subgrains indicative of a component of prism [c] slip (e.g., Mainprice et al. 1986) and large, highly irregular 'dissection' grains (e.g., Jessel 1987; Stipp et al. 2002a). Kruhl (1996) has suggested that prism [c] slip indicates deformation in the beta quartz field and thus can be used as a geothermobarometer; however, quartzite shear experiments in the beta field show no evidence of chessboard subgrains or a [c] maximum sub-parallel to the shear direction (Heilbronner and Tullis 2002).

In experimental samples the temperatures of the recrystallization mechanism transitions depend on the water content of the samples (Hirth and Tullis 1992), which is not surprising since dislocation creep flow strengths depend on water fugacity. There is some evidence that the temperature of the brittle-plastic transition, as well as those of the dislocation creep regime transitions, are higher in quartzose rocks naturally deformed at 'dry' conditions, where the fluid is CO_2-rich (e.g., Passchier 1995; Selverstone 2001) or where extreme grain size reduction may have decreased the thickness of the grain boundary fluid phase (e.g., Pennacchioni and Cesare 1997).

Microstructural recognition of the different recrystallization mechanisms allows one to infer the relative temperatures of a suite of samples; more quantitative constraints require additional experimental information such as calibrated piezometer relations. Of the potential microstructural piezometers of dislocation density, subgrain size and recrystallized grain size, the latter is probably the most useful, since dislocation density can only be evaluated using TEM and is easily changed by events after the deformation of interest, and the subgrain size varies greatly depending on where it is measured within porphyroclasts and how it is measured (optical vs. TEM). For quartz, the three distinctly different mechanisms of dynamic recrystallization probably have different piezometer relations, as already shown for two different mechanisms in halite (Guillope and Poirier 1979) and in calcite (Rutter 1995). As mentioned above, early published piezometers for quartz (e.g., Christie et al. 1980) did not take account of different recrystallization mechanisms and had significant stress errors; for samples deformed in a MSC in regime 2-3, Gleason and Tullis (1993) found that the Twiss (1977) relation gave a reasonable match to the experimental data, although there are questions about its theoretical basis (e.g., Poirier 1985; de Bresser et al. 2001).

Possible complications in using quartz recrystallization microstructures as indicators of deformation conditions could result from post-deformation static annealing, involving grain growth, or deformation continuing as temperatures decrease, involving overprinting of the higher temperature microstructures. Early studies of quartzites naturally deformed at high temperatures sometimes attributed complete dynamic recrystallization in regime 3 to static annealing. However, post-deformation static annealing is only indicated if the recrystallized grains have equant shapes, straight boundary segments and 120° angles, and/or if the recrystallized grain size does not 'match' the porphyroclast microstructures, and/or if metamorphic minerals have grown across the foliation. Applications of recrystallized grain size piezometers have generally been deemed 'safe' for rocks up through greenschist grade, whereas many amphibolite grade rocks appear to show static coarsening (e.g., Dunlap et al. 1997; Zulauf 2001). Cases of lower temperature overprinting are usually easy to recognize, because a smaller recrystallized grain size (or small-scale suturing) is superposed on a larger one.

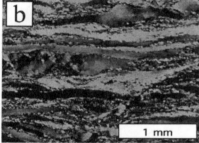

Figure 12. Optical microstructures of quartz mylonites from the Ruby Gap duplex (crossed polars). (a) Regime 1 microstructure showing patchy undulatory extinction and fine recrystallized grains along grain boundaries. (b) Regime 2 microstructure showing flattened original grains with optically visible subgrains and recrystallized grains of the same size. (c) Regime 3 microstructure with complete recrystallization and oblique foliation indicative of grain boundary migration recrystallization. [Used by permission of the editor of the *International Journal of Earth Sciences*. From Hirth et al. (2001), Fig. 1, p. 79.]

Early attempts to apply an experimental piezometer and flow law to help interpret the thermomechanical history of naturally deformed quartzites were made by Hacker et al. (1990, 1992). Their approach was sound, although the results are somewhat questionable because of their dependence on flow laws and piezometer relations based on experiments in which the stress was probably greatly in error (e.g., Gleason and Tullis 1993). Several examples of recent studies of naturally deformed quartzites and quartz veins which have documented the different recrystallization microstructures and have used the more recent flow laws are described below. All of them, however, are limited by using the Twiss (1977) piezometer relation for all three recrystallization mechanisms.

In the Heavitree quartzite of the Paleozoic mid-crustal Ruby Gap duplex of Australia, Dunlap et al. (1997) used the microstructurally identifiable regime transitions to restore a series of thrust slices. They observed microstructures ranging from bulging recrystallization (d < 20 μm) (together with pressure solution in micaceous layers), to subgrain rotation (d ~ 20-60 μm), and to grain boundary migration (d ~ 40-200 μm), all within a temperature interval of ~250 to 330°C, from sub-greenschist to very low greenschist facies (Fig. 12). Using the Twiss (1977) piezometer they inferred that the stress associated with the bulging to subgrain rotation transition was ~100 MPa and that associated with the subgrain rotation to grain boundary migration transition was ~60 MPa. Thermochronologic constraints allowed them to estimate a strain rate of $\sim10^{-15}$/s, and using this rate in the flow law of Paterson and Luan (1990) gave good agreement with their inferred temperatures and stresses.

In a major strike-slip shear zone of Oligocene age in the Eastern Alps (Alto Adige, Italy), Stoeckert et al. (1999) used optical and TEM observations to document quartz microstructures indicating semi-brittle flow and pressure solution from ~280 to ~310°C, and dislocation creep from ~310 to 350°C. Using the piezometer of Twiss (1977) they inferred a stress of 160 MPa for a grain size of 7 µm at ~280°C, and a stress of 60 MPa for a grain size of 30 µm at ~330°C. Using these stresses and temperatures in the flow laws of Paterson and Luan (1990) gave a strain rate of 10^{-13}/s to 10^{-14}/s. The deformation conditions and the recrystallized grain sizes in the studies of Dunlap et al. (1997) and Stoeckert et al. (1999) are thus quite similar.

In the Bohemian Massif, Zulauf (2001) examined quartz veins in metagraywackes deformed during Variscan shortening over a range of conditions from lowermost greenschist facies (~300°C) to amphibolite facies (>570°C). He observed bulging recrystallization (d<30 µm) accompanied by cataclasis and pressure solution at lowermost greenschist grade, a transition to subgrain rotation recrystallization (d ~ 25 to 80 µm) at mid greenschist grade (~500°C), and a transition to grain boundary migration recrystallization (d ~ 100 to 200 µm) at the transition to amphibolite facies (~570°C). Using the Twiss (1977) piezometer he inferred lower bounds of 70 MPa for the stress at the regime 1 to 2 transition and 20 MPa for the regime 2 to 3 transition.

In the Tonale Fault Zone of the southern Alps, Stipp et al. (2002a,b) documented microstructures developed in quartz veins in metapelites across an even greater temperature range (~280 to 700°C) within a strike-slip shear zone. Cataclasis occurs at <280°C. The recrystallization microstructures range from bulging (d ~ 5-25 µm, at 300 to 400°C) to subgrain rotation (d ~ 60 to 85 µm, at 440 to 500°C) to grain boundary migration (d ~ 220 µm up to several mm, at 550 to 700°C) (Fig. 13). Similar to the rocks studied by Zulauf (2001), the sizes of the recrystallized grains produced by the three mechanisms are significantly larger than those in the experimental samples or in the Ruby Gap quartzites. The reasons for these differences are not known, but using the Twiss (1977) piezometer and the water fugacity corrected flow law of Hirth et al. (2001) indicates a strain rate of 10^{-12}/s to 10^{-13}/s (Stipp et al. 2002b).

In sum, considering that experimental samples are deformed at high pressure (and water fugacity), at constant temperature and strain rate, and are quenched, whereas natural samples are generally more variable in grain size and purity, are deformed at lower pressure, temperature and strain rate, and may have been statically annealed or overprinted by lower temperature deformation, the agreement of the dislocation creep microstructures is quite remarkable. The agreement is perhaps least good for regime 1, in part because low temperature deformation in nature will tend to occur at low pressures with fluid and possibly impurities present, causing grain-boundary bulging recrystallization to occur together with microfracturing and pressure solution (e.g., Stoeckert et al. 1999). Overall, the sequence of microstructures with decreasing stress (increasing temperature or decreasing strain rate) appears robust and useful to geologists seeking to constrain thermomechanical history. Based on the experimental and natural data, Stipp et al. (2002b) constructed a recrystallization mechanism map with respect to temperature and strain rate, assuming water-saturated conditions (Fig. 14).

Analyses of LPOs developed in naturally deformed quartzose rocks have been very useful in constraining the processes responsible for the patterns and how they vary with deformation conditions (e.g., reviews of Law 1990; Schmid 1994). For example in terms of [c] axis LPOs, studies have documented an association of symmetric crossed girdles with pure shear and an oblique single girdle with simple shear (e.g., Bouchez et al. 1983) and a switch from a maximum parallel to Y at intermediate temperatures to a maximum parallel to the stretching lineation at high temperatures, consistent with a switch from dominantly prism <a> to prism <c> slip (e.g., Schmid et al. 1981; Blumenfeld et al. 1986). X-ray determinations of the complete LPO have demonstrated the tendency for [a] axes to be aligned close to the stretching direction (e.g., Bouchez 1978), and allowed the fabric skeleton to be analyzed in terms of the operative slip systems (e.g., Schmid and

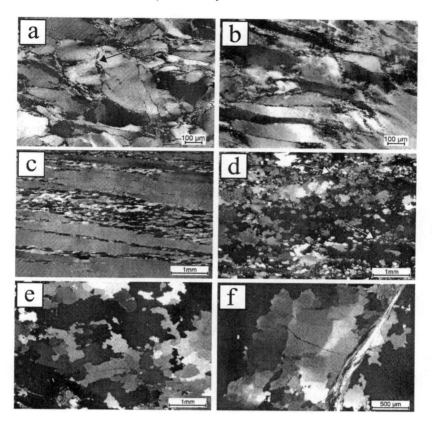

Figure 13. Optical microstructures of sheared quartz veins as a function of increasing temperature across the Tonale shear zone (crossed polars); sections are normal to foliation and parallel to stretching lineation. (a) ~300°C: bulging recrystallization (arrow). (b) ~340°C: both subgrains and bulging recrystallization. (c) ~490°C: elongate ribbon grains and subgrain rotation recrystallization. (d) ~560°C: grain boundary migration recrystallization; recrystallized grains remain smaller where pinned by biotites (arrows). (e) ~650°C: large 'amoeboid' recrystallized grains; grain boundary migration is not affected by small second phase particles. (f) ~650°C: detail showing chessboard extinction with subgrain boundaries parallel to prism and basal planes. [Used by permission of the editor of *Journal of Structural Geology*. From Stipp et al. (2002a) Fig. 7, p. 1868-1869.]

Casey 1986). There is considerable evidence that dynamic recrystallization allows achievement of a dynamic steady state LPO (Schmid 1994).

Recently there has been a remarkable confluence of results from experiments, nature and theory. The [c] axis LPOs determined for quartzites experimentally sheared over a range of conditions including all three recrystallization mechanisms (Heilbronner and Tullis 2002) match extremely well with those measured by Stipp et al. (2002a) from quartz veins naturally sheared over a wide temperature range and with microstructures indicating the same three recrystallization mechanisms. Both studies show that porphyroclasts in all three regimes develop [c] axis maxima close to the periphery of the pole figure, whereas for the recrystallized grains there is a transition from an oblique

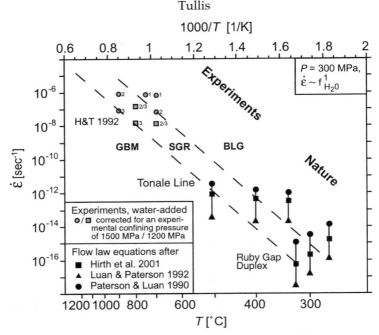

Figure 14. Strain rate vs. temperature diagram showing possible correlations of recrystallization mechanisms between the experimental data for water-added samples of Hirth and Tullis (1992), with numbers corresponding to their regimes, and data for natural samples from the Tonale shear zone (Stipp et al. 2002b) and from the Ruby Gap duplex (Dunlap et al. 1997). Flow stress data were derived using the piezometer relation of Twiss (1977); natural strain rates were calculated from three different flow laws assuming a linear dependence on water fugacity and normalizing to a pressure of 300 MPa. [Used by permission of the Geological Society of London. From Stipp et al. (2002b), Fig. 8b.]

single girdle at conditions where bulging and subgrain rotation recrystallization operate to a Y maximum at conditions where grain boundary migration recrystallization occurs. In regime 3, the fast migration evidently allows grains oriented for prism <a> slip, which accumulate less internal strain energy, to grow at the expense of other orientations (Tullis and Heilbronner 2002). Models of LPO development based only on slip do not match very well with some of the patterns commonly developed in naturally deformed quartzites, such as the Y maximum, but recent models that relax grain boundary constraints and incorporate effects of recrystallization give a much better match (e.g., Jessell and Lister 1990; Wenk 1994; Takeshita et al. 1999).

MONOMINERALIC AGGREGATES: FELDSPAR

Alkali and plagioclase feldspars constitute the dominant phases in the crust, and therefore are of major importance in controlling its deformation style and strength. However because they are crystallographically and chemically complex, as well as more refractory than quartz to experimental dislocation creep, there are fewer experimental data for feldspars than for quartz. An additional problem for experimentalists is that, unlike quartz, pure and fine-grained naturally occurring feldspar aggregates are rare. Some experiments have been done on natural albite aggregates (e.g., Tullis et al. 1990) and on fine-grained anorthosites (e.g., Tullis and Yund 1992), but recently many experimentalists have utilized synthetic aggregates with controlled composition and grain size prepared by hot pressing (e.g., Rybacki and Dresen 2000). A great deal of our current understanding of feldspar deformation has come from detailed studies of naturally deformed feldspathic rocks in areas where the pressure, temperature and fluid activity can be reasonably well constrained.

Water

Hydrogen defects of various forms, including OH and molecular water, are common in feldspars (see review by Kronenberg et al. 1996), and trace amounts of water are observed to affect the dislocation creep strength of plagioclase (Tullis and Yund 1980; Rybacki and Dresen 2000). Flow law determinations have not yet evaluated the nature of this dependence, in part because small amounts of free water drastically lower the melting temperature and thus almost eliminate the experimental dislocation creep window at higher pressures. Considerable complexity is revealed by diffusion measurements on feldspars: in alkali feldspars the rate of alkali self diffusion is independent of water pressure (Yund 1983), whereas the oxygen diffusion rate is directly related to water fugacity (Farver and Yund 1990); in plagioclase the rates of Si-Al exchange (Goldsmith 1991) and of NaSi-CaAl interdiffusion (Yund and Snow 1989) depend on hydrogen fugacity but also on pressure. An indication that the dislocation creep strength of plagioclase may have a different dependence on 'water' than does quartz comes from experiments at 10^{-5}/s, 1200°C and fH_2O= 300 MPa (Ji et al. 2000); anorthite aggregates undergo dislocation creep with a flow stress of only 20 MPa even at this low water fugacity, whereas quartz aggregates are semi-brittle with a yield strength twice the confining pressure.

Slip systems

Feldspars have a lower melting temperature than quartz and therefore might be expected to be weaker in dislocation creep; however both in experiments and nature they are stronger at almost all conditions. Their greater strength may in part reflect their large unit cell; dislocations in feldspars are commonly dissociated, inhibiting recovery (e.g., Olsen and Kohlstedt 1984). Alternatively the difference in dislocation creep strength may be due to differences in the types of water-related defects and how they affect glide and/or recovery.

The easiest glide planes in feldspars should be those across which there are the fewest T-O bonds per area of that plane in the unit cell, e.g., (010) and (001), and the easiest slip directions should be those with the shortest repeat distance, e.g., [001], [110] and [100]. Slip systems identified from optical and TEM analyses of experimentally deformed single crystals were summarized by Marshall and McLaren (1977), Tullis (1983) and Gandais and Willaime (1984), and those determined from optical and TEM analyses of naturally deformed feldspars have been summarized by Olsen and Kohlstedt (1984) and Montardi and Mainprice (1987). Information concerning dominant slip systems can also be inferred from LPO patterns in deformed aggregates. The LPOs measured from naturally deformed feldspathic rocks indicate that (010)[001] is the dominant slip system at most conditions (e.g., Kruhl 1987; Ji and Mainprice 1988, 1990) but TEM indicates there may be contributions from other systems including (010)[100], (001)[110], and (001)[1-10] (H. Stunitz, pers. comm. 2002).

Recrystallization mechanisms and microstructures

Experiments on fine-grained natural aggregates of albite and of labradorite have documented transitions, with increasing pressure and temperature, from cataclastic flow to semi-brittle flow to recrystallization-accommodated (regime 1) dislocation creep (Tullis and Yund 1985, 1987, 1992) (Fig. 15). Experiments on albite have not yet achieved climb-accommodated (regime 2) dislocation creep, largely due to its low melting temperature especially when water is present, although preliminary experiments document a transition to regime 2 in intermediate plagioclase (Tullis and Yund 2000). Evidence for higher temperature dislocation creep regimes in feldspar, characterized by subgrain rotation and grain boundary migration recrystallization, has mostly come from microstructural observations of naturally deformed feldspars.

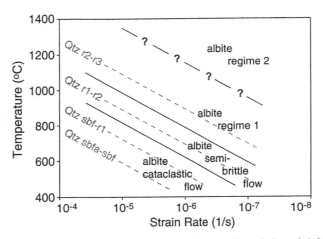

Figure 15. Plot of temperature vs. strain rate showing the relation of deformation mechanisms observed in experimentally deformed albite aggregates (solid lines) to those observed in as-is quartzites (dashed lines; taken from Fig. 8a). For albite, the deformation mechanisms plotted include cataclastic flow (distributed microcracking with no dislocation motion; see Tullis and Yund 1992), semi-brittle flow, and regime 1 dislocation creep. In albite a full transition to regime 2 dislocation creep has not been achieved experimentally; in intermediate plagioclase it appears to occur at somewhat lower temperatures (Tullis and Yund 2000).

Regime 1: bulging recrystallization. Experiments on albite aggregates (d ~ 150-200 μm) document that in the lowest temperature portion of the dislocation creep regime, the deformation process and recrystallization mechanism are very similar to those for regime 1 (recrystallization-accommodated) dislocation creep in quartz. Slow diffusion rates prevent dislocation climb; work hardening leads to very high densities of dislocation tangles and cells which drive local bulging recrystallization; and development of an interconnected grain boundary layer of very fine (1-3 μm) and initially dislocation-free recrystallized grains (Fig. 16a,c) results in strain weakening (Tullis and Yund 1985). For deformation at 1.5 GPa and 10^{-6}/s, this dislocation creep regime for albite occurs over a much wider range of temperatures (~900-1100°C) than is true for quartz (~700-750°C) (Fig. 15), and it produces greater strain weakening. Also in contrast to quartz, pre-existing zones of finer grain size, such as formed by earlier faulting or cataclasis, are able to rapidly achieve an equilibrium recrystallized grain size and highly localize further strain, leaving the coarser-grained 'wallrock' almost undeformed (e.g., Tullis et al. 1990) (Fig. 16b). The fine-grained recrystallized zones undergo a dynamic steady state of work hardening and recovery by grain boundary bulging. Although a component of grain boundary sliding cannot be ruled out, TEM observations of recrystallized regions show a variable but high dislocation density and bulging rather than planar grain boundaries (Tullis et al. 1990), and X-ray measurements show a moderately strong LPO (Heidelbach et al. 2001).

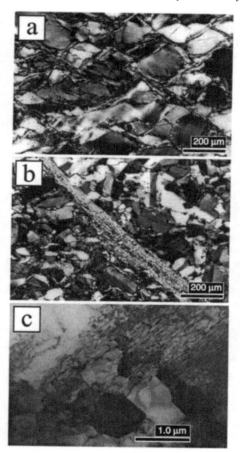

Figure 16. Microstructures produced by regime 1 dislocation creep in albite experimentally deformed in axial compression at 1.2 GPa and 10^{-6}/s. (a) Optical micrograph (crossed polars) of sample shortened 50% (parallel to short edge of photo) at 1000°C; original grains have become separated into several rhomb-shaped regions with strong and patchy undulatory extinction, bounded by thin zones of very fine recrystallized grains. (b) Optical micrograph (crossed polars) of sample pre-faulted at low pressure and temperature, then shortened 45% (parallel to short edge of photo) at 900°C; former gouge zone has recrystallized by grain boundary bulging and has localized all ductile strain. (c) TEM micrograph from sample shown in (a), illustrating part of a highly work-hardened original grain with a high density of tangled dislocations, and small dislocation-free recrystallized grains formed by bulging.

The optical microstructures characteristic of deformation in this low temperature dislocation creep regime are similar to those described for quartz; however, the two easy cleavage planes in feldspars allow microcracking and grain-scale fracturing to occur during the low strain, work-hardening phase of deformation. The fractures, twin and kink band boundaries provide internal boundaries which concentrate stress, help to nucleate dislocations (e.g., McLaren and Pryer 2001) and facilitate bulging recrystallization; in addition minor cataclasis along fractures provides 'nuclei' for additional recrystallized grains (e.g., Stunitz et al. 2001). With increasing strain original grains tend to become separated into a number of smaller lozenge-shaped porphyroclasts, characterized by strong undulatory extinction and bounded by zones of extremely fine (<3 μm) recrystallized grains which develop into long tails (Fig. 16a). The hardened porphyroclasts are gradually reduced in size by progressive bulging recrystallization, and become isolated within a fine-grained recrystallized matrix.

Transitions. The transition from semi-brittle flow to regime 1 dislocation creep is broader in feldspars than in quartz, due to their two excellent cleavages; deformation of coarse-grained aggregates includes microcracking as well as grain-scale faulting during the work hardening stage, before an interconnected layer of recrystallized grains is developed (e.g., Tullis and Yund 1985, 1987). A transition from regime 1 to 2 dislocation creep has been observed in general shear experiments on fine-grained synthetic aggregates of intermediate plagioclase, at 900°C and 1.0 GPa when small amounts of water are available (Tullis and Yund 2000). These shear experiments on a range of plagioclase compositions show a general trend from regime 1 dislocation creep in albitic aggregates, to regime 2 (climb-accommodated) dislocation creep in intermediate compositions, to diffusion creep in anorthite aggregates. The greater ease of dislocation climb in the intermediate compositions may reflect the increasing proportion of weaker Al-O bonds and faster diffusion rates resulting from non-ideality; the greater ease of diffusion creep in anorthite is consistent with the faster grain boundary diffusion rates in anorthite compared to albite (Yund and Farver 1999).

A transition from dislocation to diffusion creep has been observed in fine-grained aggregates deformed at low stress. For anorthite aggregates (d~1-7 μm) deformed in creep at 300 MPa and ~1100-1200°C, Rybacki and Dresen (2000) found this transition to occur at stresses <120 MPa. For alkali feldspar aggregates (d ~ 4-10 μm) deformed with ~0.5 wt % added water at 1.2 GPa and 900°C, where dry aggregates deform by regime 1 dislocation creep, Tullis et al. (1996) observed a switch to fluid-assisted grain boundary diffusion creep with a strength below the resolution of the Griggs apparatus (<50 MPa). The distribution of the fluid changes from isolated pores under hydrostatic conditions to mostly wetted grain boundaries during deformation; the isolated pore distribution is rapidly regained during annealing following deformation. For the water-added deformed samples, diffusion measurements document an increase in bulk transport rate of well over an order of magnitude compared to dry or hydrostatic samples. This deformation-enhanced fluid distribution is observed in albite and orthoclase, but not in anorthite (or quartz); it is not known whether it is an experimental oddity or might be important in fine-grained natural shear zones.

LPOs

LPOs in feldspar aggregates are very tedious to measure optically using a U-stage, but they cannot be measured by conventional X-ray goniometry due to the complexity of the diffraction pattern. Good results even for small volumes and fine grain sizes can be obtained with X-ray diffraction using a high intensity synchrotron source (e.g., Heidelbach et al. 2001) or with EBSD (e.g., Prior and Wheeler 1999), and larger volumes can be analyzed using neutron diffraction (e.g., Wenk et al. 1986), but most of the current data come from U-stage measurements. X-ray pole figures of albite samples experimentally deformed at 1.5 GPa and 900°C in general shear by regime 1 dislocation creep show maxima consistent with (001) [100] as the dominant slip system (Heidelbach et al. 2001), whereas EBSD measurements on anorthite aggregates experimentally deformed at 300 MPa and 1100°C in axial compression by the same process indicate that (010)[100] was dominant (Ji et al. 2000). It is not yet clear whether the difference is due to the different composition, or the difference in deformation conditions, or both.

Flow laws and piezometers

Preliminary flow laws for feldspar aggregates were reported by Shelton and Tullis (1981) but they should be ignored, for a number of reasons. First, the sample assemblies included relatively high strength components, so the measured stresses were undoubtedly higher than the true sample strengths. Second, the samples were deformed in regime 1 dislocation creep and did not come close to microstructural or mechanical steady state.

Flow laws have been determined for fine-grained synthetic anorthite aggregates with two different water contents, using creep experiments in a gas apparatus at 300 MPa (Rybacki and Dresen 2000). Their study documents a transition with increasing flow stress, from grain boundary diffusion creep with a stress exponent of 1 at stresses <120

MPa to dislocation creep with a stress exponent of 3 at higher stresses. TEM on the dislocation creep samples shows high dislocation densities and grain boundary bulging, with little evidence of dislocation climb, consistent with regime 1. Wet samples (~0.07 wt % H_2O) are weaker than dry ones (~0.004 wt % H_2O); the chief difference in the flow law is that the former have a lower activation energy (356 kJ/mol) than the latter (648 kJ/mol). Extrapolation of the dislocation creep flow law to natural strain rates is somewhat tenuous due to the limited temperature range and low water fugacity of the experiments, but indicates a transition from frictional sliding to crystal plastic flow at temperatures of ~450-500°C for a strain rate of 10^{-14}/s (Rybacki and Dresen 2000), in reasonable agreement with inferences based on microstructural observations in naturally deformed rocks (e.g., Pryer 1993). Extrapolations of the diffusion and dislocation creep flow laws indicate that for bulk strain rates of 10^{-11}/s to 10^{-13}/s, the rates from these two mechanisms will be equal for grain sizes of ~15-70 μm (Rybacki and Dresen 2000), in reasonable agreement with grain sizes found in natural shear zones.

A preliminary experimental calibration has been made for the recrystallized grain size piezometer in regime 1, where bulging recrystallization is the dominant mechanism (Post and Tullis 1999). The stress exponent (-0.66) is significantly different from that for subgrain rotation and grain boundary migration recrystallization (-1.0 to 1.3) in other materials (e.g., van der Wal et al. 1993), and very similar to that for bulging recrystallization in quartz (Bishop, unpubl. data, cited in Post and Tullis 1999). Applications of this piezometer may be limited, however, if low temperature recrystallization in nature includes compositional changes or if the deformation of such fine-grained aggregates at slower natural strain rates includes a component of grain size sensitive creep.

Comparisons with naturally deformed feldspars

There is microstructural evidence for regime 1, 2 and 3 dislocation creep in naturally deformed feldspars, although the temperatures required for each regime are significantly higher than for quartz, and they depend somewhat on fluid availability. However there are additional complexities associated with exsolution substructure as well as with the compositional changes and reactions which commonly accompany feldspar deformation. The paragraphs below give a brief summary of some of the reported evidence for crystal plasticity in naturally deformed feldspathic rocks; see Table 1 in Fitz Gerald and Stunitz (1993) for a complete summary up to that date. Further descriptions of dislocation creep microstructures of feldspars are given in a later section on naturally deformed granitic rocks.

Microstructures very similar to those observed in samples experimentally deformed by recrystallization-accommodated (regime 1) dislocation creep are observed in upper greenschist to amphibolite facies feldspathic rocks (~400-650°C), and commonly are associated with strain localization into thin ductile shear zones; however in most cases the new grains have a composition different from that of the host grains (see below). An important criterion for 'pure' strain-induced bulging recrystallization is that the recrystallized grains have the same composition as the porphyroclasts. Such cases appear to be rare, but the sequence of microstructures with increasing strain is very similar to that observed in experimental samples. For example, Shigematsu (1999) has reported the following sequence from a greenschist grade ~10 cm ductile shear zone in granite. At low strain the oligoclase grains develop grain-scale fractures and strong undulatory extinction, with a thin layer of very fine grains along their margins. At moderate strain, porphyroclasts are somewhat smaller, the volume percent of fine grains is larger and they form tails extending parallel to foliation. TEM shows that porphyroclasts have very high densities of tangled dislocations and tiny bulges along their margins. At high strain

($\gamma \sim 5$) porphyroclasts are few and small; the matrix consists of grains <5 μm in diameter, with a dislocation density that varies from very low to very high. It appears that recrystallization-accommodated dislocation creep produced the highly localized ductile shear zone, just as it does in experimental samples.

There are a number of complications in interpreting the deformation microstructures of feldspars from low-grade rocks. First, due in part to the relatively low pressure of natural low-grade deformation, fractures are common and appear to be important in facilitating dislocation generation as well as recrystallization. A patchy undulatory extinction that may superficially resemble subgrains can result from healed fractures and/or arrays of microcracks or tangles of dislocations (e.g., Tullis and Yund 1987). Second, the recrystallized grains formed in plagioclase deformed at greenschist to amphibolite grade commonly have a somewhat different composition from the host, indicating a contribution of chemical free energy to the driving potential for bulging. Stunitz (1998) has shown that the chemical free energy term can approach the same magnitude as that due to dislocation density differences. Conversely, it should be mentioned that strain-induced bulging is very effective in facilitating compositional re-equilibration (Yund and Tullis 1991). Third, and even more importantly, most feldspars deformed at low metamorphic grade undergo retrograde neomineralization, reacting with fluids to form quartz and albite with white mica and/or epidote-zoisite along grain boundaries and fractures (e.g., Mitra 1978; Gapais, 1989; Fitz Gerald and Stunitz 1993; Pryer 1993). These fine-grained, mixed-phase assemblages commonly form highly localized ductile shear zones, probably due to a switch in deformation mechanism from regime 1 dislocation creep to grain size sensitive creep (grain boundary sliding and diffusion creep) which can persist because the presence of phase boundaries prevents grain growth (e.g., Stunitz and Fitz Gerald 1993).

Microstructures indicative of climb-accommodated (regime 2) dislocation creep are observed in naturally deformed feldspars only at high temperatures, in upper amphibolite, granulite or eclogite grade feldspathic rocks (e.g., Olsen and Kohlstedt 1985; White and Mawer 1988; Altenberger and Wilhelm 2000; Kruse et al. 2001). These microstructures include elongate porphyroclasts with sweeping undulatory extinction, which contain subgrains and a mantle of recrystallized grains of approximately the same size (~50-70 μm) (Fig. 17). For plagioclase in deformed anorthosites, it is common for subgrains to be developed throughout most of the porphyroclasts; recrystallized grains may have the same composition as the porphyroclasts (e.g., Kruse and Stunitz 1999; Lafrance at al. 1998) or a different composition (e.g., Brown et al. 1980), but commonly have a LPO consistent with slip on (010)[001] (e.g., Olsen and Kohlstedt 1985; Ji and Mainprice 1988). In the K-feldspar from amphibolite grade granitic rocks, myrmekite commonly forms on the high stress margins of porphyroclasts (e.g., Simpson and Wintsch 1989) and facilitates dynamic recrystallization to a fine-grained, mixed-phase aggregate (e.g., Schulmann et al. 1996). In granitic rocks sheared at granulite grade, recrystallization of K-feldspar involves a readjustment of exsolution boundaries into larger domains and then into subgrains; continued deformation leads to recrystallization by subgrain rotation, generally over a very narrow margin of the porphyroclast, producing separate K-feldspar and plagioclase grains (e.g., White and Mawer 1986). There is a tendency for relic porphyroclasts of K-feldspar to remain when plagioclase is completely recrystallized (White and Mawer 1986), but in quartzo-feldspathic granulites both phases have approximately the same strength and the same recrystallized grain size (e.g., Martelat et al. 1999). If even a trace of water is present, the high pressure of most upper amphibolite to granulite grade deformation will result in high water and hydrogen fugacities; the resulting fast diffusion rates facilitate not only the crystal plastic recovery processes of dislocation climb and grain boundary migration, but also may allow a component of

Figure 17. Optical microstructures (crossed polars) of anorthosite deformed at ~700°C and 900 MPa by regime 2 dislocation creep (from shear zone within the Jotun nappe, Norway). The original grains are highly elongate; the recrystallized grains have approximately the same size as subgrains, and the same composition as the host grains. From Tullis et al. (2000), Photo 53.

diffusion creep, as indicated by phase boundary adjustments (e.g., Gower and Simpson 1992; Rosenberg and Stunitz 2002).

There are a number of possible complications in interpreting feldspar deformation microstructures from these high grade rocks. In upper amphibolite grade anorthosites, for example, highly elongate porphyroclasts in a recrystallized matrix may not result from extreme internal strain, but may be remnants of porphyroclasts which have recrystallized along kink band boundaries (e.g., Olesen 1987). In addition, due to the strong anisotropy of feldspar, differently oriented porphyroclasts may have different strengths and even different recrystallization mechanisms (e.g., Ji and Mainprice 1990); for example Kruse et al. (2001) found that porphyroclasts oriented for easy slip on (010)[001] underwent subgrain rotation recrystallization whereas those with a 'hard' orientation underwent fracturing and bulging recrystallization. Evidence of transient fracturing has been observed even in granulite grade anorthosites, and it may serve to localize subsequent recrystallization (e.g., Lafrance et al. 1998). Another complication is the possibility that layers and zones of fine-grained plagioclase in amphibolite grade rocks, which at first appear to be dynamically recrystallized, may contain zoning patterns inconsistent with recrystallization as well as a random LPO, strongly suggestive of dissolution and reprecipitation (e.g., Steffen et al. 2001; Lapworth et al. 2002). Finally, it is not uncommon for high temperature microstructures indicative of regime 2 dislocation creep, with subgrain rotation recrystallization, to be overprinted by later, lower temperature deformation involving grain boundary bulging.

Microstructures indicative of regime 3 dislocation creep, with grain boundary migration recrystallization, are observed only rarely and at very high temperatures. In anorthosite deformed at temperatures of ~900°C, close to the solidus and in the presence of melt, Lafrance et al. (1998) observed very large (~700 μm) 'dissection' grains with lobate boundaries, engulfing small grains of pyroxene and other phases. In plagioclase from a tonalite deformed at ~670°C, also close to the solidus, Rosenberg and Stunitz (2002) observed similar microstructures, including recrystallized grains with lobate grain boundaries which are larger than co-existing subgrains.

POLYPHASE AGGREGATES

The continental crust consists dominantly of polyphase aggregates, whose constituent phases have quite different strengths and possibly different deformation mechanisms at any given set of conditions, resulting in strain partitioning on a variety of scales. Even for a polyphase aggregate in which all phases deform by dislocation creep, the bulk strength (flow law) will depend on the flow laws of the constituent phases, their volume proportions and their geometric arrangement, and all of those factors may change with time and/or with strain. An important conceptual distinction is whether the aggregate is closer to having an interconnected strong or weak phase, as discussed by Handy (1990, 1994) in terms of a load-bearing framework (LBF) or interconnected weak layers (IWL). There is a strong tendency for progressive deformation of granitic rocks to result in a switch from LBF to IWL, accompanied by strain weakening and localization, with important implications for crustal strength.

Experimental investigations of crystal plasticity in 'granitic' aggregates are hindered by the narrow windows of conditions where the phases of interest undergo dislocation creep without melting or reacting. A second problem is the difficulty of achieving steady state flow, especially with the low strains available in axial compression; the high shear strains possible in the Paterson torsion apparatus greatly improve the likelihood of achieving steady state (for materials which undergo dislocation creep at low pressure). Experiments on synthetic polyphase aggregates of non-silicates have provided important insights (e.g., Jordan 1987), as have high strain 'see through' experiments on analog materials (e.g., Bons 1994; Herwegh and Handy 1998). Despite the difficulties, experiments on polyphase aggregates of quartz, feldspar and mica have been very useful in assessing the effects of the geometric arrangement of the phases and of phase boundaries as compared to grain boundaries.

The sections below briefly summarize some experimental results for granitic aggregates, including 2-phase aggregates of quartz-feldspar (natural aplite) and quartz-mica (synthetic), and a 3-phase mixture of quartz-plagioclase-mica (natural gneiss). These sections are followed by a brief summary of deformation mechanisms and microstructures observed in naturally deformed granitic rocks.

Quartz-feldspar aggregates

A suite of axial compression experiments on a natural fine-grained aplite (Dell'Angelo and Tullis 1996) has been useful in illustrating the important effects of recrystallization-accommodated (regime 1) dislocation creep in one of the phases of a polyphase aggregate, in terms of its role in both strain hardening and weakening. This aplite has an average grain size of 150 μm, and consists of ~30% quartz, 27% oligoclase, 40% microcline, and ~3% biotite and magnetite. The biotite does not noticeably affect the deformation and the two feldspars have almost the same mechanical behavior, so the material can be approximated as a two-phase LBF aggregate in which feldspar forms the interconnected, stress-supporting phase and quartz forms the isolated phase. The deformed aplite samples were deformed to different shortening strains up to 85% at two different temperatures, and compared with a natural quartzite and a natural albite aggregate deformed over the same range of conditions.

For deformation at lower temperatures, where quartzite deforms by regime 1 dislocation creep and is weaker than albite aggregates which deform by semi-brittle flow, the aplite has a strength close to that of pure albite and quartz grains remain almost undeformed within a highly deformed feldspar matrix (Fig. 18a). This difference in the behavior of quartzite and isolated quartz grains is a consequence of the recovery process for this low temperature dislocation creep regime: bulging recrystallization requires the migration of quartz-quartz grain boundaries, but in the aplite almost all quartz grains are surrounded by feldspar. Phase boundaries do not allow the same recovery process, and thus the quartz grains work harden until they are stronger than feldspar.

For deformation at higher temperatures, where quartzite deforms by regime 2 dislocation creep and remains weaker than albite aggregates which deform by regime 1 dislocation creep, the aplite samples strain weaken from a peak strength close to that for albite aggregates to a strength at 70% shortening close to that for quartzite. The weakening is a consequence of recrystallization-accommodated dislocation creep in the feldspars: bulging recrystallization along grain boundaries and intragranular shear zones produces interconnected weak layers of cyclically strain-free grains, and inhomogeneous deformation of the work-hardened porphyroclasts results in microboudinage, allowing the weaker and homogeneously deforming quartz grains to become increasingly interconnected (Fig. 18b,c,d). The LPO developed by the quartz grains in the aplite is almost the same as that in a quartzite (Dell'Angelo and Tullis 1986). Thus with increasing strain there is a reversal in phase contiguity, such that the weaker phase becomes interconnected even though it constitutes only 30% by volume. This reversal in phase connectivity should be even more efficient in a shearing geometry.

Effects of mica

Quartz-mica aggregates. The dislocation creep strength of mica is much lower than that of other common crustal silicates (e.g., Kronenberg et al. 1990), and micas are therefore of great importance in causing strain partitioning. Experiments have been done on a variety of mica-bearing rocks at relatively low temperatures and pressures where the framework silicates are brittle, to examine the influence of volume and alignment of micas on the rock failure mode and strength (e.g., Shea and Kronenberg 1993). However there have been few studies at higher temperatures and pressures where all phases undergo dislocation creep; one problem is that temperatures must be restricted to $<\sim 800°C$ in order to avoid melting or reactions, limiting the range of crystal plasticity achievable in quartz and especially feldspar.

In order to investigate the effects of increasing amounts of muscovite on the dislocation creep of quartz aggregates, Tullis and Wenk (1985) prepared synthetic aggregates with different volume proportions, and deformed them in axial compression at 800°C and 1.2 GPa, at two different strain rates. At the faster strain rate, where pure quartz aggregates deform by recrystallization-accommodated (regime 1) dislocation creep, addition of about 15% muscovite was enough to form interconnected grain boundary layers and significantly weaken the aggregate; further additions caused relatively little additional weakening. The presence of muscovite on most quartz boundaries prevented the recovery process of grain boundary bulging recrystallization, thus effectively hardening the quartz grains so that they remained almost undeformed. However the muscovite was so weak it could accommodate almost all of the sample strain, although its strain rate was much faster than the bulk rate. In the slower strain rate experiments, where pure quartz aggregates deform by climb-accommodated (regime 2) dislocation creep, additions of muscovite cause less weakening. Muscovite did not inhibit the quartz recovery process of dislocation climb, but it was still enough weaker to

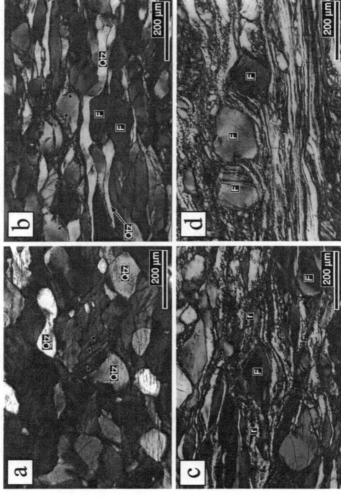

Figure 18. Optical micrographs (crossed polars) of fine-grained (d ~ 100 μm) aplite experimentally deformed ('as-is') in axial compression at 1.2 GPa and 10^{-6}/s. (a) to (c), compression direction parallel to short edge of photos; (d) dextral shear parallel to long edge. (a) Sample shortened 48% at 750°C: the dispersed quartz grains (Qtz) are not able to recover by bulging recrystallization and thus are stronger than the interconnected feldspar grains, which deformed by semi-brittle flow (arrows mark grain-scale faults). (b) Sample shortened 60% at 900°C: quartz grains (Qtz) are highly flattened with sub-grain rotation recrystallization (arrows), and are beginning to become interconnected, whereas feldspar grains (F), deforming by regime 1 dislocation creep, have started to separate into relatively undeformed augen. (c) Sample shortened 83% at 900°C: feldspar grains (F) have separated into augen with thin, finely recrystallized tails; quartz grains are highly interconnected and more coarsely recrystallized (r). (d) Sample sheared at 900°C: feldspar grains (F) form augen with asymmetric recrystallized tails; quartz grains are highly elongate and substantially recrystallized.

significantly partition the sample strain, so that the quartz LPO remained much weaker than that of a quartzite shortened by the same amount.

Quartz-plagioclase-mica aggregates. Due to their extreme anisotropy, micas are likely to be most effective in strain partitioning during a shearing deformation. Recent general shear experiments on a fine-grained (d ~ 100 μm) gneiss have documented the processes by which initially isolated biotite grains can become interconnected, and the way in which subsequent strain weakening and localization depend on its strength contrast with the quartz-plagioclase (QP) matrix (Holyoke and Tullis 2001). The gneiss consists of 55% quartz, 32% plagioclase, and 13% biotite, aligned but not interconnected; the foliation was oriented parallel to the shear plane. Samples were deformed at 800°C, 1.5 GPa and two strain rates; at the faster strain rate quartzite deforms by recrystallization-accommodated (regime 1) dislocation creep and plagioclase by semi-brittle flow, whereas at the slower strain rate quartzite deforms by climb-accommodated (regime 2) dislocation creep and plagioclase by regime 1. The biotite strength (Kronenberg et al. 1990) is much less than that of the QP matrix at both sets of conditions, although the strength contrast is lower at the slower strain rate.

Gneiss samples deformed at the faster strain rate develop a single highly localized ductile shear zone, accompanied by pronounced strain weakening. Optical and TEM observations of samples deformed to several different shear strains (γ=1.3 to 5) show that local interconnections of biotite grains occur where stress concentrations in quartz or plagioclase grains between adjacent biotite grains cause narrow zones of cataclasis followed by grain-scale offsets, allowing the easily slipping basal planes of the biotite grains to link up. These local offsets and interconnections in turn transfer the stress concentrations, and lead to further biotite interconnections. By $\gamma \sim 4$ a single highly localized shear zone cuts across the entire sample (Fig. 19a,c); minor amounts of biotite reaction products occur only in the high strain zone.

The interconnection of the weak biotite plus minor reaction to a fine-grained mixed-phase assemblage produces weakening (by a factor of ~2) as well as strain and strain rate partitioning, which in turn cause changes in the operative deformation mechanisms in both the shear zone and the host rock. The shear zone material experiences decreasing stress, but a strain rate about an order of magnitude faster than the bulk imposed strain rate. The quartz and plagioclase which initially underwent cataclasis recover, by bulging recrystallization, and form very thin and fine-grained layers with no phase mixing. The host rock also experiences decreasing stress, but a strain rate about an order of magnitude slower than the bulk imposed rate. Quartz grains flatten and develop subgrains, consistent with a switch from regime 1 to regime 2 dislocation creep, and plagioclase grains develop fine-grained recrystallized margins, consistent with a switch from semi-brittle flow to regime 1 dislocation creep. In both regions the microstructures indicative of the original deformation and interconnection processes are completely overprinted.

Gneiss samples deformed at the slower strain rate, where there is a lower contrast in strength between the biotite and the QF matrix, develop multiple interconnected biotite zones and a more homogeneous SC fabric (Fig. 19b,d). The yield strength is much lower, although strain weakening is again a factor of ~2. Initial yielding occurs by local regime 1 dislocation creep at stress concentrations in quartz or feldspar grains between adjacent biotite grains. The resulting inhomogeneous grain strain and microboudinage in turn allow biotites to become interconnected on more of a grain scale.

Comparisons with naturally deformed granitic rocks

Studies of the microstructures of naturally deformed granitic rocks over a wide range

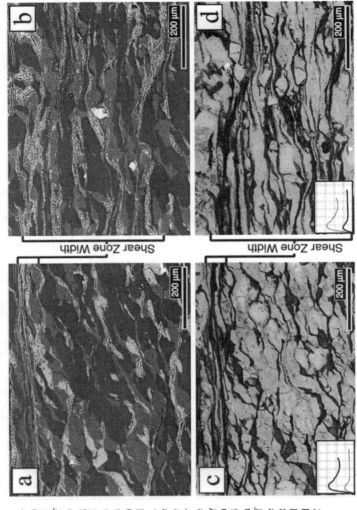

Figure 19. Micrographs of fine-grained (d ~ 100μm) quartz-plagioclase-biotite (QPB) gneiss samples experimentally sheared at 1.5 GPa and 800°C, illustrating different styles of deformation-induced interconnection of the biotite. Dextral shear is parallel to the long edge of photo. (a) and (b) are SEM backscattered images; biotite and reaction products are lightest color; plagioclase is medium gray; quartz is dark gray. (c) and (d) are optical micrographs (plane polarized light) of the same areas as in (a) and (b), respectively; biotite and reaction products are dark, plagioclase and quartz are white; on the inset plots of differential stress vs. shear strain, the darker curve is for the sample illustrated. (a) and (c) Sample deformed to γ = 4.3 at 10^{-5}/s; large contrast in strength between B and QP and yield strength of 1.4 GPa; most sample strain is localized within a single, narrow through-going shear zone of interconnected biotites. (b) and (d) Sample deformed to γ = 5.3 at 10^{-6}/s; smaller contrast in strength between B and QP and yield strength of 500 MPa; multiply interconnected biotites form a wider shear zone with an SC fabric.

of metamorphic grades provide valuable comparisons with the experimental results. They show many similarities with the experimental dislocation creep microstructures, as well as evidence for additional deformation processes. They provide information on the relative strengths of the different phases, as well as indicating what lithologies and deformation conditions lead to strain localization. Following is a brief and very general description of deformation microstructures and inferred processes for granitic rocks as a function of increasing depth in the crust, assuming average strain rates (10^{-12}/s to 10^{-14}/s) and geothermal gradient; most examples are taken from rocks in which ductile shear zones provide information about progressive deformation. This section is not intended as a complete review, but rather is meant to illustrate some of the variety of behavior that is observed. One important generalization is that the deformation behavior of granitic rocks depends strongly on the chemical stability of the feldspars. The summary below includes examples where feldspars were stable as well as examples where they were unstable; for the latter cases, syn-deformational reactions commonly lead to changes in deformation mechanism and formation of localized ductile shear zones.

At sub-greenschist facies conditions (~250-300°C) progressive deformation tends to produce phyllonites. Although the dislocation glide strength of micas is very low, they may develop cracks on cleavage planes and kink band boundaries and thus undergo some cataclasis (e.g., Behrmann 1984). Plagioclase and K-feldspar are pervasively fractured and, due to fluid access, are altered to muscovite. Quartz grains are more resistant to fracturing and show only limited crystal plasticity. Low strain granitic rocks consist of a matrix of cataclastic feldspars with admixed muscovite and quartz porphyroclasts; the high strain end-products are strongly localized shear zones of phyllonite with quartz porphyroclasts (e.g., Mitra 1978; Janecke and Evans 1988).

At lower to middle greenschist facies conditions (~300-400°C) deformation of granitic rocks tends to be quite heterogeneous, resulting in highly localized ductile shear zones. Micas undergo easy glide but also some cataclasis (e.g., Goodwin and Wenk 1990). Depending somewhat on strain rate, quartz grains may be inhomogeneously deformed, with strong undulatory extinction and very small recrystallized grains formed by bulging (regime 1), or they may be homogeneously flattened with subgrains and recrystallized grains of the same size (regime 2). At low strains plagioclase and especially K-feldspar develop grain-scale fractures which reduce the grain size and allow the weaker quartz and mica to become interconnected (e.g., Simpson 1985). The fractures also allow fluid access, and breakdown reactions accompany further deformation; new albite grains nucleate primarily along fractures and clast margins, together with quartz and mica (e.g., Fitz Gerald and Stunitz 1993). The resulting fine-grained (d ~ 10 μm) mixed-phase assemblage of quartz and albite ± white mica ± CaAl-silicates appears to deform by granular flow (grain boundary sliding, diffusion creep and/or dissolution-reprecipitation) and is weaker than coarser (d ~ 60 μm) recrystallized quartz layers deforming by climb-accommodated dislocation creep (Stunitz and Fitz Gerald 1993). Thus reactions accompanying progressive strain allow the strongest phase (feldspar) to become the weakest material, resulting in highly localized ductile shear zones.

At middle to upper greenschist facies conditions (~400-500°C) progressive strain again leads to highly localized ductile shear zones. The micas are the weakest phase, and undergo easy glide but also some cataclasis (e.g., Goodwin and Wenk 1990). Quartz is also weak and deforms by regime 2 or 3 dislocation creep, depending on strain rate and fluids. Both plagioclase and especially K-feldspar develop grain-scale fractures at low strain. Plagioclase grains commonly develop deformation and kink bands, with very fine and more albitic recrystallized grains along clast margins (e.g., Simpson 1985). K-feldspar may develop myrmekite and flame perthite on high stress boundaries (Pryer

1993). Breakdown reactions of feldspars commonly accompany deformation and result in grain-size reduction to a mixed-phase assemblage (e.g., Mitra 1992). At high strain, adjacent layers of pure fine-grained quartz and of mica show no phase mixing, and quartz layers are boudinaged between layers of feldspar reaction products, indicating that quartz deforming by dislocation creep is stronger than the fine-grained mixed-phase aggregates which presumably deform by granular flow (e.g., Gapais 1987; Stunitz and Fitz Gerald 1993; Fliervoet et al. 1997). Again at these conditions feldspar reactions accompanying progressive strain lead to strain weakening and localization.

At upper greenschist to lower amphibolite facies conditions (~500-600°C) progressive deformation of granitic rocks results in SC mylonites and ultramylonites. Micas undergo easy glide and recrystallization. Quartz grains are completely recrystallized even at low strain; the recrystallized grains may contain smaller but optically visible subgrains (e.g., Simpson 1985), indicative of regime 3 dislocation creep. At low strain both feldspars commonly develop grain-scale fractures, but generally these have no offsets. Myrmekite on K-feldspar may recrystallize to fine grains of quartz and plagioclase, and the boundaries of plagioclase porphyroclasts tend to be serrated, with very small new grains (e.g., Pryer 1993). At higher strain SC mylonites develop when mica, quartz and new feldspar grains wrap around feldspar porphyroclasts (e.g., Berthe et al. 1979). The quartz grains are recrystallized (d ~ 50 μm) and have a strong LPO (e.g., Wenk and Pannetier 1985). At very high strain ultramylonites are formed, with small porphyroclasts of oligoclase and minor K-feldspar set in a fine-grained foliated matrix of mica, quartz and feldspar (e.g., Pryer 1993). When the fine quartz grains in the ultramylonite become intimately mixed with fine mica and feldspar, they have a random LPO (e.g., Berthe et al. 1979; Wenk and Pannetier 1985). The process of phase mixing presumably involves grain boundary sliding, and the lack of LPO suggests a switch from dislocation creep to granular flow ('superplasticity' of Bouillier and Gueguen 1977).

At lower to middle amphibolite facies conditions (~600-650°C) progressive strain of granitic rocks again involves retrograde reactions, and results in SC or banded mylonites. Micas undergo easy glide and recrystallization. Quartz grains are completely recrystallized to a relatively coarse size (~150 μm) and they contain smaller subgrains (e.g., Simpson 1985). At moderate strain plagioclase is partially recrystallized with a grain size significantly smaller than that for quartz and a composition different from that of the host; some biotite grains nucleate within the aggregates of new plagioclase grains (e.g., Rosenberg and Stunitz 2002). Myrmekite developed on K-feldspar porphyroclasts recrystallizes to fine-grained (d ~ 10 μm) aggregates of quartz and plagioclase, and an SC fabric may develop by the interconnection of weak micas, quartz and new feldspar grains around K-feldspar porphyroclasts (e.g., Shulmann et al. 1996). In a granodioritic SC orthogneiss Wintsch and Yi (2002) observed compositional zoning and truncation and concluded that plagioclase, K-feldspar and epidote underwent significant dissolution and reprecipitation, whereas quartz and biotite deformed dominantly by crystal plasticity. At high strain Shulmann et al. (1996) describe a relatively coarse-grained banded mylonite, with quartz layers one grain wide (d ~ 200 μm), finer-grained plagioclase layers (d ~ 50-100 μm), and K-feldspar layers having a core of recrystallized grains (d ~ 50-100 μm) bordered by a fine-grained mixture of quartz, plagioclase and K-feldspar. The recrystallized quartz, plagioclase and K-feldspar all appear to have a strong LPO, suggesting that all phases deformed dominantly by dislocation creep, and the different layers are not boudinaged, indicating they all had approximately the same strength. In other high strain mylonites, both feldspars have completely reacted to form fine-grained (d<10 μm) mixed-phase assemblages of quartz-plagioclase-orthoclase-mica which appear to deform by granular flow; these bands sometimes appear to have the same strength as adjacent pure (d ~ 40-100 μm) quartz layers (e.g., Behrmann and Mainprice 1987) which have deformed by regime 3 dislocation creep, and sometimes appear to be weaker (e.g., Fliervoet et al. 1997).

At middle to upper amphibolite facies conditions (~650-700°C), progressive strain of granitic rocks again results in banded mylonites. At low strain quartz is completely recrystallized by grain boundary migration (d ~ 100 to 300 µm). In a low strain granite Pryer (1993) observed that both plagioclase and K-feldspar develop flattened porphyroclasts with a core and mantle structure; the porphyroclasts have a thin zone, only a few subgrains wide, along their margins and a sharp boundary with recrystallized grains (d ~ 50 µm) which may have the same or a slightly different composition. In a low strain tonalite Rosenberg and Stunitz (2002) inferred that plagioclase recrystallized by subgrain rotation with some grain boundary migration; however, biotite grains dispersed within the recrystallized plagioclase aggregates appear to allow a component of grain boundary sliding, as indicated by aligned grain boundaries and weak plagioclase LPO, and the aggregates have no competency contrast with pure quartz layers.

At granulite facies conditions (~700-750°C) localized ductile shear zones may develop, but deformation tends to be more homogeneous and syndeformational reactions are less important. Quartz is completely recrystallized, and occurs as large grains with prismatic subgrain boundaries or a chessboard microstructure with both basal and prism subgrain boundaries (e.g., Kruhl 1996). In low strain rocks, plagioclase porphyroclasts have sweeping undulatory extinction and subgrains; recrystallized grains form by subgrain rotation with some grain boundary migration, and they have the same composition as the porphyroclasts (e.g., Egydio-Silva and Mainprice 1999). Boundaries between mesoperthitic K-feldspar grains are lobate, and phase boundaries between quartz and K-feldspar are mobile, with cusps pointing toward quartz (e.g., Gower and Simpson 1992). At intermediate strains feldspar domains have the same aspect ratio as quartz (e.g., Martelat et al. 1999). K-feldspar grains undergo subgrain rotation recrystallization, involving the reorganization and coarsening of exsolved domains and the development of discrete K-feldspar and plagioclase subgrains and recrystallized grains (e.g., White and Mawer 1986). The recrystallized grains have approximately the same LPO as the porphyroclasts (e.g., Martelat et al. 1999), consistent with subgrain rotation. In high strain rocks, both feldspars are fully recrystallized (d ~ 0.5 mm) and form an equigranular matrix surrounding strongly elongated platy quartz grains. Quartz does not become interconnected, even in a rock where it constitutes 40%, and thus its dislocation creep strength must be very similar to that of the feldspars (Martelat et al. 1999). Although dislocation creep of both feldspars and quartz is pronounced, consistently cuspate phase boundaries indicate a component of diffusion creep as well (e.g., Gower and Simpson 1992).

APPLICATIONS AND IMPLICATIONS

The similarity of the dislocation creep microstructures in experimentally and naturally deformed granitic aggregates indicates that dislocation creep is an important deformation mechanism in the crust and that the microstructures it produces can be useful in constraining the thermomechanical history of many rocks. However, field studies have demonstrated the important contributions of other microstructurally less obvious deformation mechanisms, especially to the formation of localized ductile shear zones, indicating that dislocation creep strengths provide only an upper bound to the strength for most of the crust below the brittle-plastic transition.

Information from dislocation creep microstructures

Evidence from experiments indicates that in the absence of fluids and metamorphic reactions, so that no compositional changes are possible, plagioclase deforming by dislocation creep develops a similar sequence of recrystallization mechanisms with decreasing flow stress (increasing temperature) as observed in quartz, although at significantly higher temperatures (e.g., Tullis 1990; Tullis and Yund 2001). However, the situation in nature is more complicated, and in most feldspathic rocks deformed at greenschist to mid-amphibolite grade, feldspars either recrystallize by nucleation and growth of new grains with a different composition or participate in breakdown reactions; only at high temperatures is recrystallization by subgrain rotation and grain boundary migration observed (e.g., Rosenberg and Stunitz 2002). At present, dislocation creep microstructures in quartz are more useful, given that three distinct recrystallization mechanisms can be recognized and that there is a reliable flow law. For example, recognition of rotation recrystallization and grain boundary migration recrystallization should allow reasonable confidence in applying the flow laws of Luan and Paterson (1992) and Gleason and Tullis (1995) with the modifications suggested by Hirth et al. (2001). However, there is a need for quartzite flow laws to be determined with a free fluid phase present, so that the water fugacity is known and controlled, as well as a need for separate calibrations of the recrystallized grain size piezometer for the three different mechanisms of recrystallization.

LPOs are another product of dislocation creep which already provide useful information but deserve further experimental and modeling work. Recent results from quartzites experimentally sheared to high strain demonstrate that grain boundary migration recrystallization significantly changes the LPO, although only after high strain ($\gamma \sim 4$) (Heilbronner and Tullis 2002). These results need to be followed up with detailed TEM and EBSD measurements on samples of increasing shear strain to document the sequence of steps and the orientation-dependent strain energy variations governing the selected growth and consumption of grains. Similar shear experiments on feldspar aggregates of varying composition and over a range of conditions, followed by EBSD determinations of LPO, should be helpful in determining the role of slip and recrystallization in that phase.

Flow laws for crustal rocks

Dislocation creep is a common and important deformation mechanism for silicates in the mid to lower crust, and its microstructures can be used to help constrain deformation conditions, but can the strength of the crust be well approximated by dislocation creep flow laws? Answering this question requires a consideration of the mineralogy and deformation processes in ductile shear zones.

Even for monophase aggregates there is some question as to the strain required for steady state dislocation creep. One question concerns the extent of geometric softening or hardening, that is the influence of a developing LPO on the flow strength. Predictions based on modeling studies have been made about such effects for pure quartz aggregates deformed in axial compression or extension (e.g., Takeshita and Wenk 1988), although not in shear and not taking account of recrystallization. There is some evidence from shear experiments on analogue materials that steady state LPOs and microstructures require $\gamma \sim 7$ (e.g., Herwegh and Handy 1996), and evidence from torsion experiments on marble and anhydrite that $\gamma > 2$ may be required for steady state flow (e.g., Pieiri et al. 2001). Similarly, Tullis and Heilbronner (2002) have found that a γ of ~4 is required for a steady state recrystallization LPO to be achieved in regime 3 dislocation creep in quartz, although within the resolution of the solid NaCl assembly in the Griggs apparatus no change in flow stress was detected.

Another question about steady state flow in monophase aggregates concerns the effect of grain size reduction accompanying dynamic recrystallization. The common association of ductile shear zones with fine grain size has raised the question of whether reduction in grain size by dynamic recrystallization alone can change the deformation mechanism from dominantly dislocation creep to dominantly diffusion creep and grain boundary sliding. Behrmann (1985) reported that quartz mylonites recrystallized to a grain size <10 μm appeared to have a random LPO, indicative of a switch to grain size sensitive creep, although grain boundary impurities may have played a role. In contrast, Fliervoet and White (1995) found no evidence of grain boundary sliding in a quartz mylonite with d<4 μm; it had TEM microstructures, including LPO, indicative of dislocation creep, consistent with predictions based on experimental studies (e.g., Brodie and Rutter 2000).

Experimental evidence for quartz and feldspar indicates that although dynamic recrystallization does not produce a switch to grain size sensitive creep, low temperature bulging recrystallization can produce fine-grained and highly localized ductile shear zones. In regime 1 dislocation creep of quartzite and especially plagioclase aggregates, the very fine (~1-2 μm) recrystallized grains are much weaker than the work-hardened porphyroclasts; they maintain a variable but high dislocation density (up to $10^{16}/m^2$), develop a moderate LPO, and do not mix in other phases (Dell'Angelo and Tullis 1996), indicating little or no grain boundary sliding. Although a flow law has not yet been determined for regime 1 dislocation creep in quartz or feldspar at high pressure, where most of the microstructural studies have been done, results for anorthite aggregates at 300 MPa indicate a stress exponent of 3 (Rybacki and Dresen 2000), consistent with 'pure' dislocation creep. Thus for feldspathic rocks deforming at conditions where bulging recrystallization (not neomineralization) occurs, dislocation creep will tend to become highly localized along zones of initially finer grain size, which may have resulted from cataclasis (e.g., Tullis et al. 1990). Because the fine recrystallized grains can achieve a steady state flow stress much lower than that in the coarser-grained and work-hardened wall rocks, such shear zones should persist to high strain.

For polyphase aggregates there are several questions about the relevant flow law: First, is steady state dislocation creep of all phases in the aggregate ever achieved, in experiments or nature? Second, even if only one (weaker) phase deforms by dislocation creep, can the aggregate strength be well approximated by the flow law for that phase? And third, if neither of these statements is true, what is the best way to proceed in modeling the strength of the crust? These questions are addressed below.

Only at high grade (upper amphibolite to granulite) conditions can all of the major phases in a granitic rock deform by dislocation creep without reactions and strain localization. At such conditions quartz and feldspars may have sufficiently similar dislocation creep strengths that high strain occurs without phase segregation and strain partitioning (e.g., Martelat et al. 1999). For these situations a dislocation creep flow law for the aggregate could be reasonably approximated using the finite element modeling approach of Tullis et al. (1991); however note that the input flow laws they used (Shelton and Tullis 1981) are now known to be incorrect, and there is no available flow law for climb-accommodated dislocation creep of feldspar.

At most low to moderate grade (greenschist to amphibolite) conditions the initial straining of granitic rocks results in strain weakening due to disaggregation of the stronger feldspar matrix and interconnection of weaker quartz and micas. The aggregate strength may then be 'controlled' by that of the interconnected weak phases (e.g., Handy et al. 2001), and the flow law for quartz (Hirth et al. 2001) or biotite (Kronenberg et al. 1990) may provide a good approximation to the strength of granitic aggregates at such conditions.

For deformation over a wide range of crustal conditions, strain weakening and localization in granitic rocks result from progressive reactions of one or both feldspars. At mid to high-grade conditions, recrystallization of myrmekite, perthites and antiperthites produces fine-grained aggregates of quartz and plagioclase. At low to mid grade conditions, grain-scale fractures at low strain facilitate fluid access and reaction of both feldspars. In some cases the reactions directly create micas in sufficient volume to form interconnected networks, producing phyllonites (e.g., Mitra 1978; Janecke and Evans 1988). In other cases the multiple phases (albite + quartz ± white micas ± CaAl-silicates) that are nucleated ensure a very fine and stable grain size which allows a switch in deformation mechanism to grain size sensitive creep ('superplasticity' of Bouillier and Gueguen 1977), such that even thin zones of reaction products significantly weaken the rock. This process has been documented experimentally for the reaction of plagioclase + water to zoisite + quartz + kyanite + more albitic plagioclase (Stunitz and Tullis 2001) and in naturally deformed granitic rocks (e.g., Fitz Gerald and Stunitz 1993; Stunitz and Fitz Gerald 1993; Fliervoet et al. 1997). Similarly, polyphase aggregates deforming and reacting under water-saturated conditions may undergo dissolution and replacement creep (e.g., Wintsch and Yi 2002), with an effective flow stress lower than that for dislocation creep of any of the major phases.

Thus the strength of massive, low strain granitic rocks may well be controlled by the yield strength of feldspar, but rarely by its steady state dislocation creep strength, and the strength of high strain rocks will be controlled by the reaction products of feldspars which most likely undergo grain size sensitive creep. Although the flow law for quartz is commonly used on 'pine tree diagrams' (e.g., Fig. 1) to approximate the strength of the mid to lower crust, it almost certainly underestimates the strength of granitic rocks at low strain and overestimates their strength at high strain.

Assessing the strength of the crust

Field observations make it abundantly clear that the crust is both heterogeneous and anisotropic on a wide variety of scales. Thus one cannot discuss the strength of the crust without specifying the lithology, the depth interval, and the tectonic setting, including considerations of the fluid environment and whether the rocks are massive or have been previously deformed and thus have an inherited fabric.

The development of more accurate models for the strength of the crust will require a combination of field observations, experimental work and theoretical modeling. We depend on field geologists to characterize the rock compositions and fabrics in the tectonic settings of interest, and how they change with progressive strain. This characterization should include a determination of the mineralogy, grain size, phase arrangement, and fluid activity. In recent years many such careful studies have been made of crustal rocks deformed at different depths and in different tectonic settings, very briefly summarized in earlier sections, such that the greater need at present is for additional experiments and modeling studies.

To date most experimental studies of quartz, feldspar and mica have focused on monomineralic aggregates; considerable progress has been made in our understanding of flow laws, microstructures and LPO development, including considerations of grain size and fluid environment. Experiments on polyphase aggregates are needed to determine the effects of phase boundaries. For aggregates in which at least one phase is deforming by low temperature (regime 1) dislocation creep, phase boundaries have a significant strengthening effect, because the recovery process of bulging recrystallization cannot occur at phase boundaries (e.g., Dell'Angelo and Tullis 1996).

Phase boundaries are also likely to affect the strength of fine-grained aggregates deforming by grain size sensitive creep (e.g., Fliervoet et al. 1997), and several experimental studies have documented the importance of phase boundaries involving mica. The presence of clays and micas on quartz boundaries greatly enhances the rate of solution transfer (e.g., Hickman and Evans 1995). Similarly, Farver and Yund (1999) found that the rates of (wet) oxygen grain boundary diffusion in biotite-quartz-feldspar aggregates are ~5 orders of magnitude faster than those in monophase aggregates of quartz or of feldspar, or in quartz-feldspar mixtures, suggesting that the field of grain boundary diffusion creep in fine-grained micaceous aggregates should be greatly expanded relative to that of dislocation creep. As pointed out by Handy et al. (2001), there is a need for experimentally determined flow laws for fine-grained mixed phase aggregates, such as the quartz-albite-white mica and/or CaAl-silicates resulting from breakdown reactions of feldspars. However it may be difficult to find experimental conditions allowing high strain without melting.

Determination of the effective flow law for specific lithologies and tectonic settings of interest must be done by modelers using appropriate experimental flow laws and geometrical arrangement of rock units, as reviewed recently by Paterson (2001) and Handy et al. (2001). It may be difficult to actually model the gradual change of these parameters with progressive strain, and the resulting weakening and strain partitioning. Rutter (1999) has proposed a strain-dependent term in the flow law for calcite to account for weakening resulting from grain size reduction accompanying dynamic recrystallization. However it should be possible to model the phase geometry, grain size and resulting strength for several different stages in the strain history, and thus to evaluate the magnitude of the change.

In sum, great progress has been made in the last 10 or 15 years in experimental characterization of crystal plasticity in quartzo-feldspathic aggregates and the resulting microstructures, and also in the documentation of the same or similar processes occurring in naturally deformed rocks. What are needed now are better flow law data and recrystallized grain size piezometer calibrations for the different dislocation creep regimes in quartz and feldspars, as well as flow law data for fine-grained, mixed-phase assemblages deforming by grain size sensitive creep. Such data would allow modelers to make use of rock descriptions provided by field geologists to more accurately model the strength of the crust in specific tectonic settings of interest.

ACKNOWLEDGMENTS

For fruitful collaborations and stimulating discussions concerning deformation processes in crustal rocks over many years I am grateful to numerous colleagues and students, including John Christie, Lisa Dell'Angelo, Almar de Ronde, John Farver, Gayle Gleason, Harry Green, Renee Heilbronner, Greg Hirth, Caleb Holyoke, Steve Kirby, Andreas Kronenberg, Alice Post, Glen Shelton, Arthur Snoke, Michel Stipp, Holger Stunitz, and especially Dick Yund. Many thanks to Bill Collins for his superb thin sections over many years, and to Caleb Holyoke for his help with the figures. This review has benefited by constructive comments from Michael Stipp, Holger Stunitz, Jane Selverstone, and Rudy Wenk.

REFERENCES

Altenberger U, Wilhelm S (2000) Ductile deformation of K-feldspar in dry eclogite facies shear zones in the Bergen Arcs, Norway. Tectonophysics 320:107-121
Baeta RD, Ashbee KHG (1969) Slip systems in quartz: I—Experiments; II—Interpretation. Am Mineral 54:1551-1582
Bailey JE, Hirsch PB (1962) The recrystallization process in some polycrystalline metals. Proc Royal Soc London A267:11-30

Behrmann GHEE (1985) Crystal plasticity and super-plasticity in quartzite: A natural example. Tectonophysics 115:101-129

Behrmann JH, Mainprice D (1987) Deformation mechanisms in a high temperature quartzo-feldspar mylonite: evidence for superplastic flow in the lower crust. Tectonophysics 140:297-305

Bell TH, Johnson SE (1989) The role of deformation partitioning in the deformation and recrystallization of plagioclase and K-feldspar in the Woodroffe thrust mylonite zone, central Australia. J Metamor Geology 7:151-168

Berthe D, Choukroune P, Jegouzo P (1979) Orthogneiss, mylonite and non coaxial deformation of granites: the example of the South Armorican Shear Zone. J Struc Geol 1:31-42

Bons P (1994) Experimental deformation of two-phase rock analogues. Mater Sci and Engin A175:221-230

Borges FS, White SH (1980) Microstructural and chemical studies of sheared anorthosites, Roneval, South Harris. J Struc Geol 2:273-280

Bouchez JL (1978) Preferred orientations of quartz a axes in some tectonites: kinematic inferences. Tectonophysics 49:T25-T30

Bouchez JL, Lister GS, Nicholas A (1983) Fabric asymmetry and shear sense in movement zones. Geol Rundsch 72:401-419

Bouillier AM, Gueguen Y (1975) SP-mylonites: origin of some mylonites by superplastic flow. Contrib Mineral Petrol 50:93-104

Brodie KH, Rutter EH (2000) Deformation mechanisms and rheology: why marble is weaker than quartzite. J Geol Soc London 157:1093-1096

Brown, WL, Macaudiere J, Ohnenstetter D, Ohnenstetter M (1980) Ductile shear zones in a meta-anorthosite from Harris, Scotland: textural and compositional changes in plagioclase. J Struc Geol 2:281-287

Carter NL, Christie JM, Griggs DT (1964) Experimental deformation and recrystallization of quartz. J Geol 72:687-733

Christie JM, Ord A, Koch PS (1980) Relationship between recrystallized grain size and flow stress in experimentally deformed quartzite. (abstr) EOS Trans, Am Geophys Union 61:377

Cordier P, Doukhan JC (1989) Water solubility in quartz and its influence on ductility. Eur J Mineral 1:221-237

De Bresser JHP, Ter Heege JH, Spiers CH (2001) Grain size reduction by dynamic recrystallization: can it result in major rheological weakening? Intl J Earth Sci 90:28-45

Dell'Angelo LN, Tullis J (1986) A comparison of quartz c-axis preferred orientations in experimentally deformed aplites and quartzites. J Struc Geol 8:683-692

Dell'Angelo LN, Tullis J (1989) Fabric development in experimentally sheared quartzites. Tectonophysics 169:1-21

Dell'Angelo LN, Tullis J (1996) Textural and mechanical evolution with progressive strain in experimentally deformed aplite. Tectonophysics 256:57-82

den Brok B, Spiers CJ (1991) Experimental evidence for water weakening of quartzite by microcracking plus solution-precipitation creep. J Geol Soc London 148:541-548

den Brok B, Meinecke J, Roller K (1994) Fourier transformer IR-determination of intragranular water content in quartzites experimentally deformed with and without added water in the ductile deformation field. J Geophys Res 99:19821-19828

Drury MR, Urai JL (1990) Deformation-related recrystallization processes. Tectonophysics 172:235-253

Dunlap W, Hirth G, Teyssier C (1997) Thermomechanical evolution of a ductile duplex. Tectonics 16:983-1000

Egydio-Silva M, Mainprice D (1999) Determination of stress directions from plagioclase fabrics in high grade deformed rocks (Alem Paraiba shear zone, Ribeira fold belt, southeastern Brazil). J Struc Geol 21:1751-1771

Farver JR, Yund RA (1990) The effect of hydrogen, oxygen, and water fugacity on oxygen diffusion in alkali feldspar. Geochim Cosmochim Acta 54:2953-2964

Farver JR, Yund RA (1991) Oxygen diffusion in quartz: Dependence on temperature and water fugacity. Chem Geol 90:55-70

Farver JR, Yund RA (1995) Grain boundary diffusion of oxygen, potassium and calcium in natural and hot-pressed feldspar aggregates. Contrib Mineral Petrol 118:340-355

Farver JR, Yund RA (1999) Oxygen bulk diffusion measurements and TEM characterization of a natural ultramylonite: implications for fluid transport in mica-bearing rocks. J. Metamor Geol 17:669-683

Fitz Gerald JD, Stunitz H (1993) Deformation of granitoids at low metamorphic grade. I: Reactions and grain size reduction. Tectonophysics 221:269-297

Fliervoet TF, White SH (1995) Quartz deformation in a very fine grained quartzo-feldspathic mylonite: A lack of evidence for dominant grain boundary sliding deformation. J Struc Geol 17:1095-1109

Fliervoet TF, White SH, Drury MR (1997) Evidence for dominant grain-boundary sliding deformation in greenschist- and amphibolite-grade polymineralic ultramylonites from the Redbank Deformed Zone, Central Australia. J Struc Geol 19:1495-1520

Gapais D (1989) Shear structures within deformed granites: mechanical and thermal indicators. Geology 17:1144-1147

Gleason GC, Tullis J (1993) Improving flow laws and piezometers for quartz and feldspar aggregates. Geophys Res Lett 20:2111-2114

Gleason GC, Tullis J (1995) A flow law for dislocation creep of quartz aggregates determined with the molten salt cell. Tectonophysics 247:1-23

Gleason GC, Tullis J, Heidelbach F (1993) The role of dynamic recrystallization in the development of lattice preferred orientations in experimentally deformed quartz aggregates. J Struc Geol 15:1145-1168

Goldsmith J (1991) Pressure-enhanced Al/Si diffusion and oxygen isotope exchange. *In* Diffusion, atomic ordering and mass transport. Ganguly J (ed) Advances in Phys Geochem 8:248-285

Goodwin L, Wenk H-R (1990) Intracrystalline folding and cataclasis in biotite of Santa Rosa mylonite zone: HVEM and TEM observations. Tectonophysics 172:201-214

Gower RJ, Simpson C (1992) Phase boundary mobility in naturally deformed, high-grade quartzo-feldspathic rocks: evidence for diffusional creep. J Struc Geol 14:301-313

Green HW (1992) Analysis of deformation in geological materials. Rev Mineral 27:425-454

Green HW, Borch RS (1989) A new molten salt cell for precision stress measurement at high pressure. Eur J Mineral 1:213-219

Griggs DT, Blacic JD (1964) Quartz: anomalous weakness of synthetic crystals. Science 147:292-295

Griggs DT (1967) Hydrolytic weakening of quartz and other silicates. Geophys J 14:19-31

Guillope M, Poirier J-P (1979) Dynamic recrystallization during creep of single-crystalline halite: An experimental study. J Geophys Res 84:5557-5567

Hacker BR, Yin A, Christie JM, Davis GA (1992) Stress magnitude, strain rate, and rheology of extended middle continental crust inferred from quartz grain sizes in the Whipple Mountains, California. Tectonics 11:36-46

Hacker, BR, Yin A, Christie JM, Snoke AW (1990) Differential stress, strain rate, and temperatures of mylonitization in the Ruby Mountains, Nevada: implications for the rate and duration of uplift. J Geophys Res 95:8569-8580

Handy MR (1990) The solid state flow of polymineralic rocks. J Geophys Res 95:8647-8661

Handy MR (1994) Flow laws for rocks containing two non-linear viscous phases: a phenomenological approach. J Struc Geol 16:287-301

Handy MR, Braun J, Brown M, Kukowski N, Paterson MS, Schmid SM, Stoeckert B, Stuwe K, Thompson AB, Wosnitza E (2001) Rheology and geodynamic modelling: the next step forward. Intl J Earth Sci 90:149-156

Hanmer S (2000) Matrix mosaics, brittle deformation and elongate porphyroclasts: granulite facies microstructures in the Striding-Athabasca mylonite zone, western Canada. J Struc Geol 22:947-967

Heidelbach F, Post A, Tullis J (2001) Crystallographic preferred orientation in albite samples deformed experimentally by dislocation and solution precipitation creep. J Struc Geol 22:1649-1661

Heilbronner RP, Pauli C (1993) Integrated spatial and orientation analysis of quartz c-axes by computer-aided microscopy. J Struc Geol 15:369-382

Heilbronner RP, Pauli C (1994) Orientation and misorientation imaging: integration of microstructural and textural analysis. *In* Textures of geological materials. Bunge HJ, Siegesmund S, Skrotski W, Weber K (eds) DGM Informationsgesellschaft, Oberusel, p 1-18

Heilbronner R, Tullis J (2002) The effect of static annealing on microstructures and crystallographic preferred orientations of quartzites experimentally deformed in axial compression and shear. Geol Soc London Spec Pub 200:191-218

Herwegh M, Handy MR (1996) The evolution of high-temperature mylonitic microfabrics: evidence from simple shearing of a quartz analogue (norcamphor). J Struc Geol 18:689-710

Hickman SH, Evans B (1995) Kinetics of pressure solution at halite-silica interfaces and intergranular clay films. J Geophys Res 100:13113-13132

Hirth G, Tullis J (1992) Dislocation creep regimes in quartz aggregates. J Struc Geol 14:145-159

Hirth G, Tullis J (1994) The brittle-plastic transition in experimentally deformed quartz aggregates. J Geophys Res 99:11731-11747

Hirth G, Teyssier C, Dunlap WJ (2001) An evaluation of quartzite flow laws based on comparisons between experimentally and naturally deformed rocks. Intl J Earth Sci 90:77-87

Hobbs B (1984) Point defect chemistry of minerals under a hydrothermal environment. J Geophys Res 89:4026-4038

Holyoke C, Tullis J (2001) Initiation of ductile shear zones. Geol Soc Am Abstr Progr 33:A324-A325

Janecke SU, Evans JP (1988) Feldspar-influenced rock rheologies. Geology 16:1064-1067

Jessel, MW (1987) Grain boundary migration microstructures in a naturally deformed quartzite. J Struc Geol 9:1007-1014

Jessel MW, Lister GS (1990) A simulation of the temperature dependence of quartz fabrics. Geol Soc London Spec Pub 54:353-362

Ji S, Zhao P (1994) Strength of two-phase rocks: A model based on fiber-loading theory. J Struc Geol 16:253-262

Ji S, Mainprice D (1988) Natural deformation of plagioclase: implications for slip systems and seismic anisotropy. Tectonophysics 147:145-163

Ji S, Mainprice D (1990) Recrystallization and fabric development in plagioclase. J Geol 98:65-79

Ji S, Wirth R, Rybacki E, Jiang Z (2000) High-temperature plastic deformation of quartz-plagioclase multilayers by layer-normal compression. J Geophys Res 105:16,651-16,664

Jordan PG (1987) The deformational behavior of bimineralic limestone-halite aggregates. Tectonophysics 137:185-197

Jordan PG (1988) The rheology of polymineralic rocks—an approach. Geol Rundsch 77:285-294

Knipe RJ, Law RD (1987) The influence of crystallographic orientation and grain boundary migration on microstructural and textural evolution in an S-C mylonite. Tectonophysics 135:155-169

Koch PS (1983) Rheology and microstructures in experimentally deformed quartzites. PhD Dissertation, University of California at Los Angeles, Los Angeles, California

Koch PS, Christie JM, Ord A, George RP (1989) Effect of water on the rheology of experimentally deformed quartzite. J Geophys Res 94:13975-13996

Kohlstedt DL, Evans B, Mackwell SJ (1995) Strength of the lithosphere: Constraints imposed by laboratory experiments. J Geophys Res 100:17587-17602

Kronenberg AK (1994) Hydrogen speciation and chemical weakening. Rev Mineral 29:123-176

Kronenberg AK, Tullis J (1984) Flow strengths of quartz aggregates: grain size and pressure effects due to hydrolytic weakening. J Geophys Res 89:4281-4297

Kronenberg AK, Kirby SH, Aines RD, Rossman GR (1986) Solubility and diffusional uptake of hydrogen in quartz at high water pressures: implications for hydrolytic weakening. J Geophys Res 91:12723-12744

Kronenberg AK, Kirby SH, Pinkston J (1990) Basal slip and mechanical anisotropy of biotite. J Geophys Res 95:19257-19278

Kronenberg AK, Segall P, Wolf GH (1990) Hydrolytic weakening and penetrative deformation within a natural shear zone. Am Geophys Union Geophys Monogr 56:21-36

Kronenberg, AK, Yund RA, Rossman GR (1996) Stationary and mobile hydrogen defects in potassium feldspar. Geochim Cosmochim Acta 60:4075-4094

Kruhl JH (1996) Prism- and basal-plane parallel subgrain boundaries in quartz: a microstructural geothermobarometer. J Metamor Geol 14:581-589

Kruhl JH (1987) Preferred lattice orientations of plagioclase from amphibolite and greenschist facies rocks near the Insubric Line (Western Alps). Tectonophysics 135:233-242

Kruse R, Stunitz H (1999) Deformation mechanisms and phase distribution in mafic high-temperature mylonites from the Jotun Nappe, southern Norway. Tectonophysics 303:223-249

Kruse R, Stunitz H, Kunze K (2001) Dynamic recrystallization processes in plagioclase porphyroclasts. J Struc Geol 23:1781-1802

Lafrance B, John BE, Frost BR (1998) Ultra high-temperature and subsolidus shear zones: examples from the Poe Mountain anorthosite, Wyoming. J Struc Geol 20:945-955

Lapworth T, Wheeler J, Prior DJ (2002) The deformation of plagioclase investigated using electron backscatter diffraction crystallographic preferred orientation data. J Struc Geol 214:387-399

Liu M, Yund RA (1992) NaSi-CaAl interdiffusion in plagioclase. Am Mineral 77:275-283

Luan FC, Paterson MS (1992) Preparation and deformation of synthetic aggregates of quartz. J Geophys Res 97:301-320

Mainprice D, Bouchez J-L, Blumenfeld P, Tubia JM (1986) Dominant c-slip in naturally deformed quartz: implications for dramatic plastic softening at high temperature. Geology 14:819-822

Martelat J-E, Schulmann K, Lardeaux J-M, Nicollet C, Cardon H (1999) Granulite microfabrics and deformation mechanisms in southern Madagascar. J Struc Geol 21:671-687

McLaren AC, Pryer LL (2001) Microstructural investigation of the interaction and interdependence of cataclastic and plastic mechanisms in feldspar crystals deformed in the semi-brittle field. Tectonophysics 335:1-15

Means WD (1981) The concept of steady-state foliation. Tectonophysics 78:179-199

Mercier J-CC, Anderson DA, Carter NL (1977) Stress in the lithosphere: inferences from steady state flow of rocks. Pure and Appl Geophys 115:199-226

Mitra G (1978) Ductile deformation zones and mylonites: The mechanical processes involved in the deformation of crystalline basement rocks. Am J Sci 278:1057-1084

Mitra G (1992) Deformation of granitic basement rocks along fault zones at shallow to intermediate crustal levels. *In* Structural Geology of Fold and Thrust Belts. Mitra S, Fisher GW (eds) Johns Hopkins Univ. Press, Baltimore, p 123-144

Nakashima S, Matayoshi H, Yuko T, Michibayashi K, Masuda T, Kuroki N, Yamagishi H, Ito Y, Nakamura, A (1995) Infrared microspectroscopy analysis of water distribution in deformed and metamorphosed rocks. Tectonophysics 245:263-276

Okudaira T, Takeshita T, Hara I, Ando J (1995) A new estimate of the conditions for the transition from basal <a> to prism [c] slip in naturally deformed quartzite. Tectonophysics 250:31-46

Olesen NO (1987) Plagioclase fabric development in a high grade shear zone, Jotunheimen, Norway. Tectonophysics 142:291-308

Olsen TS, Kohlstedt DL (1984) Analysis of dislocations in some naturally deformed plagioclase feldspars. Phys Chem Minerals 11:153-160

Olsen TS, Kohlstedt DL (1985) Natural deformation and recrystallization of some intermediate plagioclase feldspars. Phys Chem Minerals 11:153-160

Parrish DK, Krivz AL, Carter NL (1976) Finite-element folds of similar geometry. Tectonophysics 32:183-207

Passchier CW (1985) Water-deficient mylonite zones—An example from the Pyrenees. Lithos 18:115-127

Passchier CW, Trouw RAJ (1996) Microtectonics. Springer, New York

Paterson MS (1989) The interaction of water with quartz and its influence in dislocation flow—an overview. *In* Rheology of Solids and of the Earth. Karato SI, Toriumi M (eds) Oxford University Press, Oxford, p 107-142

Paterson MS (2001) Relating experimental and geological rheology. Intl J Earth Sci 90:157-167

Paterson MS, Luan FC (1990) Quartzite rheology under geological conditions. Geol Soc London Spec Pub 54:299-307

Pennacchioni G, Cesare B (1997) Ductile-brittle transition in pre-Alpine amphibolite facies mylonites during evolution from water-present to water-deficient conditions (Mont Mary nappe, Italian Western Alps). J Metamor Geol 15:777-791

Pieri M, Kunze K, Burlini L, Stretton I, Olgaard DL, Burg J-P, Wenk H-R (2001) Texture development of calcite by deformation and dynamic recrystallization at 1000 K during torsion experiments of marble to large strains. Tectonophysics 330:119-140

Poirier JP (1985) Creep of Crystals. Cambridge University Press, Cambridge

Post A, Tullis J (1998) The rate of water penetration in experimentally deformed quartzite: implications for hydrolytic weakening. Tectonophysics 295:117-137

Post A, Tullis J (1999) A recrystallized grain size piezometer for experimentally deformed feldspar aggregates. Tectonophysics 303:159-173

Post AD, Tullis J, Yund RA (1996) Effects of chemical environment on dislocation creep of quartzite. J Geophys Res 101:22143-22155

Prior DJ, Wheeler J (1999) Feldspar fabrics in greenschist facies albite-rich mylonite from electron backscatter diffraction. Tectonophysics 303:29-49

Pryer LL (1993) Microstructures in feldspars from a major crustal shear zone: The Grenville Front, Ontario, Canada. J Struc Geol 15:21-36

Rosenberg CL, Stunitz H (2002) Deformation and recrystallization of plagioclase along a temperature gradient: an example from the Bergell tonalite. J. Struc Geol (in press)

Rutter EH (1995) Experimental study of the influence of stress, temperature and strain on the dynamic recrystallization of Carrara marble. J Geophys Res 100:24651-244663

Rybacki E, Dresen G (2000) Dislocation and diffusion creep of synthetic anorthite aggregates. J Geophys Res 105:26017-26036

Schmid SM (1994) Textures of geological materials: computer model predictions versus empirical interpretation based on rock deformation experiments and field studies. *In* Textures of Geological Materials. Bunge HJ, Skrotski W, Siegesmund S, Weber K (eds) DGM, Oberursel, p 278-301

Schmid SM, Casey M (1986) Complete fabric analysis of some commonly observed c-axis patterns. Am Geophys Union Geophys Monogr 36:263-286

Schmid SM, Casey M, Starkey J (1981) An illustration of the advantages of a complete texture analysis described by the orientation distribution function (ODF) using quartz pole figure data. Tectonophysics 78:101-117.1986

Schulmann K, Micoch B, Melka R (1996) High-temperature microstructures and rheology of deformed granite, Erzgebirge, Bohemian Massif. J Struc Geol 18:719-733

Selverstone J (2001) Microcracking induced by changes in fluid composition during decompression of graphitic schists. (abstr) Geol Soc Am Abs Prog 33:A51

Sharp ZD, Giletti BJ, Yoder HS Jr (1991) Oxygen diffusion rates in quartz exchanged with CO_2. Earth Planet Sci Lett 107:339-348

Shea W, Kronenberg AK (1993) Strength and anisotropy of foliated rocks with varied mica contents. J Struc Geol 2:135-142

Shelton G, Tullis J (1981) Experimental flow laws for crustal rocks. (abstr) EOS Trans, Am Geophys Union 62:396

Shigematsu N (1999) Dynamic recrystallization in deformed plagioclase during progressive shear deformation. Tectonophysics 305:437-452

Simpson C (1985) Deformation of granitic rock across the brittle-ductile transition. J Struc Geol 7:503-511

Simpson C, Wintsch RP (1989) Evidence for deformation-induced K-feldspar replacement by myrmekite. J Metamor Geol 7:261-275

Steffen K, Selverstone J, Brearly A (2001) Episodic weakening and strengthening during synmetamorphic deformation in a deep-crustal shear zone in the Alps. Geol Soc London Spec Pub 186:141-156

Stipp M, Stunitz H, Heilbronner R, Schmid S (2002a) The Eastern Tonale Fault Zone: A 'natural laboratory' for crystal plastic deformation of quartz over a temperature range from 250° to 700°C. J Struc Geol 24:1861-1884

Stipp M, Stunitz H, Heilbronner R, Schmid S (2002b) Dynamic recrystallization of quartz: Correlation between natural and experimental conditions. Geol Soc London Spec Pub 200:171-190

Stoeckert B, Brix MR, Kleinschrodt R, Hurford AJ, Wirth R (1999) Thermochronometry and microstructures of quartz—a comparison with experimental flow laws and predictions on the temperature of the brittle-plastic transition. J Struc Geol 21:351-369

Stunitz H (1998) Syndeformational recrystallization—dynamic or compositionally induced? Contrib Mineral Petrol 131:219-236

Stunitz H, Fitz Gerald JD (1993) Deformation in granitoids at low metamorphic grade. II: Granular flow in albite-rich mylonites. Tectonophysics 221:299-324

Stunitz H, Tullis J (2001) Weakening and strain localization produced by syn-deformational reaction of plagioclase. Intl J Earth Sci 90:136-148

Stunitz, H., Fitz Gerald, J. Tullis, J., and Kruse, R., 2001, Slip systems, dislocation generation and dynamic recrystallization in plagioclase in nature and experiments. EUG XI, Terra Nova Abstr 619

Takeshita T, Wenk H-R, Lebensohn R (1999) Development of preferred orientation and microstructure in sheared quartzite: comparison of natural data and simulated results. Tectonophysics 312:133-155

Takeuchi S, Argon AS (1976) Steady state creep of single-phase-crystalline matter at high temperatures. J Mater Sci 11:1542-1566

Tharp T (1983) Analogies between the high temperature deformation of polyphase rocks and the mechanical behavior of porous powder metal. Tectonophysics 96:T1-T11

Tullis J (1983) Deformation of feldspars. Rev Mineral 2 (2nd edn), p 297-323

Tullis J, Heilbronner R (2002) Microstructural evolution in quartzites experimentally deformed to high shear strain. (abstr) Geol Soc Am Abstr Progr 34

Tullis J, Tullis T (1972) Preferred orientation of quartz produced by mechanical Dauphine twinning: thermodynamics and axial experiments. Am Geophys Union Geophys Monogr 16:67-82

Tullis J, Wenk H-R (1994) Effect of muscovite on strength and lattice preferred orientations of experimentally deformed quartz aggregates. Mater Sci Engin A175:209-220

Tullis J, Yund RA (1980) Hydrolytic weakening of Westerly granite and Hale albite. J Struc Geol 2:439-451

Tullis J, Yund RA (1985) Dynamic recrystallization of feldspar: A mechanism for ductile shear zone formation. Geology 13:238-241

Tullis J, Yund RA (1987) Transition from cataclastic flow to dislocation creep of feldspar: Mechanisms and microstructures. Geology 15:606-609

Tullis J, Yund RA (1989) Hydrolytic weakening of quartz aggregates: The effects of water and pressure on recovery. Geophys Res Lett 16:1343-1346

Tullis J, Yund RA (1992) The brittle-ductile transition in feldspar aggregates: An experimental study. *In* Fault mechanics and transport properties of rocks. Evans B, Wong T-f (eds) Academic Press, San Diego, p 89-118

Tullis J, Yund RA (2000) Effect of composition on deformation of plagioclase. (abstr) EOS Trans, Am Geophys Union 81:F1208

Tullis J, Christie JM, Griggs DT (1973) Microstructures and preferred orientations of experimentally deformed quartzites. Bull Geol Soc Am 84:297-314

Tullis J, Dell'Angelo LN, Yund RA (1990) Ductile shear zones from brittle precursors in feldspathic rocks: the role of dynamic recrystallization. Am Geophys Union Geophys Monogr 56:67-81

Tullis J, Stunitz H, Teyssier C, Heilbronner R (2000) Deformation microstructures in quartzo-feldspathic rocks. J Virtual Explorer 2

Tullis T, Horowitz FG, Tullis J (1991) Flow laws of polyphase aggregates from end member flow laws. J Geophys Res 96:8081-8096

Twiss RJ (1977) Theory and applicability of a recrystallized grain-size paleopiezometer. Pure Appl Geophys 15:227-244

Urai JL, Means WD, Lister GS (1986) Dynamic recrystallization of minerals. Am Geophys Union Geophys Monogr 36:161-199

van der Wal D, Chopra P, Drury M, Fitz Gerald J (1993) Relationships between dynamically recrystallized grain size and deformation conditions in experimentally deformed olivine rocks. Geophys Res Lett 20:1479-1482

Wenk H-R, Pannetier J (1990) Texture development in deformed granodiorites from the Santa Rosa mylonite zone, southern California. J Struc Geol 1:177-184

Wenk H-R, Bunge HJ, Jansen E, Pannetier J (1986) Preferred orientation of plagioclase—neutron diffraction and U-stage data. Tectonophysics 126:271-284

White SH (1975) The effects of strain on the microstructures, fabrics and deformation mechanisms in quartzites. Phil Trans Roy Soc London A283:69-86

White JC, Mawer CK (1986) Extreme ductility of feldspars from a mylonite, Parry Sound, Canada. J Struc Geol 8:133-143

White JC, Mawer CK (1988) Dynamic recrystallization and associated exsolution in perthites: evidence of deep crustal thrusting. J Geophys Res 93:325-337

Wintsch R, Yi K (2002) Dissolution and replacement creep: a significant deformation mechanism in mid-crustal rocks. J Struc Geol 24:1179-1193

Yund RA (1983) Diffusion in feldspars. Rev Mineral 2 (2nd edn), p 203-222

Yund RA, Farver JR (1999) Si grain boundary diffusion rates in feldspar aggregates. (abstr) EOS Trans, Am Geophys Union 80:F1077

Yund RA, Snow E (1989) Effects of hydrogen fugacity and confining pressure on the interdiffusion rate of NaSi-CaAl in plagioclase. J Geophys Res 94:10662-10668

Yund RA, Tullis J (1991) Compositional changes of minerals associated with dynamic recrystallization. Contrib Mineral Petrol 108:346-355

Zulauf G (2001) Structural style, deformation mechanisms and paleodifferential stress along an exposed crustal section: constraints on the rheology of quartzofeldspathic rocks at supra- and infrastructural levels (Bohemian Massif). Tectonophysics 332:211-237

4 Laboratory Constraints on the Rheology of the Upper Mantle

Greg Hirth

Department of Geology and Geophysics
Woods Hole Oceanographic Institution
Woods Hole, Massachusetts 02543

INTRODUCTION

The dynamics of convection and the mechanical behavior of the lithosphere are controlled by the rheology of upper mantle rocks. For this reason, experimental and theoretical studies on the rheology of olivine aggregates have been fields of active research for at least the last 35 years. In this short-course paper, I briefly review some experimental and theoretical constraints on the rheology of upper mantle rocks and minerals and then discuss the application of these data for understanding the rheology of the upper mantle in different tectonic environments. There is an expansive literature on the subject of mantle rheology; extensive reviews and background can be found in several recent articles and books (Poirier 1985; Ranalli 1995; Karato and Wu 1993; Kohlstedt et al. 1995; Drury and FitzGerald 1998; Hirth and Kohlstedt 2003), as well as numerous references made throughout this chapter.

I focus the discussion of this chapter on the application of experimental data for constraining the rheology of the oceanic lithosphere and mantle. Constraints on mantle rheology based on extrapolation of laboratory experiments to deformation conditions in the oceanic lithosphere and asthenosphere are illuminating for several reasons. First, the composition of the mantle is relatively well constrained by analyses of peridotites from ophiolites and mid-ocean ridges, as well as chemical analyses of basalts. Second, the oceanic lithosphere is comprised of rock that cooled from high temperature conditions at high pressure, and is therefore not previously fractured. In this way the lithosphere is similar to the "ideal" rocks we use in our experiments. Third, the temperature of the lithosphere is constrained by a number of geophysical observations. Fourth, microstructural observations on naturally deformed mantle rocks justify applying experimental flow laws at geologic conditions. Finally, several independent geophysical observations, such as constraints on viscosity based on analysis of the geoid and the depth distribution of seismicity, can be compared to predictions derived from extrapolation of laboratory data.

I first provide some background information on the fracture and frictional behavior of mantle rocks. I then overview viscous deformation mechanisms including low-temperature plasticity and high temperature creep processes such as dislocation creep and diffusion creep. In the discussion, I outline several important caveats regarding the extrapolation of laboratory data to natural conditions. Finally, I review the implications of laboratory data for understanding the depth distribution of seismicity in the oceanic lithosphere, strain localization and spatial variations in the viscosity of convecting regions of the upper mantle.

BACKGROUND

Brittle deformation and low-temperature plasticity

Fracture strength and frictional properties of peridotite at low temperature. Brittle deformation of mantle rocks occurs in a wide range of tectonic environments, including oceanic transform faults, the fore-arc of subduction zones and during the formation of continental rifts. Surprisingly, there are not many data on the brittle behavior of mantle rocks and minerals. Perhaps this is a manifestation of the fact that much of the rock mechanics community's effort in studying brittle deformation has been focused on understanding earthquakes in continental regions where no seismicity is observed in the upper mantle (e.g., the San Andreas Fault). However, the temperature in many parts of the upper mantle is too low for ductile deformation mechanisms to operate. Thus, an understanding of the brittle deformation of peridotite is important for understanding the rheology of the lithosphere, especially in oceanic environments.

Data for the brittle deformation of dunite at room temperature are shown in a Mohr-Coulomb diagram in Figure 1. Similar to many other silicate rocks, a strong pressure dependence on fracture strength is observed up to extremely high pressures, indicating that a transition to fully plastic flow is not achieved at room temperature (Shimada et al. 1983). Because all of these experiments are conducted on natural rocks, the possibility that the strengths are partly influenced by the presence of alteration phases, such as serpentine, must be considered. As illustrated in Figure 1, the fracture strength of serpentinite is considerably lower than that of dunite. The fracture strength of peridotite is strongly affected by even small amounts of serpentine. For example, the strength of peridotite with ~10% serpentine is the same as that of a completely serpentinized peridotite (Escartin et al. 2001).

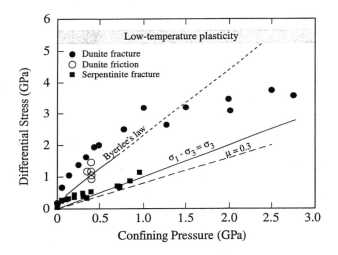

Figure 1. Plot of differential stress versus confining pressure showing fracture (Shimada et al. 1983) and frictional (Stesky et al. 1974) strengths of dunite at room temperature. The fracture strength of serpentinite (Escartin et al. 1997a) and the frictional strength predicted by Byerlee's law are also shown. A dashed line shows the extrapolation of Byerlee's law to stresses above where it is defined experimentally. The gray band shows the plastic yield strength of olivine single crystals determined from indentation tests (Evans and Goetze 1979). For comparison, a line showing where differential stress is equal to confining pressure (labeled $\sigma_1 - \sigma_3 = \sigma_3$, i.e., Goetze's criterion) and the frictional strength predicted for a coefficient of friction of 0.3 (the value measured for lizardite/chrysotile serpentinites) are also shown.

There are very few data on the frictional properties of olivine or peridotite. As a result, for most geophysical applications, Byerlee's law (Byerlee 1978) has been assumed to be appropriate. The lowest temperature data (T < 100°C) from Stesky et al. (1974) are indeed in agreement with Byerlee's law (Fig. 1).

There is considerable curvature in the fracture envelope for dunite (Fig. 1). Such curvature is often inferred to indicate onset of crystal plasticity (for example, see p. 178 in Paterson 1978). However, as illustrated in Figure 1, the stresses in these samples are considerably lower than that required to initiate crystal plasticity. Microstructural observations indicate that the high-pressure samples are brittle, but that the style of cracking changes with increasing pressure. While failure is observed on planes at an angle of ~30° at low pressure, fractures form at angle of around 45° at high pressure (Shimada et al. 1983). Similar observations have been made for quartzite (Hirth and Tullis 1994) and serpentinite (Escartin et al. 1997a). These observations suggest that the dominant mode of cracking changes from mode I to mixed mode II-III with increasing pressure, with a concomitant decrease in the pressure-dependence of brittle deformation in the high-pressure regime. In the case of serpentinite, measurements of volumetric strain also indicate that deformation in this brittle regime is nominally non-dilatant (Escartin et al. 1997a).

Based on the observation that the fracture strength of dunite becomes less than the stress predicted by the extrapolation of Byerlee's law, it appears that Byerlee's law is not valid at the highest normal stresses shown in Figure 1. The data used to define Byerlee's law were collected at normal stresses less than ~1.7 GPa, which corresponds to confining pressures less than ~0.8 GPa. While this pressure range encompasses the depths where the majority of earthquakes occur in the continental crust, the relationships illustrated in Figure 1 suggest that extrapolation of Byerlee's law to depths at the base of the oceanic lithosphere is not valid.

Low-temperature plasticity. While fully plastic flow is not observed at room temperature for dunite, experiments on olivine single crystals indicate that plasticity is possible. However, as illustrated by the gray band in Figure 1, the stresses required to activate dislocation motion in olivine at room temperature are on the order of 5-6 GPa (Evans and Goetze 1979), a value in the range of the theoretical strength. The theoretical strength, which is ~1/15 the shear modulus (Hirth 1982), represents the stress required to nucleate a dislocation in a perfect crystal.

With an increase in temperature, the stress required to activate dislocation motion decreases. The most comprehensive data set for deformation in the low-temperature plasticity regime comes from indentation tests on olivine single crystals (Evans and Goetze 1979). These data show that yield stress decreases with decreasing strain rate and increasing temperature and follows a "Peierls Law" with the form (e.g., Goetze 1978)

$$\dot{\varepsilon} = B \exp\left[-\frac{H}{RT}\left(1 - \sigma/\sigma_p\right)^q\right] \tag{1}$$

where B and q are constants, H is the activation enthalpy, σ is differential stress, σ_p is the Peierls stress, R is the gas constant and T is absolute temperature. Values of $B = 5.7 \times 10^{11}$ s^{-1}, $\sigma_p = 8.5 \times 10^3$ MPa, $H = 540$ kJ/mol and $q = 2$ have been determined for olivine (Goetze 1978). The influence of temperature on the yield stress for low-temperature plasticity is shown for a constant strain rate of 10^{-5} s^{-1} in Figure 2. The comparison of the low-temperature plasticity data with high-temperature creep data is discussed below.

High-temperature creep

At higher temperatures, where thermally activated recovery processes such as dislocation climb become effective due to relatively rapid diffusion rates, a large amount of data indicate that strain rate ($\dot{\varepsilon}$) is related to differential stress (σ) via a flow law with the form

$$\dot{\varepsilon} = A\sigma^n d^{-p} f_{H_2O}^r \exp\left(-\frac{E^* + PV^*}{RT}\right) \tag{2}$$

where A, n, p and r are constants that depend on the operative deformation mechanism,

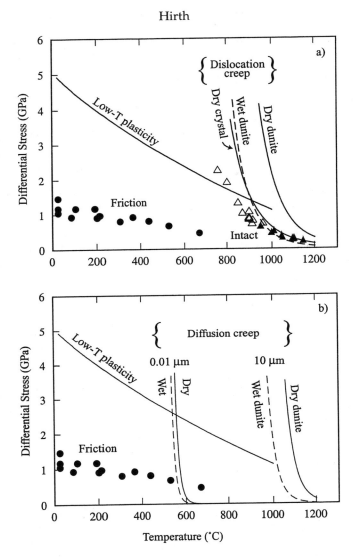

Figure 2. Plots of differential stress versus temperature comparing the high-temperature frictional behavior of olivine aggregates with strengths determined for both low-temperature plasticity and high-temperature creep. (a) The frictional strength, which was determined at a confining pressure of 400 MPa (Stesky et al. 1974) is considerably lower than both the yield strength for low-temperature plasticity (Evans and Goetze 1979) and the extrapolated strength of both wet and dry dunite in the dislocation creep field. The strength of olivine single crystals deformed on the easiest slip system (010)[100] is also shown. All of the flow laws are plotted for a strain rate of 10^{-5}/s. Data from the study of Post (1977), normalized to a strain rate of 10^{-5}/s using a stress exponent of 5.1, are shown as triangles; the open and filled triangles represent samples that were deformed at strain rates slower and faster than 10^{-5}/s, respectively. The dislocation creep flow laws are from the compilation of Hirth and Kohlstedt (2003). The olivine crystal data are from Bai et al. (1991). (b) The frictional strength is also considerably smaller than that predicted by extrapolation of the diffusion creep flow law to low temperature, unless the grain size is in the range of 0.01 μm. Even for this very small grain size, which may approach to size of some of the gouge, the Ahrrenius temperature dependence of the diffusion creep flow law results in very high stresses at temperatures less than ~550°C.

f_{H2O} is the water fugacity, E^* is the activation energy, P is pressure and V^* is the activation volume, R is the gas constant and T is absolute temperature. The values for these parameters used in this paper can be found in Table 1 of Hirth and Kohlstedt (2003). For deformation in the diffusion creep regime, where strain is accommodated by a grain boundary sliding process limited by diffusion (e.g., Raj and Ashby 1971), the strain rate is linearly dependent on stress and thus $n_{diff} = 1$. The stress dependence arises from the chemical potential gradient resulting from the applied stress field. By contrast, in the dislocation creep regime, the strain rate increases non-linearly with increasing stress. In this case, $n_{disl} \approx 3$ to 5. The strong stress dependence arises because the strain rate accommodated by the motion of dislocations depends on both the dislocation density (ρ) and the velocity of dislocations (v). The Orowan equation, $\varepsilon = \rho b v$, relates these quantities, where the Burger's vector (b) represents the displacement accommodated by one dislocation moving through the crystal lattice (see Poirier 1985 for introductory background). For simple dislocation geometries, $\rho \propto \sigma^{n'}$, with $n' \approx 2$. When dislocation motion is limited by climb, v can be related to the velocity of climb, which in turn is proportional to σ due to the changes in vacancy concentration around the dislocation (e.g., Poirier 1985).

In the laboratory, some of the most reliable rheological data have been obtained on fine-grained aggregates synthesized from natural minerals that are ground to known grain sizes. This procedure allows the analysis of a large number of experimental parameters in rocks with a controlled microstructure. Most importantly, since the diffusion creep process depends strongly on grain size, the use of fine-grained samples provides experimentalists the opportunity to run experiments in the diffusion creep regime (e.g., Schwenn and Goetze 1977; Cooper and Kohlstedt 1984; Karato et al. 1986; Hirth and Kohlstedt 1995a). When diffusion creep is limited by grain matrix diffusion (Nabarro-Herring creep), the grain-size exponent in Equation (2) is $p = 2$. If the creep rate is limited by grain boundary diffusion (Coble creep), $p = 3$. Analyses of diffusion data for olivine indicate that Coble creep dominates at all grain size expected in both the Earth and the laboratory (Hirth and Kohlstedt 2003).

An example of rheological data obtained on a fine-grained olivine aggregate is shown in Figure 3a. For this sample, with a grain size of ~15 µm, a transition from diffusion creep to dislocation creep is observed with increasing differential stress. The parameters A and n for both deformation mechanisms can be determined through a non-linear fit to the data. These data illustrate that the total strain rate of the sample is related to the sum of the strain rates accommodated by dislocation creep and diffusion creep, indicating that these processes are independent. The dislocation creep component of the total strain rate is obtained by subtracting the component of diffusion creep determined from the non-linear fit from the total strain rate.

Under hydrous conditions, the strain rate of the high-temperature creep mechanisms increases with increasing water content. Due to the large variation in water content in different tectonic environments, it is therefore important to account for the influence of water on rheology. Recent analyses indicate that the effect of water is well described by the water fugacity term in Equation (2) with $r \approx 1$. This formulation is motivated by the hypothesis that the influence of water on rheology arises from changes in point defect concentrations, and therefore diffusion kinetics, under hydrous conditions (e.g., Mackwell et al. 1985; Mei and Kohlstedt 2002b). Several observations support this hypothesis, including: (1) Above a water fugacity of approximately 50 MPa, olivine creep rates increase with increasing water fugacity (Mei and Kohlstedt 2002b). Based on the dependence of water content on water fugacity (e.g., Kohlstedt et al. 1996), this observation suggests that a change in the charge neutrality condition occurs when the olivine water content exceeds ~50 H/10^6Si. (2) The stress exponent for dislocation creep is the same, within error, under hydrous and anhydrous conditions. The observation that water also enhances creep rates in the diffusion creep regime indicates that the presence of water enhances grain boundary diffusion (Karato et al. 1986; Mei and Kohlstedt 2002a).

A comparison of the strength of dunite under wet and dry conditions at a confining pressure of 400 MPa is shown for both the dislocation creep and diffusion creep regimes in Figure 2. For these experimental conditions, the strength in the dislocation creep regime is a factor of 2-3 lower than that under dry conditions. The dislocation creep flow laws shown in Figure 2, which were determined at 1100-1300°C, are extrapolated to lower temperatures to demonstrate the strong temperature dependence of deformation associated with the Ahrrenius relationship in Equation (2).

For the flow law formulation used in Equation (2), the influence of temperature and pressure on the water fugacity (e.g., Tödheide 1972) must also be included. The effect of water fugacity on deformation is illustrated by comparing the strengths of wet and dry dunite deformed in the diffusion creep regime at relatively low and high temperatures. As shown in Figure 2b, there is little difference in the strength predicted by extrapolation of the wet and dry diffusion creep flow laws to a temperature of ~600°C and grain size of 0.01 µm. In this case, the water fugacity, and therefore the olivine water content, is much lower at 600°C than at 1100°C; this effect is solely due to the influence of temperature on the water fugacity coefficient.

When applying flow laws to deformation in the Earth, it is more convenient to consider the rheology of the mantle at a particular water content. In this case, Equation (2) can be rewritten as (Karato and Jung 2002; Hirth and Kohlstedt 2003):

$$\dot{\varepsilon} = A\sigma^n d^{-p} C_{OH}^r \exp\left(-\frac{E^*_{e\!f\!f} + PV^*_{e\!f\!f}}{RT}\right) \qquad (3)$$

with $E^*_{eff} = E^* - E_{H2O}$ and $V^*_{eff} = V^* - V_{H2O}$, where E_{H2O} and V_{H2O} define the temperature and pressure dependence of the solubility of water in olivine (Mei and Kohlstedt 2000b, Karato and Jung 2002).

Brittle-ductile/brittle-plastic transitions

The brittle-ductile transition is simple in concept. At low temperatures, thermally activated processes are not active and inelastic deformation occurs by pressure-dependent brittle processes. At higher temperatures, thermally activated viscous deformation processes dominate. However, both field and experimental observations demonstrate that the brittle-ductile transition occurs over a wide range of conditions and involves several different deformation mechanisms (e.g., Carter and Kirby 1978; Scholz 1988; Ross and Lewis 1989; Fredrich et al. 1989, Tullis and Yund 1977, 1992; Hirth and Tullis 1994).

While a constitutive law for the brittle-ductile transition has not been formulated, experimental and theoretical studies have led to important advances in our understanding of the processes responsible for its occurrence. Two empirical observations appear to be relatively robust: (1) A transition from localized to distributed deformation occurs when the strength of the intact rock is lower than the stress required for frictional sliding (e.g., Byerlee 1968). Where samples exhibit frictional behavior that does not follow Byerlee's law, the brittle-ductile transition is observed to occur where the peak strength of the intact sample becomes less than the strength of the friction behavior of the material (Escartin et al. 1997a). (2) The transition to fully plastic deformation (i.e., with no contribution from microcracking) occurs when the strength is lower than the confining pressure, referred to as the Goetze criterion (see reviews by Evans et al. 1990 and Kohlstedt et al. 1995). The depth of the transition to fully plastic deformation may be the depth limit of earthquake rupture and aftershocks (e.g., Scholz 1988; Strehlau 1986). For clarity, it is preferable to use different terms to describe these transitions (e.g., Rutter 1986; Kohlstedt et al. 1995). The brittle-ductile transition refers to a transition from macroscopically brittle (or localized) deformation to macroscopically ductile (or distributed) deformation. In this case, the ductile deformation may be accommodated by either distributed semi-brittle flow or cataclastic flow (e.g., Evans et al. 1990; Tullis and Yund 1992; Hirth and Tullis 1994). The term brittle-plastic transition is used to describe the transition from semi-brittle flow to fully plastic flow.

There are very few data to constrain the conditions where the brittle-ductile and brittle-plastic transitions occur in olivine aggregates. Over the range of conditions shown in Figure 2, the frictional strength at a confining pressure of 400 MPa (Stesky et al. 1974) remains considerably lower than the strength observed for low-temperature plasticity. The frictional strength is also lower than that predicted by the extrapolation of higher-temperature data on intact cores of olivine aggregates. No systematic study has been completed to constrain the conditions where the brittle-ductile transition occurs in olivine aggregates. However, some of the lower temperature experiments of Post (1977), shown in Figure 2a, were conducted at stress levels where either semi-brittle flow or localized deformation might be expected. Indeed, Post (1977) describes the formation of very localized ductile faults in several of his experiments.

Based on the empirical considerations described above, it is difficult to explain the decrease in frictional strength with increasing temperature. As illustrated in Figure 2a, the strong temperature dependence of the dislocation creep processes indicates that they are insignificant at temperatures less than ~800°C at experimental conditions. Even with a grain size of 0.01 μm, a possible grain size resulting from comminution of gouge in a frictional shear zone, diffusion creep cannot explain the temperature effect on friction (see Fig. 2b). It is possible that plastic flow occurs at asperities at the sliding surface or grain-to-grain contacts. For example, the ratio of the friction stress to the stress required for low-temperature plasticity indicates that creep of asperities is possible if the real area of contact is less than ~0.2. However, even if the asperities do deform by low-temperature plasticity, it is not clear that the frictional strength will decrease (see p. 62 in Scholz 1990). Alternatively, the strength of Stesky's samples could be influenced by the presence of small amounts of alteration phases.

INSIGHTS, CAVEATS AND QUESTIONS ABOUT APPLYING LABORATORY DATA TO CONSTRAIN THE RHEOLOGY OF THE OCEANIC MANTLE

Before illustrating experimental constraints on the rheology of the mantle, it is important to consider some of the problems and questions associated with extrapolating laboratory data to natural conditions. Our understanding of the rheological behavior of the Earth's crust and mantle has been greatly enhanced by microstructural observations of both naturally and experimentally deformed rocks. In conjunction with theoretical constraints, analysis of microstructures produced during deformation experiments can be used to determine the micromechanics of deformation processes. The analysis of microstructures in naturally deformed rocks therefore provides a critical link between theoretical and experimental studies and large-scale geologic processes. Many deformation processes produce diagnostic microstructures, for example the formation of lattice preferred orientations (LPO) described elsewhere is this book. Therefore, analyses of textures preserved in rocks can be used to investigate deformation mechanisms operative in the Earth. The applicability of experimental flow laws, which must be extrapolated many orders of magnitude in strain rate, can then be evaluated by combining microstructural observations with petrologic and geochemical analyses of the same rocks to estimate the conditions of deformation.

A comparison of microstructures in experimentally and naturally deformed peridotites indicates that the same slip systems are active [e.g., Nicolas 1989]. For example, the analysis of LPOs in both experimental and natural rocks demonstrates that strain is dominantly accommodated by slip on (010)[100] (e.g., Tommasi et al. 2000; Zhang et al. 2000), the easiest slip system in olivine (e.g., Durham and Goetze 1977; Bai et al. 1991).

As illustrated in Figures 3b and 3c, depending on the geological application, laboratory data must be extrapolated one to two orders of magnitude in differential stress or up to 400°C in temperature. The linear fits shown in Figures 3b and 3c illustrate the co-variation of uncertainty in various parameters in the flow law. Specifically, in Figure 3b, the dashed lines show the uncertainty in strain rate predicted by extrapolation of the laboratory data to the low stress conditions appropriate for the asthenosphere. The solid line in Figure 3b is a linear fit to the experimental data and gives a stress exponent of $n = 3.5 \pm 0.3$. The dashed lines show fits where the parameter A in Equation (2) is allowed to vary and the stress exponent is fixed to $n = 3.2$ and 3.8 (i.e., 3.5 ± 0.3). The relationships shown by the dashed lines in the Ahrrenius plot in Figure 3c were calculated in a similar way. By contrast, the dotted lines in Figure 3c are the relationships resulting from calculating strain rate while keeping A constant to the value determined from the linear fit and allowing E^* to vary by the uncertainty (i.e., $E^* = 590$ and 510 kJ/mol). Clearly, these relationships overestimate the error in extrapolation to geologic conditions; they over/underestimate the strain rates in the laboratory by approximately 2 orders of magnitude. By accounting for the co-variation of n and A, or E^* and A, the uncertainty in strain rate is considerably smaller.

When laboratory data are extrapolated to geologic conditions it is important to evaluate whether changes in the dominant deformation mechanisms may occur. For example, by extrapolating data from the higher stress dislocation creep regime to low stresses, it is possible that a transition to diffusion creep may be overlooked. Fortunately, the conditions under which changes in deformation mechanism occur can be constrained using a combination of experimental data and theoretical arguments. For example, diffusion creep processes can be studied using experimentally engineered fine-grained aggregates. A useful tool in this regard is the deformation mechanism map (see Frost and Ashby 1982 for a review).

Despite the fact that almost all rocks contain several phases, the majority of geologically applicable flow laws have been determined for monomineralic rocks. There are several good reasons for this apparent limitation, including: (1) Polyphase rocks often melt at conditions where deformation mechanisms that we want to study are activated in the lab. (2) Because the flow laws for the individual minerals are different, the large extrapolation in either temperature or stress required to apply flow laws can result in "inversions" of the hard and weak phases [e.g., Tullis et al. 1991]. For these reasons, a significant amount of experimental and theoretical work is concentrated on quantifying "mixing laws" with which the flow laws of the individual minerals can be used to study the properties of two-phase mixtures [e.g., Tullis et al. 1991; Handy 1994; Dresen et al. 1998].

The most important parameters to constrain during two-phase flow are the strength contrast of the phases, and the proportion and topology of the strong phase. For example, if the stronger phase provides the stress-supporting network, the rheology of the rock may approach a homogeneous strain condition (where the strain accommodated by the weak phase is limited by the strain accommodated by the strong framework). By contrast, if the weak phase provides the stress-supporting framework, the rheology may follow more closely to a homogeneous stress condition (where the strain is dominantly accommodated by the weak phase). Some of these effects have been observed in theoretical models for the development of LPOs in olivine+pyroxene aggregates (e.g., Wenk et al 1991). While more work is required to further document the role of pyroxenes on the rheology of peridotite, experimental work to date suggests that there is not a large difference in the rheology of olivine and olivine+pyroxene aggregates (e.g., Kohlstedt and Zimmerman 1996).

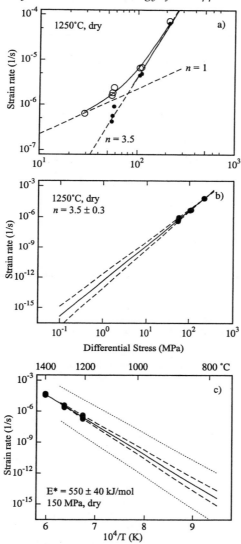

Figure 3. Example and extrapolation of high temperature creep data for olivine aggregates. (a) Experimental data from Hirth and Kohlstedt (1995a) showing a transition between diffusion creep (n = 1) to dislocation creep (n = 3.5) with increasing stress for deformation of dry olivine aggregates with a grain size of 15 μm at a confining pressure of 300 MPa. The solid line shows a non-linear fit to the entire data set. The black symbols show the dislocation creep strain rate determined after subtracting the diffusion creep component determined from the fit from the total strain rate. (b) Extrapolation of the dislocation creep data shown in (a) to lower stresses. The uncertainty in stress exponent (n = 3.5 ± 0.3) results in an increase in uncertainty of the strain rate with decreasing stress. The dashed lines show extrapolation of the data with n = 3.2 and 3.8, respectively. However, even at a mantle stress level of ~1 MPa the uncertainty is only a factor of ±6. (c) Extrapolation of dislocation creep data from the study of Chopra and Paterson (1984) to lower temperatures. The uncertainty shown by the dashed lines, which are extrapolations of the data with $E^* = 590$ and 510 kJ/mole, respectively, is only a factor of ±5 at 800°C. The gray dotted lines are calculated with $E^* = 590$ and 510 kJ/mol, but leaving A in Equation (2) constant. These lines do not "fit" the data and show how uncertainty in E^* is correlated with uncertainty in A.

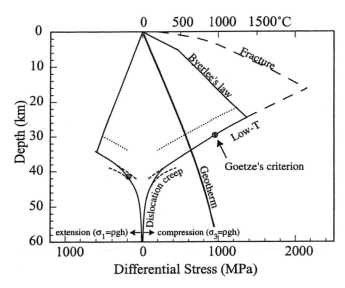

Figure 4. Maximum strength versus depth diagram for 60 Ma oceanic lithosphere with a conductive geotherm (temperature axis is on top). The strength for extension (i.e., where $\sigma_1 = \rho gh$) is considerably lower than that in compression ($\sigma_3 = \rho gh$). The strength in the brittle regime is shown for both Byerlee's law with a hydrostatic pore-fluid pressure and the room temperature fracture strength of dunite (dashed line labeled Fracture). The stress for both the low-temperature plasticity regime (labeled low-T) and the dislocation creep regime were calculated for a strain rate of 10^{-15}/s. The strength for low-temperature plasticity at a strain rate of 10^{-18}/s is shown by the dotted lines. The depths where the strength predicted by extrapolation of the low-temperature plasticity flow law is equal to the σ_3 (i.e., Goetze's criterion) are shown by the gray dots.

Strength of the lithosphere and the depth of oceanic earthquakes

A laboratory-based plot of maximum strength versus depth in the lithosphere is shown in Figure 4. Details on how to construct these types of plots, as well as caveats regarding their application, are summarized in Brace and Kohlstedt (1980) and Kohlstedt et al. (1995). In Figure 4, the strength in the brittle field is defined by two curves. A frictional curve is shown using Byerlee's law, assuming a hydrostatic pore-fluid pressure, and a fracture curve is shown based on the data of Shimada et al. (1983) illustrated in Figure 1. The strength in the plastic deformation regime is defined by extrapolation of the dislocation creep flow law (at low stresses) and the low-temperature plasticity law (at higher stresses) for a constant strain rate of 10^{-15}/s. Notice that extrapolation of the low-temperature plasticity law with the form of Equation (1) to higher temperatures (i.e., greater depths in Fig. 4) predicts unrealistically low stresses. This observation emphasizes the potential problem of extrapolating experimental flow laws many orders of magnitude in strain rate. For the low-temperature plasticity field, such extrapolations should be viewed with particular caution at stresses below ~400 MPa. For this reason, several alternate forms of a Peierls-type law have been employed for extrapolating low-temperature plasticity laws to geologic strain rates (see Evans and Goetze 1979; Liu and Yund 1995; Kameyama et al. 1999). For higher stress (lower temperature) conditions, these laws all predict similar strain rates.

The depth extent of brittle deformation can be predicted by determining where the strength predicted for low-temperature plasticity is equal to the confining pressure (i.e., Goetze's criterion). Depending on the stress state in the mantle (i.e., σ_1 vertical or σ_1 horizontal), the brittle-plastic transition is predicted to occur between depths of approximately 30-40 km for 60-Myr old lithosphere.

With our current database, the depth of the transition between localized frictional sliding and distributed semi-brittle flow (i.e., the brittle-ductile transition) is more poorly constrained (see review by Kohlstedt et al. 1995). The maximum value of stress at the brittle-ductile transition can be constrained using the transition from Byerlee's law to the low-temperature plasticity flow law. This stress should be considered a maximum for several reasons, including: (1) The observation that the brittle-ductile transition is not observed for dunite at room temperature, even though the fracture strength becomes significantly lower than the frictional strength predicted by extrapolation of Byerlee's law (i.e., Fig. 1). This observation suggests that the pressure dependence of the friction law decreases with increasing pressure. (2) Over geologically relevant time periods, the brittle strength can be reduced by sub-critical crack growth (e.g., Kohlstedt et al. 1995). Despite these concerns, it is noteworthy that in one of the only places where stresses have been measured at depths near the brittle-ductile transition, the differential stress approaches that predicted by Byerlee's law with a hydrostatic pore-fluid pressure (Brudy et al. 1997).

Based on the constraint that stress concentrations around flaws can promote localized plastic flow, Kohlstedt et al. (1995) cautiously suggested that the brittle-ductile transition occurs when the frictional strength becomes 1/5 of the flow strength. This suggestion is in conflict with the notion discussed above that there is little temperature dependence to frictional strength until the sliding surface becomes "welded" (see p. 62 in Scholz 1990). The transition to a welded sliding surface is actually another way of defining the brittle-ductile transition. Thus, while sub-critical crack growth and the onset of plasticity may decrease the fracture strength of intact lithosphere, it is not yet clear that the frictional strength will be strongly influenced by these processes.

For comparison with the rheological model shown in Figure 4, the depths of intraplate earthquakes are shown as a function of lithospheric age in Figure 5. For both compressional and extensional events, there is an increase in the depth of seismicity with increasing lithospheric age. Despite the uncertainties described above, the depth of earthquakes is confined to regions where brittle phenomenon would be predicted to occur based solely on the extrapolation of laboratory data (e.g., Chen and Molnar 1983; Wiens and Stein 1983). The depth cut-off also agrees well with the 600°C isotherm predicted for conductive cooling of the lithosphere. A similar temperature cut-off is suggested by the distribution of earthquakes along oceanic transform faults (e.g., Engeln et al. 1986). The depth to the 600°C isotherm also appears to define the limit of elastic deformation based on the analysis of the effective elastic plate thickness (e.g., Anderson 1995). In old oceanic lithosphere, the thin crustal section is too cold to deform by high-temperature creep. Thus, unlike some continental regions, there is no ambiguity in the interpretation of effective elastic plate thickness in old oceanic plates resulting from viscous deformation of the lower crust.

While the agreement between the laboratory predictions and the geophysical observations is encouraging, several additional caveats and apparent discrepancies must be noted. First, for a friction law to predict the strength of the lithosphere, the lithosphere must first be broken. The oceanic lithosphere is somewhat unique in this case, because, for a given depth, the transition from viscous deformation to elastic/brittle deformation occurs with increasing age due to cooling. Therefore, before the friction law is applicable, the rock must first be fractured. In this case, it may be more applicable to consider propagation of a fracture into the underlying viscous rock. Second, while the laboratory-based predictions indicate that brittle deformation may be possible at a depth where earthquakes occur, the limited amount of data for the high temperature frictional behavior of olivine aggregates indicate that sliding is stable at these conditions. Indeed, the data of Stesky (1977) indicate stable frictional sliding occurs at all temperatures greater than 100°C. These observations

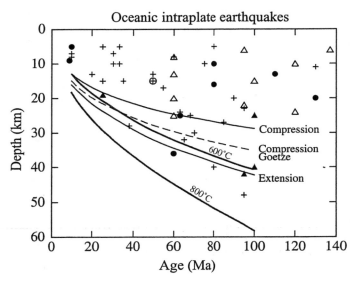

Figure 5. Plot of depth versus age of the oceanic lithosphere showing the depth and distribution of intraplate earthquakes. The lines labeled compression and extension show how the transition from frictional (i.e. Byerlee's law) to low-temperature plasticity predicted in a maximum strength versus depth diagram such as Figure 4 increases with increasing age of the lithosphere. The dashed line shows how the depth to the Goetze's criterion (i.e., the gray dots in Fig. 4) increases with increasing age. The depths to the 600 and 800°C isotherms calculated for a conductive cooling of the lithosphere are also shown. Filled (thrust earthquakes) and open (normal earthquakes) symbols are from the study of Chen and Molnar (1983). The circles show the focal depths of earthquakes in the lithosphere near oceanic trenches. The triangles are focal depths of oceanic intraplate events. The plus symbols are the focal depths of oceanic intraplate events from the study of Wiens and Stein (1983), for which no fault plane solutions were given.

again emphasize the need for more frictional sliding data on peridotites at high temperatures and also demonstrate that our understanding of the relationship between rock rheology and earthquakes may be significantly improved by further analyzing the characteristic of oceanic earthquakes.

The role of serpentinization on the strength of the oceanic lithosphere. Geological and geophysical constraints indicate that serpentinites are an important component of the oceanic lithosphere at slow-spreading ridges (e.g., Bonatti and Michael 1989; Dick 1989; Cannat et al. 1995; Karson and Lawrence 1997). Similarly, structural and geochemical observations from ophiolites indicate that serpentinization occurs at active ridge axes (e.g., Coulton et al. 1995). Serpentinites are stable below ~350-500°C at lithospheric pressures (e.g., O'Hanley 1996; and op. cit.). Lizardite, the primary polytype found in oceanic serpentinites (Aumento and Loubat 1971), is stable to ~350-400°C (O'Hanley 1996). Chrysotile is also commonly associated with lizardite in oceanic serpentinites (O'Hanley 1996).

Based on these geological observations, and motivated by experimental data on the rheology of oceanic serpentinites (e.g., Raleigh and Paterson 1965; Reinen et al. 1991, 1994; Moore et al. 1996, 1997; Escartin et al. 1997a), Escartin et al. (1997b) suggested that the strength and faulting style of the lithosphere at slow-spreading ridges could be strongly influenced by the presence of serpentinites. For example, assuming that the rheology of the lithosphere is locally controlled by the frictional properties of a mixed lizardite-chrysotile serpentinite, the strength of the lithosphere may be reduced by ~30% (relative to unserpentinized lithosphere) near ridge-transform discontinuities. In addition, a combination of the weak nature of serpentinites (relative to their unaltered equivalents) and the possibility that the nominally non-dilatant style of brittle deformation exhibited by serpentinites could promote high pore-fluid pressure conditions (Escartin et al. 1997a), suggests serpentinization can lead to significant weakening of oceanic fault zones (e.g., Escartin et al. 1997b). As an aside, frictional sliding experiments indicate that the coefficient of friction of even the weakest serpentine polymorph (chrysotile) is too large to explain the absolute weakness of the San Andreas fault (Moore et al. 1997). However, "fault sealing" resulting from serpentinization has been suggested to provide the conditions necessary to promote high fluid pressures along the San Andreas Fault (Moore et al. 1996; 1997).

While serpentinization can clearly play an important role in the rheology of shallow lithosphere near oceanic transforms and ridges, the role of serpentine at greater depths is unclear. One of the largest unknowns in this case is whether or not the fluids necessary for the serpentinization process can migrate downwards to depths at the base of the seismogenic zone. In addition, it is important to consider the role of temperature on the rheology of serpentinites. Experimental studies demonstrate that the coefficients of friction of chrysotile (e.g., Moore et al. 1996; 1997) and lizardite (e.g., Rutter and Brodie 1988; Moore et al. 1997) actually increase with increasing temperature. In contrast, experiments on intact samples of lizardite demonstrate a decrease in strength with increasing temperature (Raleigh and Paterson 1965; Murrell and Ismail 1976) over the same temperature range. These observations indicate that there are at least two effects of temperature. The increase in the coefficient of friction with increasing temperature suggests that the samples are "healing" due to lithification of the gouge. This hypothesis is supported by the observation that Moore et al.'s samples exhibited significant displacement strengthening at high temperatures, and that the displacement strengthening was most pronounced at the slowest strain rates. The weakening of intact cores of serpentine with increasing temperature indicates that a thermally activated process such as plastic flow or sub-critical crack growth may control deformation at higher temperatures. This hypothesis is supported by the positive velocity dependence of friction observed for serpentinites at "low" velocity (e.g., Reinen et al. 1991; Moore et al. 1997). However, the higher temperature experiments of Moore et al. indicate that the conditions where velocity-strengthening behavior occurs cannot be predicted using a simple trade-off between temperature and strain rate (velocity). The experimental results of Moore et al also demonstrate that adsorbed inter-layer water may play a critical role in the weakness of serpentinites.

Strength of plate boundaries in the viscous regime

Our understanding of the rheology and seismicity of plate boundaries has been significantly improved by results from a combination of modeling studies and analysis of geodetic data. For example, convection modeling studies indicate that Earth-like plate tectonics occur when lithospheric deformation is strongly localized on weak plate boundaries (e.g., Bercovici 1995; Zhong and Gurnis 1996; Bercovici 1998; Tackley 2000). Therefore, it is important to understand the deformation mechanisms that control the strength of plate boundaries, such as oceanic transform faults. At face value, extrapolation of experimental data indicates that the oceanic lithosphere is relatively strong (e.g., Fig. 4). However, analyses of microstructures in shear zones provide insight into mechanisms of weakening and ductile strain localization. Past geologic studies invariably indicate that strain localization is accompanied by grain-size reduction, leading to the hypothesis that changes in deformation mechanism (from dislocation creep to diffusion creep) facilitate weakening and localization (e.g., Kirby 1985; Rutter and Brodie 1988; Vissers et al. 1995; Jaroslow et al. 1995; Kameyama et al. 1997; Jin et al. 1998; Braun et al. 1999).

As mentioned above, there are several viscous deformation mechanisms that may occur depending on the temperature, grain size and stress state in the lithosphere. Deformation mechanism maps are a convenient way to determine which mechanism dominates at a particular condition (Frost and Ashby 1982). Deformation mechanism maps for ~25-Myr-old oceanic lithosphere at a depth of 30 km are illustrated in Figure 6. In these figures, contours of constant strain rate are shown on a plot of differential stress versus grain size. For geologically reasonable strain rates, the differential stresses predicted from Figure 6 are similar to those used in the lab. Therefore, the accuracy of these deformation mechanism maps is mostly limited by extrapolation in temperature (e.g., Fig. 3c), and thus, uncertainty in the activation energy for the various creep processes. For both the dislocation creep regime and the low-temperature plasticity regime, there is no effect of grain size on strain rate. Therefore, the transition between the two mechanisms occurs with increasing stress. By contrast, due to the strong grain-size dependence on strain rate in the diffusion creep regime, where p in Equation (2) is ~3, the field for diffusion creep increases relative to both dislocation creep and low-temperature plasticity with decreasing grain size.

The observation of very small grain sizes in naturally deformed peridotites, such as that illustrated in Figure 7, has lead to the hypothesis that grain-size reduction during deformation promotes a transition to grain-size sensitive deformation processes (e.g., Kirby 1985; Rutter and Brodie 1988; Vissers et al. 1995). The transition to grain-size sensitive deformation processes can potentially lead to extreme weakening of shear zones. However, such processes may only occur if grain growth is inhibited by the presence of second phases (e.g., Jaroslow et al. 1995). In the absence of second phases, the grain size may continually evolve towards that predicted for the boundary between dislocation creep and diffusion creep, leading to little variation in the rheology of the shear zone relative to that predicted based on dislocation creep flow laws (e.g., de Bresser et al. 2001).

One intriguing aspect of the extremely fine-grained mylonites is that the very fine-grain size appears to form at conditions within the diffusion creep regime shown in Figure 6a. Notice that for grain sizes less than ~5 mm, the recrystallized grain-size piezometer (Karato et al. 1980; van der Wal et al. 1993) plots within the diffusion creep regime. Recent studies suggest that grain-size reduction may continue into the diffusion creep regime as long as the strain rate accommodated by dislocation creep is still significant (Hall and Parmentier 2001; Montesi and Hirth 2001). However, Jaroslow et al. (1996) showed that an LPO is present in the somewhat coarser grained regions of the sample shown in Figure 7, suggesting that a dislocation creep process dominates deformation. These observations suggest that deformation by a dislocation creep process occurs at a faster rate than predicted by the dislocation creep flow laws.

Dislocation creep is generally assumed to be independent of grain size. However, at conditions near the transition from diffusion creep to dislocation creep in the laboratory, an influence of grain size on strain rate is observed for dry olivine aggregates even after the component of diffusion creep is removed from the total strain rate of the sample (Hirth and Kohlstedt 2003). In this case, the strain rate is observed to decrease with increasing grain size between 10-60 μm. By contrast, no difference in creep rate is observed between natural dunite samples with a grain size of 100 μm and those with a grain size of 900 μm. A likely explanation for these observations is that deformation of the finer grained samples is accommodated by a dislocation accommodated grain boundary sliding (GBS) process (Hirth and Kohlstedt 1995b; Drury and FitzGerald 1998; Kohlstedt and Wang 2001; Hirth and Kohlstedt 2003). Hirth and Kohlstedt (1995b) summarize several other observations consistent with this conclusion. Grain boundary sliding is implicit at conditions near the boundary between diffusion creep and dislocation creep because of the necessity for GBS in the diffusion creep regime (Raj and Ashby 1971). Constitutive laws for dislocation accommodated GBS predict $\dot{\varepsilon} \propto \sigma^n / d^p$, where $n_{gbs} \approx 2$ to 3 and $p_{gbs} \approx 2$ to 1 (e.g., Langdon 1994). Observations for olivine indicate that $n_{gbs} \approx 3.5$ and $p_{gbs} \approx 2$ (Kohlstedt and Wang

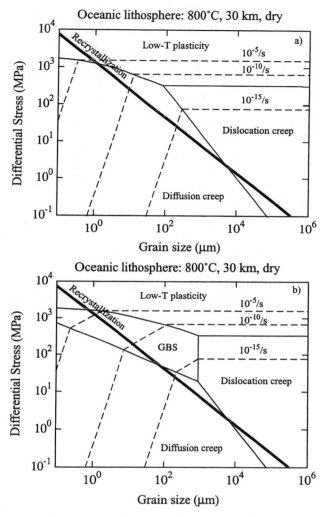

Figure 6. Deformation mechanism maps for the oceanic lithosphere. (a) The extrapolation of the recrystallized grain size piezometer falls within the diffusion creep regime at grain sizes less than ~5 mm. (b) By including the dislocation-accommodated grain boundary sliding (GBS) field, the recrystallized grain size relationship remains in a deformation regime where recrystallization is possible.

2001; Hirth and Kohlstedt 2003; Wang et al., manuscript in preparation). The deformation mechanism map illustrated in Figure 6b includes a field for GBS based on the flow law constrained by Hirth and Kohlstedt (2002). An interesting aspect of this plot is that the recrystallized grain-size piezometer remains in deformation fields dominated by dislocation processes over a much larger range in grain size.

Under conditions where dislocation climb is difficult, grain boundary migration recrystallization may also promote weakening of fine-grained shear zones relative to coarse-grained country rock (e.g., Tullis and Yund 1985; Tullis et al. 1990). During deformation by

Figure 7. Optical micrograph of extremely fine-grained peridotite mylonite from the oceanic lithosphere. This micrograph is from one of the samples from the Shaka Fracture Zone studied by Jaroslow et al. (1995).

this process, termed Regime 1 dislocation creep by Hirth and Tullis (1992), strain-free recrystallized grains are weaker than work-hardened original grains. Microstructural observations on experiments conducted on both feldspar (Tullis and Yund 1987) and quartz aggregates (Hirth and Tullis 1994) indicate that the transition from semi-brittle flow to dislocation creep with increasing temperature occurs due to enhancement of grain boundary migration.

High-temperature creep and the viscosity of the mantle

A deformation mechanism map for a depth of 30 km beneath a mid-ocean ridge is illustrated in Figure 8. The temperature at this depth is somewhat lower than that predicted for a potential temperature adiabat owing to energy expended in the MORB melting process (e.g., McKenzie and Bickle 1988; Phipps Morgan 1997). The plot in Figure 8 is calculated for dry conditions, because the melting process results in the removal of water from the residue (e.g., Hirth and Kohlstedt 1996; Karato and Jung 1998). The gray box in Figure 8 illustrates the predicted range in asthenospheric stresses (e.g., Hager and O'Connell 1981) and an estimate for the grain size based on observations of the grain size in xenoliths (e.g., Mercier 1980) and ophiolites (e.g., Dijkstra et al. 2002). The temperature of interest is similar to that used in the lab to study high-temperature creep processes. However, the stresses are two to three orders of magnitude smaller. Therefore, the primary limitation in accuracy comes from uncertainty in the stress exponent. For example, with the uncertainty in $n = 3.5 \pm 0.3$ shown in Figure 3b, there is a factor of ± 10 uncertainty in strain rate at a differential stress of 0.1 MPa. Because $\dot{\varepsilon} \propto \sigma^n$, this results in approximately a factor of ± 2 uncertainty in stress for a constant strain rate (i.e., $10^{(1/3.5)}$, with $n = 3.5$). At face value, the conditions outlined by the gray box in Figure 8 fall near the extrapolated boundary of the dislocation creep and diffusion creep flow laws, rather than on the extrapolation of the

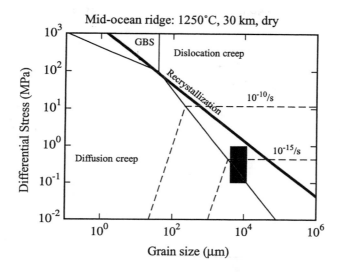

Figure 8. Deformation mechanism map for mantle beneath a mid-ocean ridge. The recrystallized grain size relationship is completely in the dislocation creep field, or dislocation accommodated GBS field, at all grain sizes. The gray box shows grain size variation in mantle xenoliths and the mantle section of ophiolites (4–12 mm) and a range of stress range estimated for convection (0.1–1 MPa). The shaded box lies almost completely in the dislocation creep field, consistent with the observation of seismic anisotropy in the mantle beneath oceanic ridges.

recrystallized grain-size piezometer, consistent with predictions based on the field boundary hypothesis (de Bresser et al. 2001). However, considering the uncertainties in extrapolation of both the creep data and the piezometer (e.g., van der Wal et al. 1993), a grain size controlled solely by the piezometer relationship cannot be ruled out. Finally, note that the GBS regime is restricted to high stresses and small grain sizes, i.e., the laboratory conditions where it is observed.

As illustrated in Figure 9, the deformation mechanism maps can also be plotted on axes of viscosity ($\eta = \sigma/\dot{\varepsilon}$) vs. grain size. In this parameter space, the diffusion creep strain rate contours all collapse onto a single line, because viscosity in this regime is independent of stress. The slope of the line illustrates the grain-size dependence of viscosity in the diffusion creep regime ($\eta \propto d^p$). The diffusion creep line also defines an upper limit for the viscosity as a function of grain size.

The plots in Figure 9 show how viscosity in the dislocation creep regime depends on stress and strain rate. In the dislocation creep regime, $\eta \propto \sigma^{1-n}$, where $n = 3.5$; thus $\eta \propto \sigma^{-2.5}$. For comparison, in the diffusion creep regime, where $n = 1$, viscosity is independent of stress. The top and the bottom of the gray boxes in Figure 9 show the viscosity in the dislocation creep regime at a constant stress of 0.1 and 1.0 MPa, respectively. The height of the box therefore spans a viscosity range of 2.5 orders of magnitude. Similarly, in terms of strain rate, $\eta \propto \dot{\varepsilon}^{(1-n)/n}$. In this case, a change in strain rate of five orders of magnitude results in 5(1−3.5)/3.5 = 3.57 orders of magnitude change in viscosity. These relationships demonstrate that viscosity is more sensitive to changes in stress than changes in strain rate in the dislocation creep regime.

The three deformation mechanism maps in Figure 9 illustrate the change in viscosity with depth beneath a mid-ocean ridge predicted by extrapolation of laboratory flow laws. In

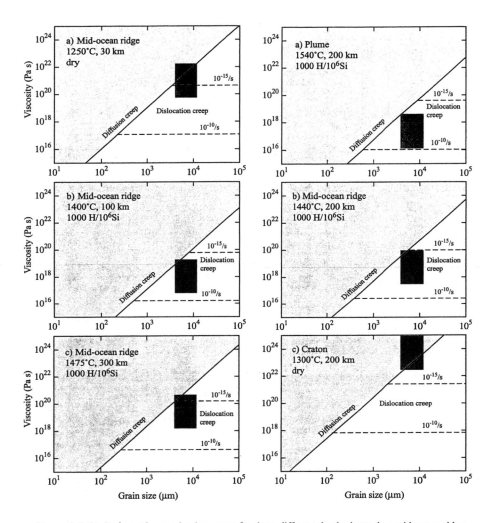

Figure 9 (left). Deformation mechanism maps for three different depths beneath a mid-ocean ridge. These maps are plotted in viscosity versus grain size space, thus strain rate contours collapse to a single relationship in the diffusion creep regime (i.e. viscosity is independent of stress). The light gray field shows a "prohibited" region. The dark gray box shows the viscosity predicted for dislocation creep at stresses of 0.1-1.0 MPa. The viscosity at 30 km (Fig. 9a) is greater than that at 100 km (Fig. 9b) for three reasons. First, water is removed from the olivine during melting at the oceanic ridge. Second, the temperature at 30 km is lower due to both the normal adiabatic gradient and the heat of fusion required to melt the mantle. Third, the effect of pressure on viscosity due to the V^* term in Equation (2) does not dominate over the E^* term at a depth of 100 km. By a depth of 300 km (Fig. 9c), the influence of V^* becomes dominant and the viscosity increases by more than an order of magnitude at a constant stress over the depth range of 100-300 km.

Figure 10 (right). Deformation maps showing variation in viscosity between different tectonic environments at a depth of 200 km. The viscosity is lowest for the high-temperature plume and highest for the continental craton, where temperature is somewhat lower and the mantle may be dry. Similar to Figure 9, the light gray field shows a "prohibited" region. The dark gray box shows the viscosity predicted for dislocation creep at stresses of 0.1-1.0 MPa.

the shallow regions beneath the ridge, the viscosity is relatively high due to the removal of water from the residue and the decrease in temperature associated with the melting process (Hirth and Kohlstedt 1996). At a depth of 100 km, the viscosity is considerably lower, due to both a significant increase in temperature and water content. A water content of 1000 H/10^6 Si is used based on the study of Hirth and Kohlstedt (1996). A water content of approximately 1000 H/10^6 Si is also consistent with independent estimates based on the analysis of the electrical conductivity of the oceanic mantle (Karato 1989; Lizarralde et al. 1995; Hirth et al. 2000). A viscosity in the range of 10^{18}-10^{19} Pa s at a depth of 100 km is consistent with independent estimates for mantle viscosity based on the analysis of the geoid (Craig and McKenzie 1986; Hager 1991). The increase in viscosity at a depth of 300 km illustrates the pressure-dependence of viscosity in the upper mantle, and demonstrates the dominance of the PV^*/RT term relative to the E^*/RT term in Equation (2) over the depth range of 100-300 km.

In addition to the uncertainty associated with extrapolation in differential stress, the accuracy of the deformation mechanism maps illustrated in Figures 9b, 9c and 10 is also influenced by uncertainty in V^*. A large range in V^*, from ~0 to 27×10^{-6} m^3/mol, has been reported in the literature based on a combination of experimental and theoretical analyses (e.g., Sammis et al. 1981; Borch and Green 1989; Karato and Rubie 1997; Karato and Jung 2002; Hirth and Kohlstedt 2003; Wang et al. (in preparation). In addition, there are very few data available to evaluate V^* for the diffusion creep regime (Kohlstedt et al. 2000). Because the majority of the highest resolution laboratory data are acquired at pressures of ~300 MPa, a small uncertainty in V^* results in considerable uncertainty in viscosity at a depth of 300 km. For example, at a constant temperature, extrapolation of data from a pressure of 300 MPa to 10 GPa (i.e., 300 km) using $V^* = 10 \times 10^{-6}$ m^3/mol or 15×10^{-6} m^3/mol results in a factor of 30 uncertainty in viscosity. The values for V^* used to construct the deformation mechanism maps in Figures 9 and 10 are based on the analyses presented in Hirth and Kohlstedt (2003).

The deformation mechanism maps shown in Figure 10 emphasize how the viscosity of the mantle can vary between different tectonic environments as a result of lateral variations in temperature and water content. Again, the positions of the gray boxes are defined by calculating the viscosity in the dislocation creep regime at a constant stress of 0.1 and 1 MPa. The lowest viscosities are predicted for the high-temperature plume environment. An upper bound for the viscosity of continental cratons is illustrated by assuming that the mantle at these depths is dry (see Pollack 1986; Hirth et al. 2000). The viscosity in these regions could be somewhat lower if there is some water present.

Observations of seismic anisotropy, from both shear wave splitting and analysis of surface waves, are used to constrain the dynamics and kinematics of the upper mantle flow. For example, the fast orientation of split shear waves is used to constrain the kinematics of plate motion (e.g., Wolfe and Solomon 1998; Fischer et al. 1998; Silver et al. 1999; Silver and Holt 2002) and in turn provide boundary conditions for dynamic models of plate tectonics (e.g., Becker and Boschi 2002). There are a wide variety of data that indicate that the fast P-wave and the fast S-wave polarization directions align parallel to the flow direction during deformation of peridotite (e.g., Nishimura and Forsyth 1989; Mainprice and Silver 1993). In addition, the observation of anisotropy provides an important constraint on the viscosity of the mantle. Specifically, since characteristics of seismic anisotropy are consistent with the development of an LPO in peridotite, and an LPO forms when deformation is accommodated by dislocation creep, the observation of anisotropy provides information on which deformation mechanisms are active in the mantle (e.g., Karato 1992). For example, the evaluation of deformation mechanism maps for the upper mantle shown in Figures 9 and 10 indicate that dislocation creep is a dominant deformation mechanism at depths less than

~200 km. This observation is consistent with the depth distribution of seismic anisotropy observed in oceanic environments (e.g., Gaherty and Jordan 1995).

ACKNOWLEDGMENTS

I am grateful to Jan Tullis, David Kohlstedt, Brian Evans, Javier Escartín, Henry Dick, Peter Kelemen, and Laurent Montesi for numerous discussions and collaborations that helped formulate the ideas presented in this chapter. In addition I am grateful to Shun Karato for his review of the manuscript and Laurent Montesi and Mike Braun for reading preliminary drafts of the manuscript. This work was partly supported by NSF grants OCE-9907244 and OCE-0099316.

REFERENCES

Anderson DL (1995) Lithosphere, asthenosphere, and perisphere. Rev Geophys 33:125-149
Aumento F, Loubat H (1971) The mid-Atlantic ridge near 45°N. Serpentinized ultramafic intrusions. Can J Earth Sci 8:631-663
Bai Q, Mackwell SJ, Kohlstedt DL (1991) High-temperature creep of olivine single crystals 1. Mechanical results for buffered samples. J Geophys Res 96:2441-2463
Becker TW, Boschi L (2002) A comparison of tomographic and geodynamic mantle models. Geochem Geophys Geosystems DOI 10.1029/2001GC000168
Bercovici D (1995) A source-sink model of the generation of plate tectonics from non-Newtonian mantle flow. J Geophys Res 100: 2013-2030
Bercovici D (1998) Generation of plate tectonics from lithosphere-mantle flow and void-volatile self-lubrication. Earth Planet Sci Lett 154:139-151
Bonatti E, Michael PJ (1989) Mantle peridotites from continental rifts to ocean basins to subduction zones. Earth Planet Sci Lett 91:297-311
Borch RS, Green HW II (1989) Deformation of peridotite at high pressure in a new molten salt cell: Comparison of traditional and homologous temperature treatments. Phys Earth Planet Inter 55:269-276
Brace WF, Kohlstedt DL (1980) Limits on lithospheric stress imposed by laboratory experiments. J Geophys Res 85:6248-6252
Braun J, Chery J, Poliakov ANB, Mainprice D, Vauchez A, Tommasi A, Daignieres M (1999) A simple parameterization of strain localization in the ductile regime due to grain size reduction: A case study for olivine. J Geophys Res 104:25167-25181
Brudy M, Zoback MD, Fuchs K, Rummel F, Baumgaertner J (1997) Estimation of the complete stress tensor to 8 km depth in the KTB scientific drill holes: implications for crustal strength. J Geophys Res 102: 18,453-18,475
Byerlee JD (1968) Brittle-ductile transition in rocks. J Geophys Res 73:4741-4650
Byerlee JD (1978) Friction of rocks. Pure Appl Geophys 116:615-626
Cannat M, Mével C, Maia M, Deplus C, Durand C, Gente P, Agrinier P, Belarouchi A, Dubuisson G, Humler E, Reynolds J (1995) Thin crust, ultramafic exposures, and rugged faulting patterns at the Mid-Atlantic Ridge (22°-24°N). Geology 23:49-52
Carter NL, Kirby SH, Transient creep and semi-brittle behavior of crystalline rocks. Pure Appl Geophys 116: 807-839
Chen WP, Molnar P (1983) Focal depths of intracontinental and intraplate earthquakes and their implications for the thermal structure and mechanical properties of the lithosphere. J Geophys Res 88:4183-4214
Chopra PN, Paterson MS (1984) The role of water in the deformation of dunite. J Geophys Res 89:7861-7876
Cooper RF, Kohlstedt DL (1984) Solution-precipitation enhanced diffusional creep of partially molten olivine basalt aggregates during hot-pressing. Tectonophysics 107:207-233
Coulton A, Harper GD, O'Hanley DS (1995) Oceanic versus emplacement age serpentinization in the Josephine ophiolite: Implications for the nature of the Moho at intermediate and slow-spreading ridges. J Geophys Res 100:22245-22260
Craig CH, McKenzie D (1986) The existence of a thin low-viscosity layer beneath the lithosphere. Earth Planet Sci Lett 78:420-426
de Bresser JHP, ter Heege JH, Spiers CJ (2001) Grain size reduction by dynamic recrystallization: can it result in major rheological weakening? Intl J Earth Sci (Geologische Rundschau) 90:28-45
Dijkstra AH, Drury MR, Frijhoff RM (2002) Microstructures and lattice fabrics in the Hilti mantle section (Oman Ophiolite): Evidence for shear localization and melt weakening in the crust-,mantle transition zone? J Geophys Res (in press)

Dick HJB (1989) Abyssal peridotites, very slow spreading ridges and ocean ridge magmatism. *In* Magmatism in the Ocean Basins. Saunders AD, Norry MJ (eds) Geol Soc Spec Publ, p 71-105

Dresen G, Evans B, Olgaard DL (1998) Effect of quartz inclusions of plastic flow in marble. Geophys Res Lett 25:1245-1248

Drury MR, FitzGerald JD (1998) Mantle rheology: insights from laboratory studies of deformation and phase transition. *In*: The Earth's Mantle: Composition, Structure and Evolution. Jackson I (ed) Cambridge University Press, Cambridge, UK, p 503-559

Durham WB, Goetze C (1977) Plastic flow of oriented single crystals of olivine, 1, Mechanical data. J Geophys Res 82:5737-5753

Engeln JF, Wiens DA, Stein S (1986) Mechanisms and depths of Atlantic transform earthquakes. J Geophys Res 91:548-577

Escartín J, Hirth G, Evans B (1997a) Non-dilatant brittle deformation of serpentinites: Implications for Mohr-Coulomb theory and the strength of faults. J Geophys Res 102:2897-2913

Escartín J, Hirth G, Evans B (1997b) Effects of serpentinization on lithospheric strength and the style of normal faulting at slow spreading ridges. Earth Planet Sci Lett 151:181-189

Escartín J, Hirth G, Evans B (2001) Strength of slightly serpentinized peridotites: Implications for the tectonics of oceanic lithosphere. Geology 29:1023-1026

Evans B, Goetze C (1979) The temperature variation of hardness of olivine and its implications for polycrystalline yield stress. J Geophys Res 84:5505-5524

Evans B, Fredrich J, Wong T-f (1990) The brittle-ductile transition in rocks: recent experimental and theoretical progress. *In* The Brittle-Ductile Transition in Rocks. Duba AG, Durham WB, Handin JW, Wang HF (eds) Geophys Monogr Ser 56:1-20

Fischer K, Fouch MJ, Wiens DA, Boettcher MS (1998) Anisotropy and flow in Pacific subduction zone backarcs. Pure Appl Geophys 151:463-475

Fredrich JT, Evans B, Wong T-f (1989) Micromechanics of the brittle to plastic transition in Carrara marble. J Geophys Res 94:4129-4145

Frost, HJ, Ashby MF (1982) Deformation Mechanism Maps. Pergamon Press

Gaherty JB, Jordan TH (1995) Lehmann discontinuity as the base of an anisotropic layer beneath continents. Science 268:1468-1471

Goetze C (1978) The mechanisms of creep in olivine. Philos Trans R Soc London A 288:99-119

Hager BH, O'Connell RJ (1981) A simple global model of plate dynamics and mantle convection. J Geophys Res 86:4843-4867

Hager BH (1991) Mantle viscosity: A comparison of models from postglacial rebound and from the geoid, plate driving forces, and advected heat flux. *In* Glacial Isostasy, Sea-Level and Mantle Rheology. Sabadini R (ed) Kluwer Academic Publishers, Dordrecht, The Netherlands, p 493-513

Hall CE, Parmentier EM (2001) The influence of grain-size evolution on convective instability. EOS Trans, Am Geophys Union 82:F1137

Handy MR (1994) Flow laws for rocks containing two non-linear viscous phases: A phenomenological approach. J Struct Geol 16:287-301

Hirth JP, Lothe J (1982) Theory of Dislocations, 2nd Edition. Wiley, New York

Hirth G, Tullis J (1992) Dislocation creep regimes in quartz aggregates. J Struct Geol 14:145-159

Hirth G, Tullis J (1994) The brittle-plastic transition in experimentally deformed quartz aggregates. J Geophys Res 99:11,731-11,748

Hirth G, Kohlstedt DL (1995a) Experimental constraints on the dynamics of the partially molten upper mantle: Deformation in the diffusion creep regime. J Geophys Res 100:1981-2001

Hirth G, Kohlstedt DL (1995b) Experimental constraints on the dynamics of the partially molten upper mantle 2. Deformation in the dislocation creep regime. J Geophys Res 100:15,441-15,449

Hirth G, Kohlstedt DL, Water in the oceanic upper mantle: implications for rheology, melt extraction and the evolution of the lithosphere. Earth Planet Sci Lett 144:93-108

Hirth G, Evans RL, Chave AD (2000) Comparison of continental and oceanic mantle electrical conductivity: Is the Archean lithosphere dry? Geochem Geophys Geosystems (G^3) Paper 2000CG000048

Hirth G, Kohlstedt DL (2003) The rheology of the upper mantle and the mantle wedge: a view from the experimentalists. *In* The Subduction Factory. Eiler J (ed) Geophys Monogr Ser, Am Geophys Union, Washington, DC (in press)

Jaroslow GE, Hirth G, Dick HJB (1996) Abyssal peridotite mylonites: implications for grain-size sensitive flow and strain localization in the oceanic lithosphere. Tectonophysics 256:17-37

Jin D, Karato SI, Obata M (1998) Mechanisms of shear localization in the continental lithosphere: inference from the deformation microstructures from the Ivrea zone, northwest Italy. J Struct Geol 20:195-209

Kameyama M, Yuen DA, Fujimoto H (1997) The interaction of viscous heating with grain-size dependent rheology in the formation of localized slip zones. Geophys Res Lett 24:2523-2526

Kameyama M, Yuen DA, Karato SI (1999) Thermal-mechanical effects of low-temperature plasticity (the Peierls mechanism) on the deformation of a viscoelastic shear zone. Earth Planet Sci Lett 168:159-172

Karato SI, Toriumi M, Fujii T (1980) Dynamic recrystallization of olivine single crystals during high-temperature creep. Geophys Res Lett 7:649-652

Karato S, Paterson MS, FitzGerald JD (1986) Rheology of synthetic olivine aggregates: Influence of grain size and water. J Geophys Res 91:8151-8176

Karato S (1990) The role of hydrogen in the electrical conductivity of the upper mantle. Nature 347:272-273

Karato S (1992) On the Lehmann Discontinuity. Geophys Res Lett 19:2255-2258

Karato S, Wu P (1993) Rheology of the upper mantle: Aa synthesis. Science 260:771-778

Karato S, Rubie DC (1997) Toward an experimental study of deep mantle rheology: a new multianvil sample assembly for deformation studies under high pressures and temperatures. J Geophys Res 102:20,111-20

Karato S-i Jung H (1998) Water, partial melting and the origin of the seismic low velocity and high attenuation zone in the upper mantle. Earth Planet Sci Lett 157:193-207

Karato S-I, Jung H (2002) Effects of pressure on high-temperature dislocation creep in olivine. Philos Mag A (in press)

Karson JA, Lawrence RM (1997) Tectonic setting of serpentinite exposures on the western median valley wall of the MARK area in the vicinity of Site 920. In Proc. ODP, Sci Results. Karson JA, Cannat M, Miller DJ, Elthon D (eds) 153:5-21

Kirby SH (1985) Rock mechanics observations pertinent to the rheology of the continental lithosphere and the localization of strain along shear zones. Tectonophysics 119:1-27

Kohlstedt DL, Evans B, Mackwell SJ (1995) Strength of the lithosphere: constraints imposed by laboratory experiments. J Geophys Res 100:17,587-17,602

Kohlstedt DL, Zimmerman ME (1996) Rheology of partially molten mantle rocks. Ann Rev Earth Planet Sci 24:41-62

Kohlstedt DL, Keppler H, Rubie DC (1996) Solubility of water in the alpha, beta and gamma phases of $(Mg,Fe)_2SiO_4$. Contrib Mineral Petrol 123:345-357

Kohlstedt DL, Bai Q, Wang ZC, Mei S (2000) Rheology of partially molten rocks. In Physics and Chemistry of Partially Molten Rocks. Bagdassarov N, Laporte D, Thompson AB (eds) Kluwer Academic Publishers, p 3-28

Kohlstedt DL, Wang Z (2001) Grain-boundary sliding accommodated dislocation creep in dunite. EOS Trans, Am Geophys Union 82:F1137

Langdon TG (1994) A unified approach to grain boundary sliding in creep and superplasticity Acta Metall 42:2437-2443

Liu M, Yund RA (1995) The elastic strain energy associated with the olivine-spinel transformation and its implications. Phys Earth Planet Inter 89:177-197

Lizarralde D, Chave AD, Hirth G, Schultz A (1995) Northeastern Pacific mantle conductivity profile from long-period magnetotelluric sounding using Hawaii to California submarine cable data. J Geophys Res 100:17,837-17,854

Mackwell SJ, Kohlstedt DL, Paterson MS (1985) The role of water in the deformation of olivine single crystals. J Geophys Res 90:11319-11333

Mainprice D, Silver PG (1993) Interpretation of SKS-waves using samples from the subcontinental lithosphere. Phys Earth Planet Inter 78:257-280

McKenzie D, Bickle MJ (1988) The volume and composition of melt generated by extension of the lithosphere. J Petrology 29: 625-679

Mei S, Kohlstedt DL (2000a) Influence of water on deformation of olivine aggregates 1. Diffusion creep regime. J Geophys Res 105:21,457-21,469

Mei S, Kohlstedt DL (2000b) Influence of water on plastic deformation of olivine aggregates 2. Dislocation creep regime. J Geophys Res 105:21471-21481

Mercier JC (1980) Magnitude of the continental lithospheric stresses inferred from rheomorphic petrology. J Geophys Res 85:6293-6303

Montesi LG, Hirth G (2001) Transient behavior of a shear zone deforming by combined dislocation and diffusion creep. Eos Trans, Am Geophys Union 82:F1145

Moore DE, Lockner DA, Summers R, Ma S, Byerlee JD (1996) Strength of chrysotile-serpentinite gouge under hydrothermal conditions: Can it explain a weak San Andreas fault? Geology 24:1041-1044

Moore DE, Lockner DA, Shengli M, Summers R, Byerlee JD (1997) Strength of serpentinite gouges at elevated temperatures. J Geophys Res 102:14,787-14,801

Murrell SAF, Ismail IAH (1976) The effect of decomposition of hydrous minerals on the mechanical properties of rocks at high pressures and temperatures. Tectonophysics 31:207-258

Nicolas A (1989) Structures of Ophiolites and Dynamics of Oceanic Lithosphere. Kluwer Academic Publishers, Dordrecht, The Netherlands

Nishimura CE, Forsyth DW (1989) The anisotropic structure of the upper mantle in the Pacific. Geophys J 96:203-229
O'Hanley DS (1996) Serpentinites. Records of Tectonic and Petrological History. Oxford University Press, New York
Paterson MS (1978) Experimental Rock Deformation-the Brittle Field. Springer-Verlag, New York
Phipps-Morgan J (1997) The generation of a compositional lithosphere by mid-ocean ridge melting and its effect on subsequent off-axis hotspot upwelling and melting. Earth Planet Sci Lett 146:213-232
Poirier J-P (1985) Creep of Crystals. Cambridge University Press
Pollack HN (1986) Cratonization and thermal evolution of the mantle. Earth Planet Sci Lett 80:175-182
Post RL (1977) High-temperature creep of Mt. Burnett dunite. Tectonophysics 42:75-110
Ranalli G (1995) Rheology of the Earth. Chapman & Hall, London
Raleigh CB, Paterson MS (1965) Experimental deformation of serpentinite and its tectonic implications. J Geophys Res 70:3965-3985
Raj R, Ashby MF (1971) On grain boundary sliding and diffusional creep. Metall Trans 2:1113-1127
Reinen LA, Weeks JD, Tullis TE (1991) The frictional behavior of serpentinite implications for aseismic creep on shallow crustal faults. Geophys Res Lett 18:1921-1924
Reinen LA, Weeks JD, Tullis TE (1994) The frictional behavior of lizardite and antigorite serpentinites: experiments, constitutive models, and implications for natural faults. Pure Appl Geophys 143:318-358
Ross JV, Lewis PD (1989) Brittle-ductile transition: semi-brittle behavior. Tectonophysics 167:75-79
Rutter EH (1986) On the nomenclature of mode of failure transitions in rocks. Tectonophysics 122:381-387
Rutter EH, Brodie KH (1988) The role of tectonic grain-size reduction in the rheological stratification of the lithosphere. Geol. Rundschau 77:295-308
Sammis CG, Smith JC, Shubert G (1981) A critical assessment of estimation methods for activation volume. J Geophys Res 86:10,707-10,718
Scholz CH (1988) The brittle-plastic transition and the depth of seismic faulting. Geol Rund 77:319-328
Scholz CH (1990) The Mechanics of Earthquakes and Faulting. Cambridge University Press, New York
Schwenn MB, Goetze C (1978) Creep of olivine during hot-pressing. Tectonophysics 48:41-60
Shimada M, Cho A, Yukutake H (1983) Fracture strength of dry silicate rocks at high confining pressures and activity of acoustic emission. Tectonophysics 96:159-172
Silver P, Mainprice D, Ben Ismail W, Tommasi A, Barruol G (1999) Mantle structural geology from seismic anisotropy. In Mantle Petrology: Field Observations and High-Pressure Experimentations. A Tribute to Francis R. (Joe) Boyd. Fei Y, Bertka CM, Mysen BO (eds) The Geochemical Society, p 79-103
Silver PG, Holt WE (2002) The mantle flow field beneath western North America. Science 295:1054-1058
Stesky RM, Brace WF, Riley DK, Robin PYF (1974) Friction in faulted rock at high temperature and pressure. Tectonophysics 23:177-203
Strehlau J (1986) A discussion of the depth extent of rupture in large continental earthquakes. In Earthquake Source Mechanics. Das S, Scholz CH (eds) Geophys Monogr Ser 37:131-146
Tackley PJ (2000) Self-consistent generation of tectonic plates in time-dependent, three-dimensional mantle convection simulation: 1. Pseudoplastic yielding. Geochem Geophys Geosystems Paper 2000GC000036
Todheide K (1972) Water at high temperature and pressure. In Water: A Comprehensive Treatise. Franks F (ed) Plenum Press, p 463-514
Tommasi A, Mainprice D, Canova G, Chastel Y (2000) Viscoplastic self-consistent and equilibrium-based modeling of olivine preferred orientations: Implications for the upper mantle seismic anisotropy. J Geophys Res 105:7893-7908
Tullis J, Yund RA (1977) Experimental deformation of dry Westerly granite. J Geophys Res 82:5705-5717
Tullis J, Yund RA (1985) Dynamic recrystallization of feldspar: A mechanism for ductile shear zone formation. Geology 13:238-241
Tullis J, Yund RA (1987) Transition from cataclastic flow to dislocation creep of feldspar: mechanisms and microstructures. Geology 15:606-609
Tullis J, Dell Angelo LN, Yund RA (1990) Ductile shear zones from brittle precursors in feldspathic rocks: the role of dynamic recrystallization. In The Brittle-Ductile Transition in Rocks. Duba AG, Durham WB, Handin JW, Wang HF (eds) Geophys Monogr Ser 56:67-81
Tullis J, Yund RA (1992) The Brittle-ductile transition in feldspar aggregates: an experimental study. In Fault Mechanics and Transport Properties of Rocks. Evans B (ed) Academic Press, p 89-117
Tullis TE, Horowitz F, Tullis J (1991) Flow laws of polyphase aggregates from end member flow laws. J Geophys Res 96:8081-8096
Van der Wal D, Chopra PN, Drury M, FitzGerald JD (1993) Relationships between dynamically recrystallized grain size and deformation conditions in experimentally deformed olivine rocks. Geophys Res Lett 20:1479-1482
Vissers RLM, Drury MR, Hoogerduijn Strating EH, Spiers CJ, Van der Wal D (1995) Mantle shear zones and their effect on lithosphere strength during continental breakup. Tectonophysics 249:155-171

Wenk H-R, Bennett K, Canova GR, Molinari A (1991) Modelling plastic deformation of peridotite with the self-consistent theory. J Geophys Res 96:8337-8349

Wiens D, Stein S (1983) Age dependence of oceanic intraplate seismicity and implications for lithospheric evolution. J Geophys Res 88:6455-6468

Wolfe CJ, Solomon SC (1998) Shear-wave splitting and implications for mantle flow beneath the MELT region of the East Pacific Rise. Science 280:1230-1232

Zhang S, Karato SI, Fitz Gerald J, Faul UH, Zhou Y (2000) Simple shear deformation of olivine aggregates. Tectonophysics 316:133-152

Zhong S, Gurnis M (1996) Interaction of weak faults and non-Newtonian rheology produces plate tectonics in a 3-D model of mantle flow. Nature 383:245-247

5   Partial Melting and Deformation

David L. Kohlstedt

Department of Geology and Geophysics
University of Minnesota
Minneapolis, Minnesota 55455

INTRODUCTION

The physical properties of partially molten rocks are directly coupled to the grain-scale as well as the broader scale distribution of the melt phase. At the grain scale, if melt forms isolated pockets in a silicate matrix, its influence on plastic flow is generally relatively minor. In contrast, if the melt wets all of the grain-grain interfaces, it may dramatically lower the viscosity of the rock. At the broader scale, deformation will localize to form shear zones in melt-rich regions that commonly develop in partially molten rocks. In turn, the melt distribution evolves during plastic deformation of a partially molten rock. Melt located in triple junctions takes on a pronounced preferred orientation. In addition, melt often segregates during plastic deformation to form melt-enriched bands separated by melt-depleted regions.

This chapter develops two interrelated themes with emphasis on the relationship between melt distribution (structure) and the plastic flow (property) of partially molten rocks. The first section concentrates on thermodynamic constraints and experimental observations on the distribution of melt in a non-deforming rock, that is, in a partially molten rock exposed simply to a hydrostatic state of stress. The second section then builds on these boundary conditions to introduce theoretically predicted and experimentally determined flow laws describing plastic deformation of partially molten rocks. Finally, the third section examines the influence of deformation (i.e., a non-hydrostatic state of stress) on melt distribution and the associated implications for the rheological properties of partially molten crustal and mantle rocks.

MELT DISTRIBUTIONS IN NON-DEFORMING ROCKS

Under a hydrostatic state of stress, the melt distribution in a partially molten rock is governed by the relative values of the solid-melt and solid-solid interfacial energies, γ_{sm} and γ_{ss}. (NB: Although the term melt is used throughout this paper, the designation fluid or liquid could equally well be substituted.) The melt distribution is thus often characterized by the dihedral angle, θ, which forms between the melt and two adjoining grains, through the relationship (Smith 1948,1964; Bargen and Waff 1986; Watson et al. 1990; Kohlstedt 1992)

$$\cos\left(\frac{\theta}{2}\right) = \frac{\gamma_{ss}}{2\gamma_{sm}}. \tag{1}$$

Based on Equation (1), the melt distribution can be divided into three categories: (i) For large dihedral angles, $\theta > 60°$ (i.e., $\gamma_{ss} < \sqrt{3}\gamma_{sm}$) melt will be isolated in pockets in grain boundaries, triple junctions and four-grain junctions in order to minimize the total amount of solid-melt interfacial area. This situation is illustrated in Figure 1. (ii) For $0 < \theta < 60°$ (i.e., $\sqrt{3}\gamma_{sm} < \gamma_{ss} \leq 2\gamma_{sm}$), melt is interconnected along triple junctions and through four-grain junctions, as shown in Figure 2. Melt along grain boundaries will be trapped in isolated pockets. (iii) For $\theta = 0°$ (i.e., $\gamma_{ss} \leq 2\gamma_{sm}$), melt wets not only the triple junctions and four-grain junctions but also the grain boundaries. It should be noted that Equation (1) applies strictly only to melt-mineral systems for which γ_{ss} and γ_{sm} are single-valued.

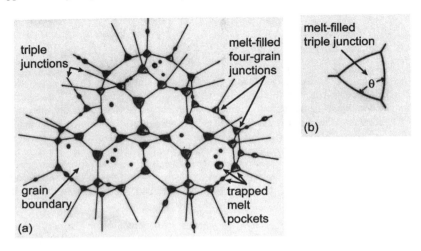

Figure 1. Melt distribution for a melt-solid system with $\theta > 60°$. (a) 3-D view illustrates the presence of melt pockets trapped in grain boundaries, triple junctions and four-grain junctions. (b) Cross-sectional view of melt pocket trapped in triple junction. Modified from Lee et al. (1991).

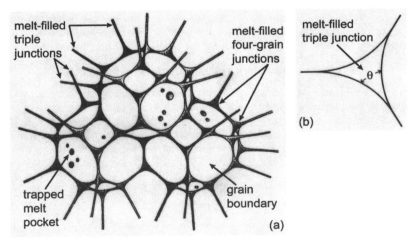

Figure 2. Melt distribution for a melt-solid system with $0 < \theta < 60°$. (a) 3-D view illustrates the presence of an interconnected melt network along triple junctions and through four-grain junctions pockets with melt pockets trapped in grain boundaries. (b) Cross-sectional view of melt-filled triple junction. Modified from Lee et al. (1991).

For the olivine + basalt system, the average dihedral angle is θ ≈ 35°. Careful examination of the scanning electron microscopy (SEM) images in Figure 3 demonstrates that every triple junction contains at least a small amount of melt, demonstrating the an interconnected network of melt exists along these triple junctions and through the four-grain junctions. A closer view of melt-filled triple junctions is presented in the transmission electron microscopy (TEM) images in Figures 4a and b.

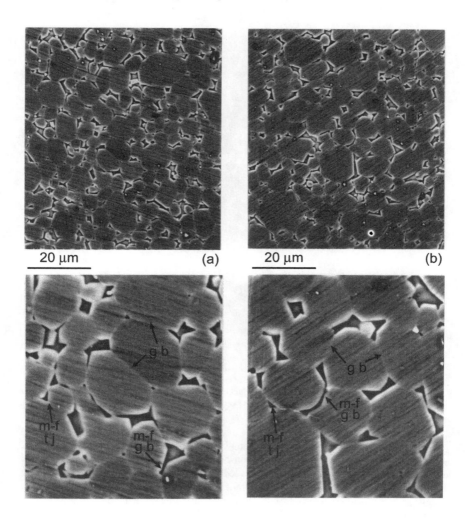

Figure 3. Backscattered electron images of partially molten sample of olivine + 7 vol % basalt. (a) and (b) Lower magnification images. (c) and (d) Higher magnification images of a region near the center of (a) and (b), respectively. In the higher magnification images, arrow mark grain boundaries (gb), melt-filled triple junctions (m-f tj) and melt-filled grain boundaries (m-f gb). Modified from Kohlstedt (1992).

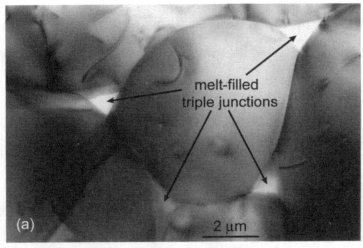

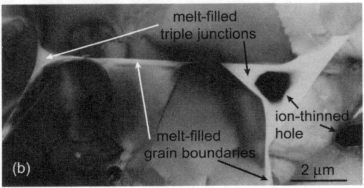

Figure 4. Dark-field transmission electron microcopy images illustrating the range of melt distribution in samples of olivine + basalt. (a) Melt fills the triple junctions formed by olivine grains. (b) Two melt-filled triple junctions are connected by a melt-filled grain boundary. Note the angular shape of the grains in contract with melt.

In real rocks, neither γ_{ss} nor γ_{sm} is single-valued. The grain boundary energy γ_{ss} increases from near zero for low-angle grain boundaries to ~1 J/m² for high-angle grain boundaries in olivine (Cooper and Kohlstedt 1982). Furthermore, γ_{sm} is highly anisotropic for the olivine-basalt and most other mineral-fluid system; melt tends to wet low-index crystallographic faces but not high-index planes. Consequently, even though the average dihedral angle is ≫0°, basaltic melt wets a portion of the grain boundaries in dunite. Examples are highlighted in Figure 3c and d in the backscattered electron SEM images of samples of olivine + basalt. At higher magnification in the TEM image in Figure 4b, two melt-filled triple junctions are clearly connected by a melt-filled (melt-wetted) grain boundary. The anisotropy in γ_{sm} is emphasized by the angular shape of many of the melt-solid interfaces. Melt does not penetrate low-angle grain boundaries, that is, the dihedral angle is relatively large. One example is shown in the TEM micrograph in Figure 5 in which strain contrast from dislocations along the low-angle boundary indicates the absence of melt.

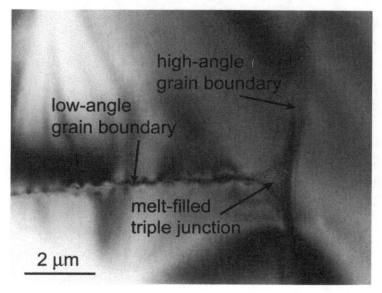

Figure 5. Bright-field image of melt-filled triple junction formed at the intersection of a high-angle grain boundary grains with a low-angle grain boundary in an olivine + basalt sample. Note the periodic strain contrast along the low-angle boundary indicative of a well-organized array of dislocations in an olivine grain.

Debate exists over the possible presence of a very thin film (~1 nm thick) of melt along all of the grain boundaries in partially molten rocks. High-resolution imaging and chemical analyses obtained with an analytical TEM have led different researchers to conflicting conclusions. Drury and Fitz Gerald (1996), Wirth (1996) and de Kloe et al. (2000) have interpreted high-resolution observations in terms of the presence of very thin melt films, while Kohlstedt and co-workers have reached the opposite conclusion also based on high-resolution TEM images and chemical examination with high spatial resolution (Vaughan et al. 1984; Kohlstedt 1990; Hiraga et al. 2002). In Figure 6, the high-resolution lattice-fringe image of the interface between two grains in an aggregate of olivine + basalt indicates that the grains are in direct contact with one another without the presence of a thin melt film. Somewhat surprisingly, however, high-resolution chemical analyses of such boundaries reveal a signature quite different from that of olivine. The X-ray intensity profiles in Figure 7 expose the presence of small amounts Ca, Al and Ti at a grain boundary between two olivine grains even though high-resolution lattice fringe images indicate that no melt film is present (Hiraga et al. 2002). The chemical composition at the grain boundary, however, is quite different from that of basalt. Hence, based on the high-resolution TEM

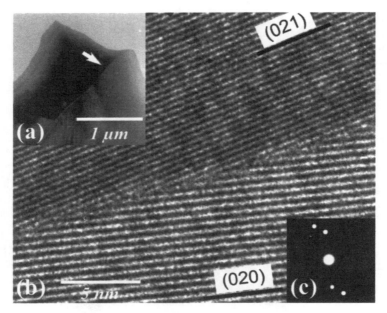

Figure 6. (a) Bright-field TEM images of grain boundary in the partially molten lherzolite sample containing 9 vol % melt. (b) Lattice fringe image of two grains meeting at an apparently melt-free grain boundary. (c) Diffraction pattern from grain boundary region. Modified from Hiraga et al. (2002).

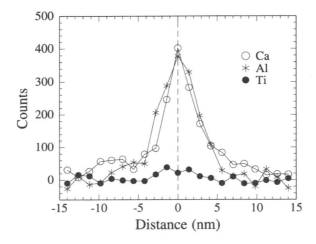

Figure 7. X-ray intensity profiles across an olivine-olivine grain boundary in a partially molten lherzolite aggregate obtained by STEM/EDX analyses. (a) Major and minor elements. (b) Ca, Al and Ti on an expanded scale. Modified from Hiraga et al. (2002).

images and chemical analyses, Hiraga et al. (2002) concluded that elements soluble in olivine only in very low concentrations have segregated to the grain boundaries, markedly altering the composition of the grain boundaries by not forming a thin amorphous layer.

DEFORMATION OF PARTIALLY MOLTEN ROCKS

Over the past twenty years, a number of researchers have investigated the high-temperature plastic (viscous) deformation of partially molten mantle rocks (Cooper and Kohlstedt 1984, 1986; Bussod and Christie 1991; Beeman and Kohlstedt 1993; Jin et al. 1994; Kohlstedt and Chopra 1994; Hirth and Kohlstedt 1995a,b; Kohlstedt and Zimmerman 1996; Zimmerman 1999; Mei and Kohlstedt 2000a,b). In a creep experiment on a partially molten rock, the strain rate $\dot{\varepsilon}$ is measured as a function of differential stress σ, grain size d, melt fraction ϕ, temperature T, and pressure P such that the flow law for steady-state deformation will have the generalized form

$$\dot{\varepsilon} = \dot{\varepsilon}(\sigma, d, \phi, T, P) . \tag{2}$$

The dependence of $\dot{\varepsilon}$ on σ and d is commonly written as a power law, while the dependence of $\dot{\varepsilon}$ on T and P is usually expressed as an exponential relationship common for thermally activated processes. The power-law equation describing the creep of a solid + melt system deformed under anhydrous (dry) conditions then has the form

$$\dot{\varepsilon}^{dry}(\sigma, d, \phi, T, P) = A_{dry} \frac{\sigma^{n_{dry}}}{d^{p_{dry}}} \mathscr{F}_{dry}(\phi) \exp\left(-\frac{E_{dry} + PV_{dry}}{RT}\right), \tag{3}$$

where A_{dry} is a material-dependent parameter, n_{dry} and p_{dry} are the stress and grain size exponents, and E_{dry} and V_{dry} are the activation energy and activation volume for creep. The function $\mathscr{F}_{dry}(\phi)$, which describes the dependence of strain rate on melt fraction, remains a subject of both experimental and theoretical exploration. A similar expression can be written to describe the creep behavior of a partially molten rock under hydrous (wet) conditions

$$\dot{\varepsilon}^{wet}(\sigma, d, \phi, C_{OH}, T, P) = A_{wet} \frac{\sigma^{n_{wet}}}{d^{p_{wet}}} \mathscr{F}_{wet}(\phi) \mathbb{C}(C_{OH}) \exp\left(-\frac{E_{wet} + PV_{wet}}{RT}\right), \tag{4}$$

where $\mathbb{C}$ is a function of water concentration C_{OH}. Recent experimental work on samples of olivine + basalt suggests that, at least to a first approximation, $\mathscr{F}_{dry}(\phi) = \mathscr{F}_{wet}(\phi) = \mathscr{F}(\phi)$, reflecting the fact that the melt distribution is similar under anhydrous and hydrous conditions (Riley and Kohlstedt 1990). Experiments also demonstrate that $\mathbb{C}(C_{OH}) \propto C_{OH}^r$ with $r \approx 1$ (Mei and Kohlstedt 2000a,b; Karato and Jung 2002).

Extrapolation of laboratory results to flow at depth in the Earth necessitates extension to significantly lower strain rates. Strain rates in laboratory experiments are typically on the order of 10^{-4} - 10^{-6} s^{-1}, while strain rates associated with plastic deformation in the Earth are closer to 10^{-14} s^{-1}. Consequently, it is necessary to carry out experiments at differential stresses larger than those appropriate for flow in the Earth (e.g., 10 MPa in the laboratory versus 0.1 MPa in the Earth) and/or temperatures that are higher than those characteristic of the Earth. If flow laws derived from laboratory experiments are to be extrapolated to geological conditions, the deformation mechanism is in the laboratory must be the same as those in the Earth. Therefore, experiments are generally carried out over as wide a range of $\dot{\varepsilon}$, σ, d, ϕ, T and P conditions as possible in order to study deformation in both the diffusional and the dislocation creep regime. If the results are expressed as viscosity, η,

$$\eta = \frac{\sigma}{\dot{\varepsilon}(\sigma)}, \tag{5}$$

then η will independent of stress in the diffusional creep regime (i.e., Newtonian with $\dot{\varepsilon} \propto \sigma^1$) but will increase with decreasing differential stress in the dislocation creep regime (i.e., non-Newtonian with $\dot{\varepsilon} \propto \sigma^n$ with $3 < n < 5$).

Laboratory experiments on partially molten olivine + basalt aggregates (Mei et al. 2002) suggest that the effect of water on strain rate can be separated from the effect of melt. Hence, Equations (3) and (4) can be combined as follows:

$$\dot{\varepsilon}^{dry/wet}(\phi) = \dot{\varepsilon}^{dry/wet}_{\phi=0} \mathcal{F}(\phi), \tag{6}$$

Again, this equation reflects the observation that the wetting behavior of basalt in a peridotite matrix is the same or very nearly the same under hydrous and anhydrous conditions, at least up to pressures of 0.3 GPa. Empirically, at the relatively small melt fractions used in our experiments, the melt-fraction dependence can be described quite well by (Kelemen et al. 1997; Mei et al. 2002; Zimmerman and Kohlstedt, submitted)

$$\mathcal{F}(\phi) = \exp(\alpha\phi), \tag{7}$$

where the value of α depends on both the creep mechanism and the melt-solid system under investigation.

Creep results for fine-grained (~15 μm) lherzolite samples deformed under anhydrous conditions are plotted in Figure 8 (Zimmerman and Kohlstedt, submitted). These samples composed of 61.8% olivine, 26.0% enstatite, 10.0% chrome-diopside and 2.2% spinel were hot-pressed from a powdered mantle xenolith from Damaping, China. In the diffusional creep regime (Fig. 8a), $\alpha = 21$, while in the dislocation creep regime (Fig. 8b), $\alpha = 30$. Somewhat larger values for α have been reported for the olivine + basalt system (Kelemen et al. 1997; Mei et al. 2002).

The exponential dependence of strain rate on melt fraction given by Equation (7) results in a greater sensitivity of creep rate to melt fraction than predicted theoretically. Cooper and Kohlstedt (1986) and Cooper et al. (1989) analyzed the contribution of melt to creep rate for the diffusion creep regime. They concluded that melt enhances creep rate in two distinct ways: (1) the presence of melt results in a local stress enhancement due to replacement of a fraction of the grain boundary contact, $\Delta d'/d'$, by melt (Fig. 2); (2) melt in triple junctions effectively shortens the path through which ions must travel in diffusing from grain boundaries under maximum compressive stress to those under smaller compressive (or tensile) stress. This analysis is based on an ideal melt distribution dictated by Equation (1) with a single-valued dihedral angle with $0 < \theta < 60°$, such that all of the melt is distributed in triple junctions and four-grain junctions but is absent along grain boundaries. The Cooper-Kohlstedt model yields

$$\dot{\varepsilon}(\phi) = \dot{\varepsilon}_{\phi=0}^{dry/wet} \left(\frac{1}{1-(\Delta d'/d')} \right)_{sc}^2 \left(\frac{1}{1-(\Delta d'/d')} \right)_{se}^2 = \dot{\varepsilon}_{\phi=0}^{dry/wet} \left(\frac{1}{1-\phi^{1/2}\mathscr{E}(\theta)} \right)^4, \quad (8)$$

where the subscripts *sc* and *se* indicate short-circuit transport and local stress enhancement, respectively, and

$$\mathscr{E}(\theta) = \frac{1.06 \sin(30° - (\theta/2))}{\left(\frac{1+\cos\theta}{\sqrt{3}} - \sin\theta - \frac{\pi}{90}(30° - (\theta/2)) \right)^{1/2}}. \quad (9)$$

For $\phi \geq 0.05$ Equation (8) underestimates the effect of melt fraction on strain rate (Hirth and Kohlstedt 1995a,b). This equation does not provide a good description of the creep of partially molten rocks in large part because of the difference between the observed melt

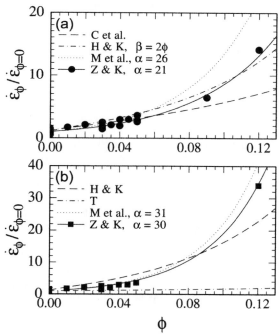

Figure 8. (a) Normalized strain rate versus melt fraction for samples of synthetic lherzolite deformed in the diffusional creep regime. C et al. = Cooper et al. (1989), Equation (8); H & K = Hirth and Kohlstedt (1995a), Equation (11); M et al. = Mei et al. (in press), Equation (7); Z & K = Zimmerman and Kohlstedt (submitted), Equation (7). (b) Normalized strain rate versus melt fraction for samples of synthetic lherzolite deformed in the dislocation creep regime. H & K = Hirth and Kohlstedt (1995b), Equation (14); T = Tharp (1983), Equation (13); M et al., Equation (7); Z & K, Equation (7). Modified from Xu et al. (2002).

distribution and that the ideal microstructure assumed in the Cooper-Kohlstedt derivation (Cooper and Kohlstedt 1986; Cooper et al. 1989). The anisotropy in solid-melt interfacial energy results in a melt distribution in which melt is not simply confined to triple junctions and four-grain junctions but also penetrates a significant fraction of the grain boundaries and separates neighboring grains, as illustrated in Figures 3 and 4. As a result, the partially molten samples have short-circuit diffusion paths and stress-enhancement geometries not considered in the derivation of Equation (8).

In their experimental investigation of diffusion creep of partially molten olivine + basalt aggregates, Hirth and Kohlstedt (1995a) accounted for this effect by quantifying the fraction of grain boundaries completely wetted by melt, β. Based on microstructural examination of samples with melt fraction between 0.02 and 0.12, they demonstrated that β increases linearly with melt fraction:

$$\beta \propto \phi . \tag{10}$$

To modify Equation (8) in order to account for the presence of wetted grain boundaries, Hirth and Kohlstedt (1995a) argued that the effective diffusion path must be shortened from $d' - \Delta d'$ to $(d' - \Delta d')(1 - \beta)$. Equation (8) thus becomes

$$\dot{\varepsilon}(\phi) = \dot{\varepsilon}^{dry/wet}_{\phi=0} \left(\frac{1}{(1 - \Delta d'/d')} \right)^2_{se} \left(\frac{1}{(1 - \Delta d'/d')(1 - \beta)} \right)^2_{sc} \tag{11}$$

$$= \dot{\varepsilon}^{dry/wet}_{\phi=0} \left(\frac{1}{1 - \phi^{1/2}\mathscr{E}(\theta)} \right)^2_{se} \left(\frac{1}{(1 - \phi^{1/2}\mathscr{E}(\theta))(1 - \beta)} \right)^2_{sc} .$$

It should be noted that, at small melt fractions, Equations (8) and (11) can be well-approximated by exponential functions, thus potentially explaining why the dependence of the strain on melt fraction can be approximated by Equation (7) (Mei et al., in press). Hirth and Kohlstedt (1995a) determined a value of $\beta = 4.5\phi$ for their aggregates of olivine + basalt deformed under anhydrous conditions, while Mei et al. (in press) determined a value of $\beta = 3\phi$ for similar samples deformed under hydrous conditions. Zimmerman and Kohlstedt (submitted) have shown that Equation (11) fits the experimental data for partially molten lherzolite samples with a value of $\beta = 2\phi$, as illustrated in Figure 8a. The smaller value of β reported for partially molten lherzolite is consistent with the observation that fewer grain-grain contacts are wetted in the lherzolite samples than in the dunite samples. This differences arises because melt does wet some of the olivine-olivine grain boundaries but seldom wets pyroxene-pyroxene grain boundaries or olivine-pyroxene phase boundaries.

Two models are generally considered in discussing the effect of a small amount of melt on the creep properties of partially molten rocks in the dislocation creep regime. Chen and Argon (1971) examined the stress enhancement produced by a weak second phase. Their analysis leads to the following expression for the dependence of strain rate on melt fraction:

$$\dot{\varepsilon}(\phi) = \dot{\varepsilon}^{dry/wet}_{\phi=0} \left(\frac{1}{1 - 2\phi} \right) \left(\frac{1}{1 - \phi} \right)^{n-1} . \tag{12}$$

Tharp (1983) explored the effect of a small amount of porosity and obtained

$$\dot{\varepsilon}(\phi) = \dot{\varepsilon}_{\phi=0}^{dry/wet} \frac{1}{\left((1-k\phi)^{2/3}\right)^n} \tag{13}$$

with $0.98 \leq k \leq 2.26$. Even for relatively low melt fractions, Equations (12) and (13) both significantly underestimate the influence of melt fraction on strain rate.

Hirth and Kohlstedt (1995b) exploited the stress-enhancement portion of Equation (8) in order to account for the discrepancy between these models and the experimental observations for dislocation creep of olivine + basalt aggregates. Their analysis leads to

$$\dot{\varepsilon}(\phi) = \dot{\varepsilon}_{\phi=0}^{dry/wet} \left(\frac{1}{1 - \phi^{1/2}\mathscr{E}(\theta)}\right)^{2n}_{se}. \tag{14}$$

This modified form of Equation (8) applied to dislocation creep provides a reasonable fit to the experimental data in Figure 8b. However, it should be noted that Equation (14) does overestimate the strain rate at low melt fractions and underestimates the strain rate at higher melt fractions. Some part of the weakening due to the presence of melt is likely results from melt-assisted relaxation of stresses at melt-filled triple junctions that develop as a result of grain boundary sliding.

MELT DISTRIBUTIONS IN DEFORMING ROCKS

One of the exciting advances in the past few years in the study of deformation of partially molten rocks is the discovery of the pronounced redistribution of melt that occurs during deformation. This redistribution occurs on two scales. First, at the grain scale, melt in triple junctions develops a strong melt preferred orientation (MPO). As illustrated in Figure 9, this MPO is most pronounced in samples sheared to large strains ($\gamma \geq 1$) (Kohlstedt and Zimmerman 1996; Zimmerman et al. 1999), although it is present in samples deformed in triaxial compression to smaller strains ($\varepsilon \leq 0.2$) (Daines and Kohlstedt 1997). Second, at a larger scale, melt segregates to form melt-rich bands separated by melt-depleted regions. As illustrated in Figure 10, in laboratory experiments on samples with a grain size of ~10 μm, the melt-rich bands are ~50 μm wide and are separated by 100-200 μm (Holtzman et al., in press).

Deformation-induced melt segregation does not occur in all partially molten rock samples. In the olivine + basalt system (Fig. 9), melt-rich bands do not form, even though a strong MPO develops during deformation. However, in samples of olivine + 25 vol% chromite + basalt, distinct melt-rich bands evolve rapidly (Holtzman et al., in press). Laboratory experiments demonstrate that melt segregation also takes place in the following systems: (i) olivine + Fe-S melt + basalt (Hustoft, private communication), (ii) olivine + albitic melt (Zimmerman, pers. comm.), and (iii) anorthite + melt (Ginsberg and Kohlstedt, in preparation).

The transition from the grain-scale MPO to larger scale melt segregation appears to be associated with the relative values of the sample thickness, $\mathscr{L}_{th}$, and the compaction length, δ_c. The compaction length is the natural scale length that emerges in analyses of compaction during two-phase flow of a melt (fluid) phase through a deformable rock (matrix) (e.g., McKenzie 1984). Melt flow and matrix deformation are coupled over a distance of δ_c.

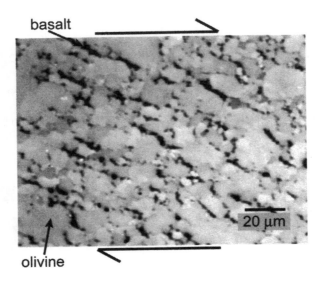

Figure 9. Reflected light optical micrograph of sample of olivine + basalt deformed in simple shear to a strain of $\gamma > 2$. Note the strong MPO developed at ~25° to the shear plane and antithetic to the shear direction. Also note that the melt remains distributed on the grain scale. Modified from Zimmerman and Kohlstedt (1996).

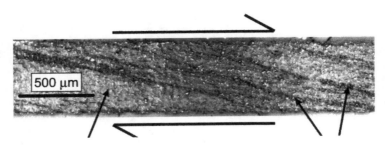

Figure 10. Reflected light optical micrograph of sample of olivine + 25 vol% chromite + 6 vol% basalt deformed to a shear strain of $\gamma > 2$. Note the melt-rich bands oriented at ~20° to the shear plane and antithetic to the shear direction.

Hence, melt migrates in response to deformation of the solid and eliminates pressure gradients in the solid. For δ_c on the order of the sample thickness, the melt distribution dynamically adjusts to the pressure gradients produced during shear deformation. The compaction length is a function of the sample permeability, k, the matrix viscosity, η, and the melt viscosity, μ:

$$\delta_c = \sqrt{k\frac{\eta}{\mu}}. \tag{15}$$

If the lack of deformation-driven melt segregation in the olivine + basalt system is taken as a reference point, then the melt segregation in the other four systems listed above can interpreted in the following way. For the olivine + basalt system, $\delta_c \geq 10$ mm, while $\mathcal{L}_{th} \leq 1$ mm, that is, $\delta_c \ll \mathcal{L}_{th}$. (i) For the olivine + chromite + basalt and the olivine + Fe-S melt + basalt systems, chromite or Fe-S melt tends to reside in the basalt channels thus reducing the permeability, k, and consequently decreasing δ_c so that $\delta_c > \mathcal{L}_{th}$. (ii) In the olivine + albitic melt system, the melt viscosity, μ, is more than three orders of magnitude smaller than that in the olivine + basalt system. Hence, $\delta_c > \mathcal{L}_{th}$. (iii) In the anorthite plus melt system, the matrix viscosity, η, is more than an order of magnitude smaller than the matrix viscosity in the olivine + basalt system. Thus again, $\delta_c > \mathcal{L}_{th}$.

The phenomenon of melt segregation during pure shear deformation of partially a molten rock was explored theoretically by Stevenson (1989). In his analysis, the driving force for deformation-driven melt segregation is determined by the pressure gradients that develop due to spatial fluctuations in melt fractions. Regions that initially have smaller than the average amount of melt will be weaker than those regions that contain more melt. Consequently, for a rock deforming at a given strain rate, the mean pressure $((\sigma_1 + \sigma_2 + \sigma_3)/3$, where σ_1, σ_2 and σ_3 are the principal stresses) will be lower in the weaker regions than in the stronger regions. Melt therefore migrates to the weaker, melt-enriched regions further lowering their strength and enhancing the pressure gradients. A positive feedback situation is thus developed. Stevenson's (1989) analysis did not yield scale lengths for the spacing and size of the resulting melt-rich bands, although he did conclude that the band spacing should be smaller than the compaction length. Further theoretical analyses and experimental studies are needed to fully characterize this important phenomenon.

Nonetheless, the presence of deformation-driven melt segregation has several important implications for the physical properties of lower crustal and upper mantle rocks. First, once melt-rich bands develop, they provide zones of relatively low shear strength such as those observed in ophiolites (Kelemen and Dick 1995; Dijkstra and Drury 2002). Deformation will therefore localize on these bands, producing a distinct anisotropy to the viscosity structure. Second, because permeability increases as the square or cube of the melt fraction, melt-rich bands will be high permeability paths. These paths will permit rapid transport of melt and will result in a highly anisotropic permeability structure. Third, the presence of melt-rich bands at depth in the Earth should affect seismic properties such as polarization and shear wave splitting. In addition, melt-rich bands will influence seismic attenuation. Finally, similar remarks apply to other physical properties such as electrical conductivity.

In summary, the presence of melt directly influences the physical properties of a partially molten rock. (i) The effect of melt on the viscosity of a partially molten rock is significantly larger than anticipated from current models because the melt distribution deviates substantially from that predicted for a simple solid + melt system. Specifically, because the solid-melt interfacial energy is highly anisotropic, a significant fraction of the grain-grain interfaces are wetted by the melt even though the dihedral angle is great than 0°. Hence, models based on an ideal melt distribution in which the melt is confined to triple junctions and four-grain junctions break down, particularly at melt fractions great than ~0.05. (ii) A pronounced segregation of melt occurs during deformation of partially molten rocks. In laboratory experiments, segregation occurs in samples for which the compaction length is smaller than the sample thickness. The melt-rich bands that develop have a profound effect on the physical properties (rheological, transport, seismic) of the rocks.

ACKNOWLEDGMENTS

Support from the National Science Foundation through grants EAR-9906986, OCE-0002463, EAR-0126277 and INT-0123224 is gratefully acknowledged. The author thanks Shun Karato for his careful reading of and insightful comments on this manuscript.

REFERENCES

Beeman ML, Kohlstedt DL (1993) Deformation of fine-grained aggregates of olivine plus melt at high temperatures and pressures. J Geophys Res 98:6443-6452

Bussod GY Christie JM (1991) Textural development and melt topology in spinel lherzolite experimentally deformed at hypersolidus conditions. In J Petrol, Special Lherzolite Issue. Vol 32. Menzies MA, Dupuy C, Nicolas A (eds) p 17-39

Chen IW, Argon AS (1979) Steady state power-law creep in heterogeneous alloys with microstructures Acta Metall 27:785-791

Cooper RF, Kohlstedt, DL (1982) Interfacial energies in the olivine-basalt system. In Advances in Earth and Planetary Sciences, Vol 12, High-Pressure Research in Geophysics. Akimoto S, Manghnani MH (eds) Center for Academic Publications, Tokyo, p 217-228

Cooper RF, Kohlstedt DL (1984) Solution-precipitation enhanced diffusional creep of partially molten olivine basalt aggregates during hot-pressing. Tectonophys 107:207-233

Cooper RF, Kohlstedt DL (1986) Rheology and structure of olivine-basalt partial melts. J Geophys Res 91:9315-9323

Cooper RF, Kohlstedt DL, Chyung K (1989) Solution-precipitation enhanced creep in solid-liquid aggregates which display a non-zero dihedral angle. Acta Metall 37:1759-1771

Daines MJ, Kohlstedt DL (1997) Influence of deformation on melt topology in peridotites, J Geophys Res 102:10257-10271

de Kloe R, Drury MR, van Roermund HLM (2000) Evidence for stable grain boundary melt-films in experimentally deformed olivine-orthopyroxene rocks. Phys Chem Minerals 27:480-494

Dijkstra AH, Drury MR (in press) Microstructures and lattice fabrics in the Hilti mantle seciton (Oman Ophiolite): evidence for shear localization and melt weakening in the crust-mantle transition zone? J Geophys Res

Drury MR, Fitz Gerald DF (1996) Grain boundary melt-films in upper mantle rocks. Geophys Res Lett 23:701-704

Hiraga T, Anderson IM, Zimmerman ME, Mei S, Kohlstedt DL (in press) Structure and chemistry of grain boundaries in deformed olivine + basalt and partially molten lherzolite aggregates: evidence for melt-free grain boundaries. Contrib Mineral Petrol

Hirth G, Kohlstedt DL (1995a) Experimental constraints on the dynamics of the partially molten upper mantle: deformation in the diffusion creep regime. J Geophys Res 100:1981-2001

Hirth G, Kohlstedt DL (1995b) Experimental constraints on the dynamics of the partially molten upper mantle: 2. Deformation in the dislocation creep regime. J Geophys Res 100:15441-15449

Holtzman BK, Groebner NJ, Zimmerman ME, Ginsberg SB, Kohlstedt DL (in press) Deformation-driven melt segregation in partially molten rocks. Geochem Geophys Geosys

Jin ZM, Green HW II, Zhou Y (1994) The rheology of pyrolite across its solidus. EOS - Trans Am Geophys Union 75:585

Karato SI, Jung H (in press) Effects of pressure on high-temperature dislocation creep of olivine. Phil Mag A

Kelemen PB, Hirth G, Shimizu N, Spiegelman M, Dick HJB (1997) A review of melt migration processes in the adiabatically upwelling mantle beneath mid-oceanic spreading ridges. Phil Trans R Soc Lond A 355:283-318

Kelemen PB, Dick HJB (1995) Focused melt flow and localized deformation in the upper mantle: juxtaposition of replacive dunite and ductile shear zones in the Josephine peridotite SW Oregon. J Geophys Res 100:423-438

Kohlstedt DL (1990) Chemical analysis of grain boundaries in an olivine-basalt aggregate using high-resolution, analytical electron microscopy. *In* The Brittle-Ductile Transition in Rocks, the Heard Volume. Duba AG, Handin JW, Wang WF (eds) American Geophysical Union, Washington DC, p 211-218

Kohlstedt DL (1992) Structure, rheology and permeability of partially molten rocks at low melt fractions. *In* Mantle Flow and Melt Generation at Mid-Ocean Ridges. Geophys Monogr 71, Phipps-Morgan J, Blackman DK, and Sinton JM (eds) American Geophysical Union, Washington DC, p 103-121

Kohlstedt DL, Chopra PN (1994) Influence of basaltic melt on the creep of polycrystalline olivine under hydrous conditions. *In* Magmatic Systems. Ryan MP (ed) Academic Press, New York, p 37-53

Kohlstedt DL, Zimmerman ME (1996) Rheology of partially molten mantle rocks. Ann Rev Earth Planet Sci 24:41-62

Lee VW, Mackwell SJ, Brantley SL (1991) The effect of fluid chemistry on wetting textures in novaculite. J Geophys Res 96:10023-10037

McKenzie DP (1984) The generation and compaction of partially molten rock. J Petrol 25:713-765

Mei S, Kohlstedt DL (2000a) Influence of water on plastic deformation of olivine aggregates: 1. Diffusion creep regime. J Geophys Res 105:21457-21469

Mei S, Kohlstedt DL (2000b) Influence of water on plastic deformation of olivine aggregates: 2. Dislocation creep regime. J Geophys Res 105:21471-21481

Mei S, Bai W, Hiraga T, Kohlstedt DL (2002) Rheology of olivine-basalt aggregates under hydrous conditions. Earth Planet Sci Lett 201:491-507

Riley GN, Kohlstedt DL (1990) An experimental study of melt migration in an olivine-melt system. *In* Magma Transport and Storage. Ryan MP (ed) John Wiley & Sons, New York, p 77-86

Tharp TM (1983) Analogies between the high-temperature deformation of polyphase rocks and the mechanical behavior of porous powder metal. Tectonophys 96:T1-T11

Vaughan PJ, Kohlstedt DL, Waff HS (1982) Distribution of the glass phase in hot-pressed, olivine-basalt aggregates: an electron microscopy study. Contrib Mineral Petrol 81:253-261

Smith CS (1948) Grains, phases and interfaces: an interpretation of microstructure. Trans AIME 197:15-51

Smith, CS (1964) Some elementary principles of polycrystalline microstructure. Metall Rev 9:1-47

Stevenson DJ (1989) Spontaneous small-scale melt segregation in partial melts undergoing deformation. Geophys Res Lett 16:1067-1970

von Bargen N, Waff HS (1986) Permeabilities, interfacial areas and curvatures of partially molten systems: results of numerical computations of equilibrium microstructures J Geophys Res 91:9261-9276

Watson EB, Brenan JM, Baker DR (1990) Distribution of fluids in the continental mantle. *In* Continental Mantle. Menzies MA (ed) Claredon Press, Oxford, p 111-125

Wirth R (1996) Thin amorphous films (1-2 nm) at olivine grain boundaries in mantle xenoliths from San Carlos, Arizona. Contrib Mineral Petrol 124:44-54

Zimmerman ME (1999) The Structure and Rheology of Partially Molten Mantle Rocks. Ph.D. dissertation, University of Minnesota, Minneapolis

6 Dislocations and Slip Systems of Mantle Minerals

Patrick Cordier

Laboratoire de Structure et Propriétés de l'Etat Solide (ESA CNRS 8008)
Université des Sciences et Technologies de Lille
59655 Villeneuve d'Ascq-Cedex, France

INTRODUCTION

This chapter focuses on dislocations and slip systems that are responsible for the plastic flow of high-pressure mantle minerals. After briefly introducing some basic concepts on crystal plasticity, we describe some recent experimental advances of the last decade that have contributed to our understanding of the rheology of mantle minerals. Among them, we describe some progress in the achievement of deformation experiments under high pressure. A more detailed review is presented by Durham et al. (this volume). We present then a novel Transmission Electron Microscopy (TEM) technique: Large Angle Convergent Beam Electron Diffraction (LACBED), which is very well adapted to the characterization of dislocations in beam sensitive materials, as well as the determination of dislocation Burgers vectors and characters from the fine analysis of X-ray diffraction peak broadening.

The present review is in the continuity of the paper entitled "Plastic Rheology of Crystals" by Poirier, published in 1995 in *"Mineral Physics & Crystallography—A Handbook of Physical Constants"* (AGU Reference Shelf 2). Poirier's paper reviewed the plasticity of some important minerals including olivine and pyroxenes. Concerning these two minerals, the reader is invited to refer to Poirier's contribution. In 1995, Poirier stated about high-pressure mantle minerals: *"Despite their importance for the rheology of the transition zone and the lower mantle, there is no information on the plasticity of the high-pressure mantle minerals in the relevant conditions of temperature and pressure."* Seven years later, the situation has significantly evolved and is still rapidly changing. It is the aim of this paper to highlight these recent advances.

BASIC CONSIDERATIONS

Plastic flow is basically a transport phenomenon controlled by the motion of defects: point defects, dislocations, or grain boundaries. The driving force is provided by the applied shear stress. In many cases, deformation is produced by the motion of dislocations and the present paper will focus on this aspect only. The strain rate then obeys the Orowan's equation:

$$\dot{\varepsilon} = \rho b v$$

where ρ is the density of mobile dislocations, b is the length of the Burgers vector, and v is the dislocation velocity. A dislocation defines the boundary between the region of the crystal which has been sheared and the unslipped region in which the displacement is zero (Fig. 1). A dislocation is characterized by a line direction (unit vector $\bar{u}$) and its Burgers vector $\bar{b}$ which is the elementary amount of plastic shear carried by the dislocation. The Burgers vector is the most invariant characteristics of a dislocation, it is the same for all parts of the line. The angle between $\bar{u}$ and $\bar{b}$ defines the dislocation character: edge, screw or mixed (Fig. 2). The plane determined by $\bar{u}$ and $\bar{b}$ is called the glide plane (or slip plane), provided the dislocation line is not of pure screw character. A dislocation loop will sweep over its glide plane (Fig. 2) under the action of an applied stress (defined by the stress tensor $\bar{\bar{\sigma}}$), the dislocation segments being subjected to the Peach-Koehler force:

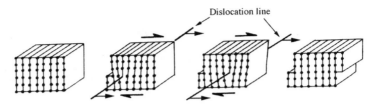

Figure 1. Slip of a crystal by propagation of an edge dislocation. The dislocation line visualized by an extra-half plane represents the border between the slipped area (on the left side) and the unslipped regions (on the right side). After Nicolas and Poirier (1976).

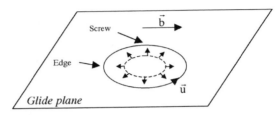

Figure 2. Expansion of a dislocation loop under the action of a shear stress applied on its glide plane. The edge and screw segments are indicated.

$$\vec{F} = (\vec{b} \cdot \overline{\overline{\sigma}}) \times \vec{u}$$

In most cases, plastic deformation is due to the glide motion of the dislocations. At high temperature, point defects become mobile and can interact with dislocation to promote climb, i.e., motion of the dislocation line out of its glide plane. However, except in some special cases where plastic deformation proceeds by pure climb, strain is generally produced by dislocation glide (although the strain-rate might be controlled by thermally activated diffusion).

The energy associated with a dislocation is mostly stored in the surrounding crystal as elastic energy. For a straight, mixed dislocation (character defined by θ, the angle between the line direction and the Burgers vector) contained in a cylinder of radius R, the energy per unit length is:

$$E_{el} = \frac{\mu b^2 (1 - \nu \cos^2 \theta)}{4\pi (1 - \nu)} \ln\left(\frac{R}{r_o}\right)$$

where μ is the shear modulus, ν is the Poisson ratio and r_0 is the core radius.

Macroscopic strain requires efficient dislocation multiplication mechanisms. A widely accepted mechanism for dislocation generation is based on the Frank-Read source (Fig. 3). In this model, a segment AB of length L bows out under an applied stress. The stress reaches a maximum σ_c when the loop is a semi-circle. If the stress exceeds σ_c the line develop into a complete loop restoring the initial segment AB (length L). The critical stress to activate a Frank-Read source is roughly:

$$\sigma_c \propto \frac{\mu b}{L}$$

It is worth noticing that this critical stress is proportional to the shear modulus μ and that any change of this constant (for instance its evolution with applied pressure) will have a direct influence on the plastic properties.

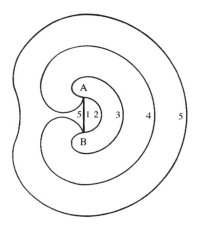

Figure 3. The Frank-Read source. A segment AB of dislocation is represented in its glide plane. A and B represent fixed pinning points where the dislocation line leave the glide plane. Under an applied shear stress, the dislocation segments bows out and moves from position 1 to positions 2, 3 and 4 where the loop is about to close leaving in 5 a large glide loop and the original segment AB.

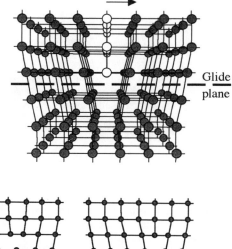

Figure 4. Schematic representation of an edge dislocation in a simple, monoatomic, cubic crystal. The Burgers vector and the glide plane are represented. The atoms of the extra-half plane are shown in white.

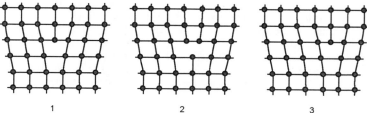

Figure 5. Movement of an edge dislocation. The dislocation originally in "1" moves by one interatomic distance to "3". In its motion, the dislocation passes through an unfavorable configuration "2" which involves bond breaking.

While the elasticity theory describes correctly the shape of a dislocation at large distances from its line, some aspects are due particularly to the crystal structure. Among those is the structure of the dislocation core. Dislocations are usually represented on simple cubic lattices. Figure 4 represents the usual simplified representation of an edge dislocation in a simple cubic crystal. The extra-half plane is shown by the white atoms. Dislocation glide involves breaking bonds at the dislocation core without transport of matter (Fig. 5). In order to glide from position "1" to an identical position "3", the line must go through an intermediary position "2" of higher energy. A stress σ_P must be applied to the dislocation to pass from "1" to "3". This stress is a function of the glide plane considered and of the Burgers vector modulus. For a close-packed plane (large d_{hkl}), the interatomic bonds are weak and not much distorted when one goes from "1" to "3". The stress σ_P is going to be small. It is large, on the other hand, for not so close packed glide planes (small d_{hkl}). A comprehensive treatment of this problem has been proposed by Peierls (1940) and Nabarro (1947), so σ_P is now called Peierls-Nabarro stress, or Peierls stress in short. Its expression is given by:

$$\sigma_P \propto \frac{2\mu}{K} \exp\left(-\frac{2\pi}{K}\frac{d_{hkl}}{b}\right)$$

where K depends on the dislocation character (K = 1 or 1-ν for screw and edge dislocations, respectively).

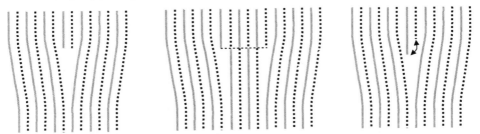

Left: **Figure 6.** Schematic of an edge dislocation in a structure with two types of atomic planes. The dislocation core is made of two extra-half planes of both types.
Center: **Figure 7.** The dislocation of Figure 6 is dissociated into two colinear partial dislocations separated by a stacking fault (glide dissociation).
Right: **Figure 8.** Climb dissociation of the edge dislocation of Figure 6.

In fact, real dislocation cores are much more complicated. Complexity increases with the Burgers vector modulus (or with the number of atoms in the unit cell). When several atomic species are involved, edge dislocations exhibit, at least, two extra half-planes (Fig. 6). It might be energetically favorable that the dislocation dissociates in the glide plane into two partial dislocations separated by a stacking fault (Fig. 7). It is also possible that the dislocation dissociates by a climb process (Fig. 8). Climb dissociation occurs by conservative self-climb, which involves redistribution of matter between the partial half-planes by short range diffusion. Mitchell et al. (1984) have shown that elasticity theory predicts the climb configuration to be usually more stable than the glide configuration, but climb dissociation requires significant diffusion. Very contrasted mobilities are expected to result from these two configurations.

Glide dissociation is expected to enhance the mobility of the dislocations because their Burgers vector is smaller, and hence they are facing smaller Peierls stresses. Climb dissociation can have a marked effect on the mechanical properties of the mineral. In particular, an increase of the yield stress with increasing temperature can be expected (Fig. 9) in some cases. The curve $\sigma_Y(T)$ first decreases with increasing temperature.

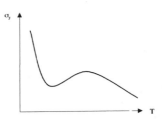

Figure 9. Possible evolution of the yield stress (σ_Y) with temperature in a material in which dislocations are dissociated in climb. At low temperatures, the yield stress decreases with increasing temperature in a usual way. At intermediate temperatures the climb dissociation width increases resulting in a decrease of the mobility of dislocations. At high temperature, the climb dissociation width has reached its equilibrium value, but diffusion is active and dislocations can move by dragging the point defects associated to the fault.

The yield stress would begin to increase when climb dissociation appears. The dislocations must then recombine to be able to glide. This process is rendered more difficult when the dissociation width increases (i.e., when T increases). Above a given temperature, the fault can diffuse and the yield stress begin to decrease. This simple example illustrates the fact that taking into account the nature of the atomic bonds is not sufficient. The dislocation core fine structure can have a tremendous influence on the plastic properties. Some information about the dislocation core can be gained from Transmission Electron Microscopy (TEM) observation (especially using high resolution TEM or the high-resolution diffraction contrast mode: weak-beam dark-field), for instance on dissociation if the dissociation width is large enough. Below that scale, the core structure can only be calculated.

More generally, common signatures of core effects are: unexpected deformation modes and slip geometries; strong and unusual dependence of flow stresses on crystal orientation and temperature; and, most commonly, a break-down of Schmid's law which states that glide on a given slip system commences when the resolved shear stress on that system reaches a critical value.

One of the more fundamental question when considering plastic deformation of a given material is: what kind of slip system is activated, and why? This section is aimed to emphasize the main parameters which govern the choice of the slip systems. A few criteria are usually proposed to predict or account for the choice of the slip systems:

1. Glide planes are chosen among close-packed planes
2. Burgers vectors are chosen among the shortest lattice repeats
3. Easy slip systems correspond to minimum values of "b/d_{hkl}" [Chalmers-Martius (1952) criterion]
4. The slip plane corresponds to the dissociation plane of the dislocations

Let us examine these different conditions.

Close-packed planes. This condition is quite straightforward in metals where all atoms have about the same sizes. In case of oxides or silicates, one notices that cations are much smaller than anions (oxygen). It seems thus reasonable to look for the possible glide planes based on the structure of the anionic sub-lattice. Indeed, some structures show a close-packing of oxygen. We will see that it is not a general rule, although it must be remembered that high-pressure structures tend to be composed of closest-packed arrays of atoms. In the case of ionic crystals, one can look for neutral planes, although it has been shown in oxides that this criterion is not always verified.

Silicates usually correspond to complex crystal chemistry involving several types of atomic bonding. The easiest glide planes are usually those in which shear breaks the weakest bonds. In silicates, at low pressure, it seems important to avoid shearing the strong SiO_4 tetrahedra. A trend in mantle mineralogy is that increasing pressure increases coordination number resulting in the transformation of SiO_4 groups into SiO_6 groups. Prewitt and Downs (1998) note that an increase in coordination number is usually accompanied by a lengthening (as well as an increase of the ionic character) of the bonds. For example, $R(Si^{IV}O) = 1.62$ Å in quartz while $R(Si^{VI}O) = 1.78$ Å in stishovite. This might have important implication on the choice of the glide system as Prewitt and Downs point out that, as a rule of thumb, short bonds are the strongest, and long bonds are the weakest. If this trend was to be confirmed, the transformation of SiO_4 groups into SiO_6 group would not correspond to a strengthening of the bonds and the role of the structural units would be less marked in the choice of the glide planes.

Short Burgers vectors. The Burgers vectors depend essentially on the Bravais lattice. The elastic energy of a dislocation is proportional to μb^2 per unit length. It follows that dislocations with Burgers vectors chosen among the shortest lattice vectors will be strongly favored from the energetic point of view. It must be kept in mind however that when dislocations are formed by plastic deformation (i.e., by mechanisms similar to the Frank-Read source for instance), it may be the most mobile rather than the less energetic dislocations which multiply more rapidly (Nabarro 1967). Indeed, the activation of dislocations with the smallest Burgers vectors is not a law of nature and many examples can be found where easy slip involves dislocations with large Burgers vectors (For instance rutile, TiO_2 which glides on <101>{101} when <001>, <100> and <110> would be shorter slip directions). In most cases, dissociation into partial dislocations is invoked to account for this behavior. Dissociated dislocations can have lower elastic energies per unit length; they might also be more mobile. However, Friedel (1964) points out that, in crystals with complex crystal chemistry, glide of partial dislocations might involve complicated atomic shuffles.

Chalmers-Martius criterion. Chalmers and Martius (1952) have observed in several simple structures that easy slip systems correspond to minimum values of "b/d_{hkl}". Indeed, the atomic displacements at the dislocation core are then smaller, because they are spread along close-packed planes which exhibit larger d_{hkl}. It is interesting to note that this criterion corresponds to the smallest Peierls stresses, that is, to dislocations which are expected to be more mobile in glide. The Chalmers-Martius criterion is more delicate to apply in minerals: which value must we consider for d_{hkl}? Is it the actual interplanar spacing (d_{hkl} *sensu stricto*) or the distance between occupied atomic layers? Once again, it can be anticipated that the validity of this criterion might be affected by dissociation and the nature of atomic bonding at the dislocation core.

Dissociation plane. If dislocations are to dissociate in a given plane in a glide configuration, their mobility is expected to increase. A special attention must be given to stacking faults which do not modify the anionic sub-lattice, because they must correspond to relatively low surface energies. It has been shown in case of oxides that this criterion was difficult to assess. Moreover, oxides have revealed complicated schemes (Bretheau et al. 1979) like zonal dissociation (i.e., dissociation spread over several levels of atomic layers) or synchroshear (collective displacements of cations along directions which are different from the shear direction).

EXPERIMENTAL ADVANCES

The recent progress accomplished in characterizing the deformation mechanisms of mantle minerals are largely due to considerable technical developments. In the first place, one must cite the developments of deformation experiments under high-pressure and high-temperature. This issue is reviewed by Durham et al. (this volume). We just intend in the present section to describe the procedure that we have used to induce plastic deformation of minerals at high-P, high-T using a multianvil apparatus.

As far as mechanisms are concerned, TEM appears as a key technique to analyze the deformation microstructures. However, high-pressure minerals are often very sensitive to beam damage and detailed characterization of the dislocation Burgers vectors by the usual techniques of contrast extinction are often precluded by early amorphisation. We present here a technique for Burgers vector determination based on Large Angle Convergent Beam Electron Diffraction (LACBED), which is very well adapted to beam sensitive materials. The analysis of broadening of X-ray diffraction peaks is an alternative technique to get information on the dislocations at a mesoscopic scale. Some recent developments concerning the characterisation of dislocations are presented.

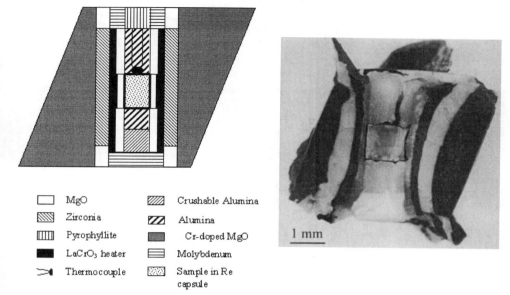

Figure 10. (a, left) Schematic cross section showing details of the octahedral pressure cell used for deformation experiments in the multianvil apparatus. Hard alumina pistons placed on both ends of the specimen induce deviatoric stresses at high pressure. (b, right) Cross-section of a recovered 10 mm pressure cell after deformation at 22 GPa and room temperature. The ringwoodite sample is located in a rhenium capsule.

High-pressure deformation experiments

Gas- and solid-medium apparatuses that permit either constant strain rate or constant load deformation under controlled thermochemical conditions are only able to operate to pressures that correspond to the crust or uppermost upper mantle (0.7 and ~4 GPa, respectively). Investigation of deep mantle minerals requires higher pressures that can be generated by devices such as multianvil or diamond-anvil apparatuses. In diamond anvil cell experiments, plastic deformation is achieved by taking advantage of the pressure gradient within the cell (Sung et al. 1977; Meade and Jeanloz 1990) or by bridging the specimen between the two diamonds (Chai et al. 1998; Wenk et al. 2000). In multianvil experiments, plastic deformation is achieved by modifying the high-pressure assembly. We have used an experimental setup used previously up to 16 GPa (Fujimura et al. 1981; Green et al. 1990; Liebermann and Wang 1992; Bussod et al. 1993; Sharp et al. 1994; Cordier et al. 1996; Durham and Rubie 1997; Voegelé et al. 1998a) and have adapted it to smaller octahedron sizes (10/5, 10/4) in order to reach higher pressures (Cordier and Rubie 2001). Two hard alumina pistons are located at each end of the sample to make the assembly mechanically anisotropic (Fig. 10). The lower piston consists of solid alumina whereas the upper piston acts as the thermocouple sleeve and contains the thermocouple wires. Thin (25 μm) Re discs were inserted between the pistons and the polished faces of the sample to avoid chemical reaction between the sample and the pistons at high temperature. Cordier and Rubie (2001) emphasize the importance of using a two-step procedure in which synthesis of the high-pressure phases is performed independently from their deformation:

1. The first step corresponds to the synthesis of the high-pressure phase in the conventional quasi-hydrostatic cell assembly. After the synthesis experiment, the

densified octahedron is broken to recover the sample capsule. The ends of the Re capsule are ground to expose the sample and it is characterized by Raman spectroscopy to ensure that the required phase has formed.

2. The second step corresponds to the deformation at high-P, high-T of the dense specimen of the high-pressure phase in the modified assembly (Fig. 10).

The deformation experiment consists of increasing the pressure first (within three to four hours) at room temperature. During this cold compression stage, the size of the MgO octahedron decreases considerably (e.g., 20% shortening) because of the collapse of its porosity and flow between the WC cubes. The extruded material together with the pyrophyllite form the gaskets that contain the pressure medium. The pistons and sample are stiffer than the octahedron materials. This results in the development of high uniaxial stresses along the piston-sample column as the octahedron size decreases.

After the required pressure is reached, temperature is increased to the desired value at a rate of 50-100°C/min. The uniaxial stresses then relax in the whole assembly, including in the sample, by plastic deformation. Following an annealing period (usually 30 min to 2 h, although much longer times are possible), the sample is quenched under pressure to <100°C in 1-2 s by cutting the power to the furnace. Quench rates are fast because of the small size of the pressure cell which is surrounded by large volumes of metal. Pressure is then decreased slowly to avoid failure of the WC cubes. After the experiment, the specimen is recovered and prepared for microstructural characterization.

Transmission electron microscopy

Dislocations are generally imaged in TEM with diffraction contrast methods. Of particular interest are the diffraction conditions where no contrast is observed because they give access to the Burgers vector. Alternatively, the Burgers vector (or at least a projection of this vector) can be obtained by counting lattice fringes along a Burgers circuit drawn around the dislocation core on a high-resolution image. This panel of techniques has been enriched recently by the addition of a defocused diffraction method: Large Angle Convergent Beam Electron Diffraction (LACBED: Tanaka et al. 1980; Cherns and Preston 1986; Cherns and Morniroli 1994). LACBED has been applied recently in mineralogy to analyze dislocations in quartz (Cordier and Doukhan 1995; Cordier et al. 1995), garnets (Cordier et al. 1996; Voegelé et al 1998a,b; Voegelé et al. 1999; Voegelé et al. 2000), stishovite (Cordier and Sharp 1998) and wadsleyite (Thurel 2001).

Convergent Beam Electron Diffraction (CBED) is obtained by focusing a convergent beam on the specimen. The diffraction pattern is made of disks containing lines (excess lines in the diffracted disks, deficient lines in the transmitted disk). These lines correspond to Bragg reflections on (hkl) planes. The diameter of the disks is directly proportional to the convergence semi-angle α. If the convergence is large enough, substantial overlap of the disks will occur such that individual lines are no longer discernible. This situation corresponds to the so-called "Kossel" pattern. This limitation can be overcome in the LACBED mode. If the specimen is moved in the column above or below the object plane, the transmitted and diffracted beams focus at different spots of the image plane (Fig. 11). The selected area aperture can thus be used to select one of those beams and a well-contrasted (bright- or dark-field) LACBED pattern will be obtained (Fig. 12).

Because LACBED is a defocused method, it brings information on both the real and the reciprocal spaces simultaneously. This allows a large number of applications including the characterization of crystal defects (Morniroli 2002).

Let us recall, briefly, the principles of Burgers vector characterization by LACBED. In this mode, when a dislocation (Burgers vector $\vec{b}$) intersects a Bragg line (i.e., the line

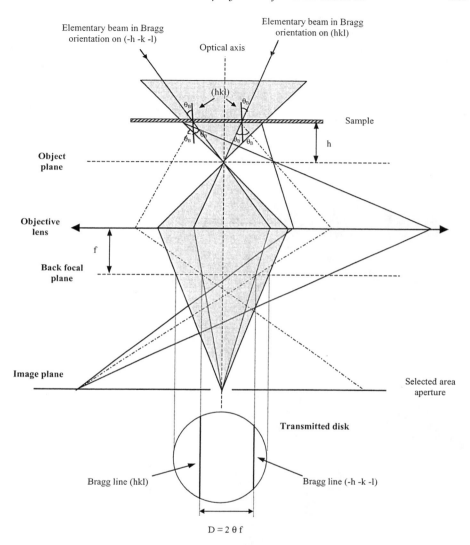

Figure 11. Ray diagram (objective lens only) showing LACBED operation in the TEM. The specimen is moved away from the object plane (it is above in case of the example presented). Bragg reflections on (hkl) and ($\overline{hkl}$) are thus located at different points of the sample. This result in a separation of the transmitted and diffracted beams in the object plane and its conjugate image plane. In the latter plane, the selected area aperture can be used to select the transmitted beam (bright-field) or a diffracted beam (dark-field) (after Morniroli 2002).

where the exact Bragg condition is satisfied) with the diffraction vector $\vec{g}$, the line rotates and splits into n nodes if $\vec{g} \cdot \vec{b} = n$ (Cherns and Preston rule, see Fig. 13). If the dislocation line is long enough, then it is possible to place it with respect to the Bragg lines so that it crosses at least three independant Bragg lines (Fig. 14). Three linear equations of the type $g_i \cdot b = n_i$ are obtained whose solution uniquely gives the Burgers vector. If the dislocation is too short, then it is successively placed on three Bragg lines and the Burgers vector is

deduced as above. The method requires knowledge of:
- n, the number of intensity minima in a dark field pattern
- the direction of the dislocation line characterized by a unit vector $\vec{u}$
- the hkl indices of the $\vec{g}$ Bragg line
- the direction of the deviation parameter s

These two last features are identified by comparison with theoretical patterns drawn by means of a computer program ("Electron Diffraction") based on the kinematical theory which was specially developed by Morniroli at the University of Lille.

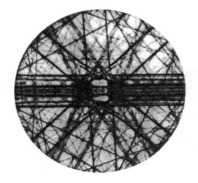

Figure 12. Example of a bright-field LACBED pattern of garnet along a [103] zone axis. Convergence semi-angle: 5°.

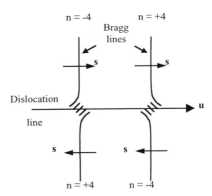

Figure 13. Cherns and Preston rules which allow to determine the sign of the $\vec{g} \cdot \vec{b}$ product from the sense of the bend of the Bragg line, the orientation of the dislocation line and the orientation of the excitation error (after Morniroli 2002).

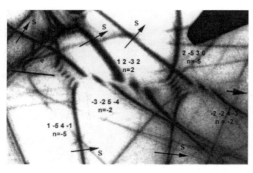

Figure 14. Example of a Burgers vector determination in quartz with one LACBED pattern. The dislocation line (sub-horizontal arrow) crosses five Bragg lines. The small arrows on each Bragg lines give the orientation of the excitation vectors. Resolution of the linear equations gives $\mathbf{b} = 1/3[\bar{1}2\bar{1}0]$.

It is also possible to determine the direction $\vec{u}$ of the dislocation line by LACBED. One has to find (at least) two Bragg lines hkl and h'k'l' which are parallel to the dislocation line. The unit vector $\vec{u}$ is then perpendicular to the corresponding $\vec{g}$ and $\vec{g}'$ vectors. This implies that $\vec{u}$ belongs to the planes hkl and h'k'l'. $\vec{u}$ is given by the intersection of these two planes (cross product $\vec{g} \times \vec{g}'$).

The knowledge on the same dislocation of the Burgers vector $\vec{b}$ and of the line direction $\vec{u}$ gives the glide plane of the dislocation (provided that the dislocation is not screw in character).

The LACBED technique has several important advantages over classical methods. In the first place, it provides very reliable results without any interpretation and without any *a priori* hypothesis on the nature of the Burgers vector. In the second place, it is a defocused method that induces very limited beam damage of the specimen. For this reason, LACBED is very well adapted to the characterization of high-pressure minerals which are often very unstable under the electron beam. The technique can be used either with perfect or partial dislocations (Cordier et al. 1995). The major drawbacks are due to the fact that dislocations must be characterized individually and to the poor visibility of the microstructure in LACBED mode. The latter point might be overcome by using a LACBED variant called Convergent Beam Imaging (CBIM: Morniroli et al. 1997).

X-ray diffraction peak broadening

X-ray diffraction techniques and instrumentation have made tremendous progress during the last two decades, including in the field of crystal defect characterization. Dislocations are characterized by a long-range strain field, therefore their diffraction effects cluster around the fundamental Bragg reflections (Fig. 15). This effect is called diffraction peak broadening. It is thus, in principle, possible to use peak broadening to get information on the dislocations. For that purpose, Ungár and Borbély (1996) have upgraded the conventional peak profile analysis methods of Williamson and Hall (1953) and Warren and Avenbach (1950) by taking into account a contrast effect of dislocations on peak broadening. The contrast factor C of dislocations depends on the relative orientation of the Burgers vector $\vec{b}$, of the line orientation $\vec{u}$ and of the diffraction vector $\vec{g}$ in a similar way as in TEM. C is thus characteristic of a given slip system. The anisotropic contrast effect of dislocations has been used to model the strain anisotropy observed on X-ray diffraction profiles of cubic and hexagonal crystals (Ungár et al. 1999; Ungár and Tichy 1999). The technique has been recently extended to orthorhombic crystals (see below). Following the Ungár and Borbely procedure, it is possible to characterize the slip systems (combination of Burgers vectors and line directions) corresponding to the dislocations stored in the sample, the relative abundance of the various dislocation types and the dislocation densities (Ungár 2001). This technique has several distinctive advantages which make it complementary to TEM characterizations: it provides information on a large number of dislocations that are more statistically relevant than information gained from TEM, it can be extended to high dislocation densities 10^{18} m^{-2}), and it can be applied to bulk materials and to specimens which are sensitive to electron irradiation.

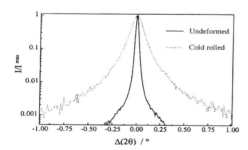

Figure 15. Typical 110 Bragg peaks of ultra-high purity iron corresponding to the underformed and cold-rolled (sample deformation: 62 %) states. The elastic energy associated with stored dislocations results in a significant peak broadening (after Borbély et al. 2000).

PLASTICITY OF MANTLE PHASES

The Earth's deep interior cannot be readily accessed for study. Mantle xenoliths and inclusions in diamonds show that down to 300-400 km, olivine, pyroxenes and garnets are the dominant minerals. Sampling becomes very limited at greater depths. However, it is possible to produce experimentally the high-pressure and high-temperature conditions of the Earth's interior and to observe their effects on these minerals. It is found that, with increasing pressure, pyroxenes and garnets undergo gradual solid solution reactions to form majoritic garnets, and finally silicate perovskite (Fig. 16). In a similar way, olivine transforms first to wadsleyite, then to ringwoodite before breaking down in an assemblage of silicate-perovskite and magnesiowüstite (Fig. 16). This mineralogical model has been found to account satisfactorily for seismological data (discontinuities, seismic velocity profiles).

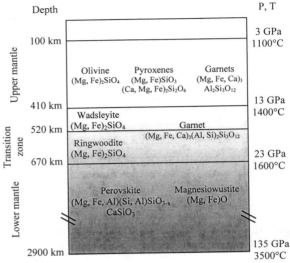

Figure 16. Simplified mineralogical model of the Earth's mantle inferred from natural sampling (xenoliths and inclusions in diamonds) and from results of high pressure and temperature experiments (after Gillet, 1995).

The aim of the following section is to review the data produced in the recent years on the plastic properties of the high-pressure minerals listed in Figure 16. Some information on high-pressure phases of silica are also presented.

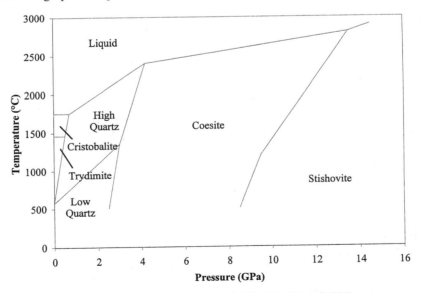

Figure 17. Simplified phase diagram for SiO_2 (after Presnall 1995).

SiO_2 system

Silica exhibits a quite complicated polymorphism (Fig. 17). At high-pressure, silica first transforms to coesite, then to stishovite. At higher pressure (above ~50 GPa), stishovite transforms to a $CaCl_2$ phase. This phase transformation is displacive, It involves just a slight tilting of the SiO_6 octahedra. Information on plasticity are restricted to coesite and stishovite.

Coesite crystallizes in the monoclinic system (space group $C2/c$) with $a = 0.72$ nm, $b = 1.25$ nm, $c = 0.72$ nm, $\alpha = \gamma = 90°$ and $\beta = 120°$ (Fig. 18).

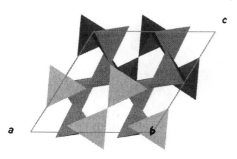

Figure 18. Structure of coesite viewed down the b-axis.

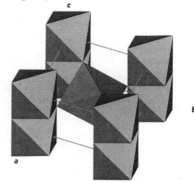

Figure 19. The unit cell of stishovite with SiO_6 octahedra represented.

The rheology of synthetic coesite polycrystals has been addressed experimentally by Renner et al. (2001) with a modified Griggs apparatus at confining pressures of 3.1 to 3.7 GPa and temperatures in the range 700-1160°C. Mechanical data follow a power law $\dot{\varepsilon} = A\sigma^n \exp[-Q/RT]$, with n = 3±1 and Q = 275±50 kJ.mol^{-1}. Although no direct determination of the dislocations and slip planes has been performed, the lattice preferred orientations appear to be consistent with the predominant slip system [001](010). Dislocations have been analyzed by TEM in relict coesite grains from the Dora Maira massif by Langenhorst and Poirier (2002). The identified Burgers vectors are [100], [001] and [110] which in a pseudo-hexagonal setting corresponds to **a** and **c+a** similar to the Burgers vectors of dislocations in quartz. The relative dislocation densities observed in coexisting quartz and coesite grains suggest (in agreement with the experiments of Renner et al. 2001) that coesite is much less deformable than quartz.

Stishovite is a high pressure polymorph of silica that is stable at pressures in excess of 10 GPa. This compound was first synthesized by Stishov and Popova (1961). Natural occurrences were later discovered in shocked specimens from terrestrial impact structures such as Meteor crater (Chao et al. 1962), Ries crater (Shoemaker and Chao 1961; Chao and Littler 1963) and the Vredefort (Martini et al. 1978). The crystal structure of stishovite, determined by Sinclair and Ringwood (1978), is a rutile-type structure with space group $P4_2/mnm$ in which silicon is octahedrally coordinated to oxygen. The Bravais lattice of stishovite is tetragonal and the lattice parameters are $a = b = 0.418$ nm and $c = 0.26678$ nm (Ross et al. 1990). The SiO_6 octahedra form edge-shared chains along the c-axis that are each connected to four other parallel chains (Fig. 19).

Microhardness measurements have been performed by Léger et al. (1996) on stishovite polycrystals synthesized at 20 GPa and 1100°C. The high shear modulus (220 GPa) and Knoop hardness (33 GPa) exhibited by stishovite at ambient pressure have classified it as a "superhard" material (immediately following sintered cubic BN and diamond). However, this characteristic shown at room temperature is difficult to extrapolate at higher temperatures.

Dislocation Burgers vectors have been characterized using LACBED by Cordier and Sharp (1998) in stishovite specimens synthesized in the multianvil apparatus at 15 GPa and 1200°C. Dislocations with Burgers vectors [001], <100>, <101> and <110> were found to be stable in stishovite (Fig. 20). No dissociations were observed.

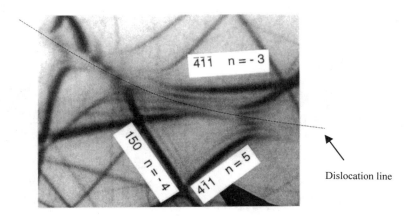

Figure 20. Stishovite grown at 15 GPa, 1200°C. Characterization by LACBED of a $[1\bar{1}0]$ dislocation.

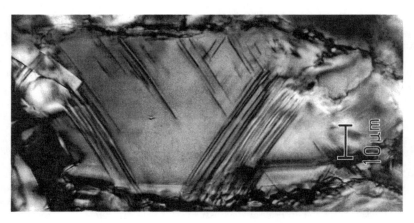

Figure 21. Deformation twins in stishovite deformed at 14 Gpa and 1300°C. Optical micrograph.

In a further study, we have performed plastic deformation experiments on stishovite in the multianvil apparatus using the procedure described above. The samples were first synthesized from wet silica glass. Deformation experiments were carried out at 14 GPa and temperatures in the range 1000-1600°C. Examination of deformed specimens at the optical microscope showed that two deformation mechanisms were activated. Mechanical twinning (Fig. 21) and dislocation activity (suggested at the optical microscope by undulatory extinction and confirmed by TEM observation). Easy slip involves dis-

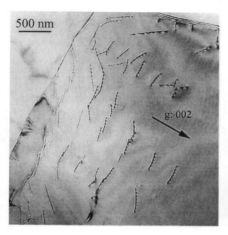
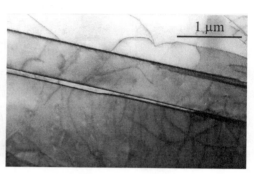

Figure 22 (left). Stishovite deformed at 14 GPa, 1300°C. [001] dislocations gliding in {110}. TEM bright-field micrograph with g: 002, foil orientation: (100).

Figure 23 (right). Stishovite deformed at 14 GPa, 1300°C. Microtwins viewed edge-on. The thichness of the twin boudaries is due to the occurrence of dislocations. TEM bright-field micrograph with g: 110.

locations with the shortest Burgers vector: [001]. They glide on {100} and {110} planes (Fig. 22). Secondary slip involving <100> dislocations is observed in grains which exhibit a large [001] dislocation density, i.e., after work hardening. This suggests that <100> slip has a higher critical resolved shear stress than [001]{100} and [001]{110}. No <101> or <110> dislocations have been observed. Comparison with results from Cordier and Sharp (1998) demonstrates that caution must be exercised when microstructures are characterized from specimens which were not unambiguously deformed. The relatively low number of easy slip systems is probably responsible for the activation of deformation twinning as an additional deformation mode. The twin law corresponds to a mirror symmetry across a {101} plane. At the TEM scale, the microtwins are shown to shear pre-existing defects like microcracks. The twin boundaries contain a high dislocation density (Fig. 23). Altogether, these observations confirm the deformation origin of the twins.

At ~50 GPa (room temperature) stishovite undergoes a second-order phase transition towards a $CaCl_2$ structure (Tsuchida and Yagi 1989; Andrault et al. 1998; Carpenter et al. 2000). This phase transformation corresponds to a symmetry change from $P4_2/mnm$ to orthorhombic Pnnm which involves slight tilting of the SiO_6 polyhedra only. No bonds are broken and the lattice parameters changes are modest (Fig. 24). From the point of view of the crystal structure only, it can be tentatively argued that the $CaCl_2$ phase would exhibit the same kind of dislocations and possibly the same slip systems as stishovite (see the case of majorite garnets below). However, the $P4_2/mnm$ to $Pnnm$ transition is also associated with a mechanical instability as suggested by *ab initio* structure calculations (Lee and Gonze 1997) and by the observed softening of the Raman B_{1g} mode (Kingma et al. 1995). The transition is then accompanied by elastic anomalies as suggested by Carpenter et al. (2000) based on the Landau theory of structural phase transformation. In particular, the shear modulus is found to decrease significantly at the transition, which should result in a remarkable softening of the material. Indeed, Shieh et al. have reported recently (2001) that the differential stress supported by stishovite in a diamond anvil cell decreases from 6 GPa at 20 GPa to less than 3 GPa close to the transition. Figure 25

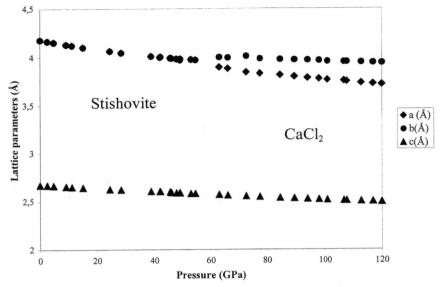

Figure 24. Pressure dependence of the lattice parameters at the stishovite-CaCl$_2$ transition (after Andrault et al. 1998).

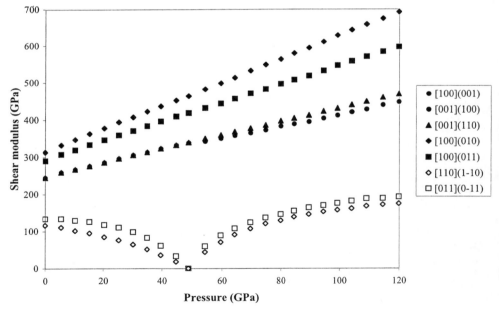

Figure 25. Variation of shear moduli of SiO$_2$ across the $P4_2/mnm$ (stishovite) to $Pnnm$ (CaCl$_2$) transition derived from the variation of the individual elastic constants proposed by Carpenter et al. (2000) and calculated in anisotropic elasticity.

presents the evolution of several shear moduli across the stishovite-CaCl$_2$ transition. These moduli have been calculated in the frame of anisotropic elasticity from the elastic constants calculated by Carpenter et al. (2000). The shear modulus $\mu_{[uvw](hkl)}$ corresponds to a sollicitation applied along the [uvw] direction, and parallel to the plane (hkl). It is shown that the slip systems [001]{hk0} or [100]{0kl} suggested from our TEM observations are not affected by the phase transformation. In contrast, we show that two moduli: $\mu_{[110](1-10)}$ and $\mu_{[011](0-11)}$ exhibit a significant softening at the transition. It can be speculated from this calculation that stishovite might undergo a change of deformation mechanism at the vicinity of the CaCl$_2$ transition with the activation of <101> and <011> slip. Indeed, the corresponding dislocations have been found to be stable by Cordier and Sharp (1998).

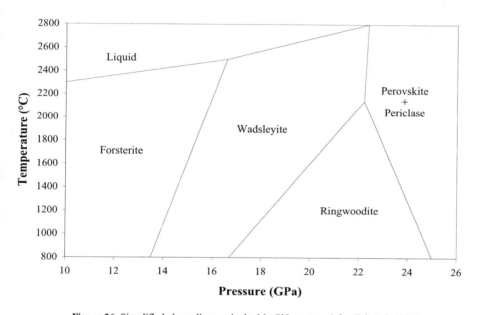

Figure 26. Simplified phase diagram in the Mg$_2$SiO$_4$ system (after Fei et al. 1990).

Olivine, wadsleyite, ringwoodite

The transition zone between the upper and the lower mantle is due to phase transformations. Among the most important ones are those which involve olivine and its high-pressure polymorphs. Their stability fields are illustrated in the simple case of Mg$_2$SiO$_4$ on Figure 26.

Olivine. Geologically relevant olivines belong to a solid-solution between two end-member phases: forsterite (Fo) Mg$_2$SiO$_4$ and fayalite (Fa) Fe$_2$SiO$_4$. The structure of olivine is based on a distorted hexagonal close-packed oxygen sublattice. The Bravais lattice is orthorhombic (Fig. 27) with, for forsterite, $a = 0.475$ nm, $b = 1.019$ nm and $c = 0.597$ nm (*Pbnm* space group). The unit cell contains four formula units.

The most common dislocations correspond to the shortest lattice repeats: [100] and [001]. Numerous deformation experiments have been performed on olivine and forsterite single crystals (Kohlstedt and Goetze 1974; Durham 1975; Durham and Goetze 1977; Durham et al 1977; Durham et al 1979; Darot and Gueguen 1981; Gueguen and Darot 1982) with a view to characterize the slip systems. In addition to usual TEM characterizations, studies on olivine have benefited from the decoration technique developed by Kohlsted et al. (1976). Annealing at 900°C in air results in the precipitation of Fe_2O_3 and/or Fe_3O_4 on dislocation cores. The dislocation lines can thus be observed in the optical microscope. This technique was extended to forsterite by Jaoul et al. (1979).

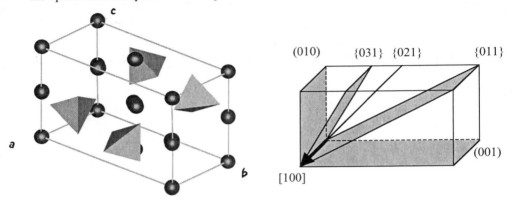

Figure 27. The forsterite unit cell. **Figure 28.** Possible glide planes of [100] dislocations in olivine.

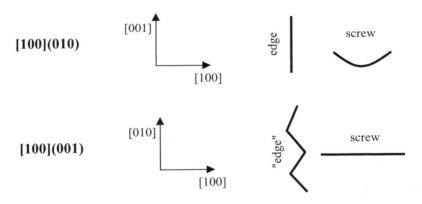

Figure 29. Most common preferred line orientations of [100] dislocations in olivine.

The common slip systems at high temperature involve [100] slip on several planes: (010), {031}, {021}, {011} and (001) (Fig. 28). Cross slip between these different planes is frequent. Dislocation lines exhibit marked crystallographic orientations characteristic of high lattice friction, even at high temperature. In (010), long edge segments are observed with short curved screw segments (Fig. 29). This is observed in natural samples (Gueguen 1979) as well as in experiments (Durham 1975; Jaoul et al. 1979). Edge [100] dislocations gliding in (001) exhibit a peculiar microstructure with zig-zag line configurations corresponding to the stabilisation of <110> segments. When several slip systems [100](0kl) are activated, one can find free screw dislocations located between (100) tilt boundaries (Durham 1975; Gueguen 1979). At low temperature and high stresses, slip occurs along [001] in (100), {110} and (010) (Fig. 30). [001] screw dislocations are always found to exhibit very marked crystallographic characters and are more pervasive that edge segments. Twist boundaries are very common; they are usually constituted of [100] and [001] screw dislocations.

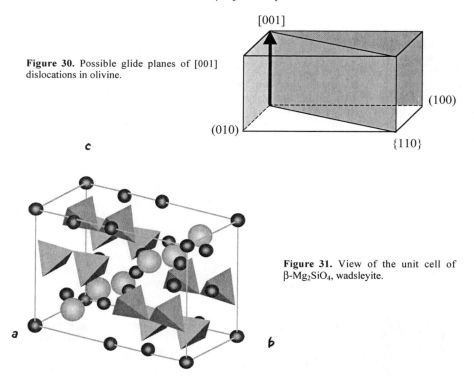

Figure 30. Possible glide planes of [001] dislocations in olivine.

Figure 31. View of the unit cell of β-Mg$_2$SiO$_4$, wadsleyite.

High temperature creep experiments performed on single crystals strained along various orientations have in common to follow a power-law creep equation with a stress exponent n ≈ 3.5 compatible with dislocation creep controlled by climb of edge dislocations.

Wadsleyite. The mineral wadsleyite (β-Mg$_2$SiO$_4$) has a spinelloid structure made of MgO$_6$ octahedra and SiO$_4$ tetrahedra. The Bravais lattice is orthorhombic (space group *Imma*) with eight formula units per unit cell (Fig. 31). The lattice parameters are $a = 0.56983$ nm, $b = 1.1438$ nm and $c = 0.82566$ nm (Horiuchi and Sawamoto 1981).

Several studies have been devoted recently to the plastic deformation of wadsleyite using the multianvil apparatus. Dupas et al. (1994) have transformed a synthetic harzburgite in the β stability field (14 GPa, 1100-1200°C). Apart from the staking faults in (010), relatively few defects are observed in wadsleyite, in contrast with remnant olivine grains which show high densities of **c** screw dislocations. The only dislocations observed are emitted at grain boundaries to relax local stress concentrations. The following slip systems have been characterized: [100]{021} (with a predominance of screw segments) and 1/2<111>{101}. In a further study, Dupas-Bruzek et al. (1998) transformed San Carlos olivine at 900°C in the β stability field (15 GPa) for 0.5 h and in the β+γ stability field (16 GPa) for 11 h. As in the previous study, dislocations are mostly found in remnant olivine grains. In wadsleyite, dislocations with [100] and 1/2<111> Burgers vectors are observed. [100] dislocations are arranged in subgrain boundaries. This led the authors to suggest that recovery by dislocation climb is active in wadsleyite at 900°C. Higher dislocation densities were introduced in wadsleyite by Sharp et al.

(1994) who transformed San Carlos olivine at pressures of 14 GPa and greater and 1450°C. Long annealing time under pressure and temperature were chosen to allow for significant creep to take place. Dislocations commonly occur in walls although tangles are also observed. The Burgers vectors are found to be [100] and orientations of dislocation segments in subgrain boundaries suggest (010) as a slip plane. Plastic deformation on pre-synthesized wadsleyite samples (of forsterite composition) have been performed by Thurel (2001) in the multianvil apparatus with the modified assembly described in Figure 10. Deformation conditions are in the pressure range 15 to 19 GPa with temperatures up to 1800°C. Complex microstructures are produced with dislocation densities in the range 10^{13} to 10^{14} m^{-2}. Dislocations are in glide configurations (Fig. 32) and many slip systems are activated. LACBED analysis confirmed the occurrence of [100] and 1/2<111> dislocations but showed that [010] (Fig. 33) and <101> slip was also activated. Characterization in the TEM of the slip planes led to the following slip systems: 1/2<111>{101}, [100](010), [100](001), [100]{011}, [100]{021}, [010](001), [010]{101} and <101>(010). Except for {021}, the slip planes are selected among those that do not shear the strong SiO$_4$ bonds (Fig. 34). The activation of slip along [010] and <101> is surprising because it corresponds to dislocations with relatively high elastic energy (Table 1) and violates the "Short Burgers vectors" rule proposed above. It is also interesting to note that [001] dislocations are almost absent. Detailed observation in weak-beam dark-field shows that [010] and <101> dislocations are dissociated. <101> dislocations are dissociated in the (010) plane (Fig. 35) following the reaction:

<101> → 1/2<101> + 1/2<101>.

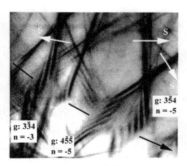

Figure 32 (left). Mg$_2$SiO$_4$ wadsleyite deformed at 18 GPa, 1600°C. TEM dark-field micrograph (g: 244) of a grain deformed by conjugate [100] and 1/2<111> slip (courtesy E. Thurel).
Figure 33 (right). LACBED characterization of a [010] dislocation in deformed wadsleyite (courtesy E. Thurel).

[010] dislocations, which are unstable with respect to the Frank criterion, decompose spontaneously into two 1/2<111> dislocations (screw segments) in a {101} plane or dissociate into four 1/4[010] partial dislocations (Fig. 36). Such relaxation events suggest a complex thermomechanical history. Indeed, experiment in which the specimens were loaded at high pressure in the deformation assembly, but not heated, show significant plastic deformation of wadsleyite during pressurization at room temperature. To overcome this problem, further deformation experiments have been performed on pre-synthesized wadsleyite samples with the shear deformation assembly designed by Karato and Rubie (1997). This experiment allows to obtain large strains at high temperature and thus high-temperature microstructures to overprint low-temperature ones. Experiments performed at 14 GPa, 1300°C on wadsleyite with Fo$_{90}$ composition confirmed that 1/2<111>{101}, [100](010) and [100]{011} are the easy slip systems of wadsleyite under these conditions (Thurel 2001; see Fig. 37).

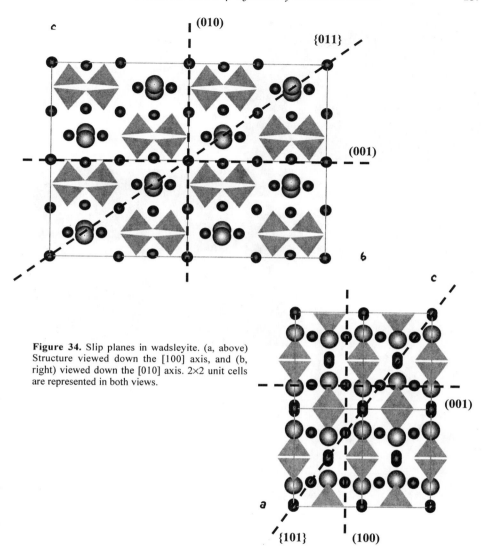

Figure 34. Slip planes in wadsleyite. (a, above) Structure viewed down the [100] axis, and (b, right) viewed down the [010] axis. 2×2 unit cells are represented in both views.

Table 1. Burgers vector modulus and elastic energy of the dislocations characterized by LACBED in Mg_2SiO_4 wadsleyite deformed in the multianvil apparatus.

Burgers vector	Modulus (Å)	μb^2 (Pa.m²)	Relative elastic energy
[100]	5.70	$4.06.10^{-8}$	1
1/2<111>	7.62	$7.25.10^{-8}$	1.8
[001]	8.27	$8.55.10^{-8}$	2.1
<101>	10.04	$1.26.10^{-7}$	3.1
[010]	11.45	$1.70.10^{-7}$	4.2

Figure 35. Mg$_2$SiO$_4$ wadsleyite deformed at 17 GPa, 1500-1600°C. TEM weak-beam dark-field (g: 211). 1/2<101> partial dislocations gliding in (010) (courtesy E. Thurel).

Figure 36. Mg$_2$SiO$_4$ wadsleyite deformed at 17 GPa, 1500-1600°C. TEM weak-beam dark-field (g: 240). Glide dissociation into four 1/4[010] of screw segments of a [010] dislocation (courtesy of E. Thurel).

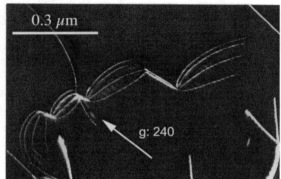

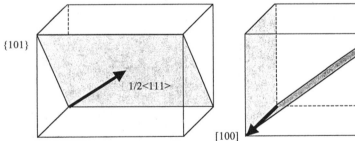

Figure 37. Easy slip systems in wadsleyite.

Ringwoodite. (Mg,Fe)$_2$SiO$_4$, ringwoodite, has the spinel structure (Fig. 38) with the Mg^{2+} or Fe^{2+} atoms occupying one-half of the octahedral sites and Si^{4+} occupying one-eighth of the tetrahedral sites within the nominal face-centered-cubic packing of the oxygen sublattice. Ringwoodite belongs to the cubic system (space group $Fd3m$). The lattice parameter of Mg$_2$SiO$_4$ ringwoodite determined by Ringwood and Major (1970) is $a = 0.8071$ nm. The unit cell contains eight formula units.

The slip direction is always observed to be parallel to the <110> close packed direction of the fcc lattice in the spinel structure (Mitchell 1999). 1/2<110> is then the shortest perfect Burgers vector. On the other hand, the observed slip plane is variable among the dense planes of the oxygen sublattice (Fig. 39). The most common glide planes are {111} and {110}, although {100} has been reported in magnetite, nickel ferrite, and chromite (Mitchell 1999 and references therein).

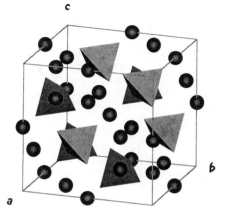

Figure 38. View of the unit cell of ringwoodite.

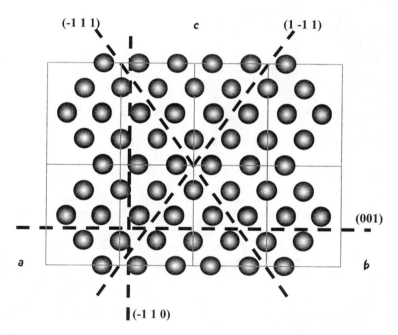

Figure 39. Structure of the oxygen sublattice of ringwoodite. 2×2×2 unit cells viewed down the [110] axis. The dense planes are represented.

The first defect characterizations of crystal defects in ringwoodite have been performed on shocked chondrites. Madon and Poirier (1980, 1983) have observed ringwoodite grains formed in shock veins from Tenham. They contain pervasive planar defects on the three families of {110} planes as well as 1/2<110> dislocations with a 60° character lying in {111} planes. The usual splitting scheme of spinel dislocations: 1/2[110] → 1/4[110] + 1/4[110] has not been observed.

Plastic deformation experiments of $(Mg, Fe)_2SiO_4$ ringwoodite have been performed at high-P, high-T by Karato et al. (1998) in the multianvil apparatus using the shear deformation assembly developed by Karato and Rubie (1997). Synthetic olivine aggregates have been transformed to ringwoodite and deformed at 16 GPa and temperatures in the range 1400-1600 K. Although small-grained samples (~0.5 μm) show evidence for grain-boundary sliding mostly, large-grained samples (above 3 μm) deform by dislocation creep. TEM investigation showed that deformation occurs mostly through slip of 1/2<110> dislocations over {111} planes although {100} was also noted. In a recent work, Thurel (2001) has studied Mg_2SiO_4 ringwoodite deformed at 22 GPa and 1000-1400°C using the compression assembly described above. Her study confirms that 1/2<110>{111} is the prominent slip systems, however, she found 1/2<110>{110} as subsidiary slip system. In both cases, dislocations appear to be in glide configurations with marked crystallographic orientations (screw, 45°, 60° or edge) and show no dissociation in weak-beam dark-field TEM micrographs (see Fig. 40).

Figure 40. Mg_2SiO_4 ringwoodite deformed at 22 GPa, 1000°C. TEM weak-beam dark-field image (g: 022). Dislocations in glide configuration. Courtesy of E. Thurel.

Perovskite, ilmenite. Perovskite is considered to be the dominant mineral in the Earth's lower mantle. This phase appears in the $MgSiO_3$ phase diagram (Fig. 41) for pressure above ca. 25 GPa. Below this pressure, one finds majorite at high temperature and akimotoite at low temperature. Majorite has a garnet structure and will be addressed in the garnet section.

Akimotoite. $MgSiO_3$ with the ilmenite structure has been named akimotoite after this mineral was identified in shocked meteorites (Sharp et al. 1997; Tomioka and Fujino 1997, 1999). It is most likely to occur at the bottom of the transition zone and top of the lower mantle. The structure of $MgSiO_3$ ilmenite has been determined by Horiuchi et al. (1982). It is trigonal (space group $R\overline{3}$) with $a = 0.47284$ nm, $c = 1.35591$ nm; it contains six formula units per cell. The structure consists of alternating layers of MgO_6 and SiO_6 octahedra stacked along the c-axis (Fig. 42).

We have carried out deformation experiments on $MgSiO_3$ akimotoite in the multi-anvil apparatus using the procedure described above and the deformation assembly of Figure 10. The samples were pre-synthesized in a conventional HP assembly from $MgSiO_3$ glass at 21 GPa, 1500°C. Two deformation experiments were carried out at 21 GPa and 1000°C and 1400°C. It is shown at the TEM scale that glide occurs mostly in the basal plane [1/3<1120>(0001); see Fig. 43], although slip in pyramidal planes is also involved.

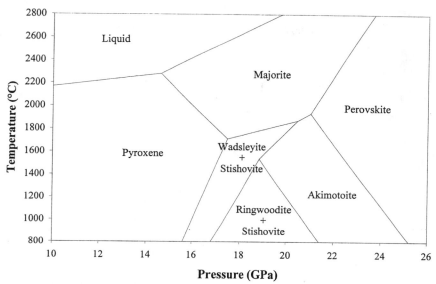

Figure 41. Simplified phase diagram for the MgSiO₃ system (after Fei et al. 1990).

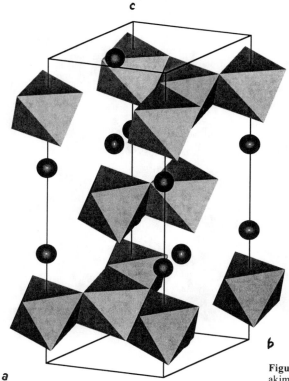

(Mg,Fe)SiO₃ perovskite.
The structure of MgSiO$_3$ perovskite consists of a corner-linked network of SiO$_6$ octahedra with Mg atoms in a cavity formed by eight octahedra. The structure is orthorhombic (space group *Pbnm*; Horiuchi et al. 1987) and the unit cell contains four formula units (Fig. 44). This structure differs from the ideal cubic perovskite (space group *Pm3m*) in that the octahedra are tilted and the Mg atoms are displaced from the centers of their octahedral sites (Fig. 45). Despite this deviation, it is usual to introduce a pseudo-cubic lattice to describe the perovskite structures. This concept is useful to discuss the dislocations and slip systems be-cause it provides a unified view for all perovskite structures.

Figure 42. The unit cell of MgSiO$_3$ akimotoite. The octahedra are represented at the silicon level only.

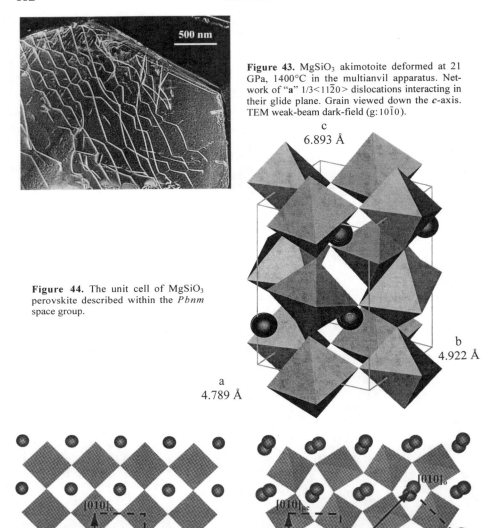

Figure 43. MgSiO$_3$ akimotoite deformed at 21 GPa, 1400°C in the multianvil apparatus. Network of "**a**" $1/3<11\bar{2}0>$ dislocations interacting in their glide plane. Grain viewed down the c-axis. TEM weak-beam dark-field (g: $10\bar{1}0$).

Figure 44. The unit cell of MgSiO$_3$ perovskite described within the *Pbnm* space group.

Figure 45. (a) Ideal (cubic) perovskite structure viewed along [001] with a unit cell representation. (b) Structure of MgSiO$_3$ perovskite viewed along the longest axis: [001]$_O$ (the subscript "o" stands for orthorhombic). An orthorhombic unit cell is represented to the right. The structure can be approximately described within a pseudo-cubic unit cell (shown on the left—labeled with subscript "pc") with $a_{pc} \approx a_o/\sqrt{2} \approx b_o/\sqrt{2} \approx c_o/2$.

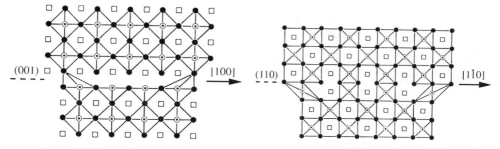

Figure 46 (left). Dissociated (unrelaxed) [001](100) dislocation in the ideal, cubic, perovskite structure. The octahedra are figured with the oxygen ions represented as circles, the Mg cations are represented as squares (after Poirier et al. 1989).

Figure 47 (right). Dissociated (unrelaxed) [1$\bar{1}$0](110) dislocation in the ideal, cubic, perovskite structure. (after Poirier et al. 1989).

The possible dislocations in ideal cubic perovskites have been discussed by Poirier et al. (1983; 1989), who pointed out that the perovskite structure could be seen as a ReO_3 structure stuffed with small cations at the center of the octahedra. Based on this analogy, they suggested that possible slip systems could be $<100>\{001\}$ and $<110>_c\{1\bar{1}0\}_c$. Possible core structures including dissociation into two colinear partial vectors have been proposed (Figs. 46 and 47). Deformation experiments have been conducted on a large number of analogue materials: $BaTiO_3$ (Doukhan and Doukhan 1986; Beauchesne and Poirier 1989), $CaTiO_3$ (Doukhan and Doukhan 1986; Wright et al. 1992; Besson et al. 1996), $SrTiO_3$ (Wang et al. 1993; Mao and Knowles 1996; Gumbsch et al. 2001), $NaNbO_3$ (Wright et al. 1992), $KTaO_3$ (Beauchesne and Poirier 1990) $KNbO_3$ (Beauchesne and Poirier 1990), $KZnF_3$ (Poirier et al. 1983), $YAlO_3$ (Wang et al. 1999). It appears from these studies that perovskites do not constitute an isomechanical group. However, they confirmed that the dislocations in perovskite have $<100>_c$ and $<110>_c$ Burgers vectors. Moreover, the $<110>_c\{1\bar{1}0\}_c$ slip system seems to be activated from room temperature and controlled by glide, whereas the $<100>_c\{001\}_c$ slip systems would be characteristic of high temperature only and controlled by climb. A strength anomaly comparable to the one described in Figure 9 has been reported recently by Gumbsch et al. (2001) in $SrTiO_3$ deformed in compression in the temperature range 100-1800 K. Microstructural investigations suggest that this behavior is related to climb dissociation of dislocations. Indeed, climb dissociation of <110> dislocations had already been reported in $SrTiO_3$ by Mao and Knowles (1996), and discussed by Poirier et al. (1983).

The first deformation experiments on $MgSiO_3$ perovskite have been performed at room temperature by Karato et al. (1990) using microindentation. They determined an average hardness $H_v = 18\pm2$ GPa. Coarse-grained polycrystals of $MgSiO_3$ perovskite have been deformed at P = 24 GPa and T = 1000-1400°C in the multianvil apparatus using the procedure described in Cordier and Rubie (2001). Optical microscopy shows dense deformation bands which interact strongly with the twins of orthorhombic perovskite (Fig. 48). Given the high sensitivity of $MgSiO_3$ perovskite to electron-irradiation, the dislocation microstructure of these samples has been analyzed by T. Ungár and colleagues using the peak broadening method described above. Preliminary results show that both $<100>_c$ and $<110>_c$ dislocations are present in these samples with densities in the range 1.8-3.6 $10^{12} m^{-2}$.

Magnesiowüstite (Mg,Fe)O. Magnesiowüstite (Mw) is likely to be the second most abundant phase of the lower mantle. It is the result of a large solid solution between periclase (MgO) and wüstite (FeO) which exhibits the same crystal structure. Although

Figure 48. MgSiO$_3$ perovskite deformed at 25 GPa, 1400°C in the multianvil apparatus. Optical micrograph showing orthorhombic twins sheared by dense deformation bands.

Figure 49. View of the unit cell of MgO.

its inferred volumetric ratio is not high enough to make Mw control the rheology of the lower mantle, one cannot exclude that there are regions of the lower mantle that are enriched in Mw. Mw is the only phase to keep the same crystal structure ("NaCl-type") from lower mantle to laboratory conditions (Duffy et al. 1995). The Bravais lattice of Mw is fcc, space group $Fm\overline{3}m$. The lattice parameters of the end-members are: 0.4211 nm for MgO (Hazen 1976) and 0.43108 nm for FeO (Foster and Welch 1956). Mg$^+$ (or Fe^{2+}) and O^{2-} are displaced by half a diagonal of the cube (Fig. 49). The predominant slip systems in NaCl-type crystals are 1/2<110>{110} (Fig. 49). Each {110} plane contains a single slip direction, resulting in 6 possibilities. Among those, some slip planes cross at 90°, some others at 60 or 120°, resulting in different dislocation interactions. Slip is also possible over other close-packed planes such as {100} or {111}. When a dislocation moves on a {110} plane, ions of opposite sign approach each other at the core of the dislocation in a "midglide" position (Fig. 50). For dislocation movement on {100} planes on the other hand, Gilman (1959) has pointed out that strong repulsive forces develop because ions of like signs are forced to approach each other (Fig. 51). Using single-crystal deformation experiments on periclase at around 0.63 T$_m$, Routbort (1979) showed that slip on 1/2<110>{110} alone was about four times easier than combined slip on 1/2<110>{110} and 1/2<110>{001}. Indeed, for MgO, the Peierls regime is not observed down to 4 K on the {110} planes, while it is operating below 1200 K for the {001} planes (Fig. 52).

Many experiments have been performed on single crystals loaded at room pressure along <100> to activate the easy slip systems. Below 1300°C (0.52 T$_m$), optical birefringence shows that slip in any given region is restricted to a single 1/2<110>{110} slip system; there is no extensive penetration between the various systems. Stress-strain

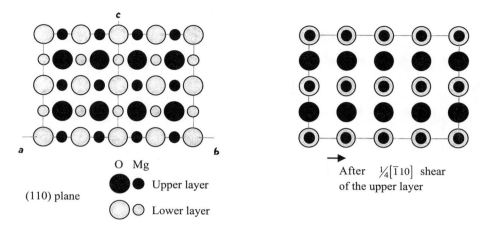

Figure 50. Slip on {110} in MgO. Left side shows two atomic layers of MgO in the unslipped position. Right side shows the same in the half-glided position along $\frac{1}{2}[\bar{1}10]$.

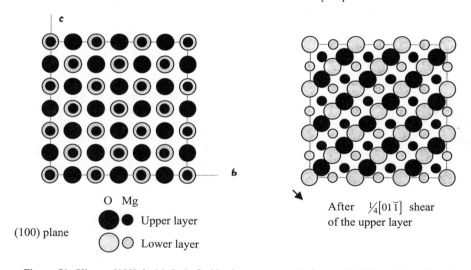

Figure 51. Slip on {100} in MgO. Left side shows two atomic layers of MgO in the unslipped position. The right side shows the same in the half-glided position along $\frac{1}{2}[01\bar{1}]$.

curves are characterised by a fairly high rate of strain-hardening. Between 1300 and 1700°C (0.52-0.65 T_m), slip on conjugate (90°) systems are able to interpenetrate whereas slip on systems making an angle of 60° are not (Day and Stokes 1964, Copley and Pask 1965). There is no dislocation reaction at the crossing of two conjugate slip systems because the dislocation reaction $1/2[\bar{1}10] + 1/2[110] \rightarrow [100]$ is not favorable. Crossing just involves mutual cutting. From a geometrical point of view, the intersection between two conjugate {110} planes is parallel to the line of edge dislocations. The primary intersection to be considered is between screw dislocations in one plane and edges or screws in the other plane. The jog on a screw cannot move forward and leaves a cusp (eventually an edge dislocation dipole) in its wake.

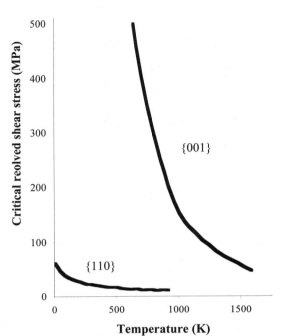

Figure 52. Critical resolved shear stress for MgO on {110} and {100}.

A jog on an edge dislocation can move conservatively with the dislocation although it is still required to move over the {100} plane. Above 1700°C (0.65 T_m), the specimens begin to harden again after 20-30% elongation in tension (Day and Stokes 1964). The shape evolution of the specimen shows that all slip systems (at 60 and 90°) can interpenetrate. This, together with the increased mobility of dislocation and point defects, permits the rearrangement and polygonization of dislocations. Dislocations belonging to slip systems at 60° can react favorably according to the reaction $1/2[110] + 1/2[0\bar{1}1] \rightarrow 1/2[101]$ to form a third dislocation lying parallel to <111>. The resulting dislocation, which is pure edge, can move only in the {112} slip plane (Day and Stokes 1964; Copley and Pask 1965; Reppich and Hüther 1974). At low temperature, slip does not normally occur over this plane and the dislocation may be regarded as sessile.

Typical dislocation structures after transient creep at ~1700°C (0.65 T_m) include dislocation walls and random network of dislocations in between (Hüther and Reppich 1973). With increasing deformation, the regions with random dislocation arrangements disappear and the parallel cell walls loose their preferred orientations and build up a nearly equiaxed cellular network. The cell boundaries act in two ways: firstly as "forest" for glide dislocations contributing to the work hardening, and as "sinks" (for dislocation annihilation) preventing the increase in flow stress. Thus, the rate-controlling processes occur in the dislocation network of the cell boundaries (Reppich and Hüther 1974).

Loading along <110> enhances the probability for multiple slip at high temperature because it results in a decrease of the resolved shear stress over the {110} planes relative to other close-packed planes. Taking τ_{011} as the resolved shear stress over the (011) plane, we have for the [001] tensile axis:

$\tau_{100} = 0$ and $\tau_{111} = 0{,}83\ \tau_{011}$

whereas for the [110] tensile axis:

$\tau_{100} = 1.41\ \tau_{011}$ and $\tau_{111} = 1.65\ \tau_{011}$

Below 1700°C (0.65T_m), the major slip systems (determined from the slip traces) are: <110>{110} (Day and Stokes 1966; Clauer and Wilcox 1976) although evidence for slip on <110>{001} and 1/2<110>{111} is also reported (Clauer and Wilcox 1976). Slip systems at 90° could interpenetrate but those at 60° could not (Day and Stokes 1966). The wavy nature of the deformation bands suggests that considerable cross-slip took place during creep (Clauer and Wilcox 1976).

Above 1700°C, slip on all systems could interpenetrate. Slip occurs over a variety of planes, but primarily over the plane of maximum macroscopic shear stress. Since this is an irrational plane, shear is accompanied by cross slip. Crystals pulled more than 60% above 1700°C completely recrystallized with a fine grain size. The grains grew on further annealing at 2000°C to yield simple polycrystalline specimens (Day and Stokes 1966).

Deformation along <111> results in an equal resolved shear stress on three of the six 1/2<110>{001} slip systems and no stress on the 1/2<110>{110} family. Yield stress for this orientation is considerably greater than for the <100> orientations and continuously decreased up to 1600°C (0.61 T_m) (Copley and Pask 1965). The ratio of the yield stresses is about 13:1 at 350°C (0.2 T_m), 4.5:1 at 600°C (0.29 T_m) and 3:1 at 1600°C (0.61 T_m).

Creep experiments performed recently on $Mg_{0.8}Fe_{0.2}O$ polycrystals at 0.44 T_m yielded CPO compatible with slip on both 1/2<110>{110} and 1/2<110>{001} although slip on {111} could not be excluded (Stretton et al. 2001).

Karki et al. (1997) have calculated the evolution of the elastic constants of MgO with pressure. They find dramatic change in elastic anisotropy with increasing pressure. We have calculated the anisotropic shear moduli $\mu_{[uvw](hkl)}$ associated with shear along 1/2<110>, parallel to {110}, {001} and {111} (Fig. 53). It is shown that 1/2<110>{110} hardens considerably from the elastic point of view compared to 1/2<110>{001}. 1/2<110>{111} has an intermediate behavior. The influence of shear modulus on plasticity has been discussed by Poirier (1982). He remarks that the critical length for activating a Frank-Read source varies as μ/σ and that a decrease in modulus increases the dislocation density (more sources are available from the Frank network), causing an extra creep rate for the corresponding slip system. Oppositely, an increase of a shear modulus could decrease the activity of a slip system. Dramatic changes in slip systems can thus be anticipated for Mw at great depths with possible implications on the crystal preferred orientations and hence on the seismic properties.

Garnets

Garnets represent an important mineral family for the rheological behavior of the upper mantle, and the mantle transition zone. They are often regarded as being very resistant to plastic flow.

The structure of garnets is based on a bcc lattice (space group $Ia3d$). The unit cell (Fig. 54) contains eight structural units $X_3Y_2Z_3O_{12}$. It can be viewed as a three-dimensional network of ZO_4 tetrahedra (SiO_4 in case of silicate garnets) and YO_6 octahedra sharing corners and forming dodecahadral cavities filled with X cations. Therefore garnets comprise a family of complex silicates with widely varying chemical composition. Garnets with no Ca in the X site and Al in the Y site are called the pyralspite series. These consist of the end members:

- Pyrope - $Mg_3Al_2Si_3O_{12}$ (a = 1.146 nm)
- Almandine $Fe_3Al_2Si_3O_{12}$ (a = 1.153 nm)
- Spessartine $Mn_3Al_2Si_3O_{12}$ (a = 1.162 nm)

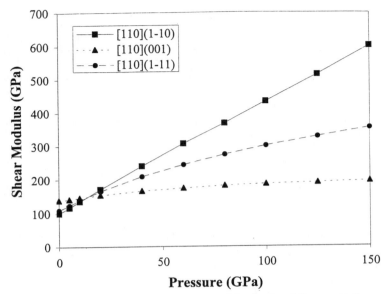

Figure 53. Evolution with increasing pressure of the shear moduli corresponding to the three slip systems 1/2<110>{110}, 1/2<110>{111} and 1/2<110>{001} of MgO. The calculation is based on anisotropic elasticity, using the elastic constants proposed by Karki et al. (1997).

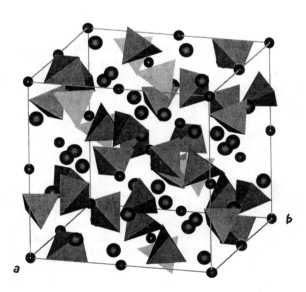

Figure 54. View of the unit cell of $Mg_3Al_2Si_3O_{12}$ pyrope.

Garnets with Ca in the A site are called the ugrandite series and consist of the end members:
- Uvarovite $Ca_3Cr_2Si_3O_{12}$ (a = 1.200 nm)
- Grossularite $Ca_3Al_2Si_3O_{12}$ (a = 1.185 nm)
- Andradite $Ca_3Fe^{+3}_2Si_3O_{12}$ (a = 1.206 nm)

In the transition zone, pyroxenes dissolve into the garnets resulting in a silicon-enriched composition called majorite: $Mg_3(Mg, Si)_2Si_3O_{12}$ (= $MgSiO_3$). The occurrence of Mg^{2+} cations (which are larger than the Al^{3+} cations) in the octahedral sites induces a distortion of the octahedron and in turn a displacement of the tetrahedra and of the dodecahedra. This results in a slight tetragonal distortion of the structure of the majorite which is described within the $I4_1/a$ space group (Angel et al. 1989). The lattice parameters of $MgSiO_3$ majorite are a = 1.1501 nm and c = 1.1480 nm with 32 formula units per cell.

In cubic garnets, the shortest lattice repeat is 1/2<111> and, as in bcc metals, one can expect plastic deformation to be dominated by 1/2<111> dislocations. One must note however that, given the large unit cell of the garnet structure, such dislocation Burgers vector exhibit an unusually large magnitude (close to 1 nm). As in bcc metals, 1/2<111> dislocations from different slip systems are likely to react and form junctions following the reaction: $1/2[1\bar{1}1]+1/2[11\bar{1}] \to [100]$.

Some earlier studies have given evidence that garnets could deform plastically in nature. Dalziel and Bailey (1968) and Ross (1973) used the distortions of Laue spots on X-ray diffraction photographs to infer for crystal plasticity in garnets from mylonitic rocks. Carstens (1969; 1971) has revealed the dislocations in pyrope garnets by etching the dislocation cores with a HF solution. The etch channels are then observed by optical microscopy. This method produced very clear pictures of the microstructures provided the dislocation density remains small enough. No information can be derived on the Burgers vectors using this method. Dislocation tangles and cell structures have been evidenced by this technique in pyropes from Norwegian and Czech peridotites (Carstens 1969). Several investigations on deformation microstructures in natural garnets have been carried out in the past few years using electron microscopy (Allen et al. 1987; Ando et al. 1993; Doukhan et al. 1994; Ji and Martignole 1994; Chen et al. 1996; Voegelé et al. 1998b, Prior et al. 2000). All these studies showed evidence for plastic deformation by dislocation creep. Characterizations of Burgers vectors lead to 1/2<111> and <100> dislocations (Ando et al. 1993; Doukhan et al. 1994; Voegelé et al. 1998b). It is impossible however to decide in these samples whether <100> slip has really been activated or if <100> dislocations are junction products. Using LACBED, Voegelé et al. (1998b) have shown that 1/2<111> dislocations glide in three kinds of planes: {110}, {112} and {123}. The ease of glide seems to be close for these three families of planes although {110} seems to be the most frequently found. Dislocations often exhibit very marked crystallographic orientations: e.g., <111> (70°, not screws! —see Fig. 55), <110> and <100> for 1/2<111> dislocations gliding in {110} planes (the easiest glide system). This marked anisotropy cannot be explained by elastic considerations, it must correspond to the structure of the core of the dislocations. LACBED studies on eclogites from Sesia Lanzo have revealed the occurrence of <110> dislocations which have an uncommonly long Burgers vector (Voegelé et al. 1998b). These dislocations always appear to be dissociated (Fig. 56).

Because they are stable at room pressure, most experimental studies on plastic deformation of garnets have been undertaken on non-silicate garnets: YAG (Hardiman et al. 1973; Parthasarathy et al. 1992; Corman 1993; Karato et al. 1994; Blumenthal and Phillips 1996), YIG (Rabier et al. 1976 a,b; Rabier 1979; Rabier et al. 1979; Rabier et al.

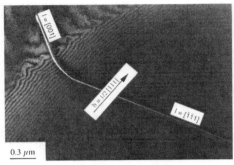 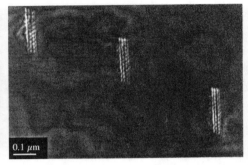

Figure 55 (left). Garnet amphibolite from Bragance. Part of a 1/2[111] dislocation loop gliding in (1 $\bar{1}$ 0). Weak-beam dark field micrograph. g: 400 (courtesy V. Voegelé).

Figure 56 (right). Eclogites from the Alps. <110> dislocation dissociated in four partials (climb dissociation). Weak-beam dark field micrograph. g: 400 (courtesy V. Voegelé).

1981), GGG (Garem et al. 1982; Garem 1983; Garem et al. 1985; Wang et al. 1996) and CGGG (Voegelé et al. 1999). However, the plastic deformation of silicate garnets has been addressed using hardness tests (Karato et al. 1995a), creep (Karato et al. 1995a; Wang and Ji 1999) and high-pressure multianvil experiments (Voegelé et al. 1998a). Garnets exhibit a brittle-ductile transition around 1000°C at the timescale of laboratory experiments (Wang and Ji 1999; Voegelé et al. 1998a). Although this transition is shifted to lower temperature at natural strain-rates, it is suggested that ductile deformation of garnets is expected under mantle conditions only. Experimentally, 1/2<111> dislocations are found to glide in {110} (Wang and Ji 1999; Voegelé et al. 1998a), {112} and {123} (Voegelé et al. 1998a). <100> dislocations compatible with the <100>{011} and the <100>{010} slip systems have also been identified (Voegelé et al. 1998a); see Figure 57.

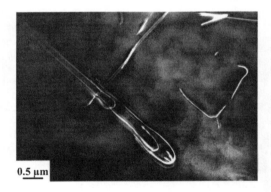

Figure 57. Garnet deformed in the multi-anvil apparatus at 6.5 GPa, 900°C. [001] dislocation loops. Weak-beam dark field micrograph. g: 040 (courtesy V. Voegelé).

Garnets represent an interesting case in crystal plasticity of a material deforming with dislocations having very large Burgers vectors (Nabarro 1984). This induces a very strong lattice-dislocation interaction (Peierls forces) as shown by the marked crystallographic orientations of the dislocation lines. Complex dislocation core structures, probably involving an extended, non-planar core, is probably responsible for the intrinsic lattice resistance to plastic shear. The core of such dislocations must expand and contract as the dislocations move from one Peierls valley to another ("breathing" of the core). Given the size expected for the core in such a structure as garnets, this renders dislocation glide very unfavorable. The brittle, Peierls regime is found to operate up to high temperature: ~1000°C. Above this temperature the dislocations become mobile. This temperature is also that for the onset for recovery (as shown by the formation of subgrain boundaries) which shows that diffusion is efficient. Voegelé et al. (1998a) suggest that dislocation glide should be assisted by diffusion. Diffusion-assisted glide is still a conservative process. Local diffusion of point defects allows displacement of the sessile component of the dislocation core. The activation energy for the glide process is the migration energy of the point defects involved. 1/2<111> and <100> slip operate in the whole temperature range.

1/2<111> and <100> have also been characterized in majorite garnets of both synthetic and meteoritic origin (Voegelé et al. 2000). The case of majorite garnets provides another example for the influence of a small lattice distortion (here the tetragonal distortion from $Ia3d$ to $I4_1/a$) on dislocations. In the present case, the Bravais mode is centered in both cases and both 1/2<111> and <100> dislocations can be expected. However, the tetragonal distortion makes [100] and [001] dislocations having slightly different Burgers vector magnitudes and hence slightly different energies. However, these differences are not important enough to govern the dislocation activity. TEM observations do not suggest significant changes (such as dissociation) of the dislocation cores either.

WHERE DO WE STAND?

A new field has emerged with the possibility of achieving deformation experiments under high-pressures exceeding ~10 GPa. Indeed, pressure appears clearly as the key parameter when dealing with the Earth's interior. The most important effect of pressure is probably the occurrence of the phase transformations that occur in the Earth's mantle. The importance of these phase transitions was recognized early on because they are critical to the understanding of the discontinuities observed by the seismologists. They also have a major effect on the rheology and hence on the dynamics of the mantle. We have seen that plastic deformation can occur through intracrystalline mechanisms (slip on crystallographic planes) that strongly depend on the crystal structure. This can have purely geometrical consequences such as the development of crystal preferred orientations that gives rise to seismic anisotropy (see Mainprice et al. 2000). The nature of dislocations and slip systems have direct consequences on the mechanical properties of the minerals and hence on the rheology of the mantle. Some structures appear very symmetrical and produce a large number of easy slip systems (garnets, MgO, ringwoodite or perovskite, for instance) when some other structures exhibit less degrees of freedom for slip (olivine, of course, which has only two orthogonal slip directions, but also wadsleyite and may be akimotoite). In some cases, the crystal structure is responsible for the activation of dislocations with very large Burgers vectors (wadsleyite, garnets). Such dislocations are clearly difficult to glide (lattice friction) and the mechanical properties are likely to depend strongly on whether glide dissociation is possible (wadsleyite) or not (garnets). Pressure also has a strong effect on the physical properties of the minerals, and among them, elasticity is one of the most important. Indeed, elasticity governs the long-range interactions between dislocations as well as the forces applied to them. We have seen in case of SiO_2 or MgO that the evolution with pressure of the elastic properties is likely to have major effects on the rheology.

The importance of the progress achieved should not lead us to forget the limitations of the studies so far conducted. In particular, it seems that high-pressure deformation experiments performed nowadays involve relatively high stress levels (compared to what is expected in the Earth's mantle). It is well known that dominant slip systems in a given material may change not only with temperature, but also with stress magnitude. This effect has been well demonstrated in case of olivine for instance ([001] vs. [100] glide). This problem raises the more general question of the extrapolation of laboratory data to natural conditions. This extrapolation involves scaling both in time and in space. There is no doubt that a detailed understanding of the physical processes involved in the plastic deformation of high-pressure minerals will be needed to address this important issue.

The study on wadsleyite has pointed out the importance of dissociation on the choice of the slip systems. More generally, most high-pressure minerals exhibit a significant lattice friction that is probably governed by the dislocation core structure. It is important to realize that the lack of information about the core structure is probably the main drawback of our field at present. The dislocation core structure is crucial because it governs to a large degree the dislocation mobility. An example has been shown with $SrTiO_3$ perovskite. In most cases the core structure is not accessible to direct characterization. It is certainly one of the major issues for the future to address the problem of dislocation core and mobility.

We have discussed so far the influence on dislocation mobility of the lattice (friction) and of other dislocations (long- or short range interactions). The interaction with point defects or impurities can also have a strong influence on the mechanical properties. Among them is water, which is likely to be present everywhere in the mantle, albeit sometimes at the parts per million (ppm) concentration (Thompson 1992). The possible influence of water-related defects on plasticity has been shown in a spectacular way in the case of quartz (see Poirier 1995). The rheology of olivine is also affected by water although the effect is less marked in that case (Mackwell 1984; Mackwell et al. 1985; Mei and Kohlstedt 2000; Karato and Jung 2002). Beyond weakening, the effect of water might be to change the crystal preferred orientations (Jung and Karato 2001) with possible implications for seismic anisotropy. It is probably naive to infer from the quartz and olivine cases that water should weaken all mantle minerals in a similar way. It appears that water weakening involves very specific mechanisms which are already found to be very different in quartz and in olivine. It seems thus necessary to assess for each phase, the possible influence of water on the rheological properties.

Some important issues have not been addressed in the present paper due to space limitations. One is related to diffusion. Although the importance of pressure has been highlighted, one must not forget that mantle conditions involve high temperatures that are likely to enhance atomic mobility. The first possible implication is the enhancement of dislocation climb, which might control the creep rate. In some case, plastic deformation can be achieved from the motion of point defects only. This mechanism is called diffusion creep. The absence of seismic anisotropy in the lower mantle has been interpreted as an evidence for deformation by diffusion creep (Karato et al. 1995b; McNamara et al. 2001). Modeling diffusion creep requires us to determine the diffusion coefficients of the relevant species at high-pressure and -temperature. The critical parameter for the relevance of diffusion creep versus dislocation creep is the grain size. Diffusion creep is likely to dominate for small grain sizes. A better knowledge of the grain size of mantle minerals is thus necessary to assess the possible role of diffusion creep in mantle deformation. The last issue we would like to address concerns the determination of mechanical data at high-pressure and -temperature. The major drawback of high-pressure deformation studies stems in the absence of mechanical data. Once again, this field has shown a rapid evolution in the past years. X-ray diffraction can be used to measure lattice strains from which elastic stresses can be deduced. This technique has been first applied to evaluate the strength of samples compressed in the diamond anvil cell (Kinsland and Bassett 1977; Meade and Jeanloz 1988a,b; 1990). Recently, it has been developed by the Stony Brook group in a multianvil apparatus connected to a synchrotron beam line, see e.g., Weidner et al. (2001) and references therein and Durham et al. (this volume). For instance, Weidner et al. (2001) present preliminary mechanical data obtained on powdered samples for most phases that exist within a subducting slab which show that higher pressure phases tend to be stronger at elevated temperatures. These data provide clues for the evolution with depths of seismicity within a subduction zone.

PERSPECTIVES: FROM ATOMIC TO THE GLOBAL SCALE

In their 1998 paper, Kubin et al. stated: "After being a major challenge in dislocation theory for decades, the multiscale modeling of plasticity is now becoming a reality." The same kind of statement can be made concerning deformation experiments under extreme pressure and temperature, which makes it possible today to investigate the plastic behavior of minerals in P,T conditions of the deep mantle. Mechanical data obtained in situ from X-ray diffraction and microstructural investigation techniques performed post-mortem provide information on the microscopic deformation mechan-isms. These new developments must be combined in a multidisciplinary approach to develop a forward modeling of the rheology of the mantle:

o *Ab initio* or semi-phenomenological atomistic approaches can be applied to high-pressure minerals to understand the core properties of dislocations. It is only at this scale that the effect of atomic bonding (and its evolution with pressure for instance) will be properly taken into account. Important constraints on the mobility of dislocations can also be expected from modeling at the atomic scale. On the other hand, simulated core properties will be constrained by TEM observations of dislocations in experimentally deformed samples.

o The mesoscopic approach combines a simplified description of the core properties with the more rigorous elastic theory of dislocations with a view to understand the formation and dynamics of the microstructures of deformed specimens. Comparison with mechanical data obtained *in situ* will provide a check or a complement on the mobility laws calculated at the atomic scale. Mesoscopic simulations of dislocation dynamics and interactions are now available for both fcc and bcc metals (Kubin et al 1998). They are being developed for other crystal symmetries such as those encountered with deep mantle minerals. As an illustration, Figure 58 presents a preliminary simulation of [100](001) glide in olivine.

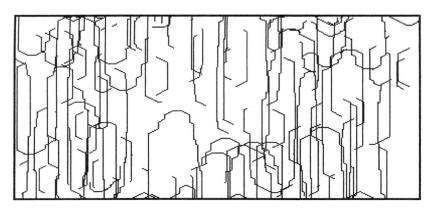

Figure 58. Simulated configuration corresponding to a crystal of forsterite deformed in single slip on [100](001) under a stress of 150 MPa applied along [101] and at 1300 K (Courtesy J. Durinck).

o The continuum theory of solids can provide the adequate formulation of boundary value problems. For instance to go from the single crystal to the polycrystal (modeling of the finite strain field can provide the crystal preferred orientations needed to interpret the anisotropic seismic properties) or to go from single crystals to multiphase assemblages. This issue is reviewed by Dawson (this volume).

Plastic deformation of mantle minerals thus appears to be one of the most exciting and challenging issue of geophysics, which involves the leading aspects of solid state physics, materials science, high pressure mineral physics on both the theoretical and experimental sides. The goal is worth this effort. One can expect this multidisciplinary approach to provide a forward modeling of the rheology of the mantle which is needed to understand mantle convection at a global scale.

ACKNOWLEDGMENTS

The author thanks S-i. Karato for an invitation to contribute to this volume and for many stimulating discussions. A large body of the work presented here has been performed at the Bayerisches Geoinstitut, Bayreuth during a sabbatical leave ("Congé thématique pour recherches") from the University of Lille and subsequent short term visits. Continuous support from the European Union ("IHP—Access to Research Infrastructures" Programme, Contract No. HPRI-1999-CT-00004 to D.C. Rubie) and from CNRS-INSU (Programme "Intérieur Terre") is gratefully acknowledged. The paper has benefited from careful reading and fruitful remarks from S-i. Karato and L. Kubin.

REFERENCES

Allen FM, Smith BK, Busek PR (1987) Direct observation of dissociated dislocations in garnet. Science 238:1695-1697

Ando J, Fujino K, Takeshita T (1993) Dislocation microstructures in naturally deformed silicate garnets. Phys Earth Planet Int 80:105-116

Andrault D, Fiquet G, Guyot F, Hanfland M (1998) Pressure-induced Landau-type transition in stishovite. Science 282(5389):720-724

Angel RJ, Finger LW, Hazen RM, Kanzaki M, Weidner DJ, Liebermann RC, Veblen DR (1989) Structure and twinning of single-crystal $MgSiO_3$ garnet synthesized at 17 GPa and 1800°C. Am Mineral 74:509-512

Beauchesne S, Poirier JP (1989) Creep of barium titanate perovskite: a contribution to a systematic approach to the viscosity of the mantle. Phys Earth Planet Int 55:187-199

Beauchesne S, Poirier JP (1990) In search of a systematics for the viscosity of perovskites: creep of potassium tantalate and niobate. Phys Earth Planet Int 61:182-198

Besson P, Poirier JP, Price GD (1996) Dislocations in $CaTiO_3$ perovskite deformed at high temperature: A transmission electron microscopy study. Phys Chem Minerals 23:337-344

Blumenthal WR, Phillips DS (1996) High-temperature deformation of single-crystal yttrium-aluminum garnet (YAG). J Am Ceram Soc 79:1047-1052

Bretheau T, Castaing J, Rabier J, Veyssière P (1979) Mouvement des dislocations et plasticité à haute température des oxydes binaires et ternaires. Adv Phys 28:835-1014

Bussod GY, Katsura T, Rubie DC (1993) The large volume multi-anvil press as a high P-T deformation apparatus. Pure Appl Geophys 141:579-599

Carpenter MA, Hemley RJ, Mao HK (2000) High-pressure elasticity of stishovite and the $P4_2/mnm-Pnnm$ phase transition. J Geophys Res 105B:10807-10816

Carstens H (1969) Dislocation structures in Pyropes from Norwegian and Czech garnet peridotites. Contrib Mineral Petrol 24:348-353

Carstens H (1971) Plastic stress relaxation around solid inclusions in pyrope. Contrib Mineral Petrol 32:289-294

Chai M, Brown JM, Wang Y (1998) Yield strength, slip systems and deformation induced phase transition of San Carlos olivine up to the transition zone pressure and room temperature. In Properties of Earth and Planetary Materials at High Pressure and Temperature. Geophys Monogr 101:483-493 American Geophysical Union, Washington, DC

Chalmers B, Martius UM (1952) Slip planes and the energy of dislocations. Proc R Soc London A 213:175-185

Chao ECT, Fahey JJ, Littler J, Milton DJ (1962) Stishovite, a new mineral from Meteor Crater, Arizona. J Geophys Res 67:419-421

Chao ECT, Littler J (1963) Additional evidence for the impact origin of the Ries basin, Bavaria, Germany. Geol Soc Am Abstr, p 127

Chen J, Wang Q, Zhai M, Ye K (1996) Plastic deformation of garnet in eclogite. Science in China (D) 39:18-25

Cherns D, Morniroli JP (1994) Analysis of partial and Stair-Rod dislocations by large angle convergent beam electron diffraction. Ultramicroscopy 53:167-180

Cherns D, Preston AR (1986) Convergent-beam diffraction of crystal defects. Proc 11th Intl Congr Electron Microscopy 1:721

Clauer AH, Wilcox BA (1976) High temperature tensile creep of magnesium oxide single crystals. J Am Ceram Soc 59:89-96

Copley SM, Pask JA (1965) Plastic deformation of MgO single crystals up to 1600°C. J Am Ceram Soc 48:139-146

Cordier P, Doukhan JC (1995) Plasticity and dissociation of dislocations in water-poor quartz. Philos Mag. A72:497-514

Cordier P, Morniroli JP, Cherns D (1995) Characterization of crystal defects in quartz by large-angle convergent-beam electron diffraction. Philos Mag A 72:1421-1430

Cordier P, Raterron P, Wang Y (1996) TEM investigation of dislocation microstructure of experimentally deformed silicate garnet. Phys Earth Planet Int 97:121-131

Cordier P, Rubie DC (2001) Plastic deformation of minerals under extreme pressure using a multi-anvil apparatus. Mater Sci Eng. A309:38-43

Cordier P, Sharp TG (1998) Large angle convergent beam electron diffraction determinations of dislocation Burgers vectors in synthetic stishovite. Phys Chem Minerals 25:548-555

Corman GS (1993) Creep of Yttrium Aluminium Garnet Single Crystals. J Mater Sci Lett 12:379-382

Dalziel IWD, Bailey SW (1968) Deformed garnets in a mylonitic rock from the greenville front and their tectonic significance. Am J Sci 266:542-562

Darot M, Gueguen Y (1981) High-temperature creep of forsterite single crystals. J Geophys Res 86(B): 6219-6234

Day RB, Stokes RJ (1964) Mechanical behavior of Magnesium Oxide at high temperatures. J Am Ceram Soc 47:493-503

Day RB, Stokes RJ (1966) Effect of crystal orientation on the mechanical behavior of Magnesium Oxide at high temperatures. J Am Ceram Soc 49:72-80

Doukhan N, Doukhan JC (1986) Dislocations in perovskites $BaTiO_3$ and $CaTiO_3$. Phys Chem Minerals 13: 403-410

Doukhan N, Sautter V, Doukhan JC (1994) Ultradeep, ultramafic mantle xenoliths—Transmission electron microscopy preliminary results. Phys Earth Planet Int 82:195-207

Duffy TS, Hemley RJ, Mao HK (1995) Equation of state and shear strength at multimegabar pressures: magnesium oxide to 227 GPa. Phys Rev Lett 74:1371-1374

Dupas C, Doukhan N, Doukhan JC, Green II HW, Young TE (1994) Analytical electron microscopy of a synthetic peridotite experimentally deformed in the β olivine stability field. J Geophys Res 99(B): 15821-15832

Dupas-Bruzek C, Sharp TG, Rubie DC, Durham WB (1998) Mechanisms of transformation and deformation in $Mg_{1.8}Fe_{0.2}SiO_4$ olivine and wadsleyite under non-hydrostatic stress. Phys Earth Planet Interiors 108(1):33-48

Durham WB (1975) Plastic flow of single crystals of olivine. PhD dissertation, M.I.T., Cambridge, Massachusetts

Durham WB, Froidevaux C, Jaoul O (1979) Transient and steady-state creep of pure forsterite at low stress. Phys Earth Planet Int 19:263-273

Durham WB, Goetze C (1977) Plastic flow of oriented single crystals of olivine. 1. Mechanical Data. J Geophys Res 82:5737-5753

Durham WB, Goetze C, Blake B (1977) Plastic flow of oriented single crystals of olivine. 2. Observations and interpretations of the dislocation structures. J Geophys Res 82:5755-5770

Durham WB, Rubie DC (1997) Can the multianvil apparatus really be used for high-pressure deformation experiments? *In* Properties of Earth and Planetary Materials at High Presure and Temperature. MH Manghnani, Y Syono (eds) American Geophysical Union, Washington DC, p 63-70

Fei Y, Saxena SK, Navrotsky A (1990) Internally consistent thermodynamic data and equilibrium phase relations for compounds in the system $MgO-SiO_2$ at high pressure and high temperature. J Geophys Res 95(B):6915-6928

Foster PK, Welch AJE (1956) Metal oxide solutions: I. Lattice constants and phase relations in ferrous oxide (wustite) and in solid solutions of ferrous oxide and manganous oxide. Trans. Faraday 52: 1626-1634

Friedel J (1964) Dislocations. Pergamon Press, Oxford.

Fujimura A, Endo S, Kato M, Kumazawa M (1981) Preferred orientation of β-Mn_2GeO_4. Programme and Abstract, The Seismological Society of Japan, p185

Garem H (1983) Dissociation des dislocations et plasticité de l'oxyde de structure grenat $Gd_3Ga_5O_{12}$. Thèse de Doctorat d'Etat, University of Poitiers

Garem H, Rabier J, Kirby SH (1985) Plasticity at crack tips in $Gd_3Ga_5O_{12}$ garnet single crystals deformed at temperatures below 950°C. Philos Mag A 51:485-497

Garem H, Rabier J, Veyssière P (1982) Slip systems in gadolinium gallium garnet single crystals. J Mater Sci 17:878-884

Gillet P (1995) Mineral physics, mantle mineralogy and mantle dynamics. C R Acad Sci Paris 320:341-356

Gilman JJ (1959) Plastic anisotropy of LiF and other rocksalt-type crystals. Acta Metall Mater 5:608-613

Green II HW, Young TE, Walker D, Scholz CH (1990) Anticrack-associated faulting at very high pressure in natural olivine. Nature 348:720-722

Gueguen Y (1979) Les dislocations dans l'olivine des péridotites. Thèse de Doctorat d'Etat, Université de Nantes

Gueguen Y, Darot M (1982) Les dislocations dans la forsterite déformée à haute température. Philos Mag A 45:419-442

Gumbsch P, Taeri-Baghbadrani S, Brunner D, Sigle W, Ruhle A (2001) Plasticity and an inverse brittle-to-ductile transition in strontium titanate. Phys Rev Lett 8708:5505-5508

Hardiman B, Bucksch R, Korczak P (1973) Observation of dislocations and inclusions in neodynium-doped yttrium aluminium garnet by transmission electron microscopy. Philos Mag 27:777-784

Hazen RM (1976) Effects of temperature and pressure on the cell dimension and X-ray temperature factors of periclase. Am Mineral 61:266-271

Horiuchi H, Hirano M, Ito E, Matsui Y (1982) $MgSiO_3$ (Ilmenite-type): Single-crystal X-ray diffraction study. Am Mineral 67:788-793

Horiuchi H, Ito E, Weidner DJ (1987) Perovskite-type $MgSiO_3$: Single-crystal X-ray diffraction study. Am Mineral 72:357-360

Horiuchi H, Sawamoto H (1981) β-Mg_2SiO_4: Single-crystal X-ray diffraction study. Am Mineral 66: 568-575

Hüther W, Reppich B (1973) Dislocation structure during creep of MgO single crystals. Philos Mag 28: 363-371

Jaoul O, Michaut M, Guegen Y, Ricoult B (1979) Decorated dislocations in forsterite. Phys Chem Minerals 5:15-19

Ji SC, Martignole J (1994) Ductility of garnet as an indicator of extremely high temperature deformation. J Struct. Geol 16:985-996

Jung HY, Karato S (2001) Water-induced fabric transitions in olivine. Science 293(:1460-1463

Karato SI (1990) Plasticity of $MgSiO_3$ perovskite: the results of microhardness tests on single crystals. Geophys Res Lett 17:13-16

Karato S, Wang Z, Fujino K (1994) High-Temperature creep of Yttrium-Aluminium garnet single crystals. J Mater Sci 29:6458-6462

Karato S, Wang Z, Liu B, Fujino K (1995a) Plastic deformation of garnets: systematics and implications for the rheology of the mantle transition zone. Earth Planet Sci Lett 130:13-30

Karato SI, Zhang S, Wenk H-R (1995b) Superplasticity in Earth's lower mantle: evidence for seismic anisotropy and rock physics. Science 270:458-461

Karato S, Rubie DC (1997) Toward an experimental study of deep mantle rheology: A new multianvil sample assembly for deformation studies under high pressures and temperatures. J Geophys Res Solid Earth 102(B):20111-20122

Karato S, Dupas-Bruzek C, Rubie DC (1998) Plastic deformation of silicate spinel under the transition-zone conditions of the Earth's mantle. Nature 395:266-269

Karato SI, Jung H (2002) Effects of pressure on dislocation creep in olivine. Philos Mag A (in press)

Karki BB, Stixrude L, Clark SJ, Warren MC, Ackland GJ, Crain J (1997) Structure and elasticity of MgO at high pressure. Am Mineral 82:51-60

Kingma KJ, Cohen RE, Hemley RJ, Mao HK (1995) Transformation of stishovite to a denser phase at lower-mantle pressures. Nature 374:243-245

Kinsland GL, Basset WA (1977) Strength of MgO and NaCl polycrystals to confining pressures of 250 kbar at 25°C. J Appl Phys 48:978-985

Kohlstedt DL, Goetze C (1974) Low-stress high-temperature creep in olivine single crystals. J Geophys Res 79:2045-2051

Kohlstedt DL, Goetze C, Durham WB (1976) New technique for decorating dislocations in olivine. Science 191:1045-1046

Kubin LP, Devincre B, Tang M (1998) Mesoscopic modelling and simulation of plasticity in fcc and bcc crystals: dislocation intersections and mobility. J Computed-Aided Mater Design 5:31-54

Langenhorst F, Poirier JP (2002) Transmission electron microscopy of coesite inclusions in the Dora Maira high-pressure metamorphic pyrope-quartzite. Earth Planet Sci Lett (submitted)

Lee CY, Gonze X (1997) SiO_2 stishovite under high pressure: Dielectric and dynamical properties and the ferroelastic phase transition. Phys Rev B Condensed Matter 56:7321-7330

Léger JM, Haines J, Schmidt M, Petitet JP, Pereira AS, Da Jornada JAH (1996) Discovery of hardest known oxide. Nature 383:401

Liebermann RC, Wang Y (1992) Characterization of sample environment in a uniaxial split-sphere apparatus. *In* High-Pressure Research: Application to Earth and Planetary Sciences. Y Syono, MH Manghnani (eds) American Geophysical Union, Washington, DC, p 19-31

Mackwell SJ (1984) Diffusion and weakening effects of water in quartz and olivine. PhD dissertation, Australian National University, Canberra

Mackwell SJ, Kohlstedt DL, Paterson MS (1985) The role of water in the deformation of olivine single crystals. J Geophys Res 90(B):11319-11333

Madon M, Poirier JP (1980) Dislocations in spinel and garnet high-pressure polymorphs of olivine and pyroxene: implication for mantle rheology. Science 207:66-68

Madon M, Poirier JP (1983) Transmission electron microscope observation of α, β and γ $(Mg,Fe)_2SiO_4$ in shocked meteorites: planar defects and polymorphic transitions. Phys Earth Planet Int 33:31-44

Mainprice D, Barruol G, Ben Ismaïl W (2000) The seismic anisotropy of the Earth's mantle: from single crystal to polycrystal. *In* Earth's deep interior: mineral physics and tomography from the atomic scale to the global scale. Geophysical Monograph Vol. 117. SI Karato, AM Forte, RC Liebermann, G Masters, L Stixrude (eds) American Geophysical Union, Washington, DC, p 237-264

Mao Z, Knowles KM (1996) Dissociation of lattice dislocations in $SrTiO_3$. Philos Mag A 73:699-708

Martini JEJ (1978) Coesite and Stishovite in the Vredeford dome, South Africa. Nature 272:715-717

McNamara AK, Karato SI, vanKeken PE (2001) Localization of dislocation creep in the lower mantle: implications for the origin of seismic anisotropy. Earth Planet Sci Lett 191:85-99

Meade C, Jeanloz R (1988a) The yield strength of MgO to 40 GPa. J Geophys Res 93:3261-3269

Meade C, Jeanloz R (1988b) The yield strength of B1 and B2 phases of NaCl. J Geophys Res 93: 3270-3274

Meade C, Jeanloz R (1990) The strength of mantle silicates at high pressures and room temperature: implications for the viscosity of the mantle. Nature 348:533-535

Mei S, Kohlstedt DL (2000) Influence of water on plastic deformation of olivine agregates 1. Diffusion creep regime. J Geophys Res 105:21457-21469

Mitchell TE, Lagerlöf KDP, Heuer AH (1984) Dislocations in ceramics. *In* Proceedings of the conference to celebrate the fiftieth anniversary of the concept of dislocations in crystals. The Institute of Metals, London, p 349-358

Mitchell TE (1999) Dislocations and mechanical properties of $MgO-Al_2O_3$ spinel single crystals. J Am Ceram Soc 82:3305-3316

Morniroli JP (2002) Large-Angle Convergent-Beam Electron Diffraction (LACBED)—Application to crystal defects. Société Française des Microscopies, Paris

Morniroli JP, Cordier P, VanCappellen E, Zuo JM, Spence J (1997) Application of the Convergent Beam Imaging (CBIM) technique to the analysis of crystal defects. Microsc Microanal Microstruct 8: 187-202

Nabarro FRN (1947) Dislocations in a simple cubic lattice. Proc Phys Soc London 59:256-272

Nabarro FRN (1967) Theory of crystal dislocations. Oxford University Press, Oxford, England.

Nabarro FRN (1984) Dislocation cores in crystals with large unit cells. *In* Dislocations 1984. P Veyssière, LP Kubin, J Castaing (eds) CNRS, Paris

Nicolas A, Poirier JP (1976) Crystalline plasticity and solid state flow in metamorphic rocks. John Wiley & Sons, Ltd, London.

Parthasarathy TA, Mah T, Keller K (1992) Creep mechanism of polycrystalline Yttrium Aluminum Garnet. J Am Ceram Soc 75:1756-1759

Peierls RE (1940) On the size of a dislocation. Proc Phys Soc London 52:34-37

Poirier JP (1982) On transformation plasticity. J Geophys Res 87:6791-6797

Poirier JP, Beauchesne S, Guyot F (1989) Deformation mechanisms of crystals with perovskite structure. *In* Perovskite: A structure of great interest to geophysics and materials science. A Navrotsky, D Weidner (eds) American Geophysical Union, Washington DC, p 119-123

Poirier JP, Peyronneau J, Gesland JY, Brebec G (1983) Viscosity and conductivity of the lower mantle: An experimental study on a $MgSiO_3$ perovskite analogue, $KZnF_3$. Phys Earth Planet Int 32:273-287

Poirier JP (1995) Plastic rheology of crystals. *In* Minerals Physics and Crystallography—A Handbook of Physical Constants. AGU Reference Shelf 2. American Geophysical Union, Washington, DC, p 237-247

Presnall DC (1995) Phase diagrams of Earth-forming minerals. *In* Minerals Physics and Crystallography—A Handbook of Physical Constants. AGU Reference shelf 2. American Geophysical Union, Washington, DC, p 248-268

Prewitt CT, Downs RT (1998) High-pressure crystal chemistry. Rev Mineral 37:283-317

Prior DJ, Wheeler J, Brenker FE, Harte B, Matthews M (2000) Crystal plasticity of natural garnet: New microstructural evidence. Geology 28:1003-1006

Rabier J (1979) Dissociation des dislocations dans les oxydes de structure grenat. Application à l'étude de la déformation plastique du grenat de fer et d'yttrium (YIG). Thèse de Doctorat d'Etat, University of Poitiers

Rabier J, Garem H, Veyssière P (1976a) Transmission electron microscopy determinations of dislocation Burgers vectors in plastically deformed yttrium iron garnet single crystals. J Appl Phys 47:4755-4758

Rabier J, Veyssière P, Grilhé J (1976 b) Possibility of stacking faults and dissociation of dislocations in the garnet structure. Phys Stat Sol (a) 35:259-268

Rabier J, Veyssière P, Garem H (1981) Dissociation of dislocation with a/2<111> Burgers vectors in YIG single crystals deformed at high temperature. Philos Mag A 44:1363-1373

Rabier J, Veyssière P, Garem H, Grilhé J (1979) Sub-grain boundaries and dissociations of dislocations in yttrium iron garnet deformed at high temperatures. Philos Mag A 39:693-708

Renner J, Stockhert B, Zerbian A, Roller K, Rummel F (2001) An experimental study into the rheology of synthetic polycrystalline coesite aggregates. J Geophys Res Solid Earth 106(B):19411-19429

Reppich B, Hüther W (1974) Formation and structure of dislocation networks developed during high temperature-deformation of MgO. Philos Mag30:1009-1021

Ringwood AE, Major A (1970) The system Mg_2SiO_4-Fe_2SiO_4 at high pressures and temperatures. Phys Earth Planet Interiors 3:89-108

Ross JV (1973) Mylonitic rocks and flattened garnets in the southern Okanagan of British Columbia. Can J Earth Sci 10:1-17

Ross N, Shu J-F, Hazen RM (1990) High pressure crystal chemistry of stishovite. Am Mineral 75:739-747

Routbort JL (1979) Work Hardening and creep of MgO. Acta Metall 27:649-661

Sharp TG, Bussod GY, Katsura T (1994) Microstructures in β-$Mg_{1.8}Fe_{0.2}SiO_4$ experimentally deformed at transition-zone conditions. Phys Earth Planet Int 86:69-83

Sharp TG, Lingemann CM, Dupas C, Stöffler D (1997) Natural occurrence of $MgSiO_3$-ilmenite and evidence for $MgSiO_3$-perovskite in a shocked L chondrite. Science 277:352-355

Shieh SR, Duffy TS, Li B (2001) Strength of SiO_2 across the stishovite-$CaCl_2$ phase boundary. International Conference on High Pressure Science and Technology (AIRAPT-18)

Shoemaker EM, Chao ECT (1961) New evidence for the impact origin of the Ries basin, Bavaria, Germany. J Geophys Res 66:3371

Sinclair W, Ringwood AE (1978) Single crystal analysis of the structure of stishovite. Nature 272:714-715

Stishov SM, Popova SV (1961) A new dense modification of silica. Geochem. (USSR) 10:923-926

Stretton I, Heidelbach F, Mackwell S, Langenhorst F (2001) Dislocation creep of magnesiowüstite ($Mg_{0.8}Fe_{0.2}O$). Earth Planet Sci Lett 194:229-240

Sung C-M, Goetze C, Mao H-K (1977) Pressure distribution in the diamond anvil press and shear strength of fayalite. Rev Sci Instrum 48:1386-1391

Tanaka M, Ueno K, Harada Y (1980) Large-angle convergent beam electron-diffraction. J Electron Microsc 29:277-277

Thompson AB (1992) Water in the Earth's upper mantle. Nature 358:295-302

Thurel E (2001) Etude par microscopie électronique en transmission des mécanismes de déformation de la wadsleyite et de la ringwoodite. PhD dissertation, University of Lille, Lille

Tomioka N, Fujino K (1997) Natural (Mg,Fe)SiO_3-ilmenite and -perovskite in the Tenham meteorite. Science 277:1084-1086

Tomioka N, Fujino K (1999) Akimotoite, (Mg,Fe)SiO_3, a new silicate mineral of the ilmenite group in the Tenham chondrite. Am Mineral 84:267-271

Tsuchida Y, Yagi T (1989) A new, post-stishovite high-pressure polymorph of silica. Nature 340:217-220

Ungár T (2001) Dislocation densities, arrangements and character from X-ray diffraction experiments. Mater Sci Eng A Struct Mater 309:14-22

Ungár T, Borbély A (1996) The effect of dislocation contrast on X-ray line broadening: a new approach to line profile analysis. Appl Phys Lett 69:3173-3175

Ungár T, Dragomir I, Révész A, Borbély A (1999) The contrast factors of dislocations in cubic crystals: the dislocation model of strain anisotropy in practise. J Appl Crystallogr 32:992-1002

Ungár T, Tichy G (1999) The effect of dislocation contrast on X-ray line profiles in untextured polycrystals. Phys Stat Sol (a) 171:425-434

Voegelé V, Ando JI, Cordier P, Liebermann RC (1998a) Plastic deformation of silicate garnets I. High-pressure experiments. Phys Earth Planet Interiors 108:305-318

Voegelé V, Cordier P, Sautter V, Sharp TG, Lardeaux JM, Marques FO (1998b) Plastic deformation of silicate garnets II. Deformation microstructures in natural samples. Phys Earth Planet Interiors 108:319-338

Voegelé V, Liu B, Cordier P, Wang Z, Takei H, Pan P, Karato SI (1999) High temperature creep in a 2-3-4 garnet: $Ca_3Ge_2Ge_3O_{12}$. J Mater Sci 34:4783-4791

Voegelé V, Cordier P, Langenhorst F, Heinemann S (2000) Dislocations in meteoritic and synthetic majorite garnets. Eur J Mineral. 12:695-702

Wang Z, Karato S, Fujino K (1996) High temperature creep of single crystal gadolinium gallium garnet. Phys Chem Minerals 23:73-80

Wang ZC, DupasBruzek C, Karato S (1999) High temperature creep of an orthorhombic perovskite—$YAlO_3$. Phys Earth Planet Interiors 110:51-69

Wang ZC, Ji SC (1999) Deformation of silicate garnets: Brittle-ductile transition and its geological implications. Can. Mineral 37:525-541

Wang ZC, Karato S, Fujino K (1993) High temperature creep of single crystal strontium titanate ($SrTiO_3$)—a contribution to creep systematics in perovskites. Phys Earth Planet Int 79:299-312

Warren BE, Avenbach BL (1950) The effect of cold work distorsion on X-ray pattern. J Appl Phys 21: 595-610

Weidner DJ, Chen JH, Xu YQ, Wu YJ, Vaughan MT, Li L (2001) Subduction zone rheology. Phys Earth Planet Int 127:67-81

Wenk HR, Matthies S, Hemley RJ, Mao HK, Shu J (2000) The plastic deformation of iron at pressures of the Earth's inner core. Nature 405:1044-1047

Williamson GK, Hall WH (1953) X-ray line broadening from filed aluminium and wolfram. Acta Metall 1: 22-31

Wright K, Price GD, Poirier JP (1992) High-temperature creep of the perovskites $CaTiO_3$ and $NaNbO_3$. Phys Earth Planet Int 74:9-22

7 Instability of Deformation

Harry W. Green, II
Department of Earth Sciences and
Institute of Geophysics and Planetary Physics
University of California
Riverside, California 92521

Chris Marone
Department of Geosciences
Pennsylvania State University
University Park, Pennsylvania 16802

INTRODUCTION

This chapter addresses the mechanisms by which instabilities can develop in rocks in the nominally ductile regime under conditions that the authors believe are relevant to Earth. We will not address brittle failure or frictional sliding processes in detail, but much of our discussion will be couched in terms of knowledge gained in studies of these processes. As a consequence, we will begin by discussing why brittle failure and frictional sliding cannot be the mechanism by which rocks become unstable at depth, despite the strong evidence that they are the underlying mechanism of earthquakes at shallow depths (e.g., Scholz 1990, 2002).

Brittle shear failure in the normal sense is fundamentally a tensile process (tensile microcracks must first nucleate and self-organize via interaction of the stress concentrations at their tips). As a consequence, brittle failure and frictional sliding are strongly inhibited by pressure because work must be done against the pressure to open the Mode I cracks. The inhibition is so strong that the stress required to create a fault or initiate sliding on an existing fault becomes greater than the room-temperature flow stress of many rocks at pressures equivalent to only a few tens of kilometers in Earth. Increasing temperature has little effect on brittle processes, but in contrast, the ductile flow stress of rocks falls exponentially with temperature. Because both pressure and temperature increase with depth in Earth, earthquakes by *unassisted* brittle fracture mechanisms or frictional sliding can only occur at depths less than ~30-50 km. Nevertheless, earthquakes occur to depths of almost 700 km in Earth and seismological evidence demonstrates unequivocally that they occur by displacement across a surface or narrow zone (i.e., they occur by faulting). This conundrum can be explained qualitatively by certain specific types of mineral reactions occurring under certain restricted conditions. The specific conditions required for faulting or sliding involve both the nature and kinetics of the mineral reactions and the nature of the way the rock that fails is loaded. That is, in order for a shearing instability to develop, certain rigorous conditions must be met involving the relative properties of the material that is faulting and its surroundings.

In this chapter, we will outline the requirements for a shearing instability to develop, followed by discussion of two different self-organizing processes associated with three different categories of mineral reactions that are currently known to lead to faulting at pressures where conventional brittle fracture is inhibited. In addition, we will discuss the popular theoretical concept of thermal runaway in shear zones as an earthquake mechanism in the context of the nature of loading in Earth and the lack of any obvious self-organizing aspect. We will show that the three known reaction-related, self-organizing, faulting mechanisms have been characterized to various degrees and probably all have relevance to faulting in Earth. However, none of them has been sufficiently investigated to allow quantitative analysis of the criteria for instability in the same way that brittle deformation has been analyzed.

SHEARING INSTABILITY

As we have seen in the foregoing discussion, the great pressures at depth mean that earthquakes under such conditions cannot represent conventional frictional shear or fracture. Nevertheless, seismic studies show that deep earthquakes have double-couple focal mechanisms and moment tensors that are consistent with shear rupture along a narrow zone (Wiens et al. 1993, 1994). Thus, it is reasonable to pursue parallels between deep earthquake instability and shallow crustal earthquakes, which are comparatively well understood (e.g., Scholz 1990, 2002).

By definition, all earthquakes radiate energy at seismic frequencies, and thus rupture must propagate at velocities approaching the elastic wave speed of the surrounding rock. Dynamic rupture velocities and stress drops of earthquakes at depth are interpreted to be in the range of 2 to 4.5 km/s and 10-100 MPa, respectively (Fukao, and Kikuchi 1987, Wiens and McGuire 1995, McGuire et al. 1997). These observations provide two important constraints on the mechanism of intermediate and deep earthquakes. First, they show that elastic strains must exist in the hypocentral region at great depths. Because ongoing plastic deformation during subduction would relieve elastic strains generated at shallow depths, the elastic strains that drive intermediate and deep earthquakes must result from local, active deformation. A second point is that these earthquakes represent a mechanical shear instability with dominantly shear components of motion. Thus, a key question is that of how the earthquakes nucleate and how strain rates increase from background levels of 10^{-15} to 10^{-12} s^{-1} to the values of 10^{-1} s^{-1} or more that are appropriate for seismic slip rates of 1-10 m/s.

Intermediate and deep earthquakes represent mechanical shear instability in the sense that some form of sudden weakening causes strength to drop below the applied stress level, resulting in a local force imbalance and dynamic acceleration. Shear instability is the result of interaction between fault zone rheology and continuum coupling with the surroundings. There are two requirements for instability. First, the rheologic response must result in weakening; a perturbation in strain rate that results in strengthening will not grow to become dynamically unstable. Second, the rate of weakening must satisfy a stability criterion governed by the continuum interactions between the nucleation region and its surroundings. In the most general case of a gradual transition from quasi-static to dynamic slip, a complete model of shearing instability requires an understanding of both the initial weakening event and how its growth leads to degradation of the local elastic stiffness (e.g., Rice and Ruina 1983). Several mechanisms have been proposed for earthquakes at depth and the exact form of initial weakening is under investigation. Nevertheless, for our purposes we can discuss instability in a generic form and examine the two primary effects.

We may start by considering a nucleation zone embedded in a homogeneous medium (Fig. 1). In terms of the mineral reaction mechanisms discussed below, the nucleation zone may be thought of as a single fluid-containing shear crack or a region containing a critical spatial-density of fluid-filled Mode I cracks (or comparable anticrack structures produced by phase transformation). At this stage, we do not consider shear heating or thermal runaway. Although one could envision Figure 1a as a region in which shear heating had reached a critical threshold via localized slip, we focus attention here on the process(es) that can lead to such initial slip localization. We assume a region of

Instability of Deformation

finite thickness W, and for simplicity, we take a circular region of radius r. We require only that the nucleation zone be mechanically weaker than the surroundings, so that its growth leads to weakening of the overall region, with stress concentration at the zone ends. The relationship between slip and weakening can be estimated using elastic dislocation theory:

$$\Delta \tau = C \frac{Gu}{r}, \tag{1}$$

where $\Delta \tau$ is the change in stress associated with slip u on a circular dislocation of radius r, G is the elastic shear modulus, and C is a constant. The effective elastic stiffness K of the region is then

$$K = C \frac{G}{r}. \tag{2}$$

Earthquake instability requires that fault zone strength fall faster than the unloading stiffness K of the surrounding region. Figure 1b is drawn in a way that emphasizes slip weakening, however equivalent plots could be constructed for strain-rate weakening, slip velocity weakening, or phase-transformation induced weakening. The energy available to drive shear rupture (including that to overcome shear resistance, create surface area, radiate seismic energy, and cause heating) is given by the shaded region in Figure 1b.

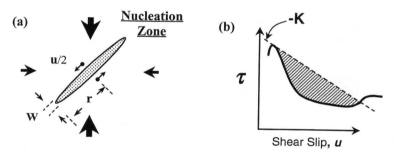

Figure 1. (a) Schematic view of the nucleation zone for a generic model of deep earthquake instability. Stippled area shows weakened region of radius r and width W. Internal shear strain is given by g = u/W, where u is slip measured from one side to the other. (b) Stress-displacement plot for the nucleation zone and surroundings. Dashed line of slope K shows the unloading stiffness (see Eqn. 2) for the region surrounding the nucleation zone. Heavy line shows hypothetical strength curve for the nucleation zone. Note that initial weakening causes strength to fall below the applied stress, leading to a dynamic force imbalance. Shaded region shows energy available to drive seismic rupture.

The instability condition is determined by the relative rates of weakening and elastic stress release. This can be quantified with reference to a generic slip-weakening model, in which we assume linear weakening over a distance D (Fig. 2). Taking τ^y as the yield strength and τ^f as the failure strength, we can write the instability condition as:

$$K < K_c, \tag{3}$$

where K_c is a critical stiffness given by:

$$K_c = \frac{(\tau^y - \tau^f)}{D}. \tag{4}$$

Because the elastic stiffness is always positive, we see that weakening is a necessary condition for instability. We may define the following cases.

Stable: $\tau^y < \tau^f, \ K_c < 0$ (5)

Conditionally Stable: $\tau^y > \tau^f, \ K > K_c$ (6)

Unstable: $\tau^y > \tau^f, \ K < K_c$ (7)

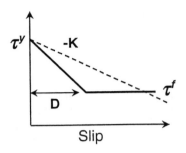

Figure 2. Generic slip weakening model for earthquake instability. Fault strength (solid line) weakens from t^y to t^f over slip distance D or shear strain $g_c = D/W$. Applied stress (dashed line) follows elastic unloading stiffness. Instability and seismic rupture are produced by the force imbalance between applied stress and fault strength.

In the first case, deformation is inherently stable because possible nucleation events, including slip or strain rate perturbations lead to strengthening. This is the standard expectation for steady-state plastic rheologies or viscous deformation in which strength increases with strain rate. The necessary condition for instability is met for the conditionally stable case, but the rate of weakening is insufficient to satisfy the instability criterion. In this case, the fault zone weakens but shear is stable and does not generate seismic radiation. In the final case (Eqn. 7), both the necessary and sufficient conditions are met and perturbations of any size could grow to become unstable. The dynamic force imbalance in this case would lead to rapid fault acceleration, shear heating and seismic radiation during slip.

In the context of a stability analysis for earthquake nucleation at high pressure, a key question is that of how rapidly a fault weakens during initial slip. For the analysis presented here we use only slip weakening, hence we are assuming that weakening is a constitutive property of a fault zone, independent of strain rate, slip or other variables. To the extent that this approximation holds, the critical weakening rate for instability can be quantified as a rheologic stiffness, as done in Equation (4).

By combining the relations for effective elastic stiffness (2) and the critical rheologic stiffness (4) we may write the stability criterion as:

$$C\frac{G}{r} < \frac{(\tau^y - \tau^f)}{D}.$$ (8)

This relation brings out an important point regarding conditionally stable shear (Fig. 3). In this case, the rate of weakening with slip is slower than the rate of stress reduction by unloading and thus shear is stable. However, growth of the nucleation region will reduce the local stiffness, which may eventually lead to instability. In the context of this approach, a critical nucleation patch size r_c can be defined:

$$r_c = C\frac{GD}{(\tau^y - \tau^f)}.$$ (9)

Figure 3 shows that as the nucleation dimension grows from r_1 to r_3, the effective stiffness falls below the critical stiffness and the instability condition is satisfied.

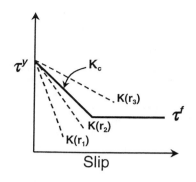

Figure 3. Slip weakening model of fault strength (solid line) together with the effective stiffness for three fault nucleation sizes ($r_1 < r_2 < r_3$). Stiffness decreases with the nucleation dimension. K_c denotes critical stiffness defined by fault strength and rheology. A critical nucleation size for instability r_c can be defined in terms of K_c.

In summary, earthquake instability requires that fault zone strength fall faster during weakening than the unloading stiffness of the surrounding region. In the context of a simplified, generic model of fault rheology, a critical stiffness for instability can be defined. A key question is that of how rapidly a fault weakens during initial slip. Clearly, much of the above analysis is highly oversimplified. More complex approaches would account for the effect of perturbation size on yield strength and failure conditions and incorporate specific mechanisms of plastic instability into relations for the critical rheologic stiffness. Also, the effect of latent heat of transformation, shear heating, and complex nucleation geometry would be included in the rheologic and continuum models. Quantitative treatment of these important effects is beyond the scope of this chapter and requires additional information about potential fault nucleation

EXPERIMENTAL HIGH-PRESSURE FAULTING MECHANISMS

Dehydration-induced embrittlement

Breakdown of hydrous minerals. Raleigh and Paterson (1965; see also Raleigh 1967) conducted deformation experiments on serpentinite under confining pressures of a few hundred MPa. They found that at elevated temperatures the rock was ductile. However, under conditions where the serpentine was breaking down to olivine + talc + H_2O, the rock failed catastrophically by faulting. This effect has been interpreted classically as a result of reduction of the effective normal stress on a fault plane or potential fault plane, resulting in the following relationship:

$$\tau = \tau_o + \mu(\sigma_n - p_f), \tag{10}$$

where τ is the shear stress, τ_o is a constant, μ is the coefficient of friction, σ_n is the normal stress on the fault, and p_f is the pressure in the pore fluid. Experimentation under conditions where the pore fluid pressure can be manipulated independently of the stress applied to the solid rock has verified this *effective stress* relationship for both brittle failure of intact rock and frictional sliding under conditions where $p_f < \sigma_3$. However, the qualitative description of the fluid as partially supporting the normal stress is unsatisfactory in a mechanistic sense because if $p_f < \sigma_3$, then the fluid can only exist in pores for which the strength of the solid matrix shelters the pores. If that is the case, how does the fluid partially reduce the normal stress on a fault? Worse, how does it partially reduce the normal stress on a fault that isn't there yet? Clearly, this relationship may provide a satisfactory continuum description of the phenomenon, but not a mechanistic explanation.

The answer comes from modern understanding that brittle shear failure is at its base tensile failure. Experimentally, it has been known for 40 years that small shear cracks cannot be made to propagate in their own plane. Application of a stress field that would be expected to make such a crack propagate causes "wing" cracks to originate at sites of greatest tensile stress concentration at the tips of the crack and propagate along the trajectory of σ_1, coming to rest when they are sufficiently far from the stress concentration that initiated them for the stress at the crack-tip to fall below the local tensile fracture strength. Thus, macroscopic shear failure initiates by a gradual process of local tensile failure at points of stress concentration (pre-existing cracks, grain boundaries, pores, grains of contrasting elastic properties, etc.), culminating in a critical population of tensile microcracks that then self-organize to initiate a shear fracture (a fault) that propagates by repetition of the "microcracks-first" process along the path of the growing fault surface.

Although perhaps counterintuitive and an unwelcome complication, this insight provides a built-in understanding of how a pressurized pore fluid facilitates faulting. It does so by enhancing the local tensile stress concentrations that lead to formation of the tiny tensile microcracks and by holding them open after they form (Fig. 4). Thus, the greater the pore pressure, the lower the applied stress necessary to bring local stress concentrations to the failure stress, and the lower the overall stress must climb to in order to generate proliferation of microcracks to the point of self-organization into an incipient fault zone that can then grow by the same pore-pressure-induced process. If the pore pressure should reach σ_3, the expected failure mode would be runaway propagation of a tensile crack, leading to sample splitting in the laboratory or a fluid-filled vein (or melt-filled dike) in Earth. In the laboratory, this is easily demonstrated to be the case. However, in a natural situation with a slowly evolving pore pressure, the more usual case should be generation of shear failure before the pore pressure reaches σ_3.

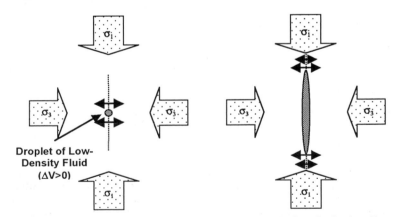

Figure 4. Fluid-assisted opening of "tensile microcracks." Nucleation of a droplet of low-density fluid (net ΔV of reaction greater than zero) at a stress concentration in a nonhydrostatic stress field induces deviatoric tensile stresses around the bubble, with maximum tangent to the plane normal to σ_1 as shown on left. Such stresses will enhance nucleation of a tensile microcrack that will grow until the stress at the crack tip produced by the combination of the applied stress and the fluid pressure within the crack falls below the local tensile fracture stress.

There is abundant laboratory evidence at moderate pressures that this mechanism can induce brittle failure at greatly reduced stresses compared to those required in the absence of a pore fluid. Importantly, as was the case in the experiments of Raleigh and Paterson

(1965), development of a pore pressure can lead to brittle failure where otherwise (i.e., in the absence of a fluid) rock behavior will be ductile. As a consequence, dehydration embrittlement is an attractive hypothesis to explain earthquakes at pressures where unassisted brittle failure is completely inhibited.

There is a problem, however. Fluids are very much more compressible than solids, hence dehydration reactions, which all produce a positive ΔV of reaction at low pressures (because of the very low density of hydrous fluids), will see the magnitude of ΔV progressively reduced as pressure rises and ΔV will become negative at some point (because the net effect of the more dense solid product phases will come to dominate). For antigorite, the serpentine mineral stable at the highest temperatures and pressures, that point is at ~750°C, 2.2 GPa (Ulmer and Trommsdorff 1995), a pressure representing a depth of only ~70 km in Earth (Fig. 5). For other hydrous minerals likely to be reasonably abundant in mantle lithologies, the conditions are comparable.

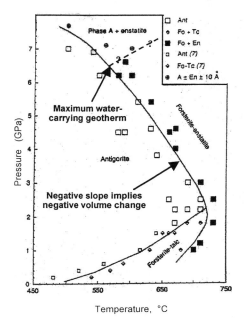

Figure 5. Antigorite phase diagram. Decomposition relations for a natural antigorite (modified after Ulmer and Trommsdorff 1995). Between ~2.2 and 6.4 GPa, dehydration yields anhydrous phases and the slope of the dehydration boundary is negative, implying $\Delta V < 0$. Above 6.4 GPa, Phase A, a dense hydrous magnesium silicate, forms rather than olivine and the slope of the dehydration curve becomes more negative.

However, if $\Delta V < 0$, the conventional rock mechanics view is that the pore-pressure effect will disappear (e.g., Wong et al. 1997). That is because when the net ΔV becomes negative, generation of a droplet of fluid results in a low-pressure pore surrounded by compressive hoop stresses, hence any stress concentrations that encourage formation of Mode I microcracks are pushed *away* from local failure rather than toward it. The greater the negative ΔV, the lower is the pore pressure produced and the greater the compressive "capsule" in which it is confined. Both effects push the material away from local tensile failure, hence this has been interpreted as inhibiting bulk shear failure. If the latter interpretation is correct, then dehydration embrittlement of serpentine and other common hydrous alteration minerals of peridotite (e.g., chlorite) will be limited to relatively shallow depths as a mechanism for triggering earthquakes.

This hypothesis has been tested in the laboratory (Reinen et al. 1998) and found that dehydration of antigorite under stress at confining pressures above 3 GPa still leads to faulting, showing that bulk shear failure can still occur with $\Delta V < 0$. However, in keeping with the predictions of brittle rock mechanics, the microstructures accompanying faulting at those high pressures show no evidence of the Mode I microcracks diagnostic of brittle shear failure and fault gouges (Fig. 6) also do not consist of angular fragments with a fractal size distribution. These observations indicate that the mode of failure is not traditional frictional failure nor embrittlement of that type assisted by dehydration.

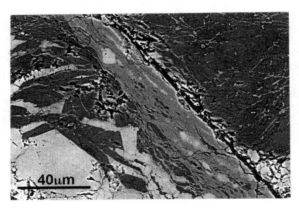

Figure 6. Faulting of antigorite when $\Delta V < 0$. Despite the fact that ΔV for the dehydration reaction becomes negative above 2.2 GPa, dehydration under stress leads to faulting. However, neither Mode I microcracks nor a fault gouge with angular fragments is observed.

Much remains to be learned about the fundamental physics of failure under these conditions. Nevertheless, at least we know that dehydration of hydrous minerals remains a viable trigger mechanism for earthquakes to depths of at least 250 km and probably much more, regardless of the sign of the ΔV of dehydration (so long as hydrous phases are present and are breaking down to form less-hydrous assemblages).

Exsolution of H_2O from nominally anhydrous minerals. A new variant of dehydration embrittlement recently has been discovered that provides additional insight into fluid-enhanced faulting (Green 2001). In a study of the rheology of eclogite at P = 3 GPa, experiments on a "wet" eclogite resulted in faulting when specimens were deformed at temperatures between the wet and dry solidi. The experimental specimens were fabricated from a natural eclogite with no hydrous phases, but with significant H_2O dissolved in both of the major, nominally anhydrous, phases (omphacitic pyroxene and garnet). Above the water-saturated solidus, H_2O exsolved from the silicates and triggered melting at grain boundaries. When this happened under stress, the specimen failed by faulting and was characterized by myriads of glass-filled Mode I cracks as well as a "normal" fault gouge except for the presence of small amounts of glass. In this case, ΔV was clearly greater than zero because the only phases being produced were fluid and melt, both of which are less dense than either omphacite or garnet; the microstructural characteristics of the specimens are as expected for $\Delta V > 0$. Thus, this failure was truly dehydration embrittlement in the original sense.

Change of micromechanism of dehydration-induced failure with change in sign of ΔV. The change of microstructures associated with dehydration-induced failure when the sign of net ΔV changes from positive to negative strongly suggests a change in failure mechanism. We defer discussion of this potential mechanism change until after discussion of transformation-induced faulting.

Transformation-induced faulting

The other experimentally established high-pressure faulting instability is the anticrack mechanism known to operate under certain restrictive conditions during the olivine $\rightarrow$ wadsleyite and olivine $\rightarrow$ ringwoodite transformations (Green and Burnley 1989,

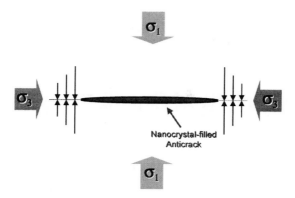

Figure 7. Nucleation and growth of anticracks is analogous to nucleation and growth of fluid-filled cracks during dehydration under stress (compare with Fig. 4). In this case, formation of a nucleus of the dense polymorph induces strong compressive stress concentrations and a small rise in T that, in turn, cause rapid-fire nucleation in the plane normal to s_1. The result is microanticracks that have high compressive stresses at their tips and are filled with a nanocrystalline aggregate of the denser phase that has fluid-like properties even at seismic slip rates.

Green et al. 1990, Burnley et al. 1991, Green & Houston 1995, Kirby et al. 1996). This is also a self-organizing instability; it requires an exothermic polymorphic phase transformation that leads to production of Mode I microanticracks (Fig. 7) filled with an extremely fine-grained aggregate of the new phase that interact via the compressive stress concentrations at their tips in an analogous way to the interactions between tensile stress concentrations at the tips of open or fluid-filled Mode I microcracks (Green and Burnley 1989).

The reason an exothermic polymorphic transformation is required to support this instability is shown in Figure 8. Cooling the high-temperature phase below the phase boundary causes it to become metastable; as the difference in Gibbs Free Energy (ΔG) between the two phases increases, so does the nucleation rate. However, when undercooling is sufficiently great, the nucleation rate becomes more dependent on the kinetics of the transformation than on the driving force, hence the nucleation rate reaches a maximum and at lower temperatures it declines at an increasingly rapid rate; at temperatures lower than defined by the left-hand branch of the curve, the reaction rate is zero. Along the near-vertical left-hand branch of the nucleation-rate curve, formation of a nucleus of the stable phase will yield a small heat release that will result in a small increase in local temperature, leading to increase in the nucleation rate. At the same time, if ΔV of the reaction is negative, the nucleus immediately will be surrounded by compressive stress concentrations that will increase the driving force for nucleation. If the material is under stress, the maximum compressive stress concentration around the initial nucleus will be located adjacent to the nucleus on planes oriented normal to σ_1, hence additional nuclei will preferentially form adjacent to the first nucleus and lie in the plane normal to σ_1. This anisotropic contraction of the transformed region increases the stress concentrations lying in the plane normal to σ_1. Therefore, if the thermal conduc-tivity of the metastable host is low and the kinetics of plastic relaxation of the stress concentrations are sufficiently slow, the combination of these two effects can lead to runaway nucleation in the plane normal to σ_1 – yielding a Mode I anticrack filled with a nanocrystalline solid with properties approximating that of a low-

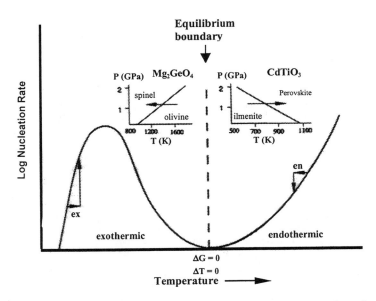

Figure 8. Thermodynamics of nucleation during polymorphic phase transformations. The topology of nucleation kinetics for transformation at constant pressure is described for exothermic reactions to the left of the vertical dashed line and for endothermic reactions to the right of that line. Examples are shown for cases where the high-pressure phase is the low-entropy phase (left) and the high-entropy phase (right). See text for discussion (modified after Green and Zhou 1996).

viscosity fluid.

As in the case with Mode I cracks, the greater the aspect ratio of the anticrack, the greater the stress concentration at the tip. Therefore, microanticracks interact with each other in a stressed material in an analogous way to microcracks, culminating in initiation of a fault and catastrophic shear failure.

In contrast, endothermic transformations have entirely different characteristics (right side of Fig. 8). As the low-temperature polymorph is progressively heated above the transformation temperature, both the driving force and the kinetics for the transformation increase monotonically. As a consequence, there is no kinetically controlled branch of the nucleation-rate curve and any thermal effects of nucleation would actually produce a small reduction of local temperature. Thus, the characteristics that can lead to positive feedback are absent; when the temperature is sufficiently high for the reaction to run, simple nucleation and growth of the high-temperature phase occurs, leading to growth of normal, blocky, crystals; no anticracks form and no instability is generated.

This argument strongly suggests that the sign of ΔS ($\geq \Delta H/T$) is critical to this instability; if the reaction is not exothermic, there will be no instability. However, nothing in the logic presented here requires that the algebraic sign of ΔV should matter. The magnitude of ΔV controls the magnitude of the stress concentration at the tips of microlenses, but there is no reason that the sign of ΔV should matter. This implication was tested in the $CdTiO_3$ system (Green and Zhou 1996); in the up-temperature (endothermic) direction, no instability developed. Under all conditions, transformation from the phase with ilmenite structure to that with perovskite structure ($\Delta V < 0$; $\Delta H > 0$) yielded blocky crystals of the daughter (high-temperature) phase growing on grain boundaries of the parent phase. In contrast, as described in the previous paragraph, when tested in the reverse direction ($\Delta V > 0$; $\Delta H < 0$), the reaction produced fine-grained lenses ("microcracks") of the ilmenite phase parallel to σ_1 and shear fracture followed. The logic for runaway growth of nanocrystalline lenses when $\Delta V > 0$ is analogous to that given above for the case $\Delta V < 0$; in this case the first nucleus is surrounded by extensile hoop stresses that prejudice the location of succeeding nuclei and the small amount of heat released leads to runaway nucleation in a plane parallel to σ_1.

Clearly, then, a critical requirement for failure is an exothermic reaction. But is that a sufficient cause? The reaction ringwoodite → perovskite + magnesiowüstite that defines the base of the upper mantle is endothermic but if olivine were to be carried metastably completely through the transition zone and then decompose into the lower-mantle assemblage, the reaction would be exothermic. It is necessary, therefore, to know whether such a disproportionation reaction could support the instability if the reaction is exothermic and ΔV is large. The reason to question this possibility is that a diffusive step is required for phase separation during this type of reaction. Is the time necessary to separate the chemical components into two phases sufficiently long that it abrogates the localized runaway nucleation of the daughter assemblage to form anticracks and lead to instability? This hypothesis was tested using the reaction

albite → jadeite + coesite ,

because it is both more exothermic and has a larger volume change than olivine → spinel in Mg_2GeO_4, the system in which we discovered transformation-induced faulting (Green and Burnley 1989), or in the olivine → wadsleyite transformation in olivine of mantle composition (Green et al. 1990). The hypothesis was confirmed; decomposition of albite under stress yielded symplectites consisting of blocky crystals of jadeite filled with wormy intergrowths of coesite; no anticracks were observed and faulting did not occur (Gleason & Green 1996).

Table 1 summarizes the systems in which the anticrack mechanism has been investigated, verifying that it can operate during a variety of exothermic polymorphic transformations but not during endothermic transformations nor during disproportionation reactions, even when ΔV is large and the heat evolved is large and positive.

Table 1. Systems tested for transformation-induced faulting.

Reaction	System	Faulting?	Pressure (Gpa)	Microstructure	Reaction type
α → γ	Mg_2GeO_4	Yes	1-2	anticracks	Exo. poly. trans
α → β	$(Mg,Fe)_2SiO_4$	Yes	14-15	"	" " "
α → β	Mn_2GeO_4	Yes	4-4.5	"	" " "
pv → il	$CdTiO_3$	Yes	0.2	"cracks"	" " "
Ice1h → II*	H_2O	Yes	0.2-0.5	"anticracks"	" " "
il → pv	$CdTiO_3$	No	0.2	Blocky xtls	Endo. " "
ab → jd + coes		No	3-3.5	Symplectite	Exo. disproportion.

Predictions for the top of the lower mantle:

γ → pv + mw		No	25	Symplectite	Endo. disproportion.
α → pv + mw		No	25	Symplectite	Exo. disproportion.

*Kirby et al. (1991); compression features similar to anticracks were observed.

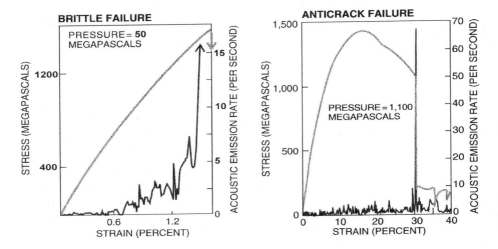

Figure 9. Comparison between low-pressure brittle failure of Westerly Granite (left) and high-pressure anticrack failure of Mg_2GeO_4 undergoing the olivine→spinel transformation. Note that brittle failure produces an exponentially increasing number of acoustic emissions that rises very high before failure, whereas anticrack failure produces significant acoustic emissions only during macroscopic failure (modified after Green, 1994).

Like normal brittle failure, anticrack failure in the laboratory generates acoustic emissions (Green et al. 1992) and thus is potentially an earthquake mechanism (Fig. 9).

"Brittle" versus "plastic" shear failure

Transformation-induced failure clearly operates by a different physical process than conventional brittle failure, yet it mimics many aspects. In particular, it achieves self-organization via Mode I nanocrystalline lenses (either parallel or perpendicular to σ_1, depending on the sign of ΔV. Nevertheless, it produces acoustic emissions only during shear failure; the anticrack growth phase before bulk failure is silent (Fig. 9). This observation plus the current interpretation of the underlying physics of anticrack growth as a runaway nucleation phenomenon indicates that this failure mechanism is perhaps better classified as a plastic instability rather than brittle failure. For dehydration-induced faulting, the change in microstructures associated with change in sign of ΔV also suggests a change in failure mechanism. In particular, change from Mode I microcracks and angular, fractal, fault gouge to microstructures indicative of mineral reactions suggests that dehydration with negative ΔV also may trigger a plastic instability. The details of this proposed switch of mechanisms remains unclear.

Thermal runaway due to shear heating

In addition to dehydration-induced faulting and transformation-induced faulting, shear failure mechanisms that have been established in the laboratory, it has been proposed a number of times on theoretical grounds that a shearing instability can develop based on thermal feedback between shear heating and thermally-induced viscosity reduction under conditions approximating constant stress (e.g., Gruntfest 1963, Griggs and Baker 1969, Hobbs and Ord 1988, Kanamori et al. 1998, Karato et al. 2001). These treatments leave little doubt that such a runaway could lead to failure. What is not clear is whether boundary conditions in Earth are such that the runaway can be supported. For example, under constant strain-rate or constant displacement conditions, the thermally-induced viscosity resulting from shear heating will lead to stress relaxation rather than runaway heating, and any thermal anomaly will dissipate rather than leading to failure.

As described above, the anticrack mechanism also appears to be a thermal runaway phenomenon. Why is this mechanism self-organizing and shear heating not? Like in brittle failure, the anticrack instability is in the development of the primary (Mode I) failure mechanism. The thermal runaway in anticrack development works off of latent heat release and therefore is proportional to reaction progress and depends on the relative rates of thermal conductivity and reaction. No energy need be supplied externally during the heating process and the volume of material transformed before failure can be extremely small. In contrast, shear heating is proportional to the work done during straining and therefore is a function of stress, strain, viscosity, thermal conductivity, and geometrical boundary conditions (the system must be pre-organized into an appropriately-oriented shear zone). Instability can be obtained only if the boundary conditions ensure that the viscosity reduction induced by increasing temperature results in increase of strain rate rather than relaxation of stress.

APPLICATION TO EARTHQUAKE MECHANISMS

Earthquake distribution with depth

Figure 10 shows the distribution of earthquakes in Earth. Virtually all earthquakes deeper than 20-30 km are generated at the interface between subducting lithosphere and the over-riding plate, or within the subducting lithosphere itself, hence this distribution shows that the generation of earthquakes with depth is generally spread out over the entire depth of subduction to the base of the mantle transition zone, where they stop abruptly. Thus, it is to be expected that a successful candidate for an earthquake trigger mechanism will naturally produce earthquakes continuously with depth, rather than bursts at particular set(s) of conditions.

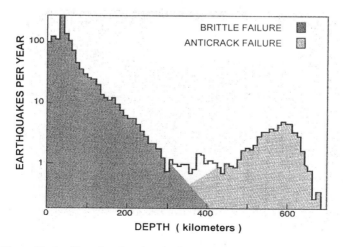

Figure 10. Semi-log plot of earthquake frequency with depth. Note exponential decline from 100 to 300 km and secondary population in mantle transition zone reaching a maximum at ~600 km before total cessation at 680 km (modified after Frohlich 1989).

Mineral reactions available to trigger earthquakes

We know from dredge hauls and heat flow that the upper portion of the oceanic lithosphere undergoes hydrothermal alteration shortly after its formation at oceanic ridges. Seismic velocities suggest that such alteration extends into the mantle beneath the crust and therefore that serpentine partially replaces olivine (and pyroxene), perhaps up to a depth of 10-12 km (Meade and Jeanloz 1991). The rare evidence for hydrous alteration in shallow mantle xenoliths from oceanic islands strongly supports this view. Thus, it has been generally assumed (e.g., Green and Houston 1995) that dehydration-induced faulting is an excellent candidate for triggering these earthquakes, but that this mechanism is probably not a candidate to explain the common occurrence of a second seismic zone deep within the lithosphere at intermediate depths (e.g., 40-200 km below Japan, Igarashi et al. 2001). However, Silver et al. (1995; see also Jiao et al. 2000) suggested that deeper hydration of the oceanic lithosphere might occur along fault zones created by great earthquakes outboard of trenches and others have pointed out the similar pattern of maximum antigorite stability in subduction zones (e.g., Ulmer and Trommsdorff 1995) and the distribution of earthquakes in intermediate-depth double seismic zones (e.g., Peacock 2001). More recently, a somewhat different mechanism has been proposed that suggests extensive serpentinization may occur at trenches (Phipps-Morgan 2001, Phipps-Morgan et al. 2002).

If it is true that the subducting lithosphere is hydrated at least sporadically to a depth of 40 km, dehydration-induced failure is a potential candidate for explanation of all earthquakes at depth. Here we will briefly examine this possibility in three depth intervals: 0-200 km, 200-300 km, 300-700 km.

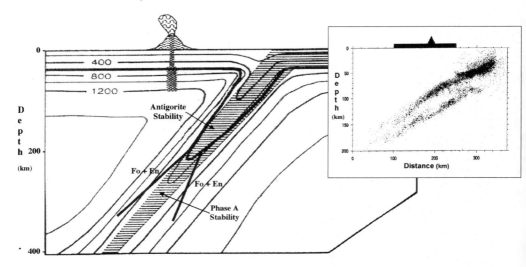

Figure 11. Stability of antigorite mapped onto a thermal model of a subduction zone. The particular model is such that antigorite stability reaches that of Phase A (~200 km depth). Hydrous phases existing in the cold core of the slab could access the antigorite→Phase A + en reaction and pass H_2O into the deeper slab. Otherwise, all H_2O will be released and could flux arc melting (heavy arrow) or trigger earthquakes, but could not be carried to great depths in significant quantities. Note similarity between limit of antigorite stability (heavy line) and earthquake distribution beneath northern Japan (inset). Hachured region represents cold lithosphere. (Modified after Ulmer and Trommsdorff 1995 and Igarashi et al. 2001).

(i) 0 - 200 km. Depending on the thermal profile of a subduction zone, antigorite can remain stable in the cold interior of the descending slab up to a maximum depth of approximately 200 km. As shown in Figure 11, the shape of the antigorite phase diagram in the thermal model chosen is remarkably similar to that of the earthquake distribution under northern Japan. As a consequence, dehydration of antigorite could explain the entire distribution of earthquakes in the deeper zone, including its approach to and merging with the shallower zone. In addition, antigorite, in combination with a myriad of other hydrous phases potentially present in the hydrated oceanic crust and shallow mantle, can also explain the shallow zone. Of course, the specific depth to which antigorite could carry water and release it to trigger earthquakes is a function of the temperature distribution in the slab. As was shown in Figure 5, the temperature of maximum stability of antigorite falls rapidly at pressures greater than ~7 GPa, hence for water to pass to greater depths, it must be passed on to hydrous phases stable to greater depths. In particular, if the minimum temperature in the slab reaches ~550-600°C before a depth of ~200 km is reached, antigorite will completely dehydrate to olivine + enstatite and the slab will be completely dehydrated except for a few rare hydrous phases. However, Figure 5 also shows that if the slab is colder, antigorite will break down into Phase A + H_2O and the door will be open to carry water into the deep interior via a succession of dense hydrous magnesium silicate phases (e.g., Angel et al. 2001). Indeed, this "eye of the needle" has profound importance for mantle dynamics and therefore for past, present, and future evolution of Earth.

(ii) 200-300 km. Deeper than 200 km, a problem arises for earthquake generation by dehydration. Examination of the data of Angel and Frost (2001) shows that any reasonable particle trajectory within a cold slab (a slab for which the minimum temperature passes below ~600°C, 7 GPa) crosses all of the phase boundaries available in directions for which water is *conserved*. That is, each hydrous phase contains less water than the more dense hydrous phase(s) that replace it. Thus, no free fluid is generated by the reactions and dehydration would appear to be unavailable for triggering earthquakes. There are, however, a few hydrous phases that would be expected to be present in minor quantities in hydrous subducting lithosphere that could potentially extend dehydration embrittlement to depths somewhat greater than 200 km. For example, titanian clinohumite is stable to depths exceeding 300 km (Ulmer and Trommsdorff 2001) and could perhaps explain the low frequency of earthquakes at that depth.

(iii) 300-700 km. The previous paragraph suggests that dehydration cannot serve as a trigger for earthquakes at depths greater than 300-400 km. In contrast, restriction of the anticrack instability to exothermic polymorphic transformations predicts that earthquakes can only be triggered by this mechanism at depths where the transformations of olivine to wadsleyite and ringwoodite occur (i.e., at depths of 300-700 km in subducting slabs). Therefore, this mechanism could explain the rise in frequency of earthquakes at the top of the transition zone and their cessation at its base (Green and Houston 1995). However, application of this mechanism to Earth requires that the central portion of descending slabs is sufficiently cold that olivine does not transform to its denser polymorphs until a critical set of conditions is reached where the kinetics of the reactions are rapid enough to run. It also may seem that the numerous special circumstances necessary for anticrack failure would make the probability small that it would happen routinely in Earth. However, in subduction zones, a given volume of material follows a particle path that is primarily up-pressure, but nevertheless is also continually encountering conditions of higher temperature as heat flows slowly into the slab from the surrounding mantle. As a consequence, if conditions are sufficiently cold in the central parts of a subducting slab for reaction to be inhibited at the top of the transition zone, a wedge of metastable olivine is created, all volumes of which must eventually encounter the conditions critical for transformation-induced failure unless they pass into the lower mantle and transform there passively to perovskite + magnesiowüstite. Unfortunately, at the present time, neither the exact temperature distribution in slabs nor the exact conditions for metastable olivine preservation are known.

Seismic evidence that such metastable olivine is present in subducting slabs traversing the transition zone has been equivocal (cf. Koper et al. 1998) and some calculations based upon experimental studies of the kinetics of the transformations and thermal models of subduction zones have suggested that metastable olivine is likely only in the coldest slabs (Mosenfelder et al. 2001). Recently, however, evidence has been collected showing slow seismic velocities in a subhorizontal, slab-like, earthquake-generating anomaly beneath Fiji that is most easily explained as a remnant slab of metastable olivine floating in the transition zone (Chen and Brudzinski 2001, Green 2001). If this interpretation is valid, the conflict between observation and prediction may indicate that subduction zones are colder than suggested by current thermal models.

However, it is premature to reach such a conclusion. In addition to the uncertainties in slab temperature distributions and olivine-spinel reaction kinetics, there are specific characteristics of certain earthquakes that are difficult to explain by either of these mechanisms. Some very deep earthquakes, most notably those in the South American slab, are interpreted from their seismic characteristics to have involved seismic slip zones of diameter several 10s of km. For example, both the 1994 Tonga and Bolivia deep earthquakes have been estimated to involve slip on regions with diameters of ~60 km. Therefore, it is difficult to explain the full slip in these earthquakes by any mechanism that is inherently restricted to cold temperatures (i.e., both transformation-induced faulting and dehydration-induced faulting). Therefore, even if slab temperatures are colder than presently envisioned, it remains questionable as to whether there could be sufficiently large volumes of metastable olivine or hydrated mantle available to support such slip. It is important, therefore, to consider possible alternative or cooperative mechanisms. It is this concern that has led to resurgence of thermal runaway models for the deepest earthquakes (e.g., Karato et al. 2001).

CONCLUSIONS AND SPECULATIONS

The discussion presented here shows that dehydration-induced faulting and transformation-induced faulting are complementary; the former appears to be restricted to depths above ~350 km and the latter to depths below ~350 km. Moreover, the waning ability of dehydration to trigger earthquakes below 200 km conveniently falls in the region of lowest earthquake production and the inability of transformation-induced faulting to trigger faulting in endothermic or disproportionation reactions offers a natural explanation for cessation of earthquakes at the base of the transition zone. It is thus possible that earthquake initiation at depth in Earth can be completely explained by a combination of these two mechanisms.

Nevertheless, it is premature to reach this conclusion for three principal reasons:

(i) It is not clear that double seismic zones at intermediate depths can be assigned to dehydration-induced faulting;

(ii) very large, very deep earthquakes seem to be inconsistent with expectations about the size and geometry of a metastable wedge of olivine;

(iii) combination of experimentally-measured kinetics of olivine breakdown reactions and thermal models suggest that subduction zones in general are too warm for metastable olivine to persist to significant depths other than in

Tonga, although if Chen and Brudzinski (2001) have truly observed metastable olivine in an older slab floating beneath Fiji, either the kinetics of transformation, the extrapolation of those kinetics to mantle conditions, or slab models may need revision.

Additional work on this problem needed at this time includes

(i) laboratory experiments to determine whether the dense hydrous magnesium silicates are capable of generation of faulting via transient release of H_2O during reaction from one to another;

(ii) quantitative determination of the conditions of failure of olivine of mantle composition as a function of pressure and development of accurate methods to extrapolate those conditions to natural time scales;

(iii) investigation of mechanisms of hydration of oceanic lithosphere at subduction zones;

(iv) a thorough analysis of the errors potentially involved in experimental measurement of the kinetics of olivine reaction at high pressures and temperatures, their extrapolation to natural conditions, and the assumptions involved in thermal modeling;

(v) evaluation of alternative mechanisms in terms of their ability to function as trigger mechanisms for earthquakes under natural boundary conditions.

REFERENCES

Angel RJ, Frost DJ, Ross NL et al. (2001) Stabilities and equations of state of dense hydrous magnesium silicates Phys Earth Planet Inter 127:181-196

Antolik M, Dreger D, Romanowicz B (1996) Finite fault source study of the great 1994 deep Bolivia earthquake. Geophys Res Lett 23:1589-1592

Burnley PC, Green HW II, Prior D (1991) Faulting associated with the olivine to spinel transformation in Mg_2GeO_4 and its implications for deep-focus earthquakes. J Geophys Res 96:425-443

Chen WP, Brudzinski, M.R (2001) Evidence for large-scale, imbricate remnant of subducted lithosphere. Science 292:2475-2479

Frohlich C (1989) The nature of deep-focus earthquakes. Ann. Rev Earth Planet. Science 17:227-254

Fukao Y, Kikuchi M (1987) Source retrieval for mantle earthquakes by iterative deconvolution of long-period P-waves. Tectonophysics 144:249-269

Gleason G, Green HW II (1996) Effect of differential stress on the albite to jadeite + coesite transition at confining pressures of >3 GPa. EOS, Trans Am Geophys Union 77:F662

Green HW II (1994) Solving the paradox of deep earthquakes. Sci Am 271:64-71

Green HW II (2001) Deep Tonga backarc: Graveyard for buoyant slabs bearing metastable olivine? Science 292:2445-2446.

Green HW II (2001) Physical mechanisms for earthquakes at intermediate depths. EOS Trans Am Geophys Union 82, Fall Mtg Suppl, Abstr S42D-03

Green HW II, Burnley PC (1989) A new self-organizing mechanism for deep-focus earthquakes. Nature 341:733-737

Green HW II, Houston H (1995) The mechanics of deep earthquakes. Ann. Rev Earth and Planet. Sci. 23:169-213

Green HW II, Scholz CH, Tingle TN, Young TE, Koczynski T (1992) Acoustic emissions produced by anticrack faulting during the olivine → spinel transformation. Geophys Res Lett 19:789-792

Green HW II, Young TE, Walker D, Scholz CH (1990) Anticrack-associated faulting at very high pressure in natural olivine. Nature 348:720-722

Green HW II, Zhou Y (1996) Transformation-induced faulting requires an exothermic reaction and explains the termination of earthquakes at the base of the mantle transition zone. *In* Avé Lallemant et al. (eds) Carter Volume. Tectonophysics 256:39-56

Griggs DT, Baker DW (1969) The origin of deep-focus earthquakes. *In* Properties of Matter Under Unusual Conditions. H Mark, S Fernbach (eds) New York: Wiley Interscience, p 23-42

Gruntfest IJ (1963) Thermal feedback in liquid flow—plane shear at constant stress. Trans Soc Rheology 7:195-207

Hobbs BE, Ord A (1988) Plastic instabilities: implications for the origin of intermediate and deep focus earthquakes. J Geophys Res 9323-42,10521-10540

Igarashi I, Matsuzawa T, Umino N, Hasegawa A (2001) Spatial distribution of focal mechanisms for interplate and intraplate earthquakes associated with the subducting Pacific plate beneath the northeastern Japan arc: A triple-planed deep seismic zone. J Geophys Res 106:2177-2191

Jiao W, Silver PG, Fei Y, Prewitt CT (2000) Do deep earthquakes occur on preexisting weak zones? An examination of the Tonga subduction zone. J Geophys Res 105:28125-38

Kanamori H, Anderson DL, Heaton TH (1998) Frictional melting during the rupture of the 1994 Bolivian earthquake. Science 279:839-842

Karato S, Riedel MR, Yuen DA (2001) Rheological structure and deformation of subducted slabs in the mantle transition zone: implications for mantle circulation and deep earthquakes. Phys Earth Planet Inter 127:83-108

Kirby SH, Durham WB, Stern L (1991) Mantle phase changes and deep earthquake faulting in subducting lithosphere. Science 252:216-25

Kirby SH, Stein S, Okal EA, Rubie DC (1996) Metastable mantle phase transformations and deep earthquakes in subducting oceanic lithosphere. Rev Geophys 34, 261-306

Koper KD et al. (1998) Modeling the Tonga Slab: can travel time data resolve a metastable olivine wedge? J Geophys Res 103:30079-30100

McGuire JJ, Wiens DA, Shore PJ, Bevis MG (1997) The March 9, 1994 deep Tonga earthquake: Rupture outside the seismically active slab. J Geophys Res 102:15163-15182.

Meade C, Jeanloz R (1991) Deep-focus earthquakes and recycling of water into the earth's mantle. Science 252:68-72

Mosenfelder JL, Marton FC, Ross CR, et al. (2001) Experimental constraints on the depth of olivine metastability in subducting lithosphere. Phys. Earth Planet. Inter 127:165-180

Ogawa M (1987) Shear instability in a viscoelastic material as the cause of deep focus earthquakes. J Geophys Res 92:13,801-13,810

Peacock SM (2001) Are the lower planes of double seismic zones caused by serpentine dehydration in subducting oceanic mantle? Geology 29:299–302

Phipps-Morgan J (2001) The role of serpentinization and deserpentinization in bending and unbending the subducting slab. EOS, Trans Am Geophys Union 82, Fall Mtg Suppl, Abstr T22D-03

Phipps-Morgan J, Ranero C, Ruepke L (2002) Serpentinization, deserpentinization, and the bending and unbending of subducting slabs (abstr) Proc Structure and Tectonics of Convergent Plate Margins. Prague, Czech Republic (in press)

Raleigh CB (1967) Tectonic implications of serpentinite weakening. Geophys J Royal Astron Soc 14: 45-51

Raleigh CB, Paterson MS (1965) Experimental deformation of serpentinite and its tectonic implications. J Geophys Res 70:3965-3985

Reinen LA, Green HW II, Nielsen SK (1998) Dehydration embrittlement of antigorite serpentinite at high pressures: Implication for intermediate depth earthquakes. EOS, Trans Am Geophys Union 79:F853

Rice JR, Ruina AL (1983) Stability of steady frictional slipping. J Appl Mech 50, 343-49

Scholz CH (1990) The Mechanics of Earthquakes and Faulting. Cambridge, UK: Cambridge University Press, 439 p

Scholz CH (2002) The Mechanics of Earthquakes and Faulting, 2nd Edition. Cambridge, UK: Cambridge University Press, 496 p

Silver PG, Beck SL, Wallace TC, Meade C, Myers S, James D, Kuehnel R (1995) The rupture characteristics of the deep Bolivian earthquake of 1994 and the mechanism of deep-focus earthquakes. Science 268:69-73

Ulmer P, Trommsdorff V (1995) Serpentine stability to mantle depths and subduction related magmatism. Science 268:858-861

Ulmer P, Trommsdorff V (1999) Phase relations of hydrous mantle subducting to 300 km. In Fei Y, Bertka CM, Mysen BO (eds) Mantle Petrology: Field Observations and High Pressure Experiments. Geochem Soc Spec Publ 6:259-281

Wiens DA, McGuire JJ (1995) The 1994 Bolivia and Tonga events: Fundamentally different types of deep earthquakes? Geophys Res Lett 22:2245-2248.

Wiens DA, McGuire JJ, Shore PJ (1993) Evidence for transformational faulting from a deep double seismic zone in Tonga. Nature 364:790-793

Wiens DA, McGuire JJ, Shore PJ, Bevis MG, Draunidalo K, Prasad G, Helus SP (1994) A deep earthquake aftershock sequence and implications for the rupture mechanism of deep earthquakes. Nature 372: 540-543

Wong T-F, Ko SC, Olgaard DL (1997) Generation and maintenance of pore pressure excess in a dehydrating system, 2. Theoretical analysis. J Geophys Res 102:841-852

8
Brittle Failure of Ice

Erland M. Schulson

Thayer School of Engineering
Dartmouth College
Hanover, New Hampshire 03755

Reprinted with modifications from *Engineering Fracture Mechanics,*
vol. 68, p. 1839-1887, copyright 2001, with permission from Elsevier Science.

INTRODUCTION

Ice exhibits either ductile or brittle behavior, depending upon the conditions under which it is loaded. Glaciers, for instance, are loaded by gravity, under deviatoric stresses of ~0.1 MPa or lower. At temperatures of interest, they flow through dislocation creep at strain rates of the order of 10^{-9} s^{-1} or lower (Patterson 1994). Sheets of sea ice—another terrestrially important ice feature—are loaded predominantly by wind, under global compressive stresses similar in magnitude to the shear stresses within glaciers (Richter-Menge and Elder 1998). These bodies deform through a combination of creep and fracture, the latter process manifested by oriented leads or open cracks and by compressive shear faults which often form as conjugate sets that traverse a large fraction of the Arctic Basin (Kwok 1999; Schulson 2002). The cold, icy crust of Europa, an extraterrestrial feature which may shield an ocean beneath within which a form of life may exist or may once have existed (Reynolds et al. 1987; Hoppa et al. 1999; Pappalardo et al. 1999; Greenberg et al. 2000; Kargel et al. 2000), is loaded by the motion of diurnal tides and by non-synchronous rotation (Greenberg and Weidenschilling 1984). The crust deforms in a brittle manner, as evident from the networks of cracks that lace through it (Greeley et al. 2000).

Of the two kinds of ice deformation, creep is probably the better known by readers of the geological and geophysical literature, and is certainly the more fully explored and better understood. The interested reader may wish to consult a number of excellent reviews of the subject (e.g., see Duval et al. 1983; Weertman 1983; Durham and Stern 2001). Fracture, in comparison, has only recently been systematically examined. The motivation with respect to terrestrial mechanics (Schulson 2001) comes largely from the growing recognition of the role it plays in limiting ice forces against off-shore engineered structures and in forming arctic leads (which serve as short-circuits paths for the transfer of heat from the ocean to the atmosphere during winter). With respect to Europa, the study of fracture is justified by the possible linking of the surface to the putative ocean. Yet, even though the field of ice fracture is relatively new compared to ice creep, much has been learned about how cracks nucleate and propagate within this material. Indeed, the study of ice has led to a new understanding of compressive failure of rock and other brittle materials (Renshaw and Schulson 2001), further justifying the effort.

This paper considers the brittle failure of polycrystalline ice Ih, called ordinary or terrestrial ice[1], at temperatures (>0.8 T_{mp}) and strain rates (~10^{-7} s^{-1} to 10^{-1} s^{-1}) of engineering interest. The work emphasizes ice as a material and thus centers on structure-property relationships. The objective is to elucidate the failure mechanisms. Tensile behavior is considered first, and is discussed in terms of the nucleation and the propagation of cracks. Brittle compressive failure, particularly under multi-axial loading,

---

[1] Ice possesses 13 different crystal structures and two amorphous states. Ice Ih forms under terrestrial conditions and possesses the hexagonal crystal structure reflected in the shape of snow-flakes. The other crystallographic forms are thermodynamically stable only under high pressure.

is addressed next, at greater length owing to its greater complexity. It is described in part by Coulomb's criterion and is discussed in terms of frictional crack-sliding and a new mechanism of shear faulting. The compressive ductile-to-brittle transition is then considered, and is explained in terms of the competition between crack-tip creep and crack propagation. The paper closes with a discussion of failure in the field and with evidence that the underlying physical processes may be independent of spatial scale. They may also operate in rock and other brittle materials. The work is based largely upon the research of the author and his students, with due regard for the literature.

Before proceeding, it is worth noting that ice is unusual in exhibiting brittle behavior right up to the melting point, at deformation rates that are well below dynamic rates. Single crystals as well as polycrystals exhibit this behavior, implying that it originates in basic deformation processes and not through the restriction to slip imposed by grain boundaries. At root are sluggish dislocation kinetics (e.g., see Ahmad and Whitworth 1988; Shearwood and Whitworth 1991; Hu et al. 1995). Ice Ih slips preferentially on basal $\{0001\}$ planes and this is impeded by the unique requirement of protonic rearrangement[2] (Glen 1968). Molecular diffusivity is low as well—about three orders of magnitude lower than atomic diffusivity through metals at an equivalent homologous temperature[3]. This factor, however, appears to play more the role of suppressing diffusional creep within the coarsely grained material (~1 to 20 mm) that occurs naturally (Duval et al. 2000) than of impeding dislocation mobility. The "degree of brittleness" of polycrystalline ice Ih is rather high (Nixon and Schulson 1987), evident from the relatively low ratio[4] of toughness (~1 J/m^2) to surface energy (~0.1 J/m^2).

FAILURE UNDER TENSION

Ice breaks in tension while bending under ships (Michel 1978) and above submarines (Bazant et al. 1995), while buckling to form pressure ridges (Hopkins 1998), while bending against the sloping sides of offshore structures (Riska and Tuhkuri 1995) and during the calving of icebergs (Hughes 1989; Rist et al. 1999). This section describes the failure characteristics and then discusses them in terms of deformation mechanisms.

Characteristics

Fresh-water ice. Equiaxed and randomly oriented aggregates of fresh-water ice (termed "granular ice") fail under uniaxial tension via transgranular cleavage (Schulson et al. 1989a), at applied stresses around 1 MPa (Hawkes and Mellor 1972; Michel 1978; Currier and Schulson 1982; Schulson et al. 1984). The strength is essentially independent of strain rate, at least over the range 10^{-7} s^{-1} to 10^{-3} s^{-1}, and is only slightly dependent upon temperature, increasing by <25% upon cooling from -5°C to -20°C (Schulson et al. 1984). Aside from cracks, grain size has the largest effect. Upon decreasing the (linear intercept) grain diameter, d, from 10 mm to 1 mm, for instance, the strength, σ_t, increases by about a factor of two (Fig. 1). The relationship is either

$$\sigma_t = \sigma_o + k_t d^{-0.5} \text{ (for } d > d_c) \tag{1}$$

---

[2] Ice Ih is possibly unique in that translation of part of the crystal relative to the rest by the Burgers vector does not exactly reproduce the atomic arrangement of its protons. The translation introduces Bjerrum defects. The stress needed to create them (of formation energy 0.68 eV) is orders of magnitude greater than can be accounted for by the flow stress, and this suggests that some kind of protonic rearrangement probably occurs (Glen 1968).

[3] The melting point diffusion coefficient of H$_2$O in ice Ih is around 10^{-15} to 10^{-14} m^2/s compared to 10^{-12} to 10^{-11} m^2/s for elemental metals.

[4] For structural ceramics and rock (see Nixon and Schulson 1987), the ratio of toughness to surface energy is generally greater than 30.

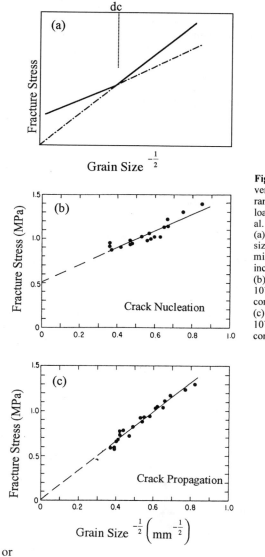

Figure 1. Graphs of tensile fracture stress versus (grain size)$^{-0.5}$ of equiaxed and randomly oriented polycrystals of ice Ih loaded under uniaxial tension (Schulson et al. 1984; Schulson 1987):
(a) showing schematically the critical grain size, d_c, below which the material exhibits a minor amount of inelastic deformation that increases with decreasing grain size;
(b) showing that at −10°C at a strain rate of 10^{-3} s^{-1} where $d_c < 1.4$ mm, the strength is controlled by crack nucleation; and
(c) showing that at −10°C at a strain rate of 10^{-7} s^{-1} where $d_c > 6.7$ mm, the strength is controlled by crack propagation.

or
$$\sigma_t = Kd^{-0.5} \text{ (for } d < d_c\text{)},\tag{2}$$

where σ_o, k_t and K are material constants and d_c is a critical grain size. Values (Schulson et al. 1984) at -10°C are $\sigma_o = 0.6$ MPa, $k_t = 0.02$ Mpa√m and $K = 0.044$ Mpa√m. d_c decreases with increasing strain rate, from >6.7 mm at 10^{-7} s^{-1} to 1.6 mm at 10^{-6} s^{-1} to <1.4 mm at 10^{-3} s^{-1} (Schulson et al. 1984; Schulson 1987). More finely grained aggregates ($d < d_c$) exhibit increasing inelastic deformation with decreasing grain size, reaching, for instance, ~0.3% at -10°C for $d = 1.6$ mm. More coarsely grained material fractures after an elongation of only ~0.02 to 0.04%, of which more than half is inelastic/viscoelastic deformation.

The notion of a "critical grain size" in ice was first suggested (Schulson 1979) in relation to a possible brittle-to-ductile transition, following Cottrell's discussion (Cottrell 1958) of a similar transition in steel. To some extent, the data support the idea, as noted. The effect, however, is small, at least within material whose grain size spans the range that is often observed in natural ice features. The transition might thus be better termed a "brittle-to-less brittle transition". Yet, there is evidence, noted below, that dynamically recrystallized ice exhibits rather extensive ductility, owing to a refined microstructure. Thus, the concept of a critical grain size is worth keeping. Within the context of brittle fracture and the mechanisms underlying Equations (1) and (2), it is given by the relationship:

$$r = \frac{K_I^2}{2\pi E^2}\left(\frac{(n+1)^2 E^n B t}{2n\alpha^{(n+1)}}\right)^{2/(n-1)} F \qquad (3)$$

The rate-dependence, it is suggested (although not proven), results mainly from the rate-dependence of σ_0.

In comparison to polycrystalline, single-plane ceramics, ice exhibits both similarities and differences. It is similar in the sense that its tensile strength increases with increasing refinement of the microstructure, exhibiting both Petch-type (Eqn. 1) and Orowan-type (Eqn. 2) relationships. The difference is the range of grain size over which the relationships apply. Alumina (Chantikul et al. 1990) and other ceramics (Zimmermann and Rodel 1998) show Petch behavior over finer grain sizes and Orowan behavior over coarser ones (Chantikul et al. 1990), while ice displays such behavior over the opposite ranges. Different mechanisms are probably operating. Ceramics contain pre-existing flaws, in which case residual stresses and stabilization mechanisms such as crack-bridging become considerations. The ice that was examined here, on the other hand, was almost certainly free from internal stresses, for it was well annealed at temperatures close to the melting point before being deformed at the same high temperatures. Also, it was initially free from cracks.

Saline ice. Salt-water ice is weaker than fresh-water ice. The reduction reflects the stress concentrating effect of brine pockets (Anderson and Weeks 1958; Assur 1958) which become entrapped within the matrix as the water freezes. The tensile strength of first-year columnar sea ice of 3 to 6 ppt salinity, for instance, loaded across the columns, varies from 0.7 to 0.8 MPa at -20°C to 0.20 to 0.3 MPa at -3°C (Dykins 1970;Richter-Menge and Jones 1993). Temperature is more important in this case, for owing to the NaCl–H_2O eutectic at -21°C the equilibrium volume fraction of brine increases above this temperature. In fact, the strength of the composite is a unique function of brine volume, v, rather than of temperature or salinity per se, decreasing linearly with either $v^{0.5}$ or $v^{0.67}$ and extrapolating to a value comparable to that for fresh-water ice (Weeks 1962). Below the eutectic, the strength is similar to that of fresh-water ice (Weeks 1962), exhibiting no further dependence upon temperature, at least to -35°C.

Unidirectionally solidified material, such as first-year arctic sea ice, possesses a microstructure in which brine pockets are arranged in a platelet-like manner on planes parallel to the crystallographic basal planes whose normals (the c-axes) are generally horizontally oriented (Weeks and Ackley 1982;Cole 2001). Its strength thus reaches a minimum in the direction perpendicular to the platelets (Shapiro and Weeks 1993). It is reduced further in the presence of networks of brine drainage channels (Shapiro and Weeks 1993). Full sheets of sea ice contain thermal cracks and other large stress concentrators which constitute an over-riding factor, lessening the tensile strength, it has been suggested (Lewis 1995), to the kPa range and raising the issue of a size effect

(Dempsey 1991; Dempsey 1996).

Failure mechanisms

Tensile failure, at least of fresh-water ice, can be understood in terms of either the nucleation or the propagation of cracks (Schulson 1979; Schulson et al. 1984; Schulson 1987). Crack nucleation limits the strength of aggregates of grain size larger than the critical size, where the stress to nucleate a crack is more than enough to propagate it. Propagation governs the strength of more finely grained material ($d < d_c$). The critical grain size thus defines the point where the stress to nucleate a crack equals the stress to propagate it (Fig. 1a). Because d_c decreases with increasing strain rate, "more coarsely grained" and "more finely grained" must be qualified with respect to the strain rate.

Crack nucleation. Nucleation control is manifested by the appearance of the ice (Schulson 1987; Cole 1990). The first crack propagates as soon as it forms, creating from tensile bars two pieces of material within which remnant cracks do not exist (e.g., see Fig. 4 of Schulson 1987). The process operates as follows: once the applied stress reaches a threshold level, denoted σ_0, stress becomes concentrated on grain boundaries, through either intragranular dislocation slip or grain boundary sliding. Both mechanisms concentrate stress in proportion to $\sqrt{L}$ where L is either the length of a dislocation pile-up impinging upon the boundaries or the distance along which the boundaries slide (Smith and Barnby 1967; Evans et al. 1980). Cracks nucleate once the inelastic strain and thus the local tensile stress reaches a level sufficiently great to overcome intermolecular cohesion. L is usually considered to scale with the grain size, and so the applied stress to nucleate cracks scales as $1/\sqrt{d}$. Equation (1) then follows, accounting for the behavior shown in Figure 1b.

Concerning the nucleation mechanism, it was suggested earlier (Schulson et al. 1984) that dislocation pileups, of the kind recently observed (Liu and Baker 1995), may be important. Accordingly, k_t was expressed by the relationship (Smith and Barnby 1967):

$$k_t = <m> (3 \pi \gamma G/8[1 - v])^{0.5} \qquad (4)$$

where γ is the surface energy (109 mJ/m^2, (Ketcham and Hobbs 1969), G is the shear modulus (initially taken to be 2300 MPa; a better value is 3500 MPa, (Gammon et al. 1983), n is Poisson's ratio (0.33, (Gammon et al. 1983) and $<m>$ is the Taylor orientation factor[5]. The problem is that the better values of G and $<m>$ yield $k_t \approx 0.14$ Mpa$\sqrt{m}$, which is about a factor of seven higher than the measured value of 0.02 Mpa$\sqrt{m}$. The current view, therefore, is that the pile-up mechanism is not the important one, at least at the higher rates of deformation where crack nucleation governs the tensile strength. Instead, grain boundary sliding probably dominates. Direct observations, in fact, have been made of crack nucleation through grain boundary sliding, albeit under compression (Nickolayev and Schulson 1993; Picu and Gupta 1995a,b; Weiss and Schulson 2000), underlining its importance.

Interestingly, the tensile strength of granular ice, when governed by crack nucleation, can double upon tension-compression cycling (Cole 1990) to stresses significantly above σ_0 ($\Delta\sigma \approx \pm 1$ MPa). The cycling evidently lessens the effectiveness of localized stress concentrators, leading to higher applied stresses for crack nucleation. Ledges on grain boundaries, for instance, have been directly observed to emit dislocations (Liu and Baker 1995; Baker 1997) under monotonic loading. Presumably under cyclic loading, such sites

---

[5] In earlier work (Schulson et al. 1984); (Schulson 1987) it was assumed that $<m> \approx 3$, the value commonly used for f.c.c. metals which exhibit a high degree of slip multiplicity. A higher value (~5) is more appropriate for hexagonal crystals which possess only two independent slip systems.

also allow strain to be accommodated, lessening the intensification of stress there. One wonders whether alumina and other ceramics might be strengthened through similar pre-treatment.

Crack propagation. Cracks within "more finely grained" ($d < d_c$) ice do not propagate immediately upon nucleation. Instead, the applied stress must increase until the mode-I stress intensity factor reaches the critical level. During the process, more inelastic deformation occurs and so, more cracks nucleate. Finely grained aggregates ($d \sim <1$ mm) thus exhibit more ductility than coarsely grained ones ($d \approx 10$ mm), although the level is still very small (<0.3%). Again, the appearance of the ice offers direct evidence of the strength-limiting process: broken tensile bars of finely grained material contain remnant cracks (see Fig. 8 in Schulson 1987). The crack size scales directly with grain size, and so crack tip stress concentration is directly proportional to (grain size)$^{1/2}$, thereby accounting for Figure 1c and for Equation (2). To account for the constant K, it is assumed that the cracks can be modeled as non-interacting, internal, penny-shaped cracks, for which

$$K = Q K_{Ic}, \tag{5}$$

where K_{Ic} is the mode-I fracture toughness and Q is a shape factor (Knott 1973) given by $Q = (\pi/2\alpha)^{0.5}$; α is the ratio of the diameter of the propagating crack to the grain size, and was measured (Lee and Schulson 1988) to be $\alpha = 3.7$. Taking the widely-used value $K_{Ic} = 0.1$ Mpa$\sqrt{m}$, Equation (5) gives $K = 0.05$ Mpa$\sqrt{m}$, in fair agreement with the measured value of 0.044 Mpa$\sqrt{m}$.

Crack propagation also controls the tensile strength of cracked ice, provided that the flaws exceed a critical size and are "sharp" or fresh. When short, or when long but "blunted", or shielded through creep deformation, the strength is again controlled by crack nucleation (Schulson et al. 1989b).

Saline ice. Salt-water ice is a more complicated material than fresh-water ice. While its microstructure can be controlled, systematic variations in factors other than brine volume/porosity are more difficult to achieve. For this reason, it is not clear whether Equations (1) and (2) are relevant in this case. The platelet spacing might be more important than the grain size, for instance. It could also be argued that crack-nucleation may never limit the tensile strength, for the brine pockets may act like cracks. Unfortunately, owing to its opacity, salt-water ice does not easily permit definitive observations of the kind made on fresh-water ice. Firm conclusions, therefore, are difficult to draw.

Ductile ice

It is noted again that ice generally exhibits little tensile ductility. However, when "preconditioned", then both fresh-water and salt-water materials can exhibit elongations greater than 10% (Schulson and Kuehn 1993; Kuehn and Schulson 1994a). In one case (Schulson and Kuehn 1993), an elongation exceeding 54% was measured (at -10°C at $10^{-7} s^{-1}$) without fracture, indicative, perhaps, of a kind of superplasticity. This increase is achieved by first pre-deforming material under compression by a few percent and then loading in tension. The processing introduces a network of short, stable cracks, as well as pockets of very fine grains (<0.1 mm) that form through dynamic recrystallizaton. The ductility is attributed to the refined microstructure, and is reminiscent of the ductility of very finely grained alumina at temperatures close to its melting point (Campbell et al. 1999). Grain boundary sliding, accommodated by a combination of dislocation creep and diffusion creep, is probably at play.

BRITTLE FAILURE UNDER COMPRESSION

Ice islands, icebergs and ice sheets too thick to buckle and too highly confined to split fail by crushing and spalling when pushed by wind and ocean currents at relatively high speeds against each other and against off-shore structures. Interactions of this kind can generate forces greater than 100-year wave forces against oil drilling platforms (API 1993), for instance, dictating engineering design. The objectives now are to describe the characteristics of brittle compressive failure, and then to discuss the failure mechanisms. Brittle compressive failure is more complicated than tensile failure, and so is discussed at greater length.

Overview

Ice exhibits two kinds of inelastic behavior under compression. At lower rates of deformation, the material is ductile (Fig. 2a). Its stress-strain curve is characterized by ascending and descending branches, and plastic strain in excess of 0.1 can be imparted without macroscopic failure. The material cracks profusely, at strain rates above $\sim 10^{-6}$ s^{-1}, but does not collapse when unconfined, at least until a kind of percolation threshold is reached. The ductile "peak" stress increases with increasing strain rate and with decreasing temperature (Fig. 2b; Michel 1978; Sinha 1982) and can be characterized ($\sigma \propto \varepsilon^m e^{mQ/RT}$) by a strain-rate sensitivity factor, m, which, although dependent upon stress-state and texture (Manley and Schulson 1997), is usually taken to be m ~ 0.3, and by an apparent activation energy, Q, which ranges between 45 kJ/mol and 90 kJ/mol (Weertman 1983). The "peak" stress decreases with increasing salinity/porosity (Peyton 1966; Schwarz and Weeks 1977), but exhibits little dependence upon grain size (Schulson and Cannon 1984; Cole 1985, 1987). Within this regime, dislocation processes are at play (Duval et al. 1983): basal glide serves as the major strain-producing mechanism, dislocation climb accounts for strain-rate hardening, both directional and non-directional interactions account for strain hardening, and internal cracking and dynamic recrystallization account for strain softening (Duval et al. 1983; Cole 1987).

At higher deformation rates—i.e., above $\sim 10^{-4}$ s^{-1} to $\sim 10^{-3}$ s^{-1}—ice exhibits brittle behavior (Schulson 1990). Under these conditions, its stress-strain curve is characterized by an ascending branch only, and failure occurs, without roll-over, after inelastic strains ~<0.003. Unconfined material fails via longitudinal splitting, while confined material fails by either faulting or spalling (Smith and Schulson 1993,1994). Post-terminal deformation (under confinement) exhibits stick/slip character and occurs at loads significantly lower than the terminal failure load. The terminal failure stress is a function of temperature, strain rate, grain size and confinement, decreasing with increasing temperature (by a factor of ~2.5 from -50°C to -10°C, (Schulson 1990), with increasing strain rate[6] (by a factor ~0.7 from 10^{-3}s^{-1} to 10^{-1}s^{-1}, (Schulson 1990) and with increasing grain size, Figure 3a (by a factor ~3 from 1 mm to 10 mm (Schulson 1990). Confinement is considered below. Unlike the tensile strength, the brittle compressive strength at -10°C appears to be independent of the salinity/porosity encountered in first-year sea ice (Schulson and Gratz 1999). However, -40°C the saline ice is weaker by a factor of about 2 than fresh-water ice of similar growth texture and grain size, under both uniaxial (Rudnicki and Rice 1975) and biaxial loading (Schulson 1990; Smith and Schulson 1993; Schulson and Nickolayev 1995; see Fig. 3b, below).

---

[6] *Strain-rate softening*, when first reported more than forty years ago (Khomichevskaya 1940; Butkovich 1954; Korzhavin 1955; Jellinek 1957; Voytkovskiy 1960), was later questioned and ascribed to an artifact of experimental technique (Hawkes and Mellor 1972). More recent experiments using careful procedures (Carter 1971; Cole 1987; Schulson 1990; Gratz and Schulson 1997a) have confirmed its existence, under both uniaxial and multiaxial loading. The effect is revisited later.

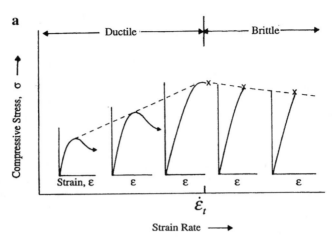

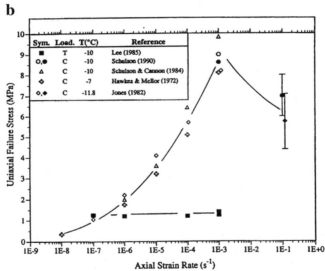

Figure 2. (a) Schematic sketch showing the effect of strain rate on the compressive stress-strain behavior of ice. At lower rates ($\dot{\varepsilon} < \dot{\varepsilon}_t$) the material exhibits strain hardening, followed by strain softening, strain-rate rate hardening and macroscopically ductile behavior. At higher rates the material exhibits strain-rate softening and at ~-10°C macroscopically brittle behavior. (b) Graph illustrating the effect of strain rate on the uniaxial compressive strength (C, upper curve) of equiaxed and randomly oriented polycrystals of ice Ih of ~1 mm grain size at approximately -10°C. The tensile strength (T) is shown for comparison. The peak in the compressive strength vs. strain rate marks the macroscopic ductile-to-brittle transition.

Discussions of brittle failure often cite the ratio of the unconfined compressive strength to the tensile strength, σ_c/σ_t. For instance, $\sigma_c/\sigma_t \approx 12$ for cold rock (Paterson 1978). Warm ice (T ≥ 0.8 Tm) is different. Because its compressive strength depends upon both temperature and strain rate while its tensile strength exhibits little dependence upon these factors, σ_c/σ_t is not constant. It varies, for instance, by an order of magnitude at -10°C (Fig. 2b) and by an even greater degree at lower temperatures.

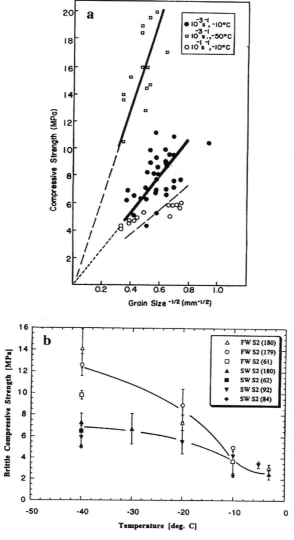

Figure 3. (a) Graph of uniaxial brittle compressive strength versus (grain size)$^{-0.5}$ of equiaxed and randomly oriented polycrystals of ice Ih at -10°C and -50°C at 10^{-3} s^{-1} and 10^{-1} s^{-1} (from Schulson 1990). (b) Graph of the uniaxial, across-column brittle compressive strength of S2 fresh-water (FW) and S2 salt-water (SW ~ 4.5 ppt salinity) ice of ~6 mm column diameter at 10^{-3} s^{-1} vs. temperature.

Is size a factor? A compilation of indentation pressures versus contact area (Sanderson 1988) indicates a decrease in strength (from ~10 MPa to ~0.01 MPa) with increasing area (from ~10^{-4} m^2 to ~10^5 m^2), suggesting some kind of "size effect". Its nature, however, is probably related more to boundary conditions (e.g., asperities leading to non-simultaneous brittle failure) and to large, resident stress concentrators (e.g., thermal cracks and refrozen leads[7]) than to the size of the ice per se. Experiments, for instance, on microstructurally similar cubes of granular ice 10 mm to 150 mm on edge (Kuehn et al. 1993) and on microstructurally similar blocks of S2 saline ice 150 mm to 1 m in size (Melton and Schulson in preparation) offer no evidence of an effect of size over the ranges noted. They do show, however, that irregular ice/platen interfaces trigger non-simultaneous failure at strain rates within the brittle regime (Kuehn et al. 1993a). The "size effect" is considered more fully by others (Jordaan 2001; Sodhi 2001; Weiss 2001).

Brittle compressive failure under multiaxial loading

Compressive failure in practice generally occurs under confinement. For instance, the contact zone that develops during ice-structure interaction is characterized by either a biaxial or a triaxial stress state that varies both spatially and temporally. Multiaxial loading has thus been studied extensively.

In exploring effects, a variety of devices has been employed. These include pressure cells (Jones 1982; Richter-Menge 1991; Rist and Murrell 1994; Gagnon and Gammon 1995), true polyaxial loading systems (Hausler 1981; Smith and Schulson 1993,1994; Schulson and Nickolayev 1995; Weiss and Schulson 1995; Gratz and Schulson 1997; Schulson and Gratz 1999) and a plane-strain device (Frederking 1977; Timco and Frederking 1983). Data from pressure cells and polyaxial systems compare favorably under lower degrees of confinement (Weiss and Schulson 1995); under higher degrees, there is no agreement because polyaxial systems induce premature failure through the destablization of material adjacent to non-loaded edges (Weiss and Schulson 1995). Polyaxial data compare favorably with those from the plane-strain device (Schulson and Nickolayev 1995), albeit along the single loading path that the device allows. The chief advantage of the polyaxial system is that it allows loads to be applied along any path in principal stress space, including biaxial ones, an ability that has proven to be useful in elucidating failure processes. The emphasis here, therefore, is on results generated using a polyaxial machine—namely, a high-stiffness (8000 MN/m), six-actuator, servohydraulic unit installed within a cold-room in the author's laboratory.

Data were obtained from plates ($150 \times 130 \times 25\text{-}40$ mm^3) and from cubes (150 mm on edge) proportionally loaded to terminal failure. To reduce end constraint, axially stiff and laterally compliant metallic brush platens (Gies 1988) were used. The brush platens, like compliant end caps (Haynes and Mellor 1977), allow length/width ratios of unity, with no apparent effect on the measured strength. Both columnar and granular ice have been studied.

S2 columnar ice. Columnar ice is the major constituent of river ice and first-year arctic sea ice. Its microstructure is reminiscent of the columnar zone of metallic ingots (Flemings 1974) and results from unidirectional heat flow during solidification. Columnar ice is an orthotropic solid: it possesses three mutually perpendicular axes about which a rotation of 180° gives an identically appearing microstructure. Three variants[8] are found in nature, S1 in which the crystallographic c-axes (normal to the basal planes) are parallel to the long axes of the columnar-shaped grains, S2 in which the c-axes are randomly oriented within a plane (the horizontal plane in natural sheets of river and sea ice) whose orientation is more or less perpendicular to the long axis of the columnar-shaped grains, and S3 ice in which the c-axes are strongly aligned within that plane. S1 ice, the least common variant, has not been studied. Little is known about S3 ice other than that its

---

[7] Leads are large cracks in floating ice covers. Their origin is considered below in the section entitled *On the Formation of "Leads" in the Arctic Sea Ice Cover.*

[8] The S1, S2 and S3 designations follow Michel's (Michel 1978) description of different types of ice Ih.

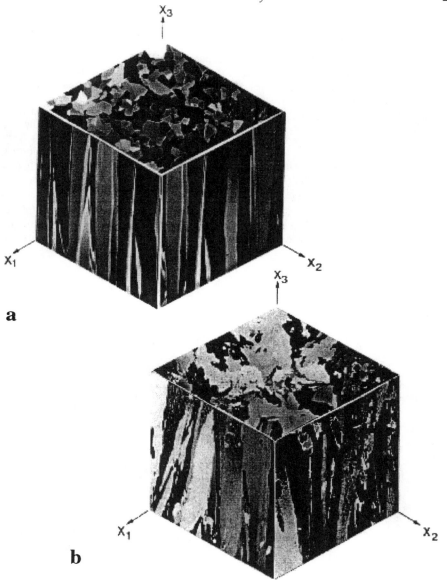

Figure 4. Composite photographs of thin-sections (~0.5 mm) viewed through cross-polarizing filters showing the X_i co-ordinate system and the macroscopic structure (a) of S2 fresh-water ice and (b) of S2 salt-water ice of 4.5 ppt salinity. Scale: width of each face is 80 mm.

strength reaches a minimum along directions at ~45° to the direction of alignment (Richter-Menge 1991). Thus, attention is focused on S2 ice (see Table 1).

To aid the discussion, an orthogonal set of coordinate axes, X_i, is defined, Figure 4: X_1 and X_2 are perpendicular to the long axis of the columnar grains and X_3 is parallel to the columns. The two normal stresses σ_{11} and σ_{22} are termed "across-column stresses," because they act on planes whose normals are perpendicular to the columns; these components of the stress tensor can be interchanged because S2 ice exhibits isotropic behavior within the X_1–X_2 plane. The normal stress σ_{33} is termed the "along-column stress" because it acts on planes whose normal is parallel to the long axis of the columnar grains. The three normal stresses are principal stresses. However, depending upon the loading path, σ_{11} is not always the largest (most compressive) principal stress nor is σ_{22} always the intermediate principal stress. To avoid confusion, the double subscript notation is used throughout this section. All compressive stresses are assigned positive values. Only loading directions either parallel or perpendicular to the columns are considered; "off axis" loading remains to be explored. The discussion addresses both fresh-water and salt-water S2 ice, Table 1. The materials were prepared in the laboratory under conditions that simulate Nature, and then proportionately loaded to terminal failure along four different kinds of path (Fig. 5).

Biaxial compressive strength: When loaded both across and along the columns such that $\sigma_{11} > \sigma_{33}$ (Fig. 5a), terminal failure occurs by irregular splitting in the loading plane; i.e., on a plane parallel to the free surface. Along these paths, the along-column confining stress has no systematic effect on the strength, taken as the across-column stress at

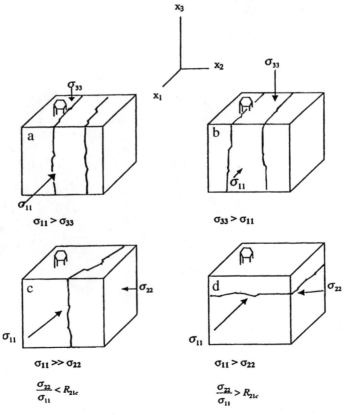

Figure 5. Schematic sketches showing four paths of proportionally loading S2 ice under biaxial compressive stresses; plus the failure mode:
 (a) $R_{21}=0$, $\sigma_{11}>\sigma_{33}$; (b) $R_{21}=0$, $\sigma_{33}>\sigma_{11}$; (c) $R_{31}=0$, $R_{21}<R_{21c}$; (d) $R_{31}=0$, $R_{21}>R_{21c}$.

terminal failure, σ_{11f} (Smith and Schulson 1994; Schulson and Nickolayev 1995; Schulson and Gratz 1999). Similarly, when $\sigma_{33} > \sigma_{11}$ (Fig. 5b), failure again occurs by splitting in the loading plane and the confining stress (this time σ_{11}) again has no effect on the strength (σ_{33f}). The strength, however, is greater along the columns than across them, by about a factor of two (Kuehn and Schulson 1994b). Along both kinds of path, the confining stress has little or no component resolved onto the macroscopic failure plane, accounting for its nil effect on the failure stress.

Table 1. Characteristics of columnar ice.

Characteristic	Fresh-water ice	Salt-water ice
Grain shape	columnar	columnar
Texture	S2	S2
Column diameter (mm)	7.6±1.6	8.5±1.8
Salinity (%)	—	4.5±0.3
Density (kg/m^3)	917.7±0.4	910±3
Porosity (%)	~0.1	3.9
Brine pocket spacing (mm)	—	0.6±0.2
Pore size (mm)	—	~0.5

When S2 ice is biaxially loaded under two across-column stresses, the confining stress (denoted σ_{22}) has a large effect. Two cases are distinguished, depending upon the ratio of the minor (i.e., confining) stress to the major stress, $R_{21} = \sigma_{22}/\sigma_{11}$, relative to a critical degree of confinement, R_{21c}:

(i) *Strengthening:* For lower confinements $R_{21} < R_{21c}$ (Fig. 5c), the strength, taken as σ_{11f}, increases sharply with increasing confinement. Figure 6 (branch AB) illustrates this effect. It can be described by the relationship:

$$\sigma_{11f} = \sigma_u + k\,\sigma_{22} = \sigma_u/(1 - k\,R_{21}) \tag{6}$$

where σ_u is the uniaxial across-column strength and k is a dimensionless constant; their values increase with decreasing temperature and are similar for both types of ice, Table 2. Terminal failure occurs by shear faulting ($R_{21} > 0$) across a plane parallel to the columns, but inclined (at -10°C) by 28±4° to the major stress (more below). That the fault does not lie in a plane parallel to the intermediate principal stress (i.e., σ_{22}) does not contradict a basic tenet (Rudnicki and Rice 1975) of faulting, because S2 ice is an anisotropic material. Within this lower confinement regime, the materials deform inelastically in essentially a plane strain manner, with little or no displacement in a direction parallel to the long axis of the columns.

(ii) *Weakening*: Under higher confinements $R_{21} > R_{21c}$ (Fig. 5d), the strength decreases with increasing confining stress; branch BB, Figure 6. This effect can be described by the relationship:

$$\sigma_{11f} = \sigma_0 - \sigma_{22} = \sigma_0/(1 + R_{21}) \tag{7}$$

where σ_0 is another constant, Table 2, which again is similar for the two types of ice. Temperature is still a factor, but a less important one. Failure now occurs by

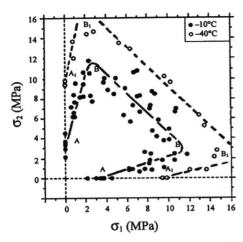

Figure 6. Brittle compressive failure envelopes for S2 fresh-water ice of ~6 mm column diameter biaxially loaded across-column at -10°C and -40°C at 10^{-2} s^{-1} (major strain rate) showing strengthening with increasing confining stress under lower degrees of confinement (branch AB, A'B') and weakening with increasing confining stress under higher degrees of confinement (branch BB, B'B'). The envelope is symmetric with respect to the principal stresses because S2 ice is isotropic in the X_1-X_2 plane. Data are thus plotted twice about the line $\sigma_{11}=\sigma_{22}$ (From (Smith and Schulson 1993).

cleavage in and then spalling out of the loading plane. Cracks propagate across the columns, through a mode analagous to that observed during higher-speed, through-thickness, across-column indentation of S2 ice where lateral confinement is provided by material surrounding the contact zone (Grape and Schulson 1992; Sodhi 2001). Note that R_{21c} decreases with decreasing temperature, Table 2.

Table 2. Experimental constants for Equations (6) and (7) describing the biaxial compressive strength of fresh-water and salt-water S2 ice at a maximum principal strain rate of 10^{-2} s^{-1} (Smith and Schulson 1993, 1994; Schulson and Nickolayev 1995; Pradeau and Schulson 1998).

Ice	T (°C)	σ_u (MPa)	k	σ_0 (MPa)	R_{21c}
fresh-water ice	-10	3.8±1	3.1	15	.25
	-40	10±1	~4	20	~0.1
salt-water ice	-2.5	2.7±0.8	1.8	11	0.45
	-10	3.5±0.7	3.1	16	.25
	-40	6.5±1	4.8	16	~0.1

Triaxial compressive strength: Confinement under triaxial loading raises the compressive strength. Three regimes are evident, depending upon the loading path (Gratz and Schulson 1997; Schulson and Gratz 1999); see Figure 7.

(i) *Regime 1*: When loaded mainly across the columns under a small across-column confining stress ($R_{21} \leq 0.25$, Fig. 7a), the strength, σ_{11f}, is independent of the degree of confinement along the columns (Fig. 8a) but increases sharply with increasing across-column confinement (Fig. 8b). Failure occurs by macroscopic shear faulting on a plane parallel to X_3, but inclined (at -10°C) by 31±6° to X_1. The behavior here is essentially identical to the behavior under lower-confinement, across-column, biaxial loading. Again, the along-column component of the confinement has little or no component resolved onto the failure plane and thus does not affect the strength.

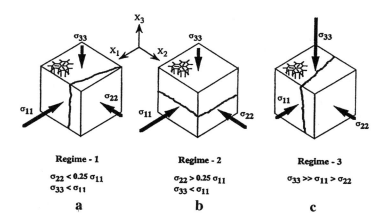

Figure 7. Schematic sketches showing three regimes of brittle compressive failure of S2 columnar ice proportionally loaded triaxially across and along the columns: (a) Regime-1; (b) Regime-2; (c) Regime-3.

(ii) *Regime 2*: When loaded mainly across the columns, but under higher degrees of across-column confining stress ($R_{21} > 0.25$; Fig. 7b), the strength increases markedly upon increasing the along-column confining stress (Fig. 8c). Failure occurs by faulting on planes that cut the columns (at -10°C) at 33±13° to their long axis. Under higher confinements ($R_{31} > 0.1$, termed Regime -2b, Fig. 8c), the strength becomes less dependent upon confining stress, exhibiting more the character of ductile failure than of brittle failure. Correspondingly, the fault is more steeply inclined, cutting the columns (at -10°C) at 41±3°. The reduction in the sensitivity to confinement corresponds to the hydrostatic component of the applied stress ($\sigma_{11} + \sigma_{22} + \sigma_{33}$)/3 reaching ~14 MPa and appears to mark the onset of dynamic recrystallization (Schulson and Gratz 1999), at least at -10°C where experiments have been performed to date.

(iii) Regime 3: When loaded mainly along the columns (Fig. 7c), the strength, σ_{33f}, increases in proportion to the smaller of the two across-column confining stresses (Fig. 8d). Failure now occurs by faulting on a plane inclined (at -10°C) 22±9° to X_3 and parallel to X_1.

Within each regime, the strength, s_f, can be described by a relationship of the form:

$$\sigma_f = \sigma_{u'} + k'\sigma_c \tag{8}$$

where $\sigma_{u'}$ and k' are constants and σ_c is the appropriate confining stress. Table 3 lists their values. Note that within Regime 1 the constant k' of Equation (8) has the same value as the constant k of Equation (6). Note also, as already mentioned, that the brittle compressive strength of the salt-water ice is essentially indistinguishable from that of fresh-water ice. The only detectable difference (Schulson and Gratz 1999) is that the shear fault is narrower in the salt-water material—i.e., less than one grain diameter vs. two to three diameters—owing to a crack-stopping effect of its pores.

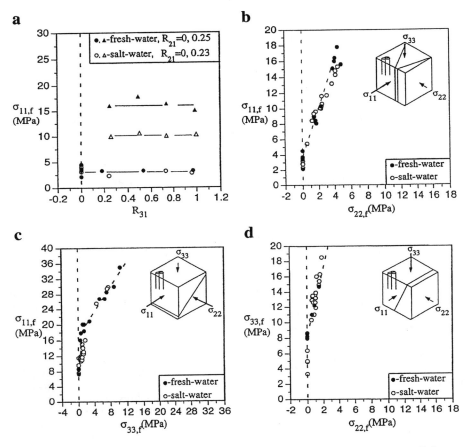

Figure 8. Graph of brittle triaxial compressive strength of S2 fresh-water and salt-water (4.5 ppt salinity) ice at -10°C at $6\times10^{-3}s^{-1}$ (maximum principal strain rate): (a) versus along-column confinement under low across-column confinement at three different levels, Regime-1; (b) versus across-column confining stress for various levels ($0<R_{31}<1$) of along-column confinement, Regime-1; (c) versus along-column confining stress for high ($R_{21}\sim1$) across-column confinement, Regime-2; (d) versus across-column confining stress ($\sigma_{22}=\sigma_{11}$), Regime-3. (From Schulson and Gratz 1999.)

Granular ice. Granular ice, also called snow ice or T1 ice in Michel's terminology (Michel 1978), is the material of icebergs, ice islands, and polar glaciers. It also constitutes the upper layer of floating sheets of columnar ice and, as such, is the material encountered by the sides of ships when moving through fields of broken ice plates. Its grain size typically ranges from ~1 mm to ~20 mm, depending upon thermal-mechanical history. Barring a deformation texture, which can develop in glaciers, for instance (Castelnau et al. 1997), the material is macroscopically isotropic.

When biaxially loaded along the path $\sigma_2 = R\sigma_1$, $\sigma_3 = 0$, where R is the proportionality constant ($R > 0$) and σ_1, σ_2 and σ_3 are the maximum (most compressive), intermediate and minimum principal stresses, respectively, terminal failure occurs by spalling out of the plane of loading at a level of stress which depends only slightly, if at all, on the confining stress (Weiss and Schulson 1995). The available data are too few to specify the trend

Table 3. Experimental constants for Equation (8) describing the triaxial strength of fresh-water and salt-water S2 ice at -10°C at a maximum principal strain rate of 6×10^{-3} s^{-1} (from (Schulson and Gratz 1999).

Regime	Definition σ_f	σ_c	Fresh-water $\sigma_{u'}$ (MPa)	k'	Salt-water $\sigma_{u'}$ (MPa)	k'
1	$\sigma_{11,f}$	$\sigma_{22,f}$	3.8 ± 1^a	3.1	3.5 ± 0.7^b	3.1
2a	$\sigma_{11,f}$	$\sigma_{33,f}$		10 ± 5	Insufficient data	4.5 ± 0.4^c
2b	$\sigma_{11,f}$	$\sigma_{33,f}$		1.6 ± 0.1 ($R_{21}=1$) 1.3 ± 0.1 ($R_{21}=0.5$)	Insufficient data	Insufficient data
3	$\sigma_{33,f}$	$\sigma_{22,f}$		4.4 ± 0.1	5.6 ± 1.7^b	7.2 ± 0.4^c

[a] Smith and Schulson (1993), [b] Smith and Schulson (1994), [c] Gratz and Schulson (1997)

with certainty. When triaxially loaded along the path $\sigma_3 = \sigma_2 = R\,\sigma_1$, failure occurs by shear faulting along a plane inclined (at -10°C) by ~30° to σ_1 (Weiss and Schulson 1995). The triaxial strength increases sharply with increasing confining stress, and the ratio of the confined to the unconfined strength increases with decreasing temperature, scaling approximately as $1/(1-3R)$ at -10°C and as $1/(1-5R)$ at -40°C (Fig. 9). More data are needed to specify these functions with greater certainty. It is noteworthy, however, that the dependence of the triaxial strength on confining stress is quantitatively similar to that of the S2 ice when failure occurs through shear faulting (i.e., $R_{21} < R_{21c}$).

When deformed under higher levels of confinement (e.g., above p = 15 MPa at -40°C at 10^{-2} s^{-1}, d = 1.7 mm), granular ice becomes ductile (Rist and Murrell 1994). This is marked by a gradual lessening of the dependence of the deviatoric stress at failure

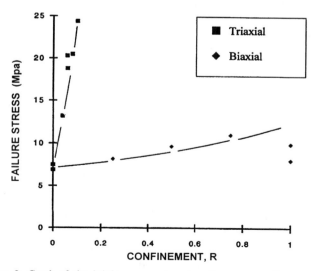

Figure 9. Graph of the brittle compressive strength versus confinement, R, of granular (d = 7 mm) fresh-water ice Ih under biaxial ($\sigma_2 = R\sigma_1$) and triaxial ($\sigma_3 = \sigma_2 = R\sigma_1$) proportional loading at -40°C at 10^{-3} s^{-1}. (From Weiss and Schulson 1995).

$(\sigma_1 - \sigma_3)_f$ on the confining pressure, p, to the point that $(\sigma_1 - \sigma_3)_f$ becomes independent of p. At high pressures ($p \geq$ 50 MPa at -11°C at 5.4×10^{-3} s^{-1}, d ≈ 0.8 mm) $(\sigma_1 - \sigma_3)_f$ decreases (Jones 1982), owing to the pressure-induced reduction in the melting temperature (~0.08K/MPa) expressed in the Clausius-Clapeyron relationship.

Orientation of shear faults. In the work cited above, shear faults were inclined by about 30° with respect to the maximum principal stress, as commonly seen in rock and concrete. However, other reports, based upon experiments using conventional triaxial pressure cells, cite orientations of 45±5° under small confining pressures (Rist et al. 1994; Rist and Murrell 1994; Rist 1997; Sammonds et al. 1998). The difference cannot be attributed to differences in macroscopic behavior, for in each case the ice exhibited the characteristics of brittle failure. Nor can it be attributed to differences in ice type: the less steeply inclined faults were found both in granular ice of 1mm and 5mm grain size and in multi-year sea ice. Instead, it has been suggested (Rist and Murrell 1994; Rist 1997) that the 45° faults result from localized plasticity that leads to unstable in-plane crack propagation.

A re-examination of the literature by Schulson (2002) has revealed two kinds of compressive shear faults in terrestrial ice when rapidly loaded under triaxial states of stress. One kind, termed a Coulombic or C- fault, develops under lower degrees of confinement and (at 263 K) is typically inclined by 28-30° to the direction of shortening. The other kind, termed a plastic or P- fault, develops under higher confinements and is inclined by 45°. The C-P transition occurs when confinement becomes sufficiently large to suppress frictional crack sliding. A criterion (upper limit) is given (Schulson 2002), in terms of the coefficient of internal friction, that defines the transition in terms of the ratio of minimum to maximum principal stress at failure. Plastic faulting, we suggest, is caused by adiabatic softening and not by in-plane crack propagation.

Failure criteria. From the effects of confinement it is clear that the brittle compressive failure stress of ice, like rock (Jaeger and Cook 1979) and concrete (Hobbs et al. 1977), depends upon the hydrostatic component of the applied stress tensor. This contrasts with the ductile compressive failure stress which is independent of that component (Melton and Schulson 1998). Different failure criteria are thus at play. Ductile failure (Melton and Schulson 1998) obeys Hill's (Hill 1950) criterion and is rooted in dislocation slip and in the fact that the critical resolved shear stress for basal glide does not depend upon the component of stress normal to the slip plane (Trickett et al. 2000). Brittle failure, on the other hand, is based upon crack mechanics and upon frictional sliding, as discussed below.

Coulomb's criterion, shear faulting and fault orientation: From the similarity in the effect of confinement on the strength of the three types of ice described above, it would appear that a single criterion governs fault-limited terminal failure. Coulomb's criterion, it is suggested, is the appropriate one. Accordingly, the shear stress, $|\tau|$ acting across the future fault plane is resisted by material cohesion, denoted S_o, and by friction, denoted by the product of the coefficient of friction, μ and the normal stress across the plane, σ:

$$|\tau| = S_o + \mu \sigma \tag{9}$$

This criterion may be more conveniently expressed in terms of the maximum shear stress $(\sigma_1-\sigma_3)/2$ and the hydrostatic component of the applied stress $(\sigma_1+\sigma_2+\sigma_3)/3$ (see (Jaeger and Cook 1979):

$$(\mu^2 + 1)^{0.5}(\sigma_1 - \sigma_3)/2 = \tau_o + \mu(\sigma_1 + \sigma_2 + \sigma_3)/3 \tag{10}$$

where τ_o is another measure of the cohesive strength. For S2 ice biaxially loaded across the columns, the criterion reduces to

$$\sigma_{11}[(\mu^2+1)^{0.5}(1-R_{21})-\mu(1+R_{21})] = 2\tau_o. \quad (11)$$

If it is assumed that μ is independent of confinement and that its value is similar to the steady-state kinetic friction coefficient for ice sliding slowly upon ice (at $v \sim 10^{-4}$ m s^{-1}), then from recent ice-on-ice measurements (Kennedy et al. 2000) $\mu = 0.3$, 0.5 and 0.8, respectively, at -2.5°C, -10°C and -40°C. These values lead to expected ratios for the confined to the unconfined biaxial strength that are in fair agreement with the measured ratios (Table 4).

Table 4. Ratio of confined/unconfined: biaxial strength of S2 ice (ratios are limited to lower confinements R_{21} such that the denominator > 0).

T (°C)	Observation (Eqn. 6, Table 2)	Coulombic criterion (Eqn. 11)
-2.5	$1/(1-1.8R_{21})$	$1/(1-1.8R_{21})$
-10	$1/(1-3.1R_{21})$	$1/(1-2.7R_{21})$
-40	$1/(1-4.8)$	$1/(1-4R_{21})$

That the failure criterion changes when confinement exceeds a critical level is not surprising. Owing to its frictional nature, sliding across the fault plane is suppressed when the normal and the shear stresses on that plane reach the level where the effective shear stress is zero. From Equation (11), this is expected to happen as R_{21} approaches the limit:

$$R_{21c} = [(\mu^2+1)^{0.5}-\mu)]/[(\mu^2+1)^{0.5}+\mu]. \quad (12)$$

Using the same friction coefficients, Equation (12) predicts that $R_{21c} = 0.55$, 0.38 and 0.25 at -2.5°C, at -10°C and at -40°C, respectively, compared with the observed values of 0.45 at -2.5°C, 0.25 at -10°C and ~0.1 at -40°C. The analysis thus captures the observed effect of temperature, but over-predicts the critical confinement. The deviation is of the correct sense, however —indeed, is expected— for upon raising confinement, the spalling mode becomes activated before the effective shear stress falls to zero.

The triaxial failure stress of ice can also be rationalized, within a factor of two or so, in terms of Coulomb's criterion. Accordingly, within Regime-1 the dependence of strength on the appropriate confining stress (σ_{22}) is expressed by the same relationship that describes biaxial failure under across-column loading at lower confinements. Thus, the expected ratio of the confined to the unconfined strength at -10°C is again expressed by $1/(1-2.7R_{21})$. This compares fairly well with the measured strengthening which scales as $1/(1-3.1R_{21})$, Table 3. That both salt-water and fresh-water ice exhibit the same scaling reflects the similarity of their friction coefficients (Kennedy et al. 2000). Within triaxial Regime-2a, Coulomb's criterion expresses the ratio of the confined to unconfined strength (at -10°C) as $1/(1-1.8R_{31}-0.4R_{21})$. This accounts reasonably well for the strengthening under full across-column confinement ($R_{21} \sim 1$). However, under more moderate across-column confinement the criterion under-predicts the strengthening by about a factor of two. For the case studied within Regime-3 ($\sigma_{22} = \sigma_{11}$) the criterion predicts a ratio of confined to unconfined strength of $1/(1 - 2.3R_{23})$ where $R_{23} = \sigma_{22}/\sigma_{33}$. Again, this under-predicts the strengthening by about a factor of two. These discrepancies suggest perhaps that in these instances the appropriate friction coefficient is greater than the one used in the analysis.

The failure of granular ice under triaxial loading can also be rationalized in terms of Coulomb's criterion. Once more, however, and possibly for the same reason, the criterion under-predicts the strength at -40°C by about a factor of two.

Consistent with Coulomb's failure criterion is the orientation of shear faults. The criterion dictates that faults form along conjugate planes whose orientation, θ, with respect to the maximum principal stress is given by the relationship (Jaeger and Cook 1979):

$$\tan^2(\pi/2 - \theta) = [(\mu^2 + 1)^{0.5} + \mu]^2. \tag{13}$$

Again using $\mu = 0.8$ at -4°C, $\mu = 0.5$ at -10°C, and $\mu = 0.3$ at -2.5°C, this gives $\theta = \pm 26°$, $\pm 29°$ and $\pm 38°$ at -40°C, -10°C, and -2.5°C, respectively, in reasonable agreement with measured values.

Hertzian criterion and spalling. Less clear is the failure criterion that governs cleavage and spalling of S2 ice biaxially loaded at $R_{21} > R_{21c}$. One suggestion (Smith and Schulson 1993) is that terminal failure in these cases results from the propagation of Hertzian cracks that are initiated along cracked grain boundaries. Accordingly, barring traction, the failure stress (from contact mechanics (Johnson 1987) can be expressed by the relationship (Smith and Schulson 1993):

$$\sigma_{11f} = \frac{6\sigma_T}{(1-2v)(1+R_{21})} \tag{14}$$

where σ_T is the tensile strength. This relationship has the same dependence upon confinement as the one found experimentally (Eqn. 7), implying that $\sigma_0 = 6\sigma_T/(1 - 2v)$. Appropriate values ($\sigma_T = 1$ MPa, ($v = 0.33$) lead to $\sigma_0 = 21$ MPa, in rough agreement with observation. The problem with the model is the implication that short cracks propagate long distances. More work is needed to clarify the spalling criterion.

Brittle failure envelope. Figure 10 shows schematically the terminal brittle failure envelope for S2 ice in biaxially loaded across the columns. The envelope combines both tensile and compressive behavior. Again, compressive stresses are positive. Loading paths within the tensile-compressive and the tensile-tensile quadrants have not been explored experimentally, and so the sections within those quadrants are drawn on the assumptions (1) that failure occurs in a purely tensile manner once one of the principal stresses reaches the tensile strength, σ_T, (point (a')), and (2) that the tensile strength is independent of the degree of biaxial tension. The larger and smaller envelopes, respectively, depict terminal failure of more finely grained and more coarsely grained ice. The slope, k, of the Coulombic segment (c-d) is assumed to be the same for both the finer-grained and the coarser-grained material, because the ice-on-ice coefficient of sliding friction is independent of grain size (Kennedy et al. 2000). The slope is lower at higher strain rates (dotted lines), because the friction coefficient decreases with increasing sliding velocity (Kennedy et al. 2000). This accounts for the strain-rate softening mentioned above and noted again below. The envelope is truncated within the compression-compression quadrant, owing to the transition from in-plane shear faulting to out-of-plane spalling. Presumably, similar envelopes describe earlier stages of failure.

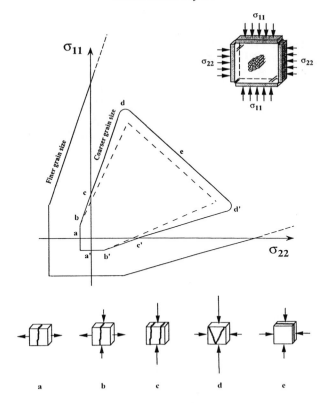

Figure 10. Schematic of the brittle failure envelope for S2 ice of two grain sizes, loaded across the columns. The broken line denotes strain-rate softening. Compressive stresses are positive.

It should be noted that the experimental brittle failure envelope (Fig. 6) and the schematic one (Fig. 10) were constructed on the basis of relatively few data, particularly the lower confinement ($R_{21} < R_{21,c}$) section, and that the data are scattered. Its true shape, therefore, may not be faceted, as drawn. Pressure-melting at contact points, for instance, could lower the friction coefficient and this could change the shape from linear to convex around the origin. More experiments are needed. It is clear, however, that the brittle compressive failure envelope is different from the ductile compressive failure envelope. The latter is semi-elliptical in shape, with its semi-major axis along the line $\sigma_{11} = \sigma_{22}$, and is defined by the relationship (Schulson and Nickolayev 1995):

$$\sigma_{11}^2 + \sigma_{22}^2 - \sigma_{11}\sigma_{22}\left[2 - \left(\sigma_{u,1}^{(d)}/\sigma_{u,3}^{(d)}\right)^2\right] = \left(\sigma_{u,1}^{(d)}\right)^2 \tag{15}$$

where $\sigma_{u,1}^{(d)}$ and $\sigma_{u,3}^{(d)}$ are the across-column and the along-column, uniaxial ductile compressive strengths, respectively. Owing to the dependence of the uniaxial strengths on strain-rate, this envelope widens and lengthens with increasing strain rate, eventually intersecting the brittle failure envelope (Fig. 11). The intersection defines the ductile-to-brittle transition, which is discussed in detail below in the section entitled *Ductile-to-Brittle Transition Under Compression*. The point to note (see also Fig. 22, below) is that over the range of strain rates marking the transition, the behavior of the ice changes with increasing confinement, from brittle (via faulting) for loading paths $0 \lesssim R_{21} < R_{21a}$ to ductile for $R_{21a} < R_{21} < R_{21b}$ to brittle (via cleavage/spalling) for $R_{21b} < R_{21} \lesssim 1$. Over the transition range of strain rates, the actual failure envelope is a composite of the two kinds, as shown by the thick curve in Figure 11.

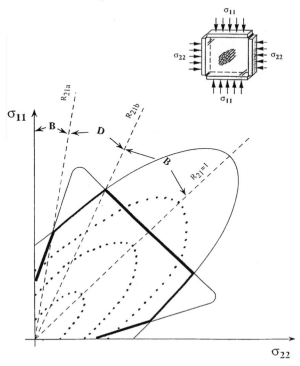

Figure 11. Schematic sketch illustrating the intersection (thick line) of brittle (faceted) and ductile (elliptical) failure envelopes for S2 ice deformed at a strain rate around the ductile (D) –to-brittle (B) transition. R_{21} denotes a loading path. D-behavior for $R_{21a} < R_{21} < R_{21b}$. As shown by dotted curves, the D-envelope widens and lengthens with increasing strain-rate until it intersects the B-envelope.

MICROSTRUCTURAL FEATURES AND MECHANISMS OF BRITTLE COMPRESSIVE FAILURE

Consider next the micromechanics of brittle compressive failure. Coulomb's criterion implies that frictional sliding is a major factor. The objective is to elucidate its nature. The discussion focuses on the mechanisms that lead up to terminal failure via longitudinal splitting and shear faulting, but does not consider cleavage/spalling further.

Longitudinal splits, material collapse and wing cracks

Longitudinal splitting, it could be argued, is not an intrinsic failure mode, but one that reflects an imperfect interface between ice and machine, as well as the smallness of test specimens and their length to width ratio. Certainly, boundaries are not perfect and radial tension can induce splitting, as happens upon the insertion of a thin latex membrane between specimen ends and loading platens (Schulson et al. 1989c). Yet, when precautions are taken to reduce asperities and specimens are creep-seated, longitudinal splitting persists (Kuehn et al. 1993). It persists also upon increasing specimen size from 0.15×0.15×0.025 to 1.0×1.0×0.15 m^3 (Melton and Schulson in preparation) and upon increasing the length/width ratio from 1.0 to 2 (Iliescu and Schulson unpublished research). Splitting thus appears to be a true failure mode, at least of test specimens. It develops as follows.

High-speed photography (Schulson 1990;Smith and Schulson 1993) shows that cracks first form within virgin ice once the applied stress reaches about 1/5 to 1/3 of the terminal failure stress. The cracks increase in number density as the load rises and are distributed more or less uniformly throughout the matrix. Closer inspection reveals that the cracks are similar in size to the grains and that they nucleate preferentially on grain boundaries inclined by about 45° to the loading direction. The nucleation mechanisms are considered elsewhere (Frost 2001). Suffice it to say that visco-elastic deformation is important, through both intragranular slip (Kalifa et al. 1989) and grain boundary sliding (Nickolayev and Schulson 1995; Picu and Gupta 1995a, 1995b; Weiss et al. 1996; Weiss and Schulson 2000).

Of particular interest are out-of-plane extensions (Cannon et al. 1990), termed *wing cracks* (Fig. 12). These features sprout from near the tips of inclined parent cracks and then lengthen as the stress increases in a macroscopically stable manner along a direction more or less parallel to the applied load. The wings interact and eventually link up, culminating in a longitudinal split. The fracture surface so formed is generally irregular, marked by inclined steps; e.g., see Figure 6 (of Smith and Schulson 1993) and Figure 7 (of Kuehn and Schulson 1994), that delineate cracked grain boundaries. Wing cracks form in both fresh-water (Smith and Schulson 1993) and salt-water S2 ice (Smith and Schulson 1994; Gratz and Schulson 1997) and in granular[9] ice (Schulson 1990), implying that they are a general feature of higher-rate compressive deformation. They are also found in deformed rock (Segall and Pollard 1983; Granier 1985; Gottschalk et al. 1990; Cruikshank et al. 1991).

Wing cracking is well understood. In essence, wings initiate in response to localized tensile stresses which develop through uniform sliding across parent cracks inclined to the maximum principal stress (McClintock and Walsh 1962; Brace and Bombolakis 1963; Kachanov 1982; Horii and Nemat-Nasser 1985; Ashby and Hallam 1986; Kemeny and Cook 1987) of the applied compressive field (Fig. 13a). Theory holds that wing cracks spout from near the tips of the parent cracks, beginning in a direction perpendicular to the maximum local tensile stress and then turning toward the direction of the most compressive stress, lengthening whenever displacement along the parent crack is sufficient to raise the mode-I stress intensity factor to the critical level; i.e., $K_I = K_{Ic}$. Observations from S2 ice show that the initial curved segment is not always present, suggesting that the early growth is influenced by local crystallography. Later growth, however, is in excellent agreement with theory (Schulson et al. 1991a), provided that the finite size of test specimens is taken into account (Fig. 13b). The wings, although macroscopically stable, lengthen in a micro-oscillated or jerky manner, as in acrylic plastic (PMMA) (Germanovich et al. 1994) and in rock (Petit and Barquins 1988), because the

---

[9] Wing cracks are more difficult to find in granular ice than in S2 columnar ice. They are entrapped more or less within single grains and do not penetrate the specimen in the way they do in columnar material loaded across-column, making less probable their capture in thin sections. Granular material exhibits 3-D as opposed to 2-D deformation. Mode-II crack sliding is thus accompanied by mode-III loading, and this impedes wing growth, just as it impedes the growth of wing cracks in other materials (e.g., silica glass and cold PMMA, (Germanovich et al. 1994). This impediment, it is suggested, accounts for the absence of wing cracks noted by others (Kalifa et al. 1989).

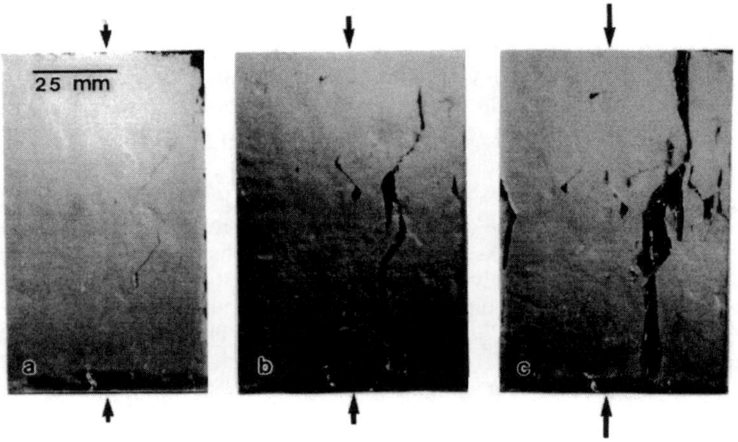

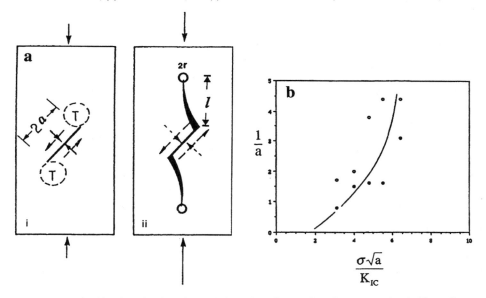

Figure 12. Photographs showing the development of wing cracks and a longitudinal split in S2 fresh-water ice (~10 mm column diameter) loaded uniaxially across the columns at –20°C at 2×10^{-2} s^{-1}. Applied stress is vertical. Viewed along the columns. Inclined segments formed at original grain boundaries and axial extensions (the "wings") run along the direction of loading. (a) After 5.17 MPa; (b) after 6.19 MPa; and (c) after failure at 6.58 MPa (from Cannon et al. 1990).

Figure 13. (a) Sketches showing the evolution of a wing crack under compressive loading: (i) Cracks nucleate on grain boundaries inclined ~45° to the maximum principal stress, once the applied stress reaches about 20-25% of the terminal failure stress. Normal stress closes cracks. Shear stress causes sliding across the crack faces and generates tension (T) at their tips on opposite sides. (ii) out-of-plane extensions or wings initiate once the critical stress intensity factor is reached for mode-I crack loading. Crack-tip creep within a region of diameter 2r. (from Schulson 1990). (b) Graph comparing the observed and calculated growth of wing cracks in S2 fresh-water ice. The length of the wing, l, is non-dimensionalized with respect to the half-length, a, of the sliding parent crack. The applied stress, σ, is non-dimensionalized with respect to the parent crack length and the critical mode-I stress intensity factor, K_{Ic}. The points show the measurements and the curve shows the calculation (from Schulson et al. 1991a).

coefficient of friction along the sliding interface falls as the parent crack slides and then rises again as sliding stops (Renshaw and Schulson 1998). Frictional resistance to sliding is a major factor, raising the applied stress for both crack initiation and growth.

Wing cracking might appear to account for terminal failure: the growth stress scales as (crack size = grain size)$^{-0.5}$, and it scales with temperature and strain rate in the correct sense, through the effect of these factors on the friction coefficient. One need only postulate that failure occurs when the wings reach a certain length. The problem with that view is that it ignores a near-surface effect. High-speed photography (Schulson et al. 1991b) shows that wing cracks that initiate near a free surface grow more rapidly than wings that initiate within the bulk of the material. For instance, Figure 14 shows that wing cracks-C (first evident in Fig. 14g) reach the ends of the specimen before wing cracks-A (first evident in Fig. 14d), even though they formed later. This rapid near-surface growth reflects a greater mode-I stress intensity factor that results from the outward bending of wing-produced, slender microcolumns. Collapse of the material is then triggered by the failure of the microcolumns. Other wing cracks adjacent to the freshly created external surface then lengthen rapidly, creating new slender microcolumns which also become unstable and fail. The process repeats and is manifested by the progressive "blowing out" of a side of the material (Fig. 14).

What destabilizes the microcolumns? Euler buckling, while a possibility, appears not to be the dominant mechanism, for two reasons. The buckling stress σ_b of a slender column of width w and length h, fixed on both ends, is given by the relationship

$$\sigma_b = \pi^2 E w^2 / 3 h^2 . \tag{16}$$

Appropriate values (E = 10 GPa (Gammon et al. 1983); w = 2mm, h = 16 mm; Fig.14) lead to a failure stress of 510 MPa, and this is two orders of magnitude greater than the measured uniaxial strength. The other reason is that Euler buckling describes an elastic instability and so cannot account for the thermal and strain-rate softening of ice. Some other kind of instability is probably at play, and this is considered in the following section.

An earlier discussion (Costin 1983) of longitudinal splitting invoked the notion of a *critical crack density*. The idea was that cracks (not necessarily wing cracks) grow along the direction of loading, under localized tensile stresses which are generated by the applied compressive stress through mismatches in materials properties. The tension is assumed to be proportional to the component of the deviatoric stress that acts normal to the surface of the crack and to be limited to a few grain diameters. The mechanism holds that material eventually splits when the global crack density reaches a critical level. One problem with this idea in the present context is that ice exhibits relatively minor mismatches in materials properties (e.g., Young's modulus varies at most by only about 30% with crystallographic direction (Fletcher 1973). Another is that on the ductile side of the ductile-brittle transition ice is far more damaged at failure (evident from the fact that transparent material becomes milky white) than it is on the brittle side, yet does not split. A third problem is that longitudinal splitting is often accompanied by very little global damage (Smith and Schulson 1993). Critical crack density is thus not a good failure criterion, at least for ice.

Shear faults and comb-cracks

Confinement prevents the development of near-surface, slender microcolumns. It thus suppresses longitudinal splitting. Instead, a different kind of instability develops, manifested by the compressive shear fault.

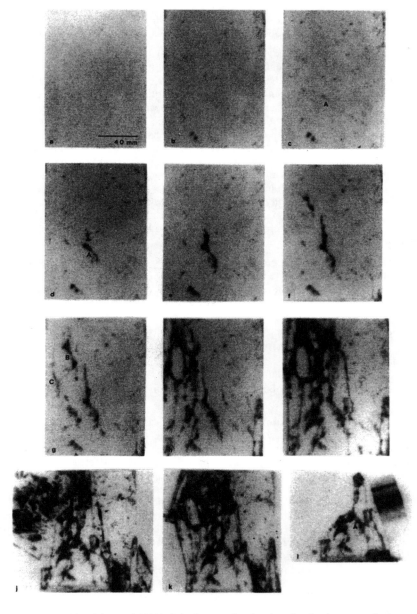

Figure 14. (a) High-speed (1000 fps) photographs showing the development of wing cracks (e.g., A,B,C) during the deformation of S2 fresh-water ice loaded uniaxially across-column at -10°C at $2\times10^{-2}\,\text{s}^{-1}$. Viewed along the columns. Applied stress vertical. From Schulson et al. (1991).

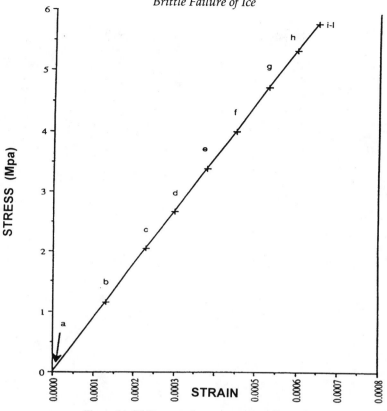

Figure 14. (b) Stress-strain curve corresponding to (a).

Figures 15a-f show the development of a typical fault, in this case within S2 freshwater ice loaded biaxially across the columns under a small degree of across-column confinement ($R_{21} \sim 0.1$). The photographs were taken after unloading the specimen, care having been taken to unload the ice at the same rate as it was loaded to suppress crack formation during unloading (see (Couture and Schulson 1994). Early in the deformation ($\sigma_{11} = 0.88$ MPa; $r = \sigma_{11}/\sigma_{11f} = 0.17$; Fig. 15a) a few grain boundaries cracked. Again, the cracked boundaries were inclined by ~45° to the major stress and were distributed more or less uniformly throughout the matrix. Shortly thereafter secondary cracks initiated from one side of some parent cracks, e.g., crack A in Figure 15b (1.25 MPa, $r = 0.25$). The secondary cracks appeared as whitish regions in thick sections, but as sets of individual cracks in thin sections (more below). Upon further deformation more damage developed and more secondary cracks formed on one side of other parent cracks; e.g., Crack B in Figure 15c (3.76 MPa, $r = 0.74$). The damage continued to increase as the load increased, but remained more or less uniformly distributed up to just below terminal failure, Figure 15d (4.85 MPa, $r = 0.96$). At this point secondary cracks developed on most of the parent cracks. Eventually, the damage linked up and an identifiable shear fault, not previously evident, was created, inclined at about $\theta = 30°$ to the major stress, Figure 15e (5.05 MPa, $r = 1.0$). Although difficult to state with certainty, the fault appears to have initiated near a free corner of the specimen. It was composed of a band of intense damage about two to three grains in width, plus fragmented material. Experiments (Schulson et al.

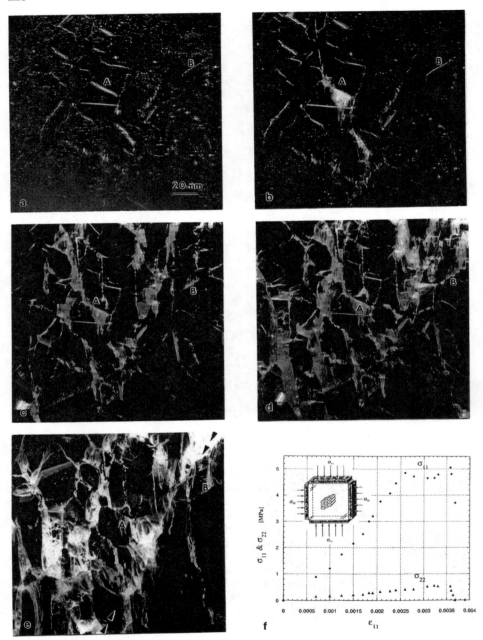

Figure 15. Photographs showing the evolution of a macroscopic shear fault in S2 fresh-water ice of ~10 mm column diameter loaded biaxially across-columns (R_{21}~0.1) at -10°C at 4.8×10^{-3} s^{-1}. Viewed along the columns. The major stress is vertical and minor stress is horizontal. The progression in terms of the percentage of the terminal failure stress is as follows: (a) 17%; (b) 25%; (c) 74%; (d) 96%; and (e) 100%. The stress-strain curves are shown in (f). The trace of the fault is noted by the double arrowheads in Figure 15e. (From Schulson et al. 1999.)

1999) using microsimilar specimens[10] showed that the evolution is remarkably reproducible, right down to the individual grains that cracked, implying that the failure process is not a random one, but one that is intimately related to the details of the microstructure.

Wing cracking is an element in fault formation. This is evident from Figure 16, which shows the structure of another shear fault, generated under the same conditions as above, but within an S2 specimen having a different set of columnar grains. Wing cracks (arrowed) are distributed across the overall field of damage. More importantly, they border the fault, as evident from its zigzag edges. Such features are reminiscent of deformed rock (e.g., see Fig. 14 of Moore and Lockner 1995). The wings are shorter than the ones created within material uniaxially loaded to failure, owing to the constraint to growth imposed by the confining stress.

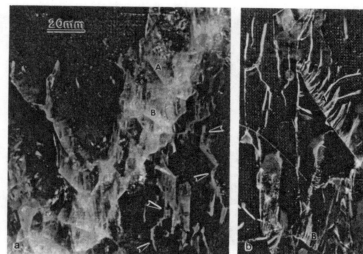

Figure 16. Photographs of a macroscopic shear fault that caused terminal failure of S2 fresh-water ice of ~10 mm column diameter loaded biaxially across-columns (R_{21}~0.1) at –10°C at 4.8×10^{-3} s^{-1}. Viewed along the columns. The major stress is vertical and minor stress is horizontal. (a) Thick section (~25 mm) showing fault running ~30° direction of greater load. Note the wing cracks (arrowed), the zigzag edges of the fault, and the whitish regions (e.g., A and B) stemming from one side of parent, intergranular cracks. (b) Thin section (~5 mm) of the same fault showing that the whitish regions A and B are comprised of sets of secondary cracks, termed "comb cracks" (from Schulson et al. 1999).

Wing cracks, however, are not the only element in fault-formation. An additional element is hidden within the milky-appearing regions that stem from one side of parent inclined cracks (e.g., Fig. 16a and 16b). When viewed in thin section, these milky regions are found to consist of sets of secondary cracks (Fig. 16, mentioned above). They form transgranularly, appear to be about one-half the grain size, and, like wings, are opening-mode cracks that generally curve toward the direction of the maximum principal stress. Secondary cracks are also seen in S2 saline ice (Smith and Schulson 1994) and in granular ice (see Schulson et al. 1991b) implying that they, too, are a general feature of the

---

[10] Microsimilar specimens are cut from the same parent block of S2 ice. They possess a set of columnar grains and grain boundaries of essentially the same orientation.

compressive deformation of ice at higher strain rates. Similar features, often termed "splay cracks", are seen in rock as well, on scales small (Conrad and Friedman 1976; Ingraffea 1981; Wong 1982; Gottschalk et al. 1990) and large (Rispoli 1981; Segall and Pollard 1983; Granier 1985; Davies and Pollard 1986; Martel et al. 1988; Cruikshank et al. 1991), implying that they are also a general feature of brittle compressive failure.

Secondary cracks, it is suggested, are the key element in triggering the instability (Schulson et al. 1999). They create, in effect, sets of slender microplates, fixed on one end and free on the other. The free ends contact the sliding, parent crack. Owing to friction, a moment is induced and this causes the plates to bend and then break, rather like the breaking of teeth in a comb under a sliding thumb (Schulson 1996). Microplates/columns near free surfaces—either external surfaces or internal ones such as the sides of holes or voids—are imagined to break first, owing to less constraint there, shedding load to ones adjacent and thereby nucleating a fault. Multiple initiations are possible. The nucleus then grows under the action of localized stress intensification, along a band of reduced shear strength. The band is imagined to be composed of "comb-cracks," of wing-cracks and of grain boundary cracks that formed prior to the initiation of the fault plus fresh cracks created just ahead of the advancing front, in the manner of a process zone—a zone, analogous to that envisaged (Reches and Lockner 1994) in rock. The front is imagined to propagate rapidly across the section, at an average velocity of ~>1 m s^{-1}, creating a band of fragmented ice or "gouge" and ending in terminal failure. Figure 17 sketches the process as it is thought to occur on the small scale. [Figure 25, below, shows evidence of its occurrence on the geophysical scale.]

The stress to trigger the fault is assumed to be the applied stress that causes a near-surface microcolumn to break. Accordingly, a first-order analysis (Schulson et al. 1999) suggests that each "tooth" in the "comb" is subjected to both an axial force P per unit depth and a moment M per unit depth, Figure 18. The moment is the more important factor. It is caused by the shear stress $\sigma_{s,eff}$ effective in causing sliding, and is equated to the difference between the applied shear stress σ_s resolved onto the sliding plane and the frictional resistance $\mu\sigma_n$ induced by the stress σ_n normal to the sliding plane. The cracks that define the fixed end of the microcolumns experience mixed-mode (I and II) loading and propagate when the mixed mode stress intensity factor reaches a critical level. Assuming that the situation is similar to the case of an edge crack in a brittle plate and that $K_{Iic} = K_{Ic}$ for ice (Shen and Lin 1986), then from the analysis of Thouless et al. 1987 it can be shown that the lower limit (under uniaxial loading) of the terminal failure stress may be approximated by the relationship (Schulson et al. 1999):

$$\sigma_f = \frac{2K_{Ic}}{(3\beta h)^{0.5}(1-\mu)} \qquad (17)$$

where $\beta = h/w$ is the slenderness ratio of the microcolumns. Under confined loading, following a derivative (Schulson et al. 1991b) of the Ashby and Hallam (1986) analysis, the failure stress may be roughly approximated by[11]:

$$\sigma_f = \frac{2K_{Ic}}{(3\beta h)^{0.5}\{(1-\mu)(1-R)-2R\mu\}} \qquad (18)$$

where R is the appropriate confining ratio and where the fuller term in the denominator expresses the impediment to parent crack sliding introduced by the confinement. By

---

[11]The denominator in Equation (18) tends to zero when μ tends to (1 − R)/(1 + R). This is not a major weakness because the model only applies to cases of low confinement. Lehner and Kachanov (1996) offer an analysis which overcomes this shortcoming.

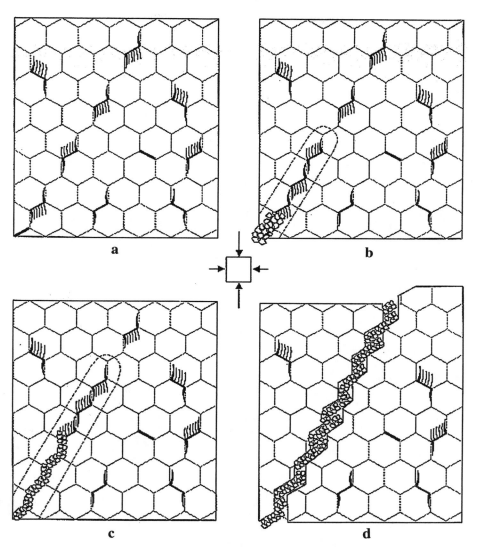

Figure 17. Schematic sketch of the initiation and growth of a macroscopic compressive shear fault, from the onset of terminal failure (i.e., from very near the peak of the stress-strain curve). The lighter, hexagonal background denotes grain boundaries. (a) Before initiation, damage is uniformly distributed throughout the matrix, and consists of a mixture of parent grain boundary cracks inclined by about 45° to the highest principal stress, parent cracks with wing-crack extensions, and sets of secondary "splay cracks" stemming from one side of parent cracks. The parent, inclined cracks plus the "splay" constitute a "comb-crack". (b) Near-surface "comb-crack" fails, owing to less constraint there, shedding load and redistributing stress to adjacent material within which new cracks initiate. The nucleus of the eventual shear fault is formed. (c) More comb-cracks fail and localized damage spreads as a band across the section, creating granulated material or gouge in its wake and forming new damage in a "process zone" ahead of the advancing fault front. (d) A recognizable macroscopic shear fault (from Schulson et al. 1999).

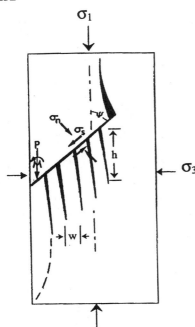

Figure 18. Schematic sketch of the "comb-crack" fault initiation mechanism. Secondary cracks initiate from one side of parent cracks which are inclined by angle ψ to the maximum (most compressive) principal stress. The secondary cracks create sets of slender microcolumns of length h and width w, fixed on one end and free on the other. Axial loading produces a longitudinal load/unit depth P on the microcolumns, and frictional sliding across the parent, inclined cracks under effective shear stress σ_s produces a moment/unit depth M on the microcolumns. M is more important than P and eventually becomes large enough to break the microcolumns, preferentially near a free surface, thereby triggering the growth of a macroscopic shear fault across the load-bearing section. From Schulson et al. (1999).

setting[12] $K_{Ic} = 0.1$ MPa$\sqrt{m}$, $\mu = 0.5$, $\beta = 5.3\pm2$ (length and width, respectively, in the case illustrated above (Fig. 16), are 6.2±4.0 mm and 1.2±0.2 mm) and h = d/2 = 5 mm (above observations) and R = 0, one finds that $\sigma_f = 1.5\pm0.5$ MPa. This estimate is of the correct order of magnitude, but is about a factor of three too small. The discrepancy reflects perhaps the exclusion of crack interactions and of additional friction that arises through the rubbing together of adjacent microcolumns within the comb.

The model captures the general character of the brittle compressive strength. It embodies (grain size)$^{-0.5}$ dependence through the dependence of the column length on grain size. It captures strain-rate softening through the friction coefficient, which, as already noted, decreases with increasing sliding speed (Kennedy et al. 2000): for instance, an increase in velocity from 10^{-4} to 10^{-2} m/s, which corresponds roughly to an increase in strain rate from 10^{-3} to 10^{-1}s^{-1}, lowers μ from 0.5 to 0.2 (at -10°C), leading to an expected reduction in strength of ~40%, in fair agreement with observation (Fig. 2). The model accounts for thermal softening, also in terms of the friction coefficient. For instance, the reduction from $\mu = 0.8$ at -40°C to $\mu = 0.5$ at -10°C leads to an expected reduction in strength of fresh-water ice of about 50%, in fair agreement with observation (Figs. 3 and 6). The friction coefficient (at least under low normal stress) of saline ice at -40°C is ≈ 15% lower than that of fresh-water ice (Kennedy et al. 2000), accounting in part for the lower strength of saline ice at the lower temperature. The model also dictates the ratio $1/(1 - 3R)$ for the confined strength to the unconfined strength of S2 ice at -10°C under lower across-column confinement ($R_{21} < R_{21c}$), in good agreement with measurement and with Coulomb's criterion [see section titled *Failure criteria*, above]. Lastly, it dictates

---

[12]The appropriate value of fracture toughness may be smaller than that assumed, owing to size effects. The lowest possible value is set by surface energy through the relationship $K_c = [2 \gamma E/(1 - v^2)]^{0.5}$. This gives $K_c \sim 0.05$ MPa$\sqrt{m}$. Whichever value is used does not change the sense of the argument.

that the critical confinement ratio $R_{21c} = (1 - \mu)/(1 + \mu) = 0.55$, 0.33 and 0.14 at -2.5°C, -10°C and -40°C, respectively, in reasonable agreement with observation [see section titled *S2 columnar ice*, above]. Although more work is needed, the comb-crack model appears to capture the essence of the faulting process.

It is interesting to note that recent modeling of faulting in rock (Gupta and Bergström 1998) invoked both a Weibull-like distribution of failure stresses and a stress enhancement factor resulting from crack interactions. In so doing, the work simulated rather closely the evolution of the process, including the "process zone" observed in granite (Lockner et al. 1991). What it did not elucidate, however, was the nature of the "trigger." Like earlier models, the work was founded upon observations (by other investigators) that were made using indirect techniques (e.g., acoustical emission, computed X-ray tomography) which have relatively coarse resolution, exacerbated by complex microstructures. An elliptical nucleus within which the shear modulus was reduced through localized cracking was thus assumed. Also, the model did not capture initial blowout at the free surface (Lockner et al. 1991), and, as its authors note, contains several free parameters. In comparison, the "comb-crack" mechanism incorporates a micromechanical trigger that is based upon direct, experimental observation, it accounts for blowout, and it has only one free parameter (the slenderness ratio of the microplates).

How are comb-cracks created? Contact, it appears, is essential across a fractured interface. This is evident from the observation that secondary cracks generally do not form on pre-engineered parent cracks, created using Teflon tape (Schulson et al. 1999). This could explain their absence in earlier modeling of brittle compressive failure, where sliding "cracks" were created in glass using slots (Brace and Bombolakis 1963) and in Columbia resin using metallic plates lined with Teflon (Horii and Nemat-Nasser 1985;1986). One possible process is Hertzian cracking, arising from stresses that develop at randomly arrayed asperities along cracked grain boundaries. The one-sided characteristic could then be explained in terms of different tensile strengths on either side of the boundary, owing to different crystallographic orientations. This kind of process is considered to be unlikely, however: the spacing of the secondary cracks seems to be too regular to have arisen from randomly arrayed points of contact; and secondary cracks are not limited to cracked grain boundaries: they also form transgranularly within S2 columnar ice (see Fig. 11; Schulson and Gratz 1999). Moreover, should their origin be similar to the origin of splay cracks in rock, the one-sided characteristic could not be explained in terms of small-scale crystallography.

A more likely explanation involves *non-uniform sliding*. The idea here, extending earlier suggestions (Martel and Pollard 1989;Cooke 1997) in connection with the fracture of rock, is that a gradient of displacement along the parent crack, $\partial u_1/\partial x_1$, creates normal stresses, of magnitude proportional to the gradient, that act across planes perpendicular to both the sliding plane and the sliding direction. Compressive stresses then develop on one side, while tensile stresses develop on the other. Figure 19 illustrates this notion. There, the uphill material that borders the upper side of the parent crack is displaced more than the downhill material, and so the upper-side material experiences a compressive stress while the underside material experiences tension. Should non-monotonic displacements develop, the tensile stress field would alternate to the other side of the sliding crack. Barring creep relaxation (see section below entitled *Ductile-to-Brittle Transition Under Compression*), the tension, once great enough, is relieved through the formation of secondary cracks. (Note that the secondary cracks in Figure 16 above and in Figure 26 below are not perpendicular to the parent, sliding crack.) The applied stress under which forms the set that constitutes a comb-crack appears to be approximately constant, given that one set formed within 10^{-3} s (i.e., in the interval between adjacent images on a high-speed film (Smith and Schulson 1993). The secondary cracks are imagined to form sequentially, starting near one tip of the parent crack and releasing proportionately more strain energy, proportional to K_{11}^{ℓ} and thus to the distance from the tip of the parent crack to the initiation site of the secondary crack. Their length is thus imagined to increase with distance from the tip. Comb-crack A of Figure 15, for instance, could have formed in this manner.

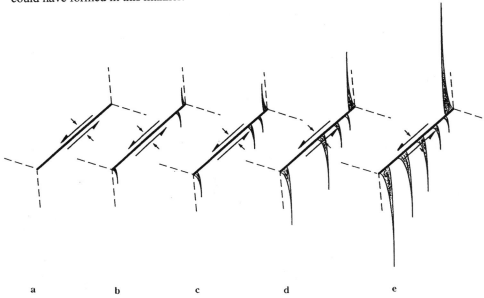

Figure 19. Schematic sketch showing the proposed sequential (a-e) creation of secondary cracks under compression through non-uniform sliding across an inclined, cracked grain boundary. The maximum (most compressive) principal stress acts along the vertical direction.

Regardless of their origin, the important point is recognition of a new frictional sliding mechanism and of the role it appears to play in destabilizing damaged material. This constitutes a new element in the understanding of brittle compressive failure. Although derived from a study of ice, the comb-crack mechanism is not limited to that material.

DUCTILE-TO-BRITTLE TRANSITION UNDER COMPRESSION

The ductile-brittle transition (Fig. 2) marks the point where the compressive strength reaches a maximum (with respect to strain rate). It thus limits the loads exerted by floating ice features against engineered and other structures. The objective now is to consider its origin.

Definition

The transition is defined macroscopically, in terms of the shape of the stress-strain curve and the behavior/appearance of the material. Accordingly, ice is considered to be ductile when its stress-strain curve exhibits either a rounded peak or a plateau and the material does not suddenly lose its load-bearing ability, even though it may become so highly damaged that initially transparent material becomes milky-white in appearance. On the other hand, ice is considered to be brittle when its stress-strain curve is pseudo-linear in shape and terminates suddenly owing to the onset of a mechanical instability. More fundamental definitions of the transition are possible, involving, for instance, the competition between the emission of dislocations from near crack tips and crack extension. However, for the present purpose, the macroscopic definition suffices. The aim is to model the strain rate at which the transition occurs. Generally, this happens over a range of strain rates (about a factor of three). For the purposes of this discussion, however, the transition is defined in terms of a specific strain rate $\dot{\varepsilon}_t$. As noted above, this is typically 10^{-4} to 10^{-3} s^{-1} (in small test specimens), but increases with decreasing grain size, increasing salinity, increasing confinement and increasing pre-deformation.

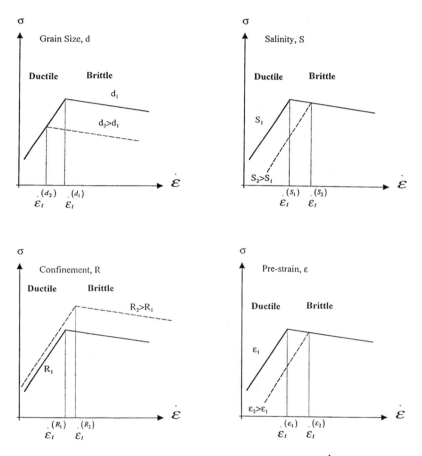

Figure 20. Schematic sketches of log strength (σ)- log strain rate ($\dot{\varepsilon}$) showing the effects of grain size, salinity, confinement and compressive pre-strain on the ductile-brittle transition strain rate, $\dot{\varepsilon}_t$.

Transition models

Two approaches have been developed (Schulson 1990): one is phenomenological and is based upon the idea that the transition occurs at the strain rate, $\dot{\varepsilon}_t$, where the ductile strength equals the brittle strength, $\sigma_d = \sigma_b$ (Fig. 20). As noted above, the ductile failure stress is characterized by strain-rate hardening and obeys Hill's (1950) criterion for the yielding of plastically orthotropic materials. The brittle strength, as already discussed, decreases with increasing strain rate. [The argument can be made as well by assuming that the brittle strength is rate independent.] The variation of ε_t with the factors noted in Figure 20 can then be understood in terms of their effects (noted above in *Overview* in the section entitled *Brittle Failure Under Compression*) on the respective strengths. Salinity/porosity (Buck et al. 1994) and pre-deformation (Kuehn et al. 1988) lower the ductile strength, but not the brittle strength, thereby increasing the transition strain rate. Grain refinement increases the brittle strength (Schulson 1990), but not the ductile strength (Schulson and Cannon 1984; Cole 1985), again raising the transition rate (Batto and Schulson 1993). Confinement increases the brittle strength more than the ductile strength (Schulson and Nickolayev 1995) and so raises the transition rate still further. These effects have been modeled (Schulson and Buck 1995; Schulson and Nickolayev 1995) and reasonable agreement with observation has been obtained.

The other approach considers micromechanical processes. An early view was that the transition occurs when the ice first begins to crack, as is the case under tension. Under compression, however, ice begins to crack, at strain rates two to three orders of magnitude below the macroscopic transition strain rate, and so crack nucleation cannot account for the transition. Another possibility is that the transition coincides, in some way, with a certain proportion of intergranular cracks. This is based upon the observation (Hallam et al. 1987; Kalifa et al. 1989; Gold 1997) that the ratio of the number density of intergranular to the transgranular cracks increases with increasing strain rate within the vicinity of the transition. The problem with this notion is that the strain rate marking any given ratio of grain-boundary cracks to total cracks increases with increasing grain size (see Fig. 2 of Gold 1997), implying that the transition strain rate should also increase with increasing grain size. The effect of grain size is just the opposite (Batto and Schulson 1993). Schulson's view (Schulson 1990) is that the transition marks the point where cracks begin to propagate, albeit in a stable manner. In this scenario, a competition is imagined between stress relaxation and stress build-up at crack tips. Relaxation through crack-tip creep is assumed to dominate at lower deformation rates, inhibiting propagation. At higher rates stress-buildup dominates, allowing the mode-I stress intensity factor to reach the critical point. Indeed, experiments (Batto and Schulson 1993) on specimens containing inclined "Teflon cracks" from whose tips wing cracks were initiated show that cracks do not propagate at strain rates below the transition, but do propagate at rates above the transition rate. Similarly, comb cracks do not develop on the ductile side of the transition (Iliescu and Schulson, unpublished), implying that the necessary tensile stresses relax. These observations support Schulson's suggestion, and so it is the one pursued.

Imagine a crack-tip creep zone of radius, r, defined as the region within which creep strain exceeds elastic strain (Fig. 13). Although brittle failure is attributed more to comb cracking than to the propagation of wing cracks, the analysis is made for wing cracks because they are more easily analyzed. It is assumed that similar stress states and creep zones characterize the near-tip region of the two kinds of crack. The model incorporates the hypothesis that cracks propagate when $r < 2\,fa$, where 2a is the length of the parent, sliding crack and $f <$ unity. The criterion for the transition may then be expressed as:

$$r = 2\,fa. \tag{19}$$

The creep zone radius can be obtained from the Riedel and Rice (1980) model of creep zones around non-interacting cracks loaded under far-field tension by invoking secondary or power-law creep ($\dot{\varepsilon} = B\sigma^n$, where n = 3). Thus:

$$r = \frac{K_I^2}{2\pi E^2}\left(\frac{(n+1)^2 E^n Bt}{2n\alpha^{(n+1)}}\right)^{2/(n-1)} F \tag{20}$$

where F and α are of order unity; t is the loading time, and is approximated as the ratio of the critical stress intensity factor and the loading rate, $t \approx K_c / \dot{K}$, where the loading rate, $\dot{K}_I$, is obtained through the product of partial derivatives $\dot{K} = (\partial K/\partial \sigma)(\partial \sigma/\partial \varepsilon)(\partial \varepsilon/\partial t)$. The derivative $\partial \sigma/\partial \varepsilon$ is essentially a modified Young's modulus E' (i.e., E reduced by damage) and $\partial \varepsilon /\partial t$ is the applied strain rate. The other derivative depends on the details of the crack geometry and here is taken from wing crack mechanics (Ashby and Hallam 1986) to be of the form (Weiss and Schulson 1995)

$$\partial K/\partial \sigma = (\pi a/3)^{0.5}\left[(1-R)(1+\mu^2)^{0.5} - \mu(1+R)\right]$$

where $2a = d$. Upon combining these relationships, upon replacing K_I in Equation (20) with K_{Ic} on the grounds that the secondary cracks are injected under load and are on the verge of growth, and upon assuming E' ≈ E, the transition strain rate may then be expressed as (Schulson and Nickolayev 1995):

$$\dot{\varepsilon}_t = \frac{BK_{Ic}^3}{fd^{1.5}\left\{[1+\mu^2]^{0.5} - \mu - R\left[\mu + (1+\mu^2)^{0.5}\right]\right\}} \qquad (21)$$

for R < (1 – μ)/(1 + μ). For larger confinement a different treatment is required owing to the change in failure mode. The transition is thus expressed in terms of independently measurable materials parameters (B, K_c, μ and d) and in terms of a fraction, f, whose value was calculated (Batto and Schulson 1993) to be f ≅ 0.015, from the Riedel-Rice (Riedel and Rice 1980) model of crack-tip creep and from separate measurements.

Schulson's model appears to capture all of the observations to date: the order of magnitude of the transition strain rate, as well as its measured dependence (Batto and Schulson 1993) on (grain size)$^{-1.5\pm0.5}$ (Fig. 21); the increase in the transition strain rate with increasing confining stress under lower confinement $R_{21} < R_{21c}$ (Schulson and Buck 1995; Schulson and Nickolayev 1995), as well as the order-of-magnitude higher transition strain rate of salt-water ice versus fresh-water ice[13] (Fig. 22). Albeit qualitatively, the model also captures the order-of-magnitude increase in the transition strain rate induced by compressive pre-deformation (Kuehn et al. 1988): the pre-strain produces damage and this raises the creep constant (Weiss 1999) without changing the exponent n. Finally, when cognizance is taken not only of the decrease in the creep, B, constant with decreasing temperature, but also of the large increase in friction coefficient (Kennedy et al. 2000) and of the small increase in fracture toughness (Nixon and Schulson 1988; Dempsey 1996), the model accounts for the order of magnitude reduction (Qi and Schulson 1998) in the transition strain rate upon reducing temperature from -10°C to -40°C.

Equation (21) also accounts for an observation not previously explained. In their early work on compressive failure, Hawkes and Mellor (1972) did not report a sudden ductile-to-brittle transition, at strain rates up to 10^{-2} s^{-1} at -7°C. The reason for its absence, it now appears, is that the grain size of their material (~0.7 mm) was so fine that it raised the transition to a strain rate of $\sim 0.7 \times 10^{-2}$ s^{-1}; i.e., to very near the highest rate they explored. Given the limited data and given the fact that the transition occurs over a

---

[13] The higher transition rate for salt-water ice stems from a higher creep rate, reflected in the creep constant: B = 5.1×10^{-6} MPa^{-3} s^{-1} for salt-water ice of 4.5 ppt salinity and B = 4.3×10^{-7} MPa^{-3} s^{-1} at -10°C (Sanderson 1988). The fracture toughness (Dempsey et al. 1999) and the coefficient of friction (Kennedy et al. 2000) are essentially the same for fresh-water and salt-water ice.

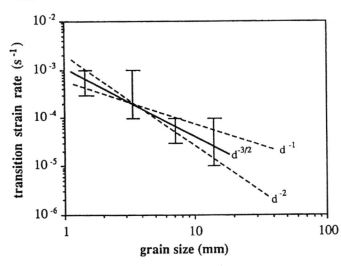

Figure 21. Graph showing the measured and the calculated effect of grain size, d, (i.e., column diameter) on the ductile-brittle transition strain rate at −10°C of fresh-water S2 ice loaded uniaxially across the columns. Calculations were made using Equation (20) and the parametric values B = 4.3×10⁻⁷ MPa⁻³ s⁻¹; K_{Ic} = 0.1 MPa√m ; μ = 0.5; f = 0.015. The dotted lines are drawn through the data and show possible d^{-1} and d^{-2} functionalities. The solid line is the one calculated from the model. From Batto and Schulson (1993).

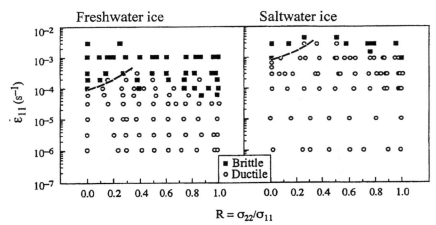

Figure 22. Graphs showing the measured and the calculated effect of across-column confining stress on the ductile-brittle transition strain rate at -10°C of both fresh-water and salt-water (of 4.5 ppt salinity) S2 ice of ~6 mm column diameter loaded biaxially across-column. The calculations (curve) were made using Equation (20) and are limited to the regime of lower confinement where failure occur via shear faulting. Parametric values: B = 4.3×10⁻⁷ (fresh-water ice) and 5.1×10⁻⁶ (salt-water ice) MPa⁻³ s⁻¹ (Sanderson 1998); K_{Ic} = 0.1 MPa√m (both materials); μ = 0.5 (both materials); d = 6 mm (both materials); f = 0.015 (both materials) (from Schulson and Buck 1995; Schulson and Nickolayev 1995).

range of strain rates, it seems likely that experiments approached, but did not fully enter, the brittle regime. Indeed, they (Hawkes and Mellor 1972) noted that the compressive strength appeared to reach a "limiting value" at 10⁻² where the specimens failed by splitting.

The agreement between theory and experiment is probably better than it should be, given the assumptions that are invoked. The Riedel-Rice analysis was developed for creep at the tips of non-interacting cracks within an isotropic, homogeneous material loaded under far-field tension. It may not be strictly applicable to crack-tip creep within tensile zones induced locally within a damaged, plastically anisotropic aggregate loaded under far-field compression. Also, stress relaxation within the crack-tip region probably occurs through a combination of primary and secondary creep, and not just secondary creep as assumed. There is also the question of friction and whether coefficients obtained from ice sliding upon ice, smoothly surfaced, are similar to those for sliding cracks. Nevertheless, in incorporating resistance to creep, to fracture and to frictional sliding, as well as the size of the stress concentrator, the model appears to capture the essential physical and microstructural elements underlying the ductile-to-brittle transition.

ON THE FORMATION OF "LEADS" IN THE ARCTIC SEA ICE COVER

Leads, or open cracks, are important features of sea ice covers at high latitudes. They serve as passageways for the movement of ships and submarines, and they act as low-resistance paths for the transfer of heat from the ocean to the atmosphere. Some leads are so long that they span a large fraction of the Arctic ocean (Kwok et al. 1992, 1999). How they form and the role they play in the subsequent deformation of the cover are issues that relate directly to polar marine transits and to global climate (Maykut 1982; Johanessen et al. 1995). The objective now is to consider their origin. Schulson and Hibler (1991) first suggested that scale-independent compressional failure processes may be at play, given wing-like cracks in the ice cover on the Beaufort Sea (Fig. 23). Although controversial, that is the path followed here.

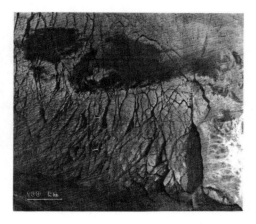

Figure 23. Satellite image of part of the sea ice cover on the Beaufort Sea off the coast of Alaska (Banks Island is just to the right of the image) 11 February 1983. The dark smudge at the top of the image is from cloud-cover. Wing-like cracks, stemming from the tips of inclined leads, are arrowed (from Schulson and Hibler 1991). Compare with wing-cracks generated in the laboratory, Figures 12 and 14.

Satellite images of high-latitude sea ice covers reveal patterns of cracks, often aligned and generally characterized by intersecting features termed "slip lines" (Marko and Thomson 1977; Erlingsson 1988; Kwok et al. 1992; Lindsay and Rothrock 1995; Overland et al. 1998) (Fig. 24a). [The figure caption describes the proposed order of shearing.] The "lines" are actually narrow (~0.5 km to 5 km) bands of damage ~50 km apart (Overland et al. 1998). They intersect over a range of angles, commonly focused around 40° (included angle), thereby forming diamond-shaped patterns. Since "slip lines" in the covers do not form orthogonal sets, they are different in character from the slip lines of metal plasticity theory (see e.g., Hill 1950). However, they do delineate diffuse boundaries diffuse boundaries across which discontinuous displacements occur, as evident from analysis of synthetic aperture radar images (Overland et al. 1998; Kwok et al. 1999). The "slip lines," it is suggested, are the geophysical analogue of the compressive shear faults that form within plates of S2 ice biaxially loaded across-column to terminal failure

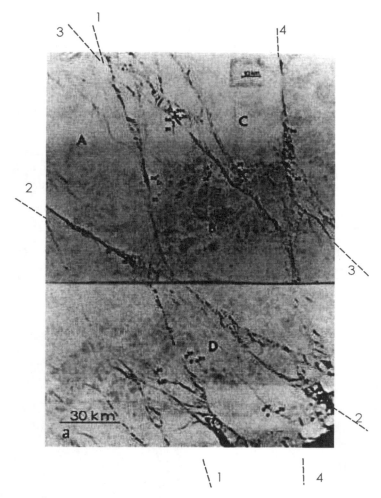

Figure 24. Photographs showing sets of intersecting faults on scales large and small: (a) Satellite image of the sea ice cover in the Beaufort sea off the coast of Alaska 25 September 1974. From Marko and Thomson (1977). Fault-1 probably formed before fault-2, and fault-3 before fault-4. Displacements of fault-1 at X probably occurred by shearing along fault-2, and displacement of fault-3 at Y by shearing along fault-4.

under moderate across-column confinement (Fig. 24b, next page). Covers of first-year arctic sea ice, it is recalled, are largely made from S2 ice and are stabilized against buckling by a potential energy gradient.

Slip lines appear to develop through a process of progressive damage to the parent ice cover, analogous to the development of shear faults in the laboratory. The development within the pack ice is expected to be more complicated, however. There the parent sliding cracks are imagined to be narrow, refrozen leads from an earlier time that failed. A recent analysis (Hibler and Schulson 2000) considered the cover to be an anisotropic continuum composed of thin and thick ice. Within each element (~10 km on edge) are located narrow, refrozen leads of random orientation. These refrozen leads

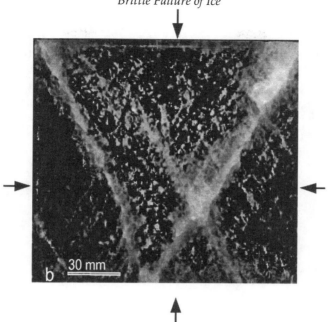

Figure 24. (b) Shear faults in S2 fresh-water ice of ~3 mm column diameter compressed biaxially across-column ($R_{21} \approx 0.1$) in the directions shown, at -10°C at 10^{-3} s^{-1}. View is along the columns.

contain thin ice and are embedded within surrounding thick pack ice. The thin, "old leads" are imagined to "fail" when the local biaxial compressive stresses satisfy the appropriate failure criterion. The criterion taken was based upon a compressive failure envelope that was generated in the laboratory (Schulson and Nickolayev 1995), similar to the ones shown in Figures 6 and 10. From a random array of such pre-existing flaws, the ones that fail preferentially are those of a particular orientation with respect to the maximum principal stress of the far-field applied stress. The critical orientation depends upon both the ratio of the minor to the major compressive stress and friction. Upon failing, the flaws redistribute stress in their vicinity, leading to fresh failures in adjacent elements. In this way is created a kind of macroscopic "process zone" within which mini-leads alternate in orientation as they form ahead of the advancing front. The mini-leads eventually link up, in a manner not yet specified but which probably depends upon the details of the local failure process, as it does on the smaller scale. Once linked, a band of damage extends all the way across the cover, thereby creating a slip line, itself oriented with respect to the maximum principal stress. The slip line, in other words, appears to be created through a process of consecutive cracking, and the damage band, which eventually defines the fault, propagates across the cover. Figure 25 shows satellite images which appear to support this proposed process. The propagation velocity is $\simeq 13$ km/day, which curiously is within an order of magnitude of the shear fault velocity in the laboratory (see section *Shear faults and comb-cracks*, above).

Slip lines divide the parent cover into individual floes whose size is set by the spacing between them. When the failure criterion of the gouge between them is satisfied, the floes slide with respect to each other, deforming little or not at all (Overland et al. 1998; Kwok et al. 1999), except perhaps by grinding away at the interface and creating additional gouge material. Slip continues, it is imagined, until either the wind stops or

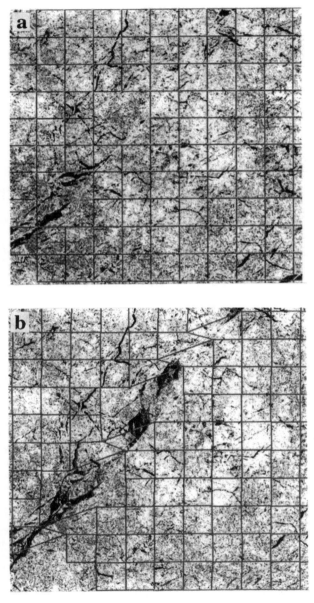

Figure 25. Synthetic aperture radar images (ERS-1) of the ice cover in the Beaufort Sea, taken by satellite (a) on day 79 (at 22:35) 1992 and (b) on day 83 (at 6:51) 1992. Note the "damage zone" ~5 km wide that started at the lower, left-hand corner and propagated ~40 km in the 3-day interval. Localization of deformation occurs within this zone. Outside the zone, deformation appears to have ceased, judging from the negligible change in the density of the distributed cracks over the 3-day interval. The damage zone defines the beginning of a shear fault, evident from the relative shear displacements (arrows) that occur across the fault. [Cell size = 5 km.]

changes direction, or until healing impedes sliding to the point that it "freezes" the process. At that stage, a new fault is expected to form, possibly of an orientation different from the first. Once formed, sliding again occurs when the failure criterion for the gouge material of that fault is satisfied. The process then repeats itself.

It is thus suggested that the ice cover exhibits a mixture two kinds of inelastic deformation (Fig. 26). One pertains to a continuum (type-I), albeit damaged, and the other (type-II), to a granular medium. Deformation of the continuum ends in terminal failure, of the kind described above. Deformation of the granular medium leads to stick/slip behavior (noted above, in the *Overview* of the section *Brittle Failure Under Compression*, for the small scale), manifested by the kind of wildly oscillating pack-ice stresses recently reported (Richter-Menge and Elder 1998). Both kinds of deformation are frictional in character, but the nature of the "friction coefficient" is probably different. The interplay between the two is thought to govern the behavior of the sheet.

Under what level of stress might the field cracks have formed? That depends upon the mechanism. The similarity in the appearance of some leads to small-scale wing cracks

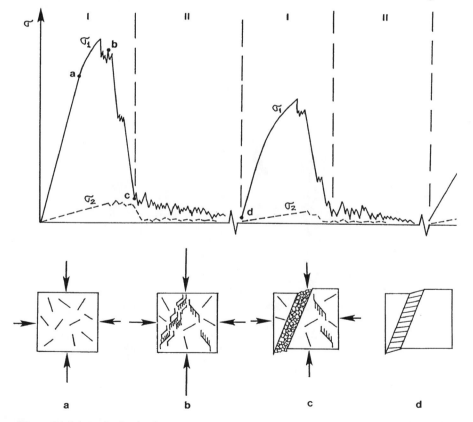

Figure 26. Schematic sketch of two types of deformation: type-I depicts ice as a continuum; type-II, as a granular material. A fault/lead develops during terminal failure (b,c), creating gouge (c) across which frictional sliding occurs at lower applied stresses. Eventually, the fault/lead heals or the wind changes direction (d) and the ice again behaves as a continuum.

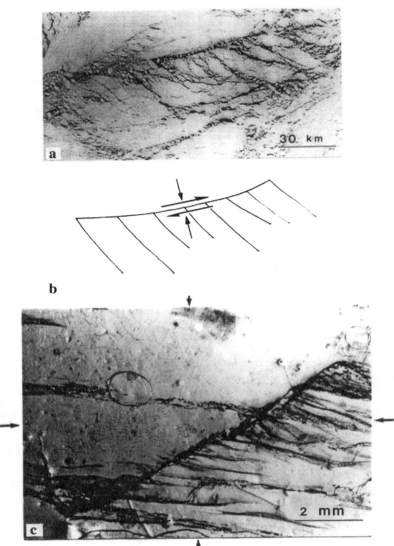

Figure 27. Photographs showing comb-cracks on scales large and small. (a) Satellite image of the sea ice cover off the coast of Greenland, 25 April 1976 (from Erlingsson 1988). (b) Suggested shear loading of the macrocolumns created by the secondary cracks. (c) Cracks in fresh-water granular ice of ~10 mm grain size loaded triaxially across-column (R ~ 0.05) at -40°C at 10^{-3} s^{-1}.

and compressive shear faults has already been noticed. Figure 27 shows another example; namely, comb-like cracks in the Greenland Sea ice field and in a small specimen of granular ice. These "look-alike" features heighten the possibility that the failure mechanisms may be scale-independent. Indeed, recent fractal analyses of crack patterns accompanying the fracture and fragmentation of ice (Weiss 2001; Weiss and Gay 1988) revealed no characteristic scale in over nine orders of magnitude of length (10^{-4} m to 10^5 m), except one corresponding to the mean grain size during the early stage of deformation. This result precludes the existence of an elementary volume within which damage is representative of the ice as a whole (Weiss 2001), and suggests universal scaling (Weiss and Gay 1998) of the fracture processes. Thus, if the wing-crack mechanism, which is scale-independent, accounts for the in-field wing-like leads of Figure 23 ($\ell/a \approx 5$; $a \approx 40$ km), and if there are no major effects of scale on the values of the physical parameters (K_{Ic} = 0.25 Mpa√m (Dempsey 1996); μ = 0.5), then from the non-dimensional quantification of that mechanism (Fig. 13b) those leads could have formed under compressive stresses of ~10 kPa to 30 kPa. Similarly, if the comb-crack mechanism, also scale-independent, accounts for the large-scale features of Figure 27 [h/w $\approx$ 3; K_c = 0.25 Mpa√m; μ = 0.5], then from Equation (17) those leads could have formed under in-field compressive stresses of ~3 kPa. Interestingly, these estimates are of the same order of magnitude as peak compressive stresses measured in pack ice, which typically range from 10 kPa to 100 kPa (Coon et al. 1989; Tucker and Perovich 1992; Richter-Menge and Elder 1998). Unfortunately, unlike the laboratory, the field has not yet offered a direct comparison between the peak stresses and the underlying mechanical events.

Implicit in the discussion so far is the idea that the arctic sea ice cover behaves in a macroscopically brittle manner. Given that covers typically deform at low strain rates, ~1%/day (10^{-7} s^{-1}), brittle behavior might appear to contradict the analysis presented above (in the section *Ductile-to-Brittle Transition Under Compression*), where the ductile-to-brittle transition strain rate for salt-water ice loaded across the columns was noted to be ~10^{-3} s^{-1}. However, the earlier discussion focused on the behavior of small pieces of material. The ice cover on the Arctic ocean is a very large body, thousands of kilometers wide and up to ~3 m thick: it contains stress concentrators of length λ—thermal cracks, for instance—that can be several orders of magnitude larger than cracked grain boundaries. In terms of the above analysis, in which the transition strain rate scales as $\lambda^{-1.5}$, one need only imagine in-field concentrators longer than a few meters to account for brittle behavior at strain rates $\leq 10^{-7}$ s^{-1}.

Another point (in relation to the ductile-brittle transition) is worth noting. In modeling ice sheet deformation, it is common practice to view the cover as a continuum, albeit an anisotropically damaged one. Hibler (2001) develops this theme. Invoked there is the notion of viscous-plastic deformation. "Viscous behavior" within that context has the same meaning as creep-based, ductile behavior within the present context; and "plastic behavior" has the same meaning as brittle failure used here. In other words, within the context of ice sheet modeling, the transition from viscous to plastic deformation is essentially the ductile-to-brittle transition.

To summarize, the evidence reveals a striking similarity in the pattern of cracking on widely different spatial scales. This suggests that the failure mechanisms are scale-independent. Owing to the concentration of stress by internal flaws of lengths, λ, the applied stress to operate the mechanisms appears to scale roughly as $\lambda^{-0.5}$. However, before firm conclusions can be drawn about scale-independence, more work is needed to relate in-field stress amplitudes to specific physical events.

ICE AND ROCK

Before closing, it is worth noting again that many of the phenomena exhibited by ice—e.g., pressure-sensitive deformation, ductile-to-brittle transition, and compressive shear faulting on scales small and large—are displayed by rock (see e.g., Duba et al. 1990, and references therein). Also, the fracture features that accompany compressive failure—distributed cracks, wing cracks and comb cracks—are also seen in geomaterials. And the evolution of a shear fault in granite (Lockner et al. 1991), for instance, is remarkably similar to its evolution in ice, even to the point of initiating at a free surface and then propagating along a narrow band of damage inclined by about 30° to the direction of the most compressive principal stress. These similarities, as already noted, strongly imply that failure mechanisms are common to many brittle or semi-brittle materials (Renshaw and Schulson 2001). Also common may be the criterion for the ductile-to-brittle transition, described by Equation (21). In other words, lessons learned from a study of ice may help to elucidate the failure of rock, and vice versa.

ACKNOWLEDGMENTS

The author acknowledges Profs. Ali Argon, Ian Baker, Bill Hibler, Harold J. Frost, Mark Kachanov, Carl Renshaw and Kaj Riska and Drs. Jerome Weiss, Paul Duval, Peter Sammonds, Devinder Sodhi, Ron Kwok, Don Nevel and Chris Heuer for helpful comments. He also recognizes again his former students whose published work has greatly contributed to this story: Rachel Batto, Stephanie Buck, Neil Cannon, Michele Couture, John Currier, Doug Fifolt, Mark Gies, Eric Gratz, Rebecca Haerle, Steven Hoxie, Daniel Iliescu, Doug Jones, Gary Kuehn, Greg Lasonde, Russell Lee, Pooi Lim, Jeff Melton, Oleg Nickolayev, Suogen Qi and Timothy Smith. Also, he acknowledges Drs. Ron Kwok and Harry Sterns for copies of Figure 25, and Dr. James Wilson for pointing out these images in the first place, following a presentation by the author. Finally, he gratefully acknowledges the J. William Fulbright Foreign Scholarship Board for a research award.

This work was supported by the Army Research Office, the Office of Naval Research, National Oceanographic and Atmospheric Administration, and Exxon and Mobil.

REFERENCES

Ahmad S, Whitworth RW (1988) Dislocation-motion in ice–a study by synchrotron X-ray topography. Phil Mag A 57:749-766
Anderson DL, Weeks WF (1958) A theoretical analysis of sea ice strength. Trans Am Geophys Union 39:632-640
API (1993) Planning, designing, constructing structures, pipelines for arctic conditions. American Petroleum Institute, Dallas, Texas
Ashby MF, Hallam SD (1986) The failure of brittle solids containing small cracks under compressive stress states. Acta Metall 34:497-510
Assur A (1958) Composition of sea ice, its tensile strength. In Arctic Sea Ice. U S Natl Acad Sci, Natl Res Council, Washington, DC, p 106-138
Baker I (1997) Observation of dislocations in ice. J Phys Chem B 101:6158-6162
Batto RA, Schulson EM (1992) A preliminary investigation of the ductile-brittle transition in columnar S2 ice under compression. IAHR Ice Symp 1021-1034
Batto RA, Schulson EM (1993) On the ductile-to-brittle transition in ice under compression. Acta metall mater 41:2219-2225
Bazant ZP, Li Y-N, Jirásek M, Li Z, Kim J-J (1995) Effect of size on distributed damage, fracture of sea ice. Northwest Research Associates, Inc., Bellevue, Washington: Sea Ice Mechanics, Arctic Modeling Workshop 1:73-83
Brace WF, Bombolakis EG (1963) A note of brittle crack growth in compression. J Geophys Res 68:3709
Buck SE, Nickolayev OY, Schulson EM (1994) Failure envelopes, the ductile to brittle transition in S2 columnar freshwater, saline ice under biaxial compression. IAHR 94, 12th Intl Symp on Ice 454-463
Butkovich TR (1954) Ultimate strength of ice. U S Snow, Ice, Permafrost Research Establishment, Res Paper 15
Campbell J, Fahmy Y, Conrad H (1999) Plastic deformation kinetics of fine-grained alumina. Metall Mater Trans A 30A:2809-2816
Cannon NP, Schulson EM, Smith TR, Frost HJ (1990) Wing cracks, brittle compressive fracture. Acta metall mater 38:1955-1962
Carter D (1971) Lois et mechanismes de l'apparente fracture fragile de la glace de riviere et de lac. University of Laval, France

Castelnau O, Canova GR, Lebensohn RA, Duval P (1997) Modeling viscoplastic behavior of anisotropic polycrystalline ice with a self-consistent approach. Acta Mater 45:4823-4834

Chantikul P, Bennison SJ, Lawn BR (1990) Role of grain size in the strength, r-curve properties of alumina. J Am Ceram Soc 73:2419-2427

Cole DM (1985) Grain size, the compressive strength of ice. Am Soc Mech Engin, 4th Intl Symp Offshore Mechanics, Arctic Engin, p 220

Cole DM (1987) Strain rate, grain size effects in ice. J Glaciol 33:274-280

Cole DM (1990) Reversed direct-stress testing of ice: Initial experimental results, analysis. Cold Regions Sci Techn 18:303-321

Cole DM (2001) The microstructure of ice, its influence on mechanical properties. Engin Fracture Mech 68:1797-1822

Conrad RE, II, Friedman M (1976) Microscopic feather fractures in the faulting process. Tectonophysics 33:187-198

Cooke ML (1997) Fracture localization along faults with spatially varying friction. J Geophys Res 102:24425-24434

Coon MD, Knoke GS, Echert DC, Pritchard RS (1989) The architecture of an anisotropic elastic-plastic sea ice mechanics constitutive law. J Geophys Res 103:21915-21925

Costin LS (1983) A microcrack model for the deformation, failure of brittle rock. J Geophys Res 88: 9485-9492

Cottrell AH (1958) Theory of brittle fracture in steel, similar metals. Trans Metall Soc AIME 212:192-203

Couture ML, Schulson EM (1994) The cracking of ice under rapid unloading. Phil Mag Lett 69:9-14

Cruikshank KM, Zhao G (1991) Analysis of minor fractures associated with joints, faulted joints. J Struct Geol 13:865-886

Currier JH, Schulson EM (1982) The tensile strength of ice as a function of grain size. Acta Metall 30:1511-1514

Davies RK, Pollard DD (1986) Relations between left-lateral strike-slip faults, right-lateral kink bands in granodiorite, Mt. Abbot Quadrangle, Sierra Nevada, California. Pure Appl Geophys 124:177-201

Dempsey JP (1991) Fracture toughness of ice. *In* Ice Structure Interactions. SJ Jones et al. (eds) Springer-Verlag, Berlin, p 109-125

Dempsey JP (1996) Scale effects on the fracture of ice. The Minerals, Metals, Materials Society, The Johannes Weertman Symposium 19:351-361

Dempsey JP, Adamson RM, Mulmule SV (1999) Scale effects on the *in situ* tensile strength, fracture of ice. Part ii: First-year sea ice at resolute, N.W.T. Intl J Fracture 95:347-366

Duba AG, Durham WB, Handin JW, Wang HF (eds) (1990) The brittle-ductile transition in rocks: *The Heard Volume*. Geophys Monogr 56, Am Geophys Union, Washington, DC

Durham WB, L.A.Stern (2001) Rheological properties of water ice, Annual Rev. Earth Planet. Science 29:295-330

Duval P, Ashby MF, Anderman I (1983) Rate-controlling processes in the creep of polycrystalline ice. J Phys Chem 87:4066-4074

Duval P, Arnaud L, Brissaud O, Montagnat M, de la Chapelle S (2000) Deformation, recrystallization processes of ice from polar ice sheets. Ann Glaciol 30:83-87

Dykins JE (1970) Ice engineering: Tensile properties of sea ice grown in a confined system. Naval Civil Engineering Laboratory, Port Hueneme, California

Erlingsson B (1988) Two-dimensional deformation patterns in sea ice. J Glaciol 34:301-308

Evans AG, Rice JR, Hirth JP (1980) Suppression of cavity formation in ceramics: Prospects for superplasticity. J Am Ceram Soc 63:368-375

Flemings MC (eds) (1974) Solidification Processing. McGraw-Hill, New York

Fletcher NH (1973) The surface of ice. *In* Physics, Chemistry of Ice. Whalley E, Jones SJ, Gold LW (ed) Royal Soc of Canada, Ottawa, p 132-136

Frederking R (1977) Plane-strain compressive strength of columnar-grained, granular-snow ice. J Glaciol 18:505-516

Frost HJ (2001) Crack nucleation in ice. Engin Fracture Mechanics 68:1823-1837

Gagnon RE, Gammon PH (1995) Triaxial experiments on iceberg, glacier ice. J Glaciol 41:528-540

Gammon PH, Kiefte H, Clouter MJ, Denner WW (1983) Elastic constants of artificial ice, natural ice samples by brillouin spectroscopy. J Glaciol 29:433-460

Germanovich LN, Salganik RL, Dyskin AV, Lee KK (1994) Mechanisms of brittle failure of rock with pre-existing cracks in compression. Pure Appl Geophys 19:197-212

Gies MC (1988) The development of brush-type platens for loading ice specimens in compression. Master of Engineering thesis, Dartmouth College, Hanover, New Hampshire

Glen JW (1968) The effect of hydrogen disorder on dislocation movement, plastic deformation in ice. Phys kondens Mater 7:43-51

Gold LW (1997) Statistical characteristics for the type, length of deformation-induced cracks in columnar-grain ice. J Glaciol 43:311-320

Gottschalk RR, Kronenberg AK, Russel JE, Handin J (1990) Mechanical anisotropy of gniess: Failure criterion, textural sources of directional behavior. J Geophys Res 95:613-621

Granier T (1985) Origin, damping, pattern development of faults in granite. Tectonics 4:721-737

Grape JA, Schulson EM (1992) Effect of confining stress on brittle indentation failure of columnar ice. J Offshore, Polar Engineering 2:212-221

Gratz ET, Schulson EM (1997) Brittle failure of columnar saline ice under triaxial compression. J Geophys Res 102:5091-5107

Greeley R, Figueredo PH, Williams DA et al. (2000) Geologic mapping of Europa. J Geophys Res 105:22559-22578

Greenberg R, Geissler P, Tufts BR, Hoppa GV (2000) Habitability of Europa's crust: the role of tidal-tectonic processes. J Geophys Res 105:17551-17562

Greenberg RP, Weidenschilling SJ (1984) How fast do Galilean satellites spin? Icarus 58:186-196

Gupta V, Bergström JS (1998) Compressive failure of rocks by shear faulting. J Geophys Res 103:23875-23895

Hallam SD, Duval P, Ashby MF (1987) A study of cracks in polycrystalline ice under uniaxial compression. J Physique 48:303-311

Hausler FU (1981) Multiaxial compressive strength tests on saline ice using brush-type loading platens. IAHR Ice Symp, p 389-398

Hawkes I, Mellor M (1972) Deformation, fracture of ice under uniaxial stress. J Glaciol 11:103-131

Haynes FD, Mellor M (1977) Measuring the uniaxial compressive strength of ice. J Glaciol 19:213-223

Hibler WD, Schulson EM (2000) On modeling the anisotropic failure, flow of flawed sea ice. J Geophys Res 105:17105-17120

Hibler WDI (2001) Sea ice fracturing on the large scale. Engineering Fracture Mechanics 68:2013-2044

Hill R (1950) The Mathematical Theory of Plasticity. Oxford University Press, New York

Hobbs DW, Pomeroy CD, Newman JB (1977) Design stresses for concrete structures subject to multi-axial stresses. Struct Engin 55:151-164

Hopkins MA (1998) Four stages of pressure ridging. J Geophys Res 103:21883-21891

Hoppa GV, Tufts BR, Greenberg R, Geissler PE (1999) Formation of cycloidal features on Europa. Science 285:1899-1902

Horii H, Nemat-Nasser S (1985) Compression-induced microcrack growth in brittle solids: Axial splitting, shear failure. J Geophys Res 90:3105-3125

Horii H, Nemat-Nasser S (1986) Brittle failure in compression: Splitting, faulting, brittle-ductile transition. Phil Trans R Soc A319:337-374

Hu X, Jia K, Baker I, Black D (1995) Dislocation mobility in HCl-doped ice. Mater Res Soc 375:287-292

Hughes T (1989) Calving ice walls. Ann Glaciol 12:74-80

Iliescu D, Schulson EM (1998) Macroscopic compressive shear faults in S2 columnar ice. A.A. Balkema Publishers, 14th Intl Symp on Ice, p 553-558

Ingraffea AR (1981) Mixed-mode fracture initiation in Indiana limestone, Westerly granite. Proc U S Symp Rock Mechanics 22:196-191

Jaeger JC, Cook NGW (1979) Fundamentals of Rock Mechanics. Chapman-Hall, London

Jellinek HHG (1957) Tensile strength properties of ice adhering to stainless steel. U S Snow, Ice, Permafrost Research Establishment Research Report 23

Johanessen OM, Miles MW, Bjorgo E (1995) The arctic shrinking sea-ice. Nature 376:126-127

Johnson KL (1987) Contact mechanics. Press Syndicate of the University of Cambridge, Cambridge, UK

Jones SJ (1982) The confined compressive strength of polycrystalline ice. J Glaciol 28:171-177

Jordaan IJ (2001) Mechanics of ice-structure interaction. Engineering Fracture Mechanics 68:1923-1960

Kachanov ML (1982) A microcrack model of rock inelasticity, part II: Propagation of microcracks. Mech Mater 1:29-41

Kalifa P, Duval P, Ricard M (1989) Crack nucleation in polycrystalline ice under compressive stress states. American Society of Mechanical Engineers, New York, Eighth Intl Conference on Offshore Mechanics: Arctic Engineering 4:13-20

Karge JS, Kaye JZ, Head (III) JW, Marion GM, Sassen R, Crowley JK, Ballseteros OP, Grant SA, Hogenboom DL (2000) Europa's crust and ocean: origin, composition, and the prospects for life. Icarus 148:226-265

Kemeny JM, Cook NGW (1987) Crack models for the failure of rock under compression. 2nd Intl Conf Constitutive Laws for Engineering Materials 2:879-889

Kennedy FE, Schulson EM, Jones D (2000) Friction of ice on ice at low sliding velocities. Phil Mag A 80:1693-1110

Ketcham WM, Hobbs PV (1969) An experimental determination of the surface energies of ice. Phil Mag 19:1161-1173

Khomichevskaya IS (1940) O vremennom soprotivlenii szhatiyu vechnomerzlykh gruntov i l'da yestestvennoy struktury. (The ultimate compressive strength of permafrost, ice in their natural states). Trudy Komiteta po Vechnoy Merzlote 10:37-83

Knott JF (1973) Fundamentals of Fracture Mechanics. John Wiley & Sons, New York

Korzhavin KN (1955) Vliyaniye skorosti deformirovaniya na velichinu predela prochnosti rechnogo l'da priodnoosnom szhatii [the effect of the speed of deformation on the ultimate strength of river ice subject to uniaxial compression]. Trudy Novosibirskogo Instituta Inzhenerov Zheleznodorozhnogo Transporta 11:205-216

Kuehn G, Schulson E, Jones D, Zhang J (1993a) The compressive strength of ice cubes of different sizes. Transactions of the ASME, 11th Intl Symp, Exhibit on Offshore Mechanics, Arctic Engineering, p 142-148

Kuehn GA, Schulson EM (1994a) The mechanical properties of saline ice under uniaxial compression. Ann Glaciol 19:39-48

Kuehn GA, Schulson EM (1994b) Ductile saline ice. J Glaciol 40:566-568

Kuehn GA, Schulson EM, Nixon WA (1988a) The effect of pre-strain on the compressive ductile-to-brittle transition in ice. IAHR Symp 1988 Sapporo, p 109-117

Kuehn GA, Schulson EM, Nixon WA (1988b) The effects of pre-strain on the compressive ductile-to-brittle transition in ice. IAHR Ice Symp, p 109-117

Kuehn GA, Schulson EM, Jones DE, Zhang J (1993b) The compressive strength of ice cubes of different sizes. Transactions of the ASME, 11th Intl Symp, Exhibit on Offshore Mechanics, Arctic Engineering 115:142-148

Kwok R, Rignot E, Holt B, Onstott R (1992) Identification of sea ice types in space-borne synthetic aperture radar data. J Geophys Res-Oceans 97:2391-2402

Kwok RE, Cunningham GF, LaBelle-Hamer N, Holt B, Rothrock D (1999). Ice thickness derived from high-resolution radar imagery. EOS Trans Am Geophys Union, October 19, 1999, p 494-497

Lee RW, Schulson EM (1988) The strength, ductility of ice under tension. J Offshore Mechanics, Arctic Engin 110:187-191

Lehner F, Kachanov M (1996) On modelling of "winged" cracks forming under compression. Intl J Fracture 77:R69-75

Lewis JK (1995) A conceptual model of the impact of flows on the stress state of sea ice. J Geophys Res 100:8819-8825

Lindsay RW, Rothrock DA (1995) Arctic sea-ice leads from advanced very high-resolution radiometer images. J Geophys Res–Oceans 10:4533-4544

Liu F, Baker I (1995) Dislocation-grain boundary interactions in ice crystals. Phil Mag A 71:15-42

Lockner DA, Byerlee JD, Kuksenko V, Pnomarev A, Sidorin A (1991) Quasi-statis fault growth, shear fracture energy in granite. Nature 350:39-42

Manley ME, Schulson EM (1997) On the strain-rate sensitivity of columnar ice. J Glaciol 43:408-410

Marko JR, Thomson RE (1977) Rectilinear leads, internal motions in the ice pack of the western arctic ocean. J Geophys Res 82:979-987

Martel SJ, Pollard DD, Segall P (1988) Development of simple strike-slip fault zones in granitic rock, Mount Abbott Quadrangle, Sierra Nevada, California. Geol Soc Am Bull 100:1451-1465

Martel SM, Pollard DD (1989) Mechanics of slip, fracture along small faults, simple strike-slip fault zones in granitic rock. J Geophys Res–Solid Earth, Planets 94:9417-9428

Maykut GA (1982) Large-scale heat exchange, ice production in the central arctic. J Geophys Res 87:7971-7984

McClintock FA, Walsh JB (1962) Frictions in Griffith's cracks in rock under pressure. Proc 4th U S Natl Congress Appl Mech, p 1015

Melton JS, Schulson EM (1998) Ductile compressive failure of columnar saline ice under triaxial loading. J Geophys Res 103:21759-21766

Michel B (1978) Ice Mechanics. Laval University Press, Quebec

Moore DE, Lockner DA (1995) The role of microcracking in shear-fracture propagation in granite. J Struct Geol 17:95-114

Nickolayev OY, Schulson EM (1993) Failure envelopes, the ductile to brittle transition in columnar saline ice under biaxial compression. 1st Joint Mechanics Meeting of ASME, ASCE, SES Meeting '93 163:61-69

Nickolayev OY, Schulson EM (1995) Grain-boundary sliding, across-column cracking in columnar ice. Phil Mag Letters 72:93-97

Nixon WA, Schulson EM (1987) A micromechanical view of the fracture toughness of ice. J Physique 48:313-319

Nixon WA, Schulson EM (1988) Fracture toughness of ice over a range of grain sizes. J Offshore Mech Arctic Eng 110:192-196

Overland JE, McNutt SL, Salo S, Groves J, Li SS (1998) Arctic sea ice as a granular plastic. J Geophys Res–Oceans 103:21845-21867

Pappalardo, RT, Belton MJS, Breneman HH, et al. (1999) Does Europa have a subsurface ocean? Evaluation of the geological evidence. J Geophys Res 104:24015-24055

Paterson MS (1978) Experimental Rock Deformation—The Brittle Field. Springer-Verlag, New York

Patterson WSB (1994) The physics of glaciers. Oxford University Press, Oxford, UK

Petit J-P, Barquins M (1988) Can natural faults propagate under mode ii conditions? Tectonics 7: 1243-1256

Peyton HR (1966) Sea Ice Strength. University of Alaska, Geophysical Institute, Fairbanks

Picu RC, Gupta V (1995a) Observations of crack nucleation in columnar ice due to grain boundary sliding. Acta metall mater 43:3791-3797

Picu RC, Gupta V (1995b) Crack nucleation in columnar ice due to elastic anisotropy, grain boundary sliding. Acta metall mater 43:3783-3789

Pradeau SI, Schulson EM (1998) Failure of columnar saline ice under biaxial compression at -2.5°C (unpublished). Thayer School of Engineering, Dartmouth College, Hanover, New Hampshire

Qi S, Schulson EM (1998) The effect of temperature on the ductile-to-brittle transition in columnar ice. A.A. Balkema Publishers, 14th Intl Symp on Ice, p 521-527

Reches Z, Lockner DA (1994) Nucleation, growth of faults in brittle rocks. J Geophys Res 99: 18159-18173

Renshaw CE, Schulson EM (1998) Non-linear rate dependent deformation under compression due to state variable friction. Geophys Res Letters 25:2205-2208

Renshaw CE, E.M.Schulson (2001) Universal behavior in compressive failure of brittle materials. Nature 412: 897-900

Reynolds RT, McKay CP, Casting JF (1987) Europa, tidally heated oceans, and habitable zones around giant planets. Adv Space Res 7:5125-5132

Richter-Menge J (1991) Confined compressive strength of horizontal first-year sea ice samples. J Offshore Mech Arctic Engin 113:344-351

Richter-Menge JA, Jones KF (1993) The tensile strength of first-year sea ice. J Glaciol 39:609-618

Richter-Menge JA, Elder BC (1998) Characteristics of pack ice stress in the Alaska's Beaufort Sea. J Geophys Res–Oceans 103:21817-21829

Riedel H, Rice JR (1980) Tensile cracks in creeping solids. ASTM-STP 7700:112-130

Riska K, Tuhkuri J (1995) Application of ice cover mechanics in design, operations of marine structures. Northwest Research Associates, Inc., Bellevue, Washington: Sea Ice Mechanics, Arctic Modeling Workshop 2:123-134

Rispoli R (1981) Stress fields about strike-slip faults from stylolites, tension gashes. Tectonophysics 75:T29-T36

Rist MA (1997) High stress ice fracture, friction. J Phys Chem B 101:6263-6266

Rist MA, Murrell SAF (1994) Ice triaxial deformation, fracture. J Glaciol 40:305-318

Rist MA, Jones SJ, Slade TD (1994) Microcracking, shear fracture in ice. Ann Glaciol 19:131-137

Rist MA, Sammonds PR, Murrell SAF, Meredith PG, Doake CSM, Oerter H, Matsuki K (1999) Experimental, theoretical fracture mechanics applied to antarctic ice fracture, surface crevassing. J Geophys Res 104:2973-2987

Rudnicki JW, Rice JR (1975) Conditions for the localization of deformation in pressure-sensitive dilatant materials. J Mech Phys Solids 23:371-394

Sammonds PR, Murrell SAF, Rist MA (1998) Fracture of multi-year sea ice. J Geophys Res 100: 21795-21815

Sanderson TJO (1988) Ice mechanics: Risks to offshore structures. Graham & Trotman, London

Schulson EM (1979) An analysis of the brittle to ductile transition in polycrystalline ice under tension. Cold Regions Sci Techn 1:87-91

Schulson EM (1987) The fracture of ice 1h. J Physique C1-207–C-220

Schulson EM (1990) The brittle compressive fracture of ice. Acta metall mater 38:1963-1976

Schulson EM (1996) The failure of ice under compression. TMS Annual Meeting, Weertman Symposium, p 363-374

Schulson E.M (2001) Fracture of ice. Engg Fracture Mech 68:1793-1795

Schulson E.M (2002) Compressive shear faults within the arctic sea ice cover. J Geophys Res (submitted)

Schulson EM (2002 On compressive shear faulting in ice: Coulombic vs. plastic faults. Acta Mater 50:3415-3424

Schulson EM, Cannon NP (1984) The effect of grain size on the compressive strength of ice. IAHR Ice Symp

Schulson EM, Hibler WD (1991) The fracture of ice on scales large, small: Arctic leads, wing cracks. J Glaciol 37(127):319-323

Schulson EM, Kuehn GA (1993) Ductile ice. Phil Mag Lett 67:151-157

Schulson EM, Buck SE (1995) The ductile-to-brittle transition, ductile failure envelopes of orthotropic ice under biaxial compression. Acta metall mater 43:3661-3668

Schulson EM, Nickolayev OY (1995) Failure of columnar saline ice under biaxial compression: Failure envelopes, the brittle-to-ductile transition. J Geophys Res 100:22383-22400

Schulson EM, Gratz ET (1999) The brittle compressive failure of orthotropic ice under triaxial loading. Acta Mater 47:745-755

Schulson EM, Lim PN, Lee RW (1984) A brittle to ductile transition in ice under tension. Phil Mag A 49:353-363

Schulson EM, Hoxie SG, Nixon WA (1989a) The tensile strength of cracked ice. Phil Mag A 59:303-311

Schulson EM, Jones DE, Kuehn GA (1991a) The effect of confinement on the brittle compressive fracture of ice. Ann Glaciol 15:216-221

Schulson EM, Iliescu D, Renshaw CE (1999) On the initiation of shear faults during brittle compressive failure: A new mechanism. J Geophys Res 104:695-705

Schulson EM, Geis MC, Lasonde GJ, Nixon WA (1989b) The effect of the specimen-platen interface on the internal cracking, brittle fracture of ice under compression: High-speed photography. J Glaciol 35:378-382

Schulson EM, Kuehn GA, Jones DE, Fifolt DA (1991b) The growth of wing cracks, the brittle compressive failure of ice. Acta metall mater 39:2651-2655

Schulson EM, Gratz ET, Johnson RC, Spring W (unpublished research)

Schulson EM, Baker I, Robertson CD, Bolon RB, Harnimon RJ (1989c) Fractography of ice. J Mater Sci Lett 8:1193-1194

Schwarz J, Weeks WF (1977) Engineering properties of sea ice. J Glaciol 19:499-531

Segall P, Pollard DD (1983) Nucleation, growth of strike-slip faults in granite. J Geophys Res 88:555-568

Shapiro LH, Weeks WF (1993) The influence of crystallographic, structural properties on the flexural strength of small sea ice beam. Am Soc Mech Eng, Ice Mechanics–1993; 1993 Joint ASME Applied Mechanics, Materials Summer Mtg, Appl Mech Div 163:18159-18173

Shearwood C, Whitworth RW (1991) The velocity of dislocations in ice. Phil Mag A 64:289-302

Shen W, Lin SZ (1986) Fracture toughness of bohai bay sea ice. 5th Intl Offshore Mech, Arctic Engin Symp IV:354-357

Sinha NK (1982) Constant strain-, stress-rate compressive strength of columnar-grained ice. J Mater Sci 17:785-802

Smith E, Barnby JT (1967) Crack nucleation in crystalline solids. Metall Sci J 1:56-64

Smith TR, Schulson EM (1993) The brittle compressive failure of fresh-water columnar ice under biaxial loading. Acta metall mater 41:153-163

Smith TR, Schulson EM (1994) Brittle compressive failure of salt-water columnar ice under biaxial loading. J Glaciol 40:265-276

Sodhi DS (2001) Crushing failure during ice-structure interaction. Eng Fracture Mechanics 68:1889-1921

Thouless MD, Evans AG, Ashby MF, Hutchinson JW (1987) The edge cracking, spalling of brittle plates. Acta metall 35:1333-1341

Timco GW, Frederking RMW (1983) Confined compressive strength of sea ice. 7th Intl Conf, Port and Ocean Engineering Under Arctic Conditions (POAC) I:243-253

Trickett YL, Baker I, Pradhan PMS (2000) The orientation dependence of the strength of ice single crystals. J Glaciol 46:41-44

Tucker WB, Perovich DK (1992) Stress measurements in drifting pack ice. Cold Regions Sci Techn 20:119-139

Voytkovskiy KF (1960) Mekhanicheskiye svaystva l'da [mechanical properties of ice]. Moscow, Izdatel'stvo Akademii Nauk SSSR, [U S Dept of Commerce, Office of Technical Services: transl]

Weeks WF (1962) Tensile strength of nacl ice. J Glaciol 4:25-52

Weeks WF, Ackley SF (1982) The growth, structure, properties of sea ice. U S Army Cold Regions Research, Engineering Laboratory, November 1982

Weertman J (1983) Creep deformation of ice. Ann Rev Earth Planet Sci 11:215-240

Weiss J (1999) The ductile behaviour of damaged ice under compression. Port and Ocean Engineering Under Arctic Conditions (POAC), p 10

Weiss J (2001) Fracture, fragmentation of ice: A fractral analysis of scale invariance. Engin Fracture Mech (in press)

Weiss J, Schulson EM (1995) The failure of fresh-water granular ice under multi-axial compressive loading. Acta metall mater 43:2303-2315

Weiss J, Gay M (1998) Fracturing of ice under compression creep as revealed by a multifractal analysis. J Geophys Res 103:24005-24016

Weiss J, Schulson EM (2000) Grain boundary sliding, crack nucleation in ice. Phil Mag 80:279-300

Weiss J, Schulson EM, Frost HJ (1996) The nucleation of microcracks in ice cubes compressed equally on all boundaries. Phil Mag A 73:1385-1400

Wong T-F (1982) Micromechanics of faulting in Westerly granite. Intl J Rock Mech Mining Sci 19:49-64

Zimmermann A, Rodel J (1998) Generalized Orowan-Petch plot for brittle fracture. J Am Ceram Soc 81:2527-2532

9 Seismic Wave Attenuation: Energy Dissipation in Viscoelastic Crystalline Solids

Reid F. Cooper

Department of Materials Science and Engineering
University of Wisconsin–Madison
Madison, Wisconsin 53706

INTRODUCTION

Seismic imaging as a tool to understand the structure and, perhaps, the dynamics of the planet at depth involves the spatially resolved, combined study of wave velocities and velocity dispersion, wave birefringence and wave attenuation (mechanical absorption). When integrated with insights from petrology, plus-or-minus input from magnetotellurics, specific hypotheses concerning structure at depth can be formulated (cf. Karato 1993; Karato and Karki 2001). The recent advances in tomography have allowed significant improvements in spatial resolution which also have allowed ever more exacting hypotheses of structure to be articulated. Nevertheless, as is nicely illustrated by the relatively recent seismic analyses of an accreting plate margin (the East Pacific Rise —e.g., Toomey et al. 1998; Webb and Forsyth 1998), the structures inferred at depth can vary significantly: clearly, the interpretation of seismic data is limited specifically by a lack of understanding of the physical processes by which low-frequency wave absorption occurs, particularly in the cases (a) where melt is present and/or (b) where the material is being actively plastically deformed.

This chapter emphasizes the mineral physics/materials science of mechanical absorption, specifically in dense materials at elevated temperature; the interest, then, is in employing the physical study of mechanical absorption to understand natural phenomena such as those mentioned above. Beyond questions of the basic mineral physics of absorption, the guiding interest the community of scholars pursuing the ideas is in isolating actual absorption mechanisms operative in the geological setting thereby allowing for a greater discrimination in the interpretation of seismic data.

At the outset, I note that this contribution does not seek to be a comprehensive review of the ideas and work involved in the physics of mechanical absorption. There are marvelous reviews in the literature that are both authoritative and current; I refer to these frequently in contemplating my own experimental work and so am pleased to point them out here. In the geophysics literature, Karato and Spetzler (1990) have extensively reviewed the variety of lattice-defect-based mechanical loss mechanisms described theoretically in the physics and materials science communities, with an eye towards their application to minerals, while Jackson (1993, 2000) reviewed experimental approaches geared to study these effects in mineral systems. In the materials science literature, the text by Nowick and Berry (1972) remains the standard. In the mechanics literature, the books by Findley et al. (1976; corrected and republished 1989) and, recently, by Lakes (1999) accurately tie down the phenomenology of attenuation. Additionally, in that all of the above authors point out the importance of grain boundaries and solid-state phase boundaries in effecting attenuation (at least in part), Kê's (1999) recent review of a half-century of his own work in the area is significant.

LINEAR VISCOELASTICITY: A CHEMICAL KINETICS PERSPECTIVE

The mechanical properties of solids are distinctly time dependent. When a deviatoric-stress thermodynamic potential is applied to a solid system initially at mechanical and chemical equilibrium, that system responds—specifically in the form of strain—in a manner that minimizes the new potential: LeChatlier's principle holds in mechanical properties as well as in thermochemistry. (From this point forward, I will refer to the deviatoric stress potential as the "stress potential" or simply the "stress;" "mechanical potential" is used in most thermodynamics treatises to describe pressure, which, of course, must incorporate the hydrostatic component of the applied tensorial stress. Further, hydrostatic pressure can effect mechanical absorption, too: the melt-migration absorption response of partial melts being one example.) The responses to the stress potential combine elastic and inelastic phenomena; the resultant behavior overall is labeled "viscoelasticity." Elastic response, i.e., strain in the form of the stretching and bending of atomic bonds, is considered instantaneous or time-independent (to first order, i.e., ignoring inertial effects). In contrast, inelastic responses are distinctly time dependent: in fully crystalline solids, they involve the interaction of the stress potential with point (vacancies, interstitials, substitutionals, as well as jogs and kinks on dislocations), line (dislocations) and planar (twin boundaries; grain and phase boundaries) lattice defects. Time dependence is a manifestation that the strain so produced involves the motion of these lattice defects: the motion involves the breaking and reforming of atomic bonds, the rate of which is dictated by Boltzmann statistics (and, thus, the rate depends exponentially on temperature). One can further divide inelastic responses, identifying those that produce time-dependent but fully recoverable strain ("anelastic") and those that produce time-dependent, permanent (irrecoverable, "plastic") strain. In that viscoelasticity combines elastic and inelastic responses, viscoelasticity is time dependent.

From a thermodynamics perspective, one understands elasticity as a reversible process where strain energy (i.e., the energy manifest as the distortion of bonds) is stored in the material and is recoverable without loss, while inelastic response is irreversible: strain energy is dissipated within the material, that is, it is converted irreversibly to heat. (Of course, if fracture is initiated, strain-energy dissipation sees the formation of new surfaces as well as the conversion to heat. This chapter, though, will not consider fracture behavior.) Kinetics enters the discussion—and dominates much of the consideration of mechanical response—when (i) temperature is sufficiently high and/or (ii) the application of a deviatoric stress sufficiently slow and/or (iii) the deviatoric stress persists for a sufficiently long time such that either (a) the inelastic strain response(s) occurs at a rate(s) competitive to the elastic response or (b) the accumulated strain produced by the inelastic responses outstrips that from the elastic response.

The elastic and many inelastic responses of materials act in "parallel" kinetically, meaning the responses to stress are independent physically and, as such, the strain produced by each response is additive. Inelastic responses, however, often involve a number of "serial" kinetic steps, which are physical processes that must act in sequence in order to produce strain. Some series-process phenomena are inherent in the process physics (e.g., solution-precipitation—or pressure-solution—creep involves the serial physical steps of (i) atomic dissolution into the liquid phase, (ii) atomic/ionic transport through the liquid phase and (iii) precipitation from the liquid phase; dislocation creep involves the serial process of climb (or cross-slip) and glide of the lattice dislocations), while others are based on geometrical constraints (e.g., diffusional creep acts in series with the processes of grain boundary sliding). The distinction between parallel and serial kinetic processes in mechanical response is important: for a given set of thermodynamic potentials—pressure, temperature, various chemical potentials (e.g., water fugacity, oxygen fugacity, oxide chemical potentials, etc.), and deviatoric stress—and microstructure parameters (e.g., grain size, phase percolation, etc.), the parallel kinetic mechanism that produces strain most rapidly will dominate the overall mechanical response, while the slowest serial step of that fastest parallel process will rate-limit the response. The major challenge in experimental geophysics is to understand the ranges of thermodynamic potentials giving rise to various mechanical responses in the laboratory time frame, and, also, to understand well the physics of their extrapolation: one must study in the laboratory the same physical process active in the geological environment for the data to be applicable (Paterson 1976). This point is as true for the phenomenon of seismic-frequency mechanical absorption as it is for large-strain creep phenomena.

The maximum strains involved in wave attenuation are quite small, generally $<10^{-6}$ (e.g., Romanowicz and Durek 2000; Karato and Spetzler 1990); correspondingly, the deviatoric stresses involved are small as well, ~10–100 kPa. In this regime, the viscoelasticity in crystalline matter is found to be linear, which means, specifically, two qualities characterize the mechanical behavior: (i) the magnitudes of the strain response and of its first time derivative are directly proportional to the magnitude of the applied stress (e.g., $\varepsilon, \dot{\varepsilon} \propto \sigma^1$, where ε is strain, σ is deviatoric stress and the superposed dot represents a time derivative) and (ii) the rules of mechanical (Boltzmann) superposition hold (e.g., Findley et al. 1976, p.5). Superposition means that the elastic and the (perhaps many) plastic and anelastic responses can be simply summed to discern the overall mechanical response. Superposition is important in understanding how models of dynamic mechanical response are formulated, as will be discussed below.

Wave attenuation arises due to inelastic behavior. It is useful to consider inelastic responses as the relaxation of elastic strain energy: with the application of stress, elastic energy is stored into the material (the "system") instantaneously; inelastic, time-dependent relaxation can then occur, assuming the applied stress is sustained for sufficient time. Behavior can be easily contemplated by considering the system response to two loading approaches. (1) If a fixed amount of strain is applied instantaneously, and then the system is maintained at this strain, the system will experience a relaxation of the stress—and, thus, of the elastic strain energy, which is proportional to the product of stress and strain—until the energy is fully dissipated. In nature, the simplest strain-energy-relaxation function seen—and fairly frequently (e.g., chemical diffusion in a fixed potential gradient; cf. Nowick and Berry 1972, Chapter 8)—is that involving an exponential decay, i.e.,

$$U(t) = U_0 \exp(-t/\tau_R), \tag{1}$$

where U is strain energy (with U_0 being its value after application of the instantaneous stress), t is time and τ_R is the relaxation time, i.e., that time where the energy is decreased to 1/e of its initial value. Lakes (1999, Eqn. 2.6.1) describes the *effective*, time-dependent Young's modulus with this function: if one defines Young's modulus (1-D) as $E = \sigma/\varepsilon$, the stress relaxation means that the modulus apparently decays with time. Alternatively, (2) if the stress step is applied instantaneously and then that stress is maintained, the system will experience flow. The strain energy stored in the material will remain constant, but mechanical energy will be dissipated through the specimen as strain is accumulated. This behavior—flow at constant stress—is called creep; phenomenologically, the system appears to have a time-dependent compliance, J, i.e. (cf. Lakes 1999, Eqn. 2.6.2),

$$J(t) = J_0(1 - \exp[-t/\tau_C]), \tag{2}$$

where τ_C is the characteristic creep or retardation time and, similar to Equation (1), the subscript "0" being the initial value after application of the stress step.

SPRING AND DASHPOT MODELS OF VISCOELASTICITY

In linear mechanical response, exponential "response" functions like Equations (1) and (2) can be modeled using combinations of discrete elastic elements (Hookean springs) and plastic elements (Newtonian-viscous "dashpots"); various of these elements can be arrayed in serial-kinetic or parallel-kinetic form to replicate the mechanical behavior of a system. It is useful for the non-specialist to be exposed to the use of spring-dashpot models because these models often dominate the thinking of mechanics and materials experimentalists and theoreticians—for both good and ill. Before presenting the basic ideas, then, two admonitions are offered. The first—because it is so exacting—is directly quoted from Lakes (1999, p.23):

> *Warning.* Spring-dashpot models have a pedagogic role; however, real materials in general are not describable by models containing a small number of springs and dashpots. (The emphasis is Lakes'.)

Lakes makes this warning because many investigators tend to cling too tightly to exponential-decay, spring-dashpot models as describing exactly some experimental results. The second admonition concerns the relationship between the chemical-kinetics description and the mechanics description—as represented by spring-dashpot models—of mechanical response. Independent physical responses, described above as "parallel-kinetic," end up being portrayed as spring and dashpot elements arranged in series; correspondingly, dependent, "serial-" or "series-kinetic" physical processes translate to springs and dashpots arranged in parallel. In what follows, then, I will carefully distinguish, e.g., "mechanical parallel" from "parallel-kinetic," etc.

Figure 1a presents the spring-dashpot model where one spring (with spring constant R_1) plus one dashpot (with Newtonian viscosity η_1) are arranged in mechanical series; this construct is historically known as the Maxwell Solid model. The response functions corresponding to the constant stress and constant strain conditions mentioned previously are shown as well. One sees clearly that the total strain and strain rate in the Maxwell Solid are the sums of those provided by the spring and the dashpot (i.e.,

$\varepsilon = \varepsilon_1 + \varepsilon_2$ and $\dot{\varepsilon} = \dot{\varepsilon}_1 + \dot{\varepsilon}_2$,

where, in this 1-D model with normal stresses and strains, R_1 will be equivalent to Young's modulus and η_1 to a normal viscosity) and that the elastic and plastic elements respond independently to the stress (i.e., they are kinetically parallel). Release of the stress sees instantaneous recovery of the strain in the spring, but no strain recovery of the dashpot: the Maxwell Solid displays no anelastic behavior. The Maxwell Solid response functions are easily contemplated by considering the strain rate:

$$\dot{\varepsilon} = \dot{\varepsilon}_1 + \dot{\varepsilon}_2 = (\dot{\sigma}/R_1) + (\sigma/\eta_1) ; \tag{3}$$

thus, the constant-stress (creep) response is

$$\varepsilon(t) = (\sigma_0/R_1) + (\sigma_0/\eta_1)t , \tag{4}$$

and the constant-strain, stress-relaxation response is

$$\sigma(t) = \sigma_0 \exp(-R_1 t/\eta_1) = R_1 \varepsilon_0 \exp(-R_1 t/\eta_1) . \tag{5}$$

Comparison of Equations (5) and (1) reveals that $\tau_R = \eta_1/R_1$, i.e., 1/e (~37%) of the stress is relaxed in this time.

Placing a spring and dashpot in mechanical parallel results in the Voigt/Kelvin Solid model (Fig. 1b). Here, one sees that the application of step-function (tensile) stress does not see the instantaneous extension of the spring; rather, the spring is loaded/extended at a rate dictated by the viscous-flow extension of the dashpot. Upon release of the stress,

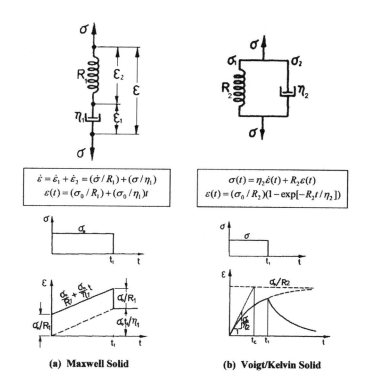

Figure 1. Simple, two-element spring-and-dashpot models of linear viscoelastic solids, including characteristic differential equations and strain responses for static loading. (a) Maxwell Solid. (b) Voigt/Kelvin Solid.

the spring relaxes, again dictated kinetically by the dashpot. The model describes plainly a serial-kinetic process for the accumulation and relaxation of strain. Further, the Voigt/Kelvin Solid is clearly anelastic. In that the responses of the elements are dependent, the Voigt/Kelvin Solid is best considered by considering the constitutive relation between stress and strain:

$$\sigma(t) = \eta_2 \dot{\varepsilon}(t) + R_2 \varepsilon(t) . \tag{6}$$

The creep function for the Voigt/Kelvin Solid is

$$\dot{\varepsilon}(t) = (\sigma_0 / R_2)(1 - \exp[-R_2 t / \eta_2]) , \tag{7}$$

and, if the stress is removed from the system at $t = t_1$, the creep function for $t > t_1$ is

$$\dot{\varepsilon}(t) = \varepsilon_{t=t_1} \exp[-R_2(t - t_1)/\eta_2] . \tag{8}$$

For the Voigt/Kelvin Solid, one sees that $\tau_C = \eta_2/R_2$.

Neither the Maxwell or Voigt/Kelvin Solid models, in themselves, accurately model real, viscoelastic solids: the Maxwell Solid has no anelastic behavior and, as such, could not produce the transients seen in plastic flow; the Voigt/Kelvin Solid cannot produce instantaneous strain, nor plastic flow, nor the rather high-rate transients associated with primary creep. Superposition of these linear models, however, can demonstrate "complete" viscoelastic behavior. The simplest of these compound models is the Burgers Solid, created by arranging the Maxwell and Voigt/Kelvin Solid models in mechanical series (Fig. 2). From the illustration, it is clear that an applied stress allows the two elements of the Maxwell Solid to strain independently of the Voigt/Kelvin Solid and of each other (a parallel-kinetic response), while serial-kinetic behavior of the Voigt/Kelvin solid remains. The creep function of the Burgers Solid model is simply the sum of Equations (4) and (7), or:

$$\dot{\varepsilon}(t) = (\sigma_0/R_1) + (\sigma_0/R_2)(1 - \exp[-R_2 t/\eta_2]) + (\sigma_0/\eta_1)t , \qquad (9)$$

where the three terms represent the elastic, anelastic and plastic components, respectively, of the model. The ε vs. t curve for creep in Figure 2 shows, too, the strain response after removal, at time t_1, of the applied stress σ_0: the elastic strain from the Maxwell Solid spring is recovered instantly, the strain from the Maxwell Solid dashpot ceases but experiences no recovery and the anelasticity of the Voigt/Kelvin Solid gives back its strain over time; the creep function for $t > t_1$ is (cf. Eqn. 8):

$$\dot{\varepsilon}(t) = (\sigma_0/R_2)(\exp[-R_2 t_1/\eta_2] - 1)(\exp[-R_2 t/\eta_2]) + (\sigma_0/\eta_1)t_1 . \qquad (10)$$

One can differentiate Equation (9) to determine strain rate in a continuing creep experiment at stress, σ_0, i.e.,

$$\dot{\varepsilon}(t) = (\sigma_0/\eta_2)(\exp[-R_2 t/\eta_2]) + (\sigma_0/\eta_1) . \qquad (11)$$

The function can be evaluated at very short, but finite time to discern an initial strain rate: $\sigma_0[(1/\eta_2) + (1/\eta_1)] = \tan \alpha$ in the figure; evaluating the function at long time sees that it asymptotically approaches $\sigma_0/\eta_1 = \tan \beta$. In curve-fitting the Burgers Solid model to

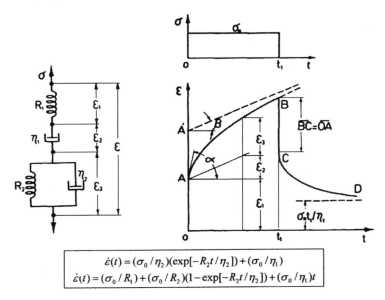

Figure 2. Burgers Solid (four-element) model, including characteristic differential equation and strain response for static loading/unloading.

high-temperature, Newtonian (diffusional) creep data (of polycrystalline material), the fit is usually dominated (not surprisingly) by the steady-state (long-time) portion of the data; what is learned is that the initial strain rate is invariably underestimated by this simple model. (This behavior will be demonstrated with some real data below.) Nevertheless, the lessons the Burgers Solid model teaches regarding attenuation remain valuable.

ATTENUATION AND THE LINEAR VISCOELASTIC MODELS

Dynamic mechanical response is often described by the "quality factor," Q, which is the ratio of the elastic energy stored to that dissipated. Q is unitless. Attenuation is Q^{-1}, and is further characterized by the nature of loading: the subscript E, G (or μ) or K is applied to discriminate Young's-modulus, shear-modulus or bulk-modulus loading modes, respectively. High-Q (low Q^{-1}) materials are said to "ring," that is, a resonant vibration sees little decay in amplitude for many, many cycles; alternatively, low-Q materials are "lossy," i.e., resonant vibrations decay rapidly. Under ambient conditions, aluminum alloys are high-Q—e.g., $Q \sim 5000$ ($Q^{-1} \sim 2\times10^{-4}$) for a 440 Hz tuning fork—while, by comparison, an identical fork made from wood is quite lossy, vibrating at $\sim$300 Hz and displaying a $Q \sim 33$ ($Q^{-1} \sim 0.03$) (e.g., Lakes 1999, Ch. 7). For very lossy materials ($Q^{-1} > 0.1$), the physical meaning of Q^{-1} is poorly defined. The most common definition is

$$Q^{-1} \equiv (1/2\pi)(\Delta\Phi/W_{MAX}), \tag{12}$$

where $\Delta\Phi$ is the elastic energy dissipated per cycle of harmonic loading and W_{MAX} is the maximum stored energy per cycle (e.g., Nowick and Berry 1972, p.22; cf. O'Connell and Budiansky 1978). If one considers application to a system of a cyclical stress, i.e., $\sigma = \sigma_0 \cos(\omega t)$, where (now) σ_0 is the stress amplitude and ω is the angular frequency (i.e., the frequency in Hz is $f = \omega/2\pi$), this definition of Q^{-1} is equivalent to the tangent of the phase angle, δ, by which the strain lags the applied stress (i.e., $Q^{-1} \equiv \tan\delta$ and, relative to the stress, the strain is given by $\varepsilon = \varepsilon_0 \cos(\omega t - \delta)$). (In subresonant attenuation measurements, it is δ that is measured directly (e.g., Jackson 1993).) Further, one sees that the viscoelastic modulus (and the viscoelastic compliance) of the system is a complex number, and so it can be demonstrated (e.g., Findley et al. 1976, p.93) that Q^{-1} equals the ratio of the imaginary component of the modulus (termed the "loss modulus") to the real component of the modulus (the "storage modulus").

With these definitions in mind, one can imagine interrogating the Maxwell, Voigt/Kelvin and Burgers Solid models with a cyclic stress function (instead of the static creep function shown in Figs. 1 and 2), and from the responses determine the attenuation spectrum (i.e., Q^{-1} vs. f, which is usually plotted as log Q^{-1} vs. log f) for each model. Given the forms of the differential equations describing the time-dependence of strain rate for the models, these can be Laplace-transformed to discern the frequency response. Table 5.1 in Findley et al. (1976) is a nice presentation of the calculations involved. In short, the Maxwell Solid model produces $Q^{-1} = R_1/(2\pi f\eta_1) = 1/(2\pi f\tau_R)$, which is a straight line with slope -1 on a log Q^{-1} vs. log f plot. Alternatively, the Voigt/Kelvin Solid model produces $Q^{-1} = (2\pi f\eta_2)/R_2 = 2\pi f\tau_C$, a straight line with slope +1 on a log Q^{-1} vs. log f plot. For the Burgers Solid model, the form of $Q^{-1}(f)$ is significantly more complex:

$$Q^{-1} = \frac{p_1 q_2 [2\pi f]^2 + \eta_1 \left(1 - p_2 [2\pi f]^2\right)}{p_1 \eta_2 [2\pi f] - q_2 [2\pi f]\left(1 - p_2 [2\pi f]^2\right)} \tag{13}$$

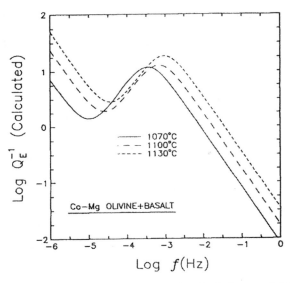

Figure 3. Dynamic response of the viscoelastic Burgers Solid model. The curves shown are for an analytical Laplace inversion of the Burgers model being first fit to experimental, flexural creep response of a Co-Mg olivine-basalt partial melt The characteristic attenuation response of the Burgers Solid includes a Debye peak related to the anelastic (Voigt/Kelvin) element. [Used by permission of American Geophysical Union, from Green et al. (1990) *Geophysical Research Letters*, Vol. 17, Fig. 4, p. 2099.]

where

$$p_1 \equiv (\eta_1/R_1) + (\eta_1/R_2) + (\eta_2/R_2), \tag{14}$$

$$p_2 \equiv (\eta_1\eta_2)/(R_1R_2), \text{ and} \tag{15}$$

$$q_2 \equiv (\eta_1\eta_2)/R_2. \tag{16}$$

The form of the Burgers Solid absorption spectrum is illustrated in Figure 3 (above), in which experimental flexural creep curves—for partially molten, nominally texturally equilibrated, Newtonian-viscous (Co-Mg) olivine-liquid "basalt" aggregates—were first curve-fit (for η_1, η_2 and R_2, with a textbook value used for R_1) to the model (Eqn. 9) and then transformed via Equations (13)-(16) (Green et al. 1990). (The creep experiments, in four-point flexure (Cooper 1990), examine the time-dependent Young's modulus behavior, thus the curves in Figure 3 are for Q_E^{-1} and R_1 equals the temperature-compensated, high-frequency (i.e., unrelaxed) Young's modulus.) The attenuation spectrum for the Burgers Solid shows Q_E^{-1} tending to infinity at low frequency, consistent with domination of energy dissipation by the Maxwell Solid dashpot (steady-state creep response with η_1) under these conditions; at high frequency, the dominant absorption behavior remains the Maxwell Solid dashpot (the spring cannot effect absorption, by definition), and the slope of -1 continues as Q_E^{-1} approaches zero. In an intermediate frequency, one sees that the anelastic Voigt/Kelvin component produces a discrete and pronounced attenuation peak (a Debye peak), with the deviation from the Maxwell Solid model spectrum occurring at $f \sim 1/2\pi\tau_C = 1/(2\pi[\eta_2/R_2])$. One sees that an increase in temperature causes the spectrum, in its entirety, to shift to higher frequency.

The primary temperature effect is on the values of η_1 and η_2, which, following Boltzmann statistics, each depend exponentially on temperature (i.e., $\eta_i \propto \exp(E_{a,i}/RT)$, where E_a is the activation energy for the physical process rate-limiting the viscosity, T is temperature and R is the gas constant). The shift of the anelastic peak to higher f as T increases is not necessarily the same shift as that for the straight-line Maxwell Solid-dominated component of the spectrum in that η_1 might not have the same E_a as does η_2.

Two points require emphasis here, justifying the time spent contemplating the attenuation spectrum produced by Burgers Solid model. First, experimentalists get very excited when a Debye peak, like those that punctuate the spectra shown in Figure 3, appears in their *measured* attenuation data. (Please note: the peaks shown in Fig. 3 were *not* produced in cyclic loading experiments; rather, the peaks shown came from fitting a model—the Burgers Solid model that, by definition, produces a Debye peak when it is transformed—to creep data.) This excitement occurs because the peak's presence suggests operation of an anelastic mechanism that has a unique τ_C, one that can perhaps be proven to correspond to a unique, physical causal mechanism in the material. Examples of Debye-peak behavior include (1) the Snoek relaxation, in which interstitial impurities diffuse from one interstitial site to an empty, adjacent one because an applied stress destroys the potential-energy degeneracy of the interstitial sites (e.g., Nowick and Berry 1972, Ch.9) and (2) the Zener (1941) grain boundary relaxation (seen infrequently) associated with uniform grain-sized aggregates having grain boundaries of nominally identical viscosity in a thermodynamic regime where mechanical energy dissipation is dominated by grain boundary sliding (e.g., Kê 1947, 1999); the review by Karato and Spetzler (1990) covers other mechanisms as well. The materials and mechanics literature on attenuation (or "damping," or "internal friction," if one is contemplating a literature search) has many, many contributions where, if a peak is discovered—even one that might be considered noise—the entire discussion section is dedicated to discerning that unique physical phenomenon giving rise to the peak, while the rest of the experimental spectrum is ignored, that is, simply extracted away as "background."

The second point justifying time on the Burgers Solid model is that the impact of temperature on absorption is better illustrated and thus contemplated. The shift in frequency—and, to first order (at least), not-at-all in height—of the Debye peak with temperature is clearly illustrated in Figure 3 and, thus, understood as the impact of Boltzmann statistics: i.e., as T is increased, the probability of a stress-driven motion of a specific lattice defect occurring within a fixed time increases exponentially. But consider, for example, those portions of the spectra in Figure 3 between $-2.5 \leq \log f \leq 0$. The power-law form of the data (i.e., $Q_E^{-1} \propto f^m$) in this region produces an array of straight lines (of slope m, which, in this example, happens to be $m = -1$) in the log-log plot. This straight-line form tempts one to think that it is Q_E^{-1} that is exponentially dependent with temperature at constant f. Indeed, in power-law attenuation data for which a Debye peak is not seen—which, as noted below, occurs frequently in high-T experimental materials studies and, too, characterizes, e.g., the absorption response of the upper mantle—this approach to temperature-effect extrapolation/interpretation is often taken, and referenced to Schoeck et al. (1964). For most situations, to do so is to ignore the physics of absorption: Nowick and Berry (1972, p.458) note explicitly that any "effective" activation energy discerned this way has no physical meaning. The frequency shifts with T for the Debye peaks remind us of this fact. Further, in Figure 3, one realizes that the frequency-shift phenomenology must remain true for those portions of the spectra located away from the peak: the Maxwell Solid dashpot represents, in this case, the diffusional (Newtonian) creep of polycrystalline olivine, which is effected by point-defect mechanisms of chemical diffusion.

To be complete, however, one can articulate two physical possibilities for a vertical shift of a log Q^{-1} vs. log f spectrum with temperature. The magnitude of specific absorption at a fixed frequency is dependent on the spatial density of lattice-defect-based absorbers. If, for entropic reasons, an increase in temperature causes an increase in density of these defects, one can envision an absorption increase as T increases. Most mechanisms of absorption are based on extrinsic defects, however; thus this possibility seems limited. (Indeed, in the case of the extrinsic-defect-effected Snoek relaxation noted above, an increase in T is found to *diminish* the height of the Debye peak: higher T promotes randomization of impurity interstitials, negating the stress-induced loss of degeneracy of interstitial site energy.) The other possibility is that an increase in temperature allows access to a new, and perhaps more potent, absorption mechanism, i.e., a change in the parallel-kinetic mechanism and/or the rate-limiting serial-kinetic process dominating the dynamic behavior. Behavior of this type—i.e., mechanism change with a change in thermodynamic and/or microstructural conditions—can perhaps be inferred from recent experimental attenuation data, as will be presented below.

THE ATTENUATION BAND/HIGH-TEMPERATURE BACKGROUND

Experimental attenuation data seen in most rock and mineral studies, as well as in many studies of engineering materials, do not match well with the simple viscoelastic models described above—which is the basis of Lakes' (1999) warning noted earlier. At high (homologous) temperature, what is usually seen is a power-law absorption, $Q^{-1} \propto f^m$, where m is in the range $-0.5 \le m \le 0$ over several orders of magnitude in frequency. This response has been referred to at the "attenuation band" or "absorption band" in the geophysics literature; in the materials literature it is known as the "high-temperature background" absorption. Most dramatic of the experimental studies of the high-temperature background in the materials literature is the work of Lakes and co-workers on low-melting-point metals, alloys and intermetallic compounds, where the power-law behavior is noted for up to ten orders of magnitude in frequency ($10^{-5} \le f[\text{Hz}] \le 10^5$) for, e.g., two-phase mixtures of $\beta\text{-In}_3\text{Sn} + \gamma\text{-InSn}_4$ at room temperature, which is ~75% of the β-γ eutectic temperature (Lakes and Quackenbush 1996). The mantle of Earth, too, shows this behavior, where shear-waves demonstrate $Q_G^{-1} \propto f^{-(0.15-0.4)}$ (i.e., $-0.40 \le m \le -0.15$; e.g., Anderson and Minster 1979; Anderson and Given 1982; Molodenskiy and Zharkov 1982). Figure 4, from Anderson and Given (1982), shows well the behavior for the mantle, giving both references for the window of frequencies of interest ($10^{-3} \le f[\text{Hz}] \le 1$) as well as absolute values of Q^{-1} (10^{-3} to 10^{-2}, although at shallow depths near accreting plate margins, Q_G^{-1} can reach ~0.1—e.g., Chan et al. 1989).

The broad absorption band requires a *distribution* of material compliances. From a perspective that emphasizes solely exponential relaxation processes, this distribution arises from (i) a variety of exponential-decay mechanisms, each with a unique τ_C, that sum to produce the response overall or (ii) a single exponential-decay mechanism in which a variation of a spatial component, e.g., grain size, allows for a variety of τ_C's or (iii) some combination—probably complex—of both. Mechanistically, one can apply the exponential-decay, linear viscoelastic models to experimental data so as to discern whether a discrete number of Voigt/Kelvin elements of differing τ_C might match well the data. Schematically, the idea is presented in Figure 5, which compares a "Multiple (Voigt/)Kelvin" approach to the Burger's Model.

Figure 6 presents some experimental creep data (Fig. 6a,c,e; the same data are in each figure) and experimental attenuation data (Fig. 6b,d,f; again, the same data are in each figure) for an enstatite glass-ceramic. This material is composed of 95 vol % of a very uniform, 0.5-μm grain-sized (ortho)enstatite (with some clinoenstatite intergrowths, though insufficient in volume to perceive in X-ray diffraction) with 5 vol % of a residual sodium-magnesium aluminosilicate glass, confined to grain triple junctions—a model

Seismic Wave Attenuation: Energy Dissipation in Crystalline Solids   263

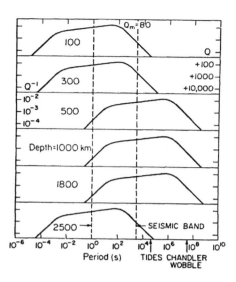

Figure 4. Absorption band model of Earth's mantle: Q (and Q^{-1}) vs. wave period ($\equiv f^{-1}$) and depth. [Used by permission of American Geophysical Union, from Anderson and Given (1982) *Journal of Geophysical Research*, Vol. 87, Fig. 3, p. 3896.]

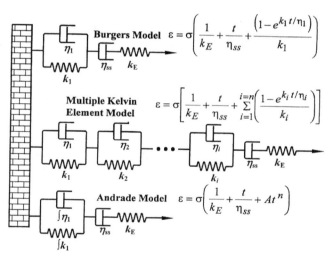

Figure 5. Alternative viscoelastic models—and their respective strain responses—to compare with the Burgers Solid model. The Multiple (Voigt/)Kelvin model employs discrete exponential-decay elements with unique τ_C's that superpose to estimate a power-law absorption response. The Andrade model assumes a continuous distribution of compliances, which can represent either many exponential-decay elements or one or more non-exponential-decay relaxation processes. [Used by permission of Plenum Press, from Gribb and Cooper (1995). In *Plastic Deformation of Ceramics*, edited by R.C. Bradt et al., Fig. 3, p. 91.]

silicate partial melt. (Details of the materials fabrication and its microstructure and thermal/mechanical stability are given in Cooper (1990)). The creep tests were performed in four-point flexure; the attenuation measurements were performed in reciprocating four-point flexure (Gribb and Cooper 1995).) The material, at the temperatures tested, had a

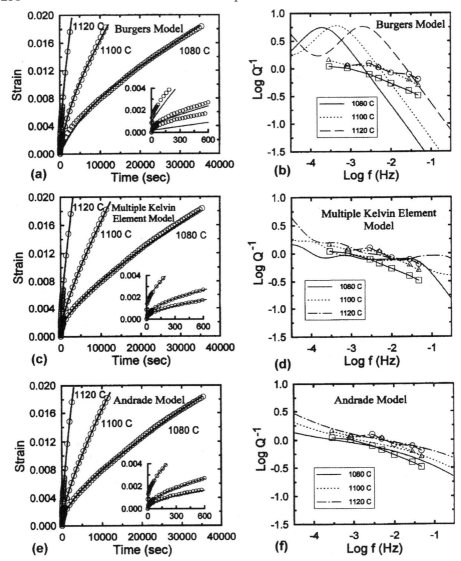

Figure 6. (a),(c),(e) Experimental, unidirectional, four-point flexural creep data (open circles) for an enstatite glass-ceramic (a model partial melt) versus the best fits of the three linear models, respectively, shown in Figure 5 (solid lines). For each graph, the short-time/small-strain response and model fit are shown in the inset. (b),(d),(f) Experimental Q_E^{-1} response for the enstatite glass-ceramic obtained from reciprocating four-point flexure (curves with symbols) versus Q_E^{-1} predictions calculated from numerical inversion of the linear models (curves without symbols). [Used by permission of Plenum Press, from Gribb and Cooper (1995). In *Plastic Deformation of Ceramics*, edited by R.C. Bradt et al., Fig. 5, p. 94.]

Newtonian steady-state viscosity; linearity was proven in both creep and attenuation modes. As shown in the constant-stress creep curves, there is a significant creep transient. This transient was experimentally demonstrated as fully recoverable; it is anelastic.

Figures 6a and 6c compare the fit of the Burgers Solid model and of a computer-arbitrated, Multiple Kelvin Solid model, respectively, to the creep data. ("Computer-arbitrated" means the algorithm selects the number of elements and the τ_C of each following the philosophy articulated by Thigpen et al. (1983); the programmer diminishes the size of discrete time steps available for the τ_C's until the result is unaffected by the size of these steps (Gribb 1992).) Clearly seen in Figure 6a is the Burgers Solid model underestimating the initial strain rate, as was noted earlier. Also evident is the far better fit to the data by the Multiple Kelvin Solid model. All of the Multiple Kelvin Solid fits to these data produced three distinct Voigt/Kelvin elements, which demonstrated appropriate frequency shifts with the changes in temperature studied. These model results are Laplace-transformed and compared to the experimental attenuation data in Figures 6b and 6d; the fits to the creep data are the data-free curves, while the experimental data (points) are connected with discrete line segments. The attenuation data demonstrates clearly the simple power-law spectrum. Equally clear is the inadequacy of the Burgers Solid model to predict the attenuation. (Gribb et al. (1994) measured, too, the Q_E^{-1} response of the Co-Mg olivine-basalt material, and the comparison of those data to the transformed creep curves shown in Figure 3 are qualitatively similar in their poorness of fit.) One sees that the Multiple Kelvin model inversion matches the data far better, but, consistent with its discrete nature, produces three Debye peaks that superpose to closely match the data. The data, in that they appear to vary monotonically with frequency, certainly do not loudly proclaim the adequacy of the "bumpy" Multiple Kelvin model fit, although their resolution could be limiting. Of course, if the fit was better, the problem remains regarding deciding what physical phenomena these discrete Voigt/Kelvin elements represent.

A third curve fit is shown in Figure 6. The Andrade Model is an empirical, phenomenological model, first used to describe the creep of polycrystalline metals:

$$\dot{\varepsilon} = \sigma[(1/E) + (At^n) + (t/\eta_{ss})], \tag{17}$$

where η_{ss} is the steady-state viscosity, A is a constant and the time exponent n is less than one. (Andrade (1910) explicitly articulated $n = 1/3$ and that η_{ss} could be an effective viscosity, i.e., the steady-state behavior was not limited to Newtonian flow.) In the framework of linear viscoelasticity, the Andrade Model can represent an infinite number of (exponential-decay) Voigt/Kelvin elements possessing a continuous distribution of relaxation times. The match of the Andrade model, both with $n = 1/2$ and with n used as a flexible parameter (the error difference between approaches was negligible), to the creep data is quite accurate, and its transformation matches the attenuation results beautifully; indeed the transformation of the Andrade Model produces a region in the attenuation spectrum where $Q^{-1} \propto f^m$ with $(1-n) \leq m \leq 0$.

Thus, the materials physics in this case demands reflection. As a creep function that possesses a continuous distribution of compliances, the Andrade Model need not represent an infinite number of linear viscoelastic elements with exponential decay behavior, for other decay functions (e.g., a "stretched" exponential; Lakes 1999, p.326) can be integrated to effect as well. For example, in the cases of the enstatite glass-ceramic (Gribb and Cooper 1995) and the Co-Mg olivine-basalt partial melt (Gribb et al. 1994), the Young's-modulus-mode relaxation could be modeled quite nicely, albeit qualitatively (A in Equation (17) being empirical and overly flexible, for example), as D'Arcy flow of the melt phase across the flexure specimen in response to the gradient in the hydrostatic component of the stress tensor. The melt-migration problem thus resembles chemical diffusion in a potential gradient that diminishes with response, which is a $t^{1/2}$ relaxation and not an exponential one. Convoluted in these data, then, are solid-state effects, some undoubtedly in shear (the Young's-modulus bending experiments convolute shear- and bulk-modulus responses (cf. Green and Cooper 1993)). How should these effects be dealt with? Further, can one *predict* the magnitude of the attenuation and the shape of the attenuation-band spectrum from a physically sound foundation?

There are lovely data in the literature for melt-free, polycrystalline aggregates tested in both shear creep and subresonant oscillatory shear. In the geophysics literature, the work out of Jackson's laboratory on olivine (e.g., Tan et al. 2001; Jackson et al. 1992, 2002) and on perovskite (Webb et al. 1999) is noteworthy for its attention to detail in specimen preparation and characterization and, particularly—because of the use of high confining pressure in their experiments—for the elimination of grain-boundary cracking, which the rock physics community considered affected negatively ambient-pressure experiments performed earlier (e.g., Berckhemer et al. 1982). In materials and mechanics, the literature is broad, quite extensive—particularly for metals—and of long standing (e.g., the many references in Zener (1948) and Nowick and Berry (1972)), but it is the recent work out of Lakes' lab on various metals/alloys that is provocative because of the large range of frequencies studied and other experimental innovation (e.g., Cook and Lakes 1995; Lakes and Quackenbush 1996; Dooris et al. 1999). The attenuation band behavior is seen in all cases, and, as already noted, for wide ranges of frequency in the case of both polyphase and single-phase polycrystalline metals. Inevitably, then, some of the data in Figure 6 must be attributed to solid-state processes in addition to melt-migration effects.

Perusal of many experimental and observational papers on absorption in crystalline materials, where the data display the attenuation band with no discernable Debye peak(s) superposed, inevitably reveals a qualitative-to-semiquantitative discussion concerning a variation/distribution in defect length scale and its consequent impact to create a distribution in τ_C in the material. Most frequently speculated are variations in grain size and/or variations in the pinning lengths of lattice dislocations (cf. Anderson and Given 1982; Lakes 1999, p.315). And yet, even as models for such broadband absorption become more complicated (e.g., D'Anna et al. (1997) for the case of variation in dislocation pinning length), they so far fail to produce an accurate prediction of the magnitude of Q^{-1} for a given set of thermodynamic and microstructural constraints, nor of the value of *m* characterizing the attenuation band.

ISOLATION/CHARACTERIZATION OF A SINGLE PHYSICAL MECHANISM PRODUCING A POWER-LAW ATTENUATION SPECTRUM: THE INTRINSIC TRANSIENT IN DIFFUSIONAL CREEP

Gribb and Cooper (1998) set out to characterize the shear attenuation behavior in polycrystalline olivine aggregates that (i) had such a uniform grain size that, to first order, one could analyze as having no grain-size variation, (ii) were prepared with such a fine grain size that no thermal-expansion-anisotropy grain-boundary cracking would occur in the process of heating or cooling the specimens (the experiments were to be performed at ambient pressure) and (iii) were fabricated and mechanically tested under differential-stress conditions that would disallow the nucleation of lattice dislocations. The idea was simple: we desired to get rid of microstructural scale distributions and see what happens to the shear absorption spectrum. The material was prepared by pulverizing Balsam Gap (Jackson Co., NC) dunite to sub-micrometer dust, classify the particle size and "gently" (low-pressure), uniaxially hot-press the material back into a polycrystalline solid. Given the three material qualities noted above, the stress/temperature conditions for hot-pressing were the critical parameter: for a 3-μm grain size aggregate, for example, conservative theory indicated that the hot-press pressure could not exceed 24 MPa at 1200°C if nucleation of lattice dislocations was to be avoided. Further, use of the Balsam

Gap material had some advantages: very-fine (≤ 1 μm in the fabricated materials) second phases of chromite and enstatite (< 1 vol % each) plus impurity Ca^{2+}, which segregates to grain boundaries, succeeded in shutting down grain growth of the dense, 3-μm grain-size aggregates during the mechanical test experiments. Test conditions included a temperature range of 1200–1285°C, maximum torsional shear stresses of 10–120 kPa (with mechanical linearity demonstrated for this range), and frequencies for the reciprocating experiments of $10^{-0.5}$–$10^{-2.5}$ Hz.

The creep curves demonstrated a distinct transient, similar to those seen, e.g., in Figure 6a; these were demonstrated to be fully anelastic. The creep behavior was demonstrated as Newtonian, that is, diffusional deformation dominated the steady-state flow—a result consistent with other studies of olivine under similar thermodynamic conditions (e.g., Karato et al. 1986; Hirth and Kohlstedt 1995). Further, the thermal activation energy for the transient matched that for the steady-state response, indicating their cause by the same physical mechanism. The attenuation data are presented in Figure 7, along with the inversions of the Andrade Model fit to the creep data. Once again, the attenuation-band behavior is discovered ($Q_G^{-1} \propto f^{-0.35}$) and its useful characterization by the inversion of the Andrade Model reconfirmed. Further, the absolute values measured for the of the attenuation, and the behavior in general, are found to be consistent with studies on similar synthetic olivine aggregates measured in Jackson's laboratory at high confining pressure (e.g., Tan et al. 1997, 2001), meaning that, as expected from theory, the fine-grained aggregates have not been affected by grain-boundary cracking. Now, however, one cannot cite a microstructural scale distribution to be the source of a distribution of compliances: the Andrade behavior and the related attenuation band are intrinsic to the diffusional creep mechanism operative in the experiments.

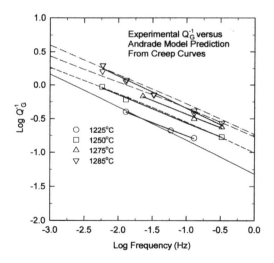

Figure 7. Q_G^{-1} of fine-grained ($d \approx 3$ μm) olivine (Fo_{88}) measured in reciprocating torsion (points with heavy curves) versus that predicted from the Laplace transform of the Andrade-model, best fit (with $n = 1/2$) to the unidirectional torsional creep response (fine curves). [Used by permission of American Geophysical Union, from Gribb and Cooper (1998) *Journal of Geophysical Research*, Vol. 103, Fig. 7, p. 27272.]

268   Cooper

Following the theoretical treatments of Lifshits and Shikin (1965) and Raj and Ashby (1971), and then extended by Raj (1975), one discovers that the apparent continuous distribution of compliances is created in the chemical diffusion response that seeks to evolve the stress distribution (normal traction) on grain boundaries from an initial state where singularities exist at grain triple junctions to the steady state that requires the traction be highest in the center of boundaries and near zero at triple junctions. The process is well characterized as chemical diffusion within a diminishing potential; its analysis allows the articulation of a universal creep response and, through its inversion, a universal attenuation spectrum, applicable to all crystalline materials having a Newtonian (diffusional) rheology.

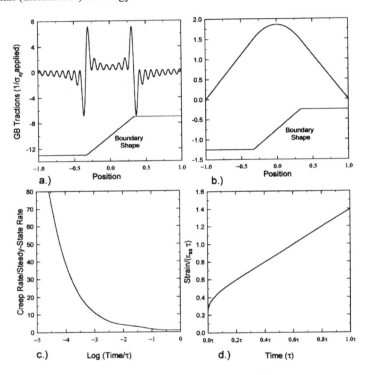

Figure 8. Calculation of a universal creep curve for all polycrystalline materials deforming by grain-boundary-diffusion creep under conditions in which the grain boundaries have no resistance to shear (after Raj 1975). (a) Initial distribution of normal traction on grain boundaries for a 2-D array of hexagonal grains: singularity-like high tractions occur at triple junctions. (The high-frequency oscillations are numerical artifacts due to the Fourier approximation to the stress state.) (b) Traction distribution consistent with a steady-state diffusive flux along grain boundaries (and, thus, a steady-state creep rate). (c) Integrated creep rate (normalized by the steady-state strain rate) from a numerical solution of the diffusion equations (Fick's laws). The characteristic time of the transient, τ, is defined based on the diffusion behavior (Eqn. 18); one sees that 99% of the transient is played out by ~0.25τ. (d) Universal creep curve obtained by integration of the curve shown in (c). One sees that, over the period τ, an additional 60% of inelastic strain is generated over that predicted by the steady-state response. The amount of elastic strain (strain at time zero) is constrained by the model to be ~$(\dot{\gamma}_{ss}\tau)/4$. [Used by permission of American Geophysical Union, from Gribb and Cooper (1998) *Journal of Geophysical Research*, Vol. 103, Fig. 8, p. 27273.]

The framework for the model is presented in Figure 8. Diffusional creep is driven by the spatial gradient of normal tractions on grain boundaries and is rate-limited either by diffusion through the crystalline lattice (Nabarro-Herring creep: Nabarro 1948; Herring 1950) or along the grain boundaries (Coble creep: Coble 1963). Material continuity, however, requires that these diffusional mechanisms act in kinetic series with some process of grain boundary sliding (Raj and Ashby 1971; cf. Courtney 1990, p.276). Because of the serial nature of the process overall, there are thermodynamic conditions where grain boundary sliding component essentially dissipates no energy, i.e., conditions where the grain boundaries have no resistance to shearing. It is this situation of boundaries inviscid to shear that is described by Raj (1975), and replicated and extended in Figure 8—for the geometrical case of a 2-D array of hexagonal grains—to produce a "universal" creep curve for diffusional creep. Figure 8a shows the initial stress distribution for the material with inviscid boundaries: traction singularities exist at all triple junctions. The traction distribution consistent with a steady-state diffusion flux, on the other hand, consists of tractions being maximized at the center of grain faces and diminish to zero at the triple junctions, a parabolic distribution in 2-D (Fig. 8b) or paraboloidal in 3-D. Evolution of the traction distribution from that shown in Figure 8a to that in Figure 8b constitutes the intrinsic transient in diffusional creep. One should note that this transient is anelastic: for example, removal of the far-field applied stress after the material has achieved steady state sees an internal stress predicated on the traction distribution in Fig 8b that will be relieved by diffusion, giving strain recovery. One can solve Fick's 2nd law for the evolution of the tractions from the initial to the steady-state conditions to calculate the transient (Fig. 8c) and then integrate this function to discern a universal creep curve (Fig. 8d). Finally, this creep curve can be transformed to an attenuation spectrum.

This "universal" attenuation spectrum is presented in Figure 9. The single mechanism produces an attenuation-band response, though three regions of isothermal behavior are noted: (i) at low frequency, Q^{-1} is dominated by the steady-state creep response, producing $Q^{-1} \propto f^{-1}$ (cf. the Maxwell Solid model); (ii) at high frequencies, the elastic contribution to the total strain becomes dominant and $Q^{-1} \propto f^{-0.5}$; and (iii) at intermediate frequencies the anelastic transient dominates behavior, with $Q^{-1} \propto f^{-(0.25-0.4)}$. One sees that the creep and attenuation response is scaled by a characteristic relaxation time τ, which, for the case of a grain-boundary-diffusion-effected Newtonian rheology, is given by (for the simplified, 2-D hexagonal grain model)

$$\tau = \frac{3\sqrt{3}\left(1-v^2\right)d^3 k T}{2\pi^3 E D_b \xi \Omega} , \qquad (18)$$

where v is Poisson's ratio, d is the grain size, E is Young's modulus, D_b is the grain boundary diffusion coefficient for the rate-limiting ionic species, ξ is the characteristic grain boundary width, Ω is the molecular volume, and kT has the usual meaning. Please note: this τ is not to be confused with the standard period ascribed to an exponential decay process in a linearly viscoelastic solid; rather, τ is a non-exponential relaxation associated with chemical diffusion within a diminishing potential gradient: with application of stress, e.g., in a creep test, the strain rate decays to within 1% of the steady-state value by $\sim 0.25\tau$. One sees a power-law grain-size dependence of τ: the value 3 for this exponent follows the grain-boundary diffusion model. The steady state strain rate calculated for this same, 2-D model is (Raj and Ashby 1971; cf. Coble 1963):

$$\dot{\gamma}_{ss} \cong \frac{132 \sigma_{xy} \Omega \xi D_b}{k T d^3} = \frac{\sigma_{xy}}{\eta_{ss}} , \qquad (19)$$

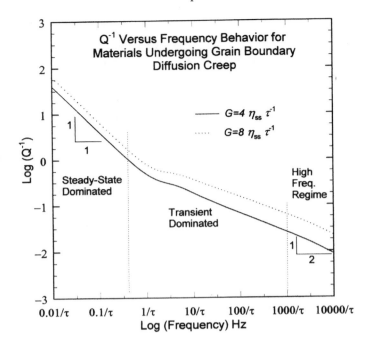

Figure 9. Universal Q^{-1} versus frequency curve for all materials deforming via grain boundary diffusional creep, predicted by numerical Laplace transform of the creep curve shown in Figure 8d. The Q^{-1} behavior displays three distinct regions of behavior: (i) at low frequencies, steady-state dominates and $Q^{-1} \propto f^{-1}$; (ii) at intermediate frequencies, the transient interacts with the steady-state response and $Q^{-1} \propto f^{-(0.25-0.4)}$; (iii) at high frequencies, the elastic behavior interacts with the transient response and $Q^{-1} \propto f^{-0.5}$. The simplifications inherent in the formulation (e.g., geometry) may result in an error on the order of a factor of two. The effect of such an error is illustrated by the dotted line that represents the attenuation behavior of the model with a reduced modulus or, equivalently, an increased steady-state creep rate. [Used by permission of American Geophysical Union, from Gribb and Cooper (1998) *Journal of Geophysical Research*, Vol. 103, Fig. 9, p. 27274.]

where $\dot{\gamma}_{ss}$ is the steady state shear strain rate and σ_{xy} is the applied, far-field shear stress. By combining Equations (18) and (19), and assuming that $\nu = {}^1/_3$, τ is defined in terms of the steady state response:

$$\tau \cong \frac{10\sigma_{xy}}{E\dot{\gamma}_{ss}} = \frac{10\eta_{ss}}{E} \cong \frac{4\eta_{ss}}{G}. \tag{20}$$

Simply stated, for an aggregate that deforms by diffusional creep, knowing the steady-state viscosity and the temperature-affected (but unrelaxed) modulus, one can determine a priori the attenuation spectrum—its magnitude and its shape—at least that portion associated with the high-T background under thermodynamic conditions where the shear-inviscid grain boundary assumption holds. Figure 10 shows just this approach: the continuous curves were calculated based on Equation (20) defining τ, with η_{ss} values coming from the creep experiments and E coming from Simmons and Wang (1971); the correlation is remarkable.

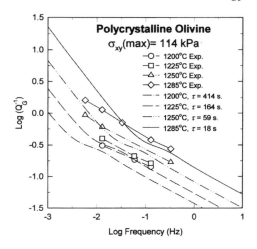

Figure 10. Experimentally measured values of Q_G^{-1} for polycrystalline olivine versus the corresponding predictions of the intrinsic creep transient model. The value of τ for each temperature was calculated via Equation (18) with η_{ss} obtained from the unidirectional creep response of the material and E obtained from Simmons and Wang (1971). The experimental data do not show the transition to the $Q_G^{-1} \propto f^{-1}$ behavior predicted by the model at higher temperatures and lower frequencies; however, the effect of errors inherent in the simplifications of the model are sufficient to explain this inconsistency (cf. Fig. 9). [Used by permission of American Geophysical Union, from Gribb and Cooper (1998) *Journal of Geophysical Research*, Vol. 103, Fig. 10, p. 27275.]

DATA EXTRAPOLATION AND APPLICABILITY OF EXPERIMENTS TO GEOPHYSICAL CONDITIONS

The interpretation of the intrinsic diffusion-creep transient in terms of a grain-boundary-diffusion-limited mechanism was based on a significant database suggestive of such behavior (e.g., Karato et al. 1986; Hirth and Kohlstedt 1995). Nevertheless, the predicted, substantial sensitivity of the dynamics to the grain size in this case (i.e., $\eta_{ss} \propto d^3$ and $\tau \propto d^3$) prompts a direct experimental analysis of the effect, which is required if the experimental data are to be evaluated relative to the geophysical situation of interest (e.g., upper mantle grain sizes, where d is estimated in the range 1 mm to 1 cm (cf. Karato and Spetzler 1990)). The impact of grain size on low-frequency absorption in polycrystalline olivine has recently been measured both in Jackson's laboratory (d in the range 3–24 μm, T in the range 1000–1200°C; Jackson et al. (2002)) as well as in my laboratory (d in the range 4–17 μm, T in the range 1175–1300°C; Bunton (2001); Bunton and Cooper, in preparation). 1200°C experimental data from both studies—the condition where the data sets directly overlap—are shown in Figure 11. Despite the materials being fairly different (Jackson et al. studied aggregates prepared by hot-pressing (i) pulverized San Carlos, Arizona, peridot as well as (ii) powder prepared entirely by sol-gel (solution) processing, while Bunton examined aggregated prepared by vacuum sintering of pulverized Balsam Gap dunite), the agreement of the data, particularly at $\log f < -1.2$, is very strong. The Q_G^{-1} measurements from both laboratories demonstrate that the power-law exponent for the spectra increases (i.e., becomes less negative) as frequency and grain size are increased (and also as temperature is decreased, though confirmation of this point is not available in Fig. 11); for Bunton's data, that slope approaches zero for some conditions. Bunton's creep measurements characterize unequivocally that $\eta_{ss} \propto d^2$, which is a surprise, given the database (already cited) for diffusional creep of olivine. The conventional way to interpret the $\eta_{ss} \propto d^2$ result is to conclude that the steady-state response is rate-limited by lattice diffusion. The activation energy measured for both the creep and attenuation in these experiments, however, is $E_a \approx 650$ kJ mol^{-1}, a value far too high compared to the thermal sensitivities directly measured for lattice diffusion of all component ion species in olivine (e.g., Brady 1995). The "mystery" here most likely involves a threshold phenomenon for diffusional creep that is effected by the distribution of point defects in and near the grain boundaries (Jamnik and Raj 1996); this effect can result in a $\eta_{ss} \propto d^2$ behavior for grain boundary diffusion creep for a very low stress potential, as employed in these experiments. This argument, though, is beyond the scope of this chapter. Extrapolation of the experimental data with grain size, however, requires use of a $\tau \propto d^2$ relationship, and Equations (18)–(20) can be modified to reflect the behavior.

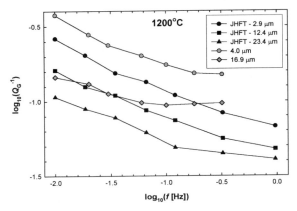

Figure 11. Impact of grain size on the measured attenuation response of polycrystalline olivine: 1200°C data from Jackson et al. (2002) (labeled "JHFT; black solid symbols; high-confining-pressure experiments on aggregates prepared from San Carlos peridot and from chemical sols) and from Bunton (2001) (gray symbols; ambient-pressure experiments on aggregates prepared from Balsam Gap dunite). The grain sizes are as noted in the key. The curves indicate an increase in the value of the power-law exponent m (i.e., it becomes less negative) with an increase in frequency.

Figure 12 presents all of Bunton's attenuation data (white circles), extrapolated in frequency based on both temperature (Boltzmann statistics, see Section 4) and grain size (i.e., overall, $\tau \propto d^2 \exp(-E_a/RT)$ with $f \propto \tau^{-1}$, cf. Fig. 9) to conditions of 1200°C and d = 1 mm; one sees clearly that a master curve results for the data, proving the efficacy of the physically based, frequency extrapolation approach outlined above. To this plot has been added (i) the 1200°C attenuation data of Jackson et al. (2002), (ii) the 1200°C attenuation data of Gribb and Cooper (1998), (iii) 1200°C attenuation data for a forsterite single crystal (minimum dimension ~1 mm) that was first plastically deformed to steady state in dislocation creep at T = 1600°C and $(\sigma_1-\sigma_3)$ = 20 MPa and then low-stress attenuation-tested in shear (Gueguen et al. 1989) and (iv) the Andrade model inversion—consistent with the universal attenuation curve—for $\tau \propto d^2$ behavior. The frequency shift of both the Jackson et al. (2002) and Gribb and Cooper data is based on $\tau \propto d^2$ (direct analysis of the Jackson et al. (2002) 1200° data gives $\tau \propto d^{(1.2-2.0)}$, depending on the Q_G^{-1} at which the measurement is made). An oblique view of the graph better reveals the curvature in the model curve. The correlation of the polycrystalline data from different investigations is again revealed as strong. Further, one sees that the extrapolation of the inviscid-boundary model to seismic/teleseismic frequencies for this mantle-like grain size gives Q_G far, far in excess of the values 100–500 cited in many mantle models (e.g., Anderson and Given 1982; Romanowicz and Durek 2000). Besides, it is clear that the polycrystalline data themselves deviate from the model at (the transformed) $f > 10^{-6}$ for this grain size and temperature. It is logical to conclude, then, that a different physical mechanism begins to affect the absorption response with the increase in frequency/drop in temperature/increase in grain size conditions associated with the high-frequency "tail" on the spectrum. The "gentleness" of the deviation from the model suggests that it is the shear-inviscid grain boundary assumption in the model that is failing, that is, with the change in f, T, d, the serial-kinetic process of grain boundary sliding begins to dissipate, in part, the mechanical energy. This point is the centerpiece of the Jackson et al. (2002) argument concerning their $Q_G^{-1}(d)$ data and its extrapolation in grain size.

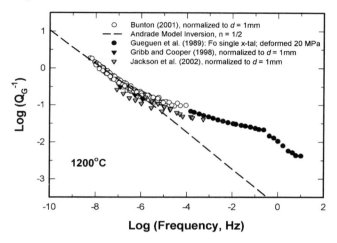

Figure 12. Comparison of a 1200°C, polycrystalline olivine absorption master curve (created from 1200°C data from Jackson et al. 2002, Gribb and Cooper 1998, and all data from Bunton 2001; see text)—for an extrapolation to $d = 1$ mm—to the 1200°C torsional attenuation response of a forsterite single crystal that was first crept to steady state at 1600°C, 20 MPa (Gueguen et al. 1989). The tail in the master curve, associated with the absorption response(s) of viscid grain boundaries, maps into the response of the deformed single crystal. The correlation suggests low-angle (subgrain) boundaries play a significant—perhaps primary—role in the absorption of the deformed single crystal.

Because it acts in kinetic series with both diffusional and dislocation flow mechanisms—meaning that, under most thermodynamic conditions, it cannot be physically isolated—grain boundary sliding is difficult to characterize. At one extreme, the oft-cited model for grain boundary sliding-effected absorption treats grain boundaries as thin films of a finite width and a constant and finite Newtonian viscosity (Zener 1941). Loading produces sliding displacements on the boundaries, at a rate dependent on their unique viscosity, and elastic loading of triple junctions (hence the description of the physics as "elastically accommodated grain-boundary sliding," e.g., Raj and Ashby 1971). From a mechanics-model perspective, one can imagine a Voigt/Kelvin Solid model placed in mechanical series with a spring (i.e., no plastic response is possible, only anelastic and elastic responses); the resultant model is one form of the "Standard Linear Solid," and its dynamic response is an exponential-decay-based Debye peak. τ_C in this case is given by (Nowick and Berry 1972, p. 437):

$$\tau_C = \eta_b d / \xi G, \tag{21}$$

where η_b is the boundary shear viscosity and G is the unrelaxed material shear modulus. The grain-size scaling for the peak location goes as $\tau_C \propto d$, in that strain is characterized as the relative displacement across the boundary normalized by the grain size. An important ramification of the model is that the magnitude of the absorption is independent of grain size: a change in d solely causes a shift in frequency of the Debye peak with no change in its height.

A nice example of the phenomenon is the absorption response of hot-pressed, polycrystalline Si_3N_4 ceramics (e.g., Mosher et al. 1976; Pezzotti et al. 1996). Si_3N_4 is one of the few materials where it is apparently all-but-impossible to form crystalline grain boundaries: in these materials, all grain boundaries have been replaced by ~1-nm thick films of amorphous silica or silicate (e.g., Clarke 1987). The chemistry (and thus the molecular structure and, with it, the viscosity) of these films is controlled by oxide additives. In general, the Newtonian viscosity of these thin films is $>10^3-10^4$ times smaller than that expected of diffusional processes in the Si_3N_4 proper. It is this extreme viscosity difference that allows experimental "access" to the boundary-sliding relaxation: to first order, the conditions envisioned by Zener (1941) (i.e., pure elastic accommodation of boundary sliding and little other time-dependent response) are met, quite clean Debye peaks are characterized and, from them, η_b and its temperature sensitivity are determined.

The Debye-peak grain-boundary absorption response matching the Zener (1941) model has been seen only rarely in metals. Kê's (1947) work on high-purity aluminum is the classical example: the study revealed the single Debye peak whose magnitude correctly matched theoretical prediction and as well as the property of being independent of grain size; in addition, the $\tau_C \propto d$ was confirmed. Other careful studies comparing polycrystalline metals with single crystals (reviewed in Nowick and Berry (1972) and Lakes (1999)) periodically reveal a broad absorption peak of low magnitude in the polycrystalline material, one barely "competing" to be seen over the power-law background absorption. The interpretation of such a response is problematic. The simplest (albeit not unique) defensible interpretation is that the constant-boundary-viscosity assumption of the Zener (1941) model is, in general, inappropriate: actual grain boundary sliding (that is, in materials with crystalline grain boundaries, in contrast to the Si_3N_4 example cited earlier) involves chemical diffusion and the glide+climb motion of primary and secondary grain boundary dislocations (cf. Ashby 1972). In that grain boundary structure is a strong function of, for example, the relative orientation of grains, the loss of degeneracy of η_b is easily contemplated.

Of course, not even a broad grain-boundary-sliding peak is evident in the polycrystalline olivine data presented in Figures 11 and 12. This is perhaps not surprising: when the selection of thermodynamic and microstructural conditions is such that two (or more) serial-kinetic processes are vying for rate control, the impact on the measured kinetics is often subtle. Crossman and Ashby (1975) have modeled the plasticity behavior of polycrystals, specifically to gage the impact of the grain boundary response on the kinetics. The impact on bulk deformation kinetics of changes in thermodynamic conditions that cause the boundaries to evolve from inviscid to (uniformly) Newtonian-viscous to rigid is demonstrated to be very minor: at constant T, the effect is a ~10% increase in the stress providing a given strain rate, with the stress transition being played out over less than a half-order of magnitude in strain rate. Not minor, however, is the spatial distribution of strain (and thus of energy dissipation) within individual grains corresponding to the transition: for an overall Newtonian-viscous aggregate, the strain associated with viscous grain boundary sliding evolves from 15% of the total to 0%, but played out over two-to-three orders of magnitude in strain rate. Because of this serial-kinetic coupling of the diffusional based transient flow with the viscous response(s) of grain boundaries, simple application of superposition, e.g., as is done with parallel kinetic processes, is not physically sound.

What can be said, then, of extrapolation of experimental data on grain-boundary effects in attenuation to the geological setting? For polycrystalline olivine, Jackson et al. (2002) have noted the complexities of physical behavior associated with the inviscid-to-viscous transformation of the grain boundaries, including the ineffectiveness of a straightforward application of superposition of the effects represented here by Equations (18) and (21), and so chose to apply the essentially empirical Schoeck et al. (1964) equation to their data, modified to include the apparent effect of grain size, i.e.,

$$Q^{-1} \propto [f^{-1}d^{-1}\exp(-E_a/RT)]^{\alpha}. \tag{22}$$

Their best-fit lines to those log Q^{-1} vs. log f data at higher-frequency and lower temperature give a value for α of 0.27 and a value of E_a ~400 kJ mol^{-1}. Jackson et al. (2002) continue by showing that a "vertical" translation (i.e., a shift in Q^{-1} with changes in grain size and/or temperature) of their data to grain sizes thought appropriate for the upper mantle produce $Q^{-1}(f)$ results in rational agreement with values measured seismologically. Jackson et al. (2002) conclude, accordingly, that the "same grain-size-sensitive processes [active in the experiments] might be responsible for much of the observed seismic-wave attenuation." Barring a better understanding of the physics of absorption of extended defects, like grain and phase boundaries, both Jackson et al.'s approach to data analysis and their conclusions must be considered reasonable.

Discerning the physics of absorption processes of extended defects, and thus how absorption is affected by microstructure, is the present "cutting edge" of experimental research in attenuation/internal friction. Of interest in both geophysics and materials science is the specific impact of deformation-induced microstructure on the absorption behavior. Figure 12, thus, presents the provocative comparison of the polycrystalline olivine data with that for the attenuation response of a deformed forsterite single-crystal (Gueguen et al. 1989). In these experiments, the single crystal was first deformed to steady state at 1600°C and 20 MPa differential stress and, subsequently, a slice with minimum dimension 1 mm was tested in reciprocating shear at 1200°C. One sees clearly that the tail of the polycrystalline master curve "maps" into the single-crystal data. Further, there exists no high-angle grain boundaries to effect absorption (crystals not "pre-deformed" demonstrated a high-temperature experimental $Q_G^{-1} \sim 10^{-2}$ that was essentially frequency independent). The result suggests strongly that low-angle (subgrain) boundaries in the single crystal, a result of the 20-MPa plastic deformation, are of primary importance in the overall attenuation response, that is, if the scale of these low-angle boundaries matches reasonably the 17–24 µm associated with the "tail" of the polycrystalline data. Such is the case: in a high-T, steady-state dislocation rheology, the single crystal must develop a polygonized microstructure, that is, there is developed a three-dimensional network of low-angle subgrain boundaries whose mean spacing (and spacing distribution) is set by the level of differential stress (and little-effected by T) (cf. Glover and Sellars 1973; Twiss 1977; Kohlstedt and Weathers 1980; Stone 1991). For forsterite at 1200°C, a 20 MPa differential stress creates a mean subgrain boundary spacing of ~20 µm.

The physical implication, then, is that a small-amplitude, oscillatory variation in stress, one that is added to a nominally constant, and significantly greater differential stress that is effecting dislocation creep in a material, is "sampling" the deformation-induced microstructure, and particularly the network of subgrains. Such a perspective on the high-temperature background absorption behavior and its interrelation with the subgrain network is consistent fully with models attempting to formulate a microstructural foundation for the phenomenological mechanical (plasticity) equation of state. The equation-of-state approach (theory: e.g., Hart (1970); experiments in ionic solids: e.g., Lerner et al. (1979); Covey-Crump (1994)) suggests that a *single*, internal state-variable can be used to describe the strain-effected "hardness" (i.e., the microstructural resistance to flow) of a material. The model of Stone (1991) correlates this state variable to the statistical distribution of low-angle boundaries and how this distribution is affected by changes in stress and by accumulated strain: the steady-state distribution is primarily dependent on the applied stress, while the evolution of the distribution—effected by diffusion-based processes within the low-angle boundaries—is strain sensitive. Recent experiments on the flow of halite (Plookphol 2001), comparing/correlating the deformation kinetics and deformation-induced microstructures from both creep experiments and load-relaxation experiments, confirm many of Stone's hypotheses. While this research correlating deformation-induced microstructure and "hardness" is young, one ramification is clear: with experimental care, the relative absorption effects of high-angle and low-angle boundaries (as well as of solid-state phase boundaries) can be deconvolved, physical models for these phenomena formulated and tested, and the results extrapolated. The possibilities of a more exacting and robust way of interpreting seismic Q^{-1} data, i.e., its use as a prospecting tool to understand structures and stresses (and perhaps strains) in active tectonic terranes, seems within the realm of reason.

SUBGRAIN ABSORPTION AND THE ATTENUATION BAND

As evidence of the possibilities, consider the room-temperature attenuation spectra (tan $\delta \equiv Q_G^{-1}$), presented in Figure 13a, for various phases in the solder system indium-tin. These data are from Lakes' laboratory, collected on his "broadband viscoelastic spectrometer" (Brodt et al. 1995); the breadth of frequencies explored is extraordinary. The data for γ-InSn$_4$ (solid squares) and β-γ eutectic (solid triangles) data were published earlier (Lakes and Quackenbush 1996); the data for single-phase β-In$_3$Sn ("×" symbol; between 10^{-1} and 10^3 Hz, these data are so dense that they appear as a solid line) represent a recently inaugurated collaboration with Lakes (McMillan et al., submitted). β-In$_3$Sn is not an intermetallic compound but rather is an endothermic solid solution having the same body-centered tetragonal (BCT) crystal structure as pure In; as a consequence, β-In$_3$Sn has a limited number of dislocation slip systems and the Peierls barriers to slip are high. Similarly, γ-InSn$_4$ is also an endothermic solid solution with a hexagonal crystal structure, again having limited slip systems and a high Peierls barrier. β-γ eutectic is, thus, a two-phase (with different crystal structures) solid; in this case, γ grains are dispersed within a continuous, polycrystalline β phase. At frequencies in the range 10^{-3} to 1 Hz, one sees that the eutectic is significantly more absorbing than is either component phase individually: clearly, one cannot apply a simple rule-of-mixtures to model the data for these frequencies. Solid-state β–γ phase boundaries are perhaps so demonstrated to be more potent mechanical absorbers than β grain boundaries or subgrain boundaries in this frequency range; affirmation of this postulate requires a better understanding of absorption in the individual phases that constitute the eutectic.

To better probe the physics of the broadband power-law absorption behavior, we have initiated a hierarchal study with single-phase β, using deformation and thermal processing to vary the dislocation (sub)microstructure. The baseline behavior—the as-cast condition (labeled "tan δ; cast β In-Sn")—is that presented in Figure 13a. The data display an attenuation plateau, $m \sim 0$, in the region 10^{-1}–10 Hz. Beyond 10 Hz, the data trend back towards the $m \sim 0.3$ slope characteristic of the frequencies below 10^{-1} Hz; as such, a description of the data above 10^{-1} Hz as a "broad hump" is perhaps appropriate. (Note, too, absorption behavior of single-phase γ has a similar plateau for frequencies ≥10 Hz. The "tail" visible in the polycrystalline olivine data presented in Fig. 12 is also consistent with this spectral feature.)

Figure 13b presents results of a parametric study in which the torsional absorption spectra (emphasizing the frequency range 10^{-1}–10^3 Hz) of β specimens were measured as functions of various specimen preparations. These preparations were serial and cumulative, meaning that specimens were mechanically tested after each step of process-

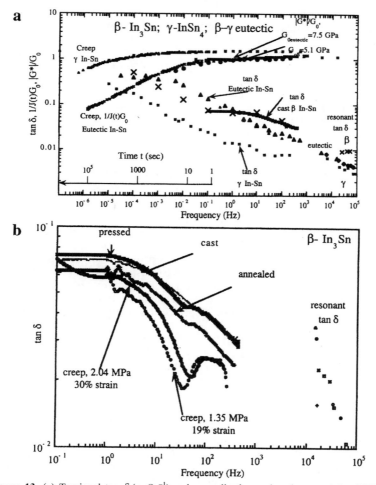

Figure 13. (a) Torsional tan δ ($\equiv Q_G^{-1}$) and normalized complex shear modulus ($|G^*|$) as functions of frequency, and creep compliance (J(t)) as a function of time, for β-In$_3$Sn (×), γ-InSn$_4$ (■) and the β–γ eutectic (solid triangles), all in the as-cast condition and at room temperature. (The tanδ data for β-In$_3$Sn in the frequency range 10^{-1}–10^3 Hz is so dense as to appear as a solid line.) One sees that the power-law absorption behavior holds for both single-phase β and γ materials at low frequency ($m \approx 0.3$); both β and γ deviate from this behavior at increasing frequency (~10^{-1} Hz for β; ~1 Hz for γ). (b) Effect of processing conditions on the tanδ of single-phase β-In$_3$Sn specimens. The processing followed a hierarchy of as-cast ("cast" in the figure); cast + pressed (high-rate, high-strain deformation; "pressed"); cast + pressed + annealed ("annealed"); cast + pressed + annealed + creep ("creep," with the stress and strain noted). One sees a significant effect of the creep treatment: despite an increase in lattice dislocation density due to the tensile creep deformation, tan δ actually decreases. In addition, an absorption peak is created at ~10^2 Hz. Both results suggest an important—perhaps dominant—role of low-angle boundaries in effecting the power-law absorption.

ing. Beyond the as-cast condition (spectrum marked "cast" in Fig. 13b, which is a replication of a portion of the spectrum shown in Fig. 13a), specimens were (i) plastically deformed at a high rate to greater than 60% compressive strain at 293 K (producing the absorption spectrum marked "pressed"), followed by (ii) annealing at 390 K (~95% of the melting temperature) for several hours (producing the spectrum marked "annealed"), followed by (iii) high-strain, plastic deformation in tension by dislocation creep (cf. Frost and Ashby 1982). Two different creep stresses were studied, 1.35 and 2.04 MPa. The absorption spectra are marked "creep;" the respective deformation stress and strain are noted as well. Optical microscopy of the specimens subsequent to various treatments revealed little effect of processing on grain shape or size: β grains remained nominally equiaxed with mean grain sizes ranging from ~500 μm for the "pressed" specimens to ~700 μm for the "annealed" specimens and, despite the large accumulated strain, creep subsequent to annealing had no statistically significant affect on the size or shape of the grains. In that torsional-specimen cross-sections were approximately 3×3 mm, the number of grains making up that cross-section was small. Torsional maximum strains in the absorption experiments were 10^{-6}; despite the presence of lattice dislocations in the "pressed" and "creep" specimens, the materials were proven linearly viscoelastic at this strain.

One sees clearly in the data (Fig. 13b) that, despite the creep deformation that must endow the specimens with a significantly higher density of mobile lattice dislocations than are remnant in an "annealed" specimen, the absorption of the crept specimens is significantly *lower*. Specifically, the spectra demonstrate a diminishment of the plateau/broad hump and demonstrate, too, the sharpening of a Debye peak near $f = 10^2$ Hz. Because of the limited slip systems in the BCT-structure β, the tensile creep must result in a significantly limited distribution of subgrain boundary structures—a distribution in dynamic equilibrium with the applied tensile stress. The spectral peak at 10^2 Hz results from the absorption of a specific boundary structure (cf. Yan and Kê 1987), one obviously now more prevalent due to strain accumulation. Further, one sees that, for the "creep" specimens, absorption is apparently increased—actually, it is shifted to higher frequency—by deforming at a higher creep stress. The amount of this frequency shift is exactly that predicted by the change in τ (Eqn. 18), when d is considered to be the mean subgrain size, the magnitude of which is calculated from the mean-subgrain-size/differential-stress semi-empirical "calibration" of Twiss (1977) (and employing the measured, high-frequency shear modulus).

An obvious question to ask, then, is whether the shallow peak near $f = 10^{-1}$ Hz in the absorption data for the deformed forsterite single crystal (Fig. 12) is related to a similar concentration of specific low-angle boundary structures. Gueguen et al. (1989) did not comment on this aspect of their data. Nevertheless, with limited slip systems, olivine, too, realizes a limitation on the structures of low-angle boundaries (e.g., Ricoult and Kohlstedt 1983; Bai and Kohlstedt 1992a).

Finally, the diminishment of absorption with the addition of lattice dislocations to the material (Fig. 13b) should give pause to any generalized, qualitative speculation that "dislocation mechanisms" (i.e., the behavior of "free" lattice dislocations as opposed to the dislocation arrays constituting low-angle boundaries) are dominating a power-law, high-temperature-background absorption response. Inevitably, dislocation "vibration" with cyclic loading occurs, and must lead to nonlinearities in absorption (e.g., Lakes 1999, Ch.8). Yet under the thermodynamic (high temperature, low deviatoric stress), microstructural (polyphase, mineral (ionic) aggregates with each phase having a high Peierls barriers to dislocation glide) and frequency conditions of interest in geological settings, such mechanisms of absorption may prove of little consequence.

IMPACT OF PARTIAL MELTING

The impact of partial melting on low-frequency mechanical absorption depends primarily on (i) the spatial distribution of the melt and crystalline phases, (ii) the relative volume fractions of these phases and (iii) the practical rheological contrast of these phases. Early geophysical analyses to correlate materials properties with their capability to absorb shear waves concentrated on the spatial distribution of a liquid phase at the scale of the grain size of the crystalline residuum (e.g., Walsh 1969; Birch 1970; Mavko 1980). For example, it is easy to comprehend that a partial-melt microstructure in which all solid-state grain and phase boundaries are replaced by a homogeneous, low-viscosity liquid will be far more absorbing of shear waves (at a fixed frequency related to the viscosity of the liquid films—cf. the Si_3N_4 example cited earlier) than a system where grain and phase boundaries remain melt-free. Waff (e.g., Waff and Bulau 1979; Waff 1980; Waff and Faul 1992) inaugurated the petrologic study of partial-melt morphology, which followed in philosophy the microstructure topology work of Smith (1952). Waff defined "textural equilibrium" as the minimization of solid-solid (grain boundary or solid-state phase boundary) and solid-liquid interfacial energy; in olivine-basalt partial melts, under hydrostatic conditions and for a small volumetric melt fraction, textural equilibrium sees the liquid phase confined to three-grain ("triple") junctions and four-grain corners, forming a 3-D interconnected melt network, with melt-free olivine grain boundaries. That the grain boundaries in these materials were indeed melt free was confirmed to a resolution of 0.5 nm using high-resolution transmission electron microscopy (HRTEM; e.g., Vaughan et al. 1982; Cooper and Kohlstedt 1982), a result confirmed by high-resolution transmission analytical electron microscopy (HRAEM; Kohlstedt 1990). The result remains controversial: for example, de Kloe et al. (2000) argue from HRTEM/AEM evidence that a deviatoric stress (≥100 MPa) changes this textural equilibrium so as to produce nm-scale melt films that replace grain boundaries. The chapter in this volume by David Kohlstedt, examining the rheology of partial melts and the physics of melt segregation, presents the breadth of microstructural evidence and associated arguments. Nevertheless, in drained mafic-to-ultramafic material at low melt fraction, the vast majority of the liquid phase is confined to triple junctions and four-grain corners.

The impact of grain-scale, partial-melt textural equilibrium on attenuation, then, involves three possible effects that are perhaps distinguishable experimentally: (i) long-range (i.e., beyond the grain scale) melt migration based on gradients in hydrostatic pressure (a bulk-modulus-mode absorption); (ii) melt "squirt" (Mavko and Nur 1975), a shear effect that adjusts the volume of adjacent triple junctions at the grain scale; and (iii) any shear effect that might arise from melt films replacing grain boundaries and/or solid-state phase boundaries.

Long-range melt migration is best contemplated following a D'Arcy flow model in which the compaction of the crystalline residuum imparts momentum to the interconnected liquid phase; in the geophysics literature, the seminal papers of McKenzie (1984), Richter and McKenzie (1984) and Scott and Stevenson (1986) present these ideas, now known as "compaction theory." The relative velocities of the crystalline residuum and the liquid are linearly proportional to the inverse gradient of the pore fluid pressure, with the constant of proportionality being the permeability of the residuum, i.e.,

$$u - U = -\left[k(\phi,d)/\eta_f\right]\nabla \Pi, \tag{23}$$

where

$$\nabla \Pi = \eta \nabla^2 U + \left(\zeta + \eta/3\right)\nabla\left(\nabla \cdot U\right) - \left(1-\phi\right)g\Delta\rho - H\left(\gamma_{sl},\phi\right)\nabla\phi. \tag{24}$$

In Equation (23), u and U are the velocities of the melt and crystalline residuum, respectively, k is the permeability (which is a function of the volumetric melt fraction, ϕ, and the grain size), η_f is the shear viscosity of the liquid and $\nabla \Pi$ is the gradient in pore fluid pressure; in Equation (24), η is the shear viscosity and ζ the bulk viscosity of the partial-melt aggregate, respectively, g is the acceleration of gravity, $\Delta\rho$ is the density difference between liquid and crystalline phases and γ_{sl} is the interfacial energy ("surface tension") of the crystal (solid)-liquid interface. The first two terms in Equation (24) represent pressure gradients affected by shear and dilatational deformation, respectively, the third term is the buoyancy effect and the last term (Stevenson 1986; Cooper 1990) is the restraining effect of capillarity. In the geophysical context, gravity drives the liquid phase upwards while the residuum compacts downward; capillarity restrains the expulsion of all of the melt. From a chemical kinetics perspective, buoyancy-driven melt migration is obviously a serial-kinetic process based on a geometrical constraint: liquid is not expelled unless the residuum compacts. Parametric analysis of the equations reveals a characteristic "compaction length," δ_c:

$$\delta_c = [k(\zeta + 4\eta/3)/\eta_f]^{1/2} \; ; \qquad (25)$$

δ_c is the instantaneous length over which the compaction is rate limited solely by the capabilities of the residuum to deform (i.e., the actual flow of the liquid phase dissipates none of the energy in compaction). Beyond δ_c, energy dissipation is convolved between the deformation of the residuum and shear flow of the viscous liquid, although at scales large compared to δ_c, the shear flow of the liquid should be rate-limiting.

Bulk-modulus anelasticity for a partial melt arises due to the restorative force of capillarity: a hydrostatic compressional force promotes compaction; capillarity (the Laplace-Young relationship, e.g., that defines the correlation of the pressure in a bubble to its surface tension) promotes swelling upon removal of the applied force. Flexural creep experiments, which optimize the melt-migration driving force while minimizing the driving force for shear creep of the aggregate, combined with both microstructural analyses and strain-relaxation measurements, proved the operation of the long-distance migration mechanism and its anelastic nature (Cooper 1990; Gribb et al. 1994). The seismic-frequency absorption response for this mechanism was measured directly for partially molten, Newtonian-viscous silicate partial melts in reciprocating flexure ((ortho)enstatite glass-ceramic, Gribb and Cooper (1995)) and reciprocating compression (Co-Mg olivine-"basalt", Gribb et al. (1994)). In both these cases of Young's-modulus-mode loading, the effective δ_c was significantly smaller than the specimen size: flow of the liquid phase rate-limited the anelastic response. The absorption spectra revealed the power-law behavior, with a very-low $m \sim 0.10$–0.14 (see Fig. 6 for the glass-ceramic data). As noted earlier, a single, non-exponential-decay mechanism related to the D'Arcy-flow (diffusion)-effected migration of the flexural-specimen's neutral axis (or to the growth of the compaction layer—from the specimen's free sides to the specimen center—in compression specimens) accurately models the absorption response for this long-range melt migration.

Given the thermodynamics of capillary forces, the long-distance-migration mechanism is not expected to be operative in pure-shear loading. As a consequence, shear experiments can isolate the other two, grain-scale partial-melt absorption effects listed above. Gribb and Cooper (2000) compared the shear absorption responses of fine-grained ($d \sim 3$ µm) olivine aggregates with and without the addition of 5 wt % (6 vol %, i.e., $\phi \sim 0.06$) of a texturally equilibrated, K^+-doped basanite melt (the melt composition was selected to be saturated with respect to olivine as well as to have a low-enough viscosity to provide a δ_c that was significantly greater than the specimen size). Experiments were done in the regime of (f,d,T) characteristic of the shear-inviscid boundary assumption being applicable to the melt-free specimens. The 1250°C result, characteristic of all the experiments, is presented in Figure 14a: the partial melt specimens showed essentially the same power-law absorption spectra as do the melt-free

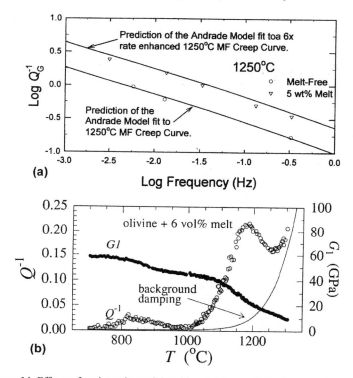

Figure 14. Effects of grain scale partial melting—at low melt fraction and the textural equilibrium corresponding to nominally hydrostatic pressure (melt-free grain boundaries)—on the shear attenuation response of olivine. (a) Sub-resonant behavior. At the high-temperature and/or low-frequency conditions where the inviscid grain boundary assumption holds, partial melting affects the absorption spectrum solely by its shifting to slightly higher frequency. The magnitude of the frequency shift is predicted by the effect of partial melting on the aggregate η_{ss}, following the model of Cooper et al. (1989). There is no apparent disaggregation of the material—with consequent dramatic drop of shear modulus—because of the presence of melt. [Used by permission of American Geophysical Union, from Gribb and Cooper (2000), *Geophysical Research Letters*, Vol. 27, Fig. 3, p. 2343.] (b) Resonant behavior at $f \sim 2$ Hz. Two Debye peaks are revealed: the one at high temperature may be related to the grain-scale "melt squirt" phenomenon envisioned by Mavko and Nur (1975); alternatively, it may represent a solid-state grain-boundary absorption. [Used by permission of Columbia University Press, from Xu et al. (2002). In *MARGINS Theoretical and Experimental Earth Science*, edited by G. Karner, Fig. 10.3, in press.]

specimens ($Q_G^{-1} \propto f^{-0.4}$), with the partial melt spectra shifted to a higher frequency that corresponds to an decrease in η_{ss} of a modest factor of 6. In the figure, a shift of the Andrade-Model inversion for the melt-free material, corresponding to this decrease in η_{ss} for the partial melt, predicts beautifully the partial melt attenuation. The important interpretation is that relative to point (iii) above: there is no melt-effected plummeting of viscosity or shear modulus, nor a dramatic increase in shear attenuation, as might be anticipated if all grain boundaries were replaced by films of a low-viscosity liquid. Further, the thermal sensitivity (activation energy) of both creep and attenuation responses were identical between partial-melt and melt-free materials, clearly suggesting that the same physical process rate-limits both creep and attenuation in both materials.

The results (i.e., the magnitude of decrease in η_{ss} and the consequent effect on the attenuation spectra) are consistent fully with the model for melt-enhanced diffusional creep in which diffusion through grain boundaries is rate limiting and the triple-junction melt network, representative of textural equilibrium, provides a partial short-circuit for diffusion (Cooper et al. 1989; see the chapter in this volume by Kohlstedt).

Shear resonance (torsion pendulum) experiments (Xu et al. 2002) performed on fine-grained ($d \sim 20$ μm) olivine-basalt aggregates ($1 \leq f[\text{Hz}] \leq 10$; $22 \leq T[°C] \leq 1350$; $0.004 \leq \phi \leq 0.13$) reveal two distinct Debye peaks, one centered on $T \approx 875°C$ and the other on $T \approx 1175°C$ (the data for $\phi = 0.06$ is presented in Fig. 14b). The high-T peak has a magnitude that is weakly sensitive to ϕ: $Q_G^{-1} \sim 0.3$ for $\phi = 0.13$; $Q_G^{-1} \sim 0.1$ for $\phi = 0.004$. The low-temperature peak is associated with the glass transition of the quenched basalt; the high-temperature peak, however, speaks either of a melt-squirt effect and/or of a grain-boundary effect in shear absorption. Melt squirt is defined as an exponential-decay absorption mechanism with a characteristic time, τ_{ms}, given by (cf. Mavko 1980):

$$\eta_f = (\tau_{ms} K_f / 40)(R/d)^2, \tag{26}$$

where K_f is the bulk modulus of the liquid and R is a measure of the cross-section of the triple-junction "tube." Xu et al. (2002) calculate that, for the characteristic τ_C of their experiments, the η_f is too high, by an order of magnitude, than that of a bulk, olivine-saturated liquid basalt; as a consequence, they interpret the high-temperature peak to be one associated with the Zener grain-boundary absorption (i.e., Eqn. 21). Their argument continues by demonstrating that an extrapolation of their data to $\phi = 0$ matches well the "plateau" result of Bunton (2001) for melt-free olivine of similar grain size. It seems premature, however, to completely rule out a melt-squirt contribution in this absorption peak. For example, the viscosity of a bulk liquid may not accurately reflect the shear resistance of a silicate melt within a ≤1 μm "diameter" triple junction tube, particularly considering that the melt flow involves, too, stress-effected chemical reaction with the tube (cf. Riley and Kohlstedt 1991). Further, the significant difference in the thermal sensitivity for their background absorption ($E_a \sim 600$–650 kJ mol^{-1}, which is related to grain boundary diffusion and, most likely, to grain boundary viscosity as well) compared to that for the measured τ_C of the Debye peak ($E_a \sim 300$–350 kJ mol^{-1}) suggests a distinctly different physical process controlling the latter phenomenon. A melt-induced dissipation peak superposed on the power-law absorption response of melt-free olivine has also been observed recently in sub-resonant torsion experiments by Jackson et al. (2001).

A critical factor not yet addressed in any attenuation experiments on partial melts is the role played in absorption by incipient melt segregation beyond the grain scale. Geochemical evidence strongly indicates that melt flow at the grain scale must be very limited in extent, that is, upon formation of melt it must be focused into larger channels so as to avoid continuous reaction with the crystalline residuum (e.g., Spiegelman and Kenyon 1992). Arguments based on the thermodynamics of irreversible processes suggests that a fractal drainage system should prove effective to this end (e.g., Hart 1993). Consistent with the fractal hypothesis, the incipient process of beyond-grain-scale phase separation of melt and crystalline residuum, driven by deformation and affected in scale by the relative values of (bulk-specimen) strain rate and δ_c, has recently been demonstrated experimentally (Holtzman et al. 2001). Correlation of this scaling with the seismic-frequency absorption response for the material overall is an experimental challenge with obvious important implications.

IMPACTS OF DEFECT CHEMISTRY (INCLUDING WATER) AND THE STRUCTURE(S) OF INTERFACES

The rates of chemical diffusion and of temperature-dependent rheology of crystalline solids depend directly on the concentrations of point defects on the crystal lattice. These point defects include vacancies (missing ions or atoms), interstitials (ions/atoms placed in the structure in interstices not occupied in the "perfect" crystal structure) and substitutionals (e.g., as occurs in a solid solution); they additionally include bonded complexes of these single defects as well as the jogs and kinks on lattice dislocations. Made thermodynamically stable by the configurational entropy of their distribution on the lattice, the concentrations of point defects are set by the temperature, pressure (in the limit, they can perhaps be set solely by temperature and pressure—these are the *intrinsic* point defects), deviatoric stress, and by the various chemical potentials defining chemical equilibrium in the system. Indeed, the concentrations of point defects are the physical manifestation of chemical activities within a crystal. In minerals—because the bonding is primarily ionic and, thus, the energy gap between the valence and conduction bands is so large—point defects carry a charge; as such, equilibrium requires that the product of concentration and valence of the positively charged defects equals that of the negatively charged defects; this condition is part of the overall charge neutrality requirement for the crystal. The large band gap also insures that, in the geologic setting, point defect concentrations are affected significantly by changes in the activities (fugacities) defining the equilibrium; this situation is defined as the *extrinsic* condition. In that most minerals contain transition-metal cations, even in very small concentrations, additional important point defects include conduction electrons and electron holes, which are (for example) transition-metal cations having either a too-small valence or a too-large valence, respectively, relative to that valence required for neutrality on the cation site of interest. The concentrations of conduction electrons and electron holes, and therefore (because of the charge-neutrality constraint) the concentrations of most of the other point defects on the lattice, are consequently affected significantly by the oxygen fugacity. (The best pedagogical treatise on point-defect thermodynamics in ionic materials and its application to dynamic processes is the text by Schmalzried (1995).)

The effect of point defect concentrations on mechanical absorption is fourfold (cf. Karato 1995): (i) the defects can perhaps affect the bond strength in the material, and so affect the modulus; (ii) at very high concentrations, the defects can bond and this polar "complex" can respond to stress in a similar manner to that producing the Snoek relaxation described earlier; (iii) the defects affect the rates of all dynamic processes including chemical diffusion and the climb and glide of lattice dislocations and of primary and secondary grain boundary (phase boundary) dislocations; and (iv) the defects affect the structure of grain boundaries and of solid-state phase boundaries, which, in ionic solids, also includes an electrostatic space charge penetrating into the grains on either side of the boundary. Because, in general (though not including substitutional solid solutions of widely ranging concentration), the total concentration of point defects is usually very small, effect (i) can be considered of little consequence. Similarly, the formation of defect complexes is prevalent at lower temperatures and the frequencies anticipated for their mechanical absorption are far higher than those of interest seismologically; effect (ii) can also be set aside. Effects (iii) and (iv), however, are significant to this discussion.

The dynamic effect of point defect concentration (point (iii) above) is often probed by measuring the impact of various chemical fugacities on, for example, viscosity (or, in non-linear systems, effective viscosity). What is discovered experimentally (and supported by point-defect thermodynamics theory) is a power-law dependence, that is, viscosity is characterized as:

$$\eta \propto f_{ox1}^{a} f_{ox2}^{b} \cdots f_{O_2}^{c} \exp(E_a / RT), \tag{27}$$

where f_j is fugacity of species j (ox1 and ox2, etc., representing different crystalline oxides that are components of the phase(s) whose viscosity is being probed). The values of power-law exponents a, b, and c (± the value of E_a) determined in experiments are compared to theoretical models to discern the specific point defects affecting/effecting the physical behavior. The careful work of Bai et al. (1991) probing the impacts of oxygen fugacity and orthopyroxene (or magnesiowüstite) activity on the dislocation creep rate of single-crystal olivine is a fine example of the application of point defect thermodynamics to mechanical response.

It has long been known that water fugacity (f_{H_2O}) has a dramatic effect on the viscosity of silicates (both crystalline and amorphous), decreasing it dramatically. Griggs (1967) described the phenomenon as "hydrolytic weakening;" the experimental rock physics community has been exploring the phenomenology of this chemical/mechanical effect ever since (the chapters by Jan Tullis and by Greg Hirth in this volume detail the current understanding). The phenomenology suggests clearly that hydrolytic weakening is best contemplated from the perspective of (extrinsic) point defect thermodynamics (i.e., relationships similar in form to Equation (27) can be defined for f_{H_2O}-affected viscosity; e.g., Hobbs (1981)). Identifying the specific point defect(s) responsible for hydrolytic weakening (which could vary for different silicate phases) remains difficult and controversial; the experimental work of Kohlstedt and co-workers, for example, on the reaction of water with various polymorphs of $(Mg,Fe)_2SiO_4$ illustrates well the confluence of experiment with point-defect theory (e.g., Mackwell and Kohlstedt 1990; Bai and Kohlstedt 1992b; Kohlstedt et al. 1996). These studies additionally suggest that the primary source of water in the mantle is not that incorporated in the ("perfect") structure of hydrous minerals, but rather as extrinsic point defects in the "anhydrous" minerals.

To zero-th order, then, the impact of chemical environment (via its effect on point-defect concentrations—including water-related extrinsic defects) on attenuation is straightforward: impacts on viscosity cause changes in the various τ_C's with consequent frequency shifts in the Q^{-1} spectra. For example, in that recent grain boundary diffusion creep studies on polycrystalline olivine demonstrate that $\eta_{ss} \propto f_{H_2O}^{-1}$ (Mei and Kohlstedt 2000; cf. Karato et al. 1986), the effect on the absorption model presented in Equations (18) and (20) (Fig. 9) is easily calculated.

There are, however, more profound issues that, while not yet explored directly experimentally, bear contemplation with regard to the impact of point defects, and particularly water-related defects, on the mechanical absorption of mantle material. In partial melting, for example, partitioning of water between the crystalline residuum and the silicate melt functions both to increase the solid-state viscosity of the olivine and lower that of the melt; calculations of the partitioning, integrated with the understanding of its effects on rheology (Karato 1986; Hirth and Kohlstedt 1996), affect the understanding of upper-mantle dynamics, particularly melt segregation. As for attenuation, then, the zeroth-order effects of crystalline point defects are convolved with the partial melt effects noted earlier.

In that the low-frequency, power-law absorption behavior seems intimately related to the density, distribution and structure of interfaces (grain boundaries, solid-state phase boundaries, subgrain boundaries, ± solid-liquid phase boundaries) the interaction of lattice point defects with the structural elements of these interfaces is of critical importance to a quantitative understanding of attenuation. In the solid state, the robust understanding of grain/phase boundaries to have a 2-D crystallographic structure based on dislocations (Sutton and Balluffi 1995) means that the jogs and kinks on these boundary dislocations (as well as those on the "free" lattice dislocations) contribute to the point-defect equilibrium of the system. Thus the kinetics of stress-induced motion of the extended defects is affected to first order by the extrinsic chemical condition. The interaction (in the case of ionic solids) of charged jogs and/or kinks with charged lattice point defects can either inhibit or enhance the motion of the extended defect—as well as the temperature sensitivity of that motion (cf. Maier 1994; Jamnik and Raj 1996; Chiang et al. 1997, p.155). (The case of rate inhibition is fully analogous to the elastic interactions between solute atoms and dislocations/grain boundaries in metals that results in "solid-solution strengthening" (e.g., Courtney 1990, p.173).) The phenomenological experimental record of the impact of f_{H_2O} on the kinetics of dislocation creep and on fabric development of olivine is an important example. An increase in f_{H_2O}, while lowering the effective viscosity overall, is known to effect *relative* changes in the resistance to dislocation motion of different slip systems in olivine (e.g., Mackwell et al. 1985); an increase in f_{H_2O} has also been correlated with an increase the mobility of grain boundaries (e.g., Jung and Karato 2001a). The combination of the two f_{H_2O} effects, in the presence of a constant deviatoric stress, has been demonstrated to result in distinctly different fabric (i.e., lattice preferred orientation) for the deformed material compared to that seen for nominally "dry" conditions (Jung and Karato 2001b). The impact on seismic responses are (i) to change significantly the wave velocity anisotropy, that is, its directionality relative to the principal stress directions, and (ii) to decrease the variety of grain boundary structures in the deformed material (a geometrical requirement consistent with the development of fabric), and so affect the absorption spectrum (cf. the case for the β-In$_3$Sn creep specimens presented in Fig. 13b, where one aspect of fabric—spatial distribution of low-angle boundary structures—has a notable, demonstrable effect on absorption).

The fabric effect presented above is based on the kinetic response of continued, accumulated strain in an polycrystalline aggregate. But, as noted earlier, the attenuation response is that associated with a very small stress pulse that adds to the overall stress potential giving rise to the large-scale deformation. Both the crystallographic structure *and* the electrostatic structure of the interfaces dictates the response: the stress-pulse-induced climb and glide of primary and secondary grain (phase)-boundary dislocations are subject both to their specific crystallography and to the electrostatic interactions with lattice point defects. The electrostatic and crystallographic structures are coupled to the extrinsic point-defect chemistry. A dramatic, and quantitatively characterized, example of the boundary-structure/extrinsic-defect relationship is the case of "twist" boundary (boundary normal and twist axis both [001]) structure in pure (BCC) α-iron and α-iron doped with 0.18 at % gold (Sickafus and Sass 1985). TEM and ion-backscattering spectroscopy proved that the impurity gold atoms preferentially segregate to the grain boundary, and the segregation results in a two-dimensional structural reaction. In pure α-iron, the boundary consists of an orthogonal array of [$\bar{1}$10] and [110] screw dislocations; with the addition of the impurity, the boundary transforms to an orthogonal array of [010] and [100] screw dislocations. It is not a great stretch to believe that similar transformations occur in ionic solids; certainly, impurity segregation to boundaries has been well characterized in variety of oxides and other ionic compounds (e.g., Chiang et al. 1997, p.156).

The impact of these chemical-segregation effects and their possible influence on boundary structure and viscosity, etc. are now being contemplated by the experimental community. For example, very small concentrations of Ca^{2+} in olivine apparently effect a significant change in the thermal sensitivity of grain boundary diffusional creep (E_a in the range 300–400 kJ mol^{-1} for aggregates prepared from pulverized San Carlos peridot, with CaO content of ~0.05 wt %; E_a in the range 600–700 kJ mol^{-1} for aggregates prepared either from a dunite with a CaO content of ~0.3-0.6 wt % or for San Carlos specimens prepared with very small amounts of CaO-bearing basaltic melt). (There are copious examples of similar behavior in the ceramic materials literature, e.g., Cho et al. (1997).) Similarly, the $\eta_{ss} \propto d^2$ result for diffusional creep in fine-grained olivine at small deviatoric stress (noted in Section 7) seems consistent with a threshold effect related to the electrostatic structure(s) of the grain boundaries. And the exact impact of water-related extrinsic defects on boundary-effected absorption have yet to be characterized. Because the stresses (and strains) involved in seismic wave attenuation are so small—and the extrapolation goals for the experimental research so grand!—these chemical effects may prove of first-order importance and so require close scrutiny in future experiments.

FINAL COMMENTS

This chapter has sought specifically to be a primer on linear viscoelasticity at high temperature and low frequency, conditions characterizing wave attenuation in geological systems. The emphasis placed on appreciating viscoelastic models and their impact on thinking with regard to attenuation is to allow the reader a base for access to the literature. The geophysical problem is a difficult one: the power-law absorption spectrum is fairly non-descript; and, despite one quantitative characterization (the shear-inviscid-boundary, diffusional creep model giving rise to Fig. 9) that can forward-predict the absorption-band behavior for a finite window of thermodynamic and microstructural conditions, the overall absorption behavior of the terrestrial planets undoubtedly involves contributions of a number of serial-kinetic mechanisms (and maybe some parallel kinetic mechanisms as well), which are difficult, indeed, to isolate experimentally.

Progress in characterizing attenuation response(s) quantitatively and with predictive capability requires advancement on a number of fronts. First, there is significant need for an improved understanding of the physical processes involved in crystalline plasticity, particularly in the areas of statistical studies of dislocations and dislocation arrays, e.g., correlating phenomenological mechanical-equation-of-state approaches to plasticity with the more classical, steady-state creep and microstructural studies, and in the application of irreversible-process thermodynamics to plasticity problems in polyphase crystalline aggregates. Second, in dynamic experiments, the role of threshold phenomena will need to be studied directly. Critical for rock and mineral studies will be the multiple impacts of various chemical potentials—including, particularly, that of H_2O—on the distributions of lattice point defects and on the structure and related viscosity of grain and phase boundaries. Nevertheless, given recent experimental and theoretical advances in the mineral physics of mechanical absorption, along with the inspiration of various geodynamic models, there is reason for optimism that an experimentally calibrated, but physically based "key" can be developed that will correlate the magnitude and frequency-sensitivity of seismic wave absorption with the stresses and accumulated strains in tectonic terranes.

ACKNOWLEDGMENTS

I have been blessed with a number of graduate students—Tye Gribb, Joe Bunton, Jeff Lee and Sam Zhang—and one postdoc—Doug Green—who have pursued with me various aspects of the attenuation problem in polycrystals and partial melts; all were outstanding collaborators and I am pleased to acknowledge them for their efforts and continued friendship. The experimental wave attenuation community in geophysics is a fairly small and collegial group, from whom I have learned a great deal; in particular, periodic discussions with Ian Jackson, Shun Karato, David Kohlstedt, Brian Bonner and

Ivan Getting have proven distinctly valuable. Likewise, in the materials and mechanics community, active collaborations with Rod Lakes and Donald Stone, and periodic discussions with Rishi Raj, continue to sharpen my perspective. The attenuation research pursued in my group at Wisconsin has been funded, in part, by the National Science Foundation Division of Earth Sciences Program in Geophysics; that support is gratefully acknowledged.

REFERENCES

Anderson DL, Given JW (1982) Absorption band Q model for the Earth. J Geophys Res 87:3893-3904
Anderson DL, Minster JB (1979) The frequency dependence of Q in the Earth and implications for mantle rheology and Chandler wobble. Geophys J R Astron Soc 58:431-440
Andrade, ENDaC (1910) On the viscous flow in metals, and allied phenomena. Proc R Soc London A84: 1-12
Ashby MF (1972) Boundary defects and atomistic aspects of boundary sliding and diffusional creep. Surf Sci 31:498-542
Bai Q, Kohlstedt DL (1992a) High-temperature creep of olivine single crystals, 2. Dislocation structures. Tectonophysics 206:1-29
Bai Q, Kohlstedt DL (1992b) Substantial hydrogen solubility in olivine and implications for water storage in the mantle. Science 357:672-674
Bai Q, Mackwell SJ, Kohlstedt DL (1991) High-temperature creep of olivine single crystals, 1. Mechanical results for buffered samples. J Geophys Res 96:2441-2463
Berckhemer H, Kampfmann W, Aulbach E, Schmeling H (1982), Shear modulus and Q of forsterite and dunite near partial melting from forced-oscillation experiments. Phys Earth Planet Inter 29:30-41
Birch F (1970) Interpretations of the low velocity zone. Phys Earth Planet Inter 3:178-181
Brady JB (1995) Diffusion data for silicate minerals, glasses and liquids. *In* Mineral Physics and Crystallography: A Handbook of Physical Constants. Ahrens TJ (ed) American Geophysical Union, Washington, p 269-290
Brodt M, Cook LS, Lakes RS (1995) Apparatus for determining the properties of materials over ten decades of frequency and time: Refinements. Rev Sci Instrum 66:5292-5297
Bunton JH (2001) The impact of grain size on the shear creep and attenuation behavior of polycrystalline olivine. MS thesis, University of Wisconsin–Madison, Madison, Wisconsin
Chan WW, Sacks IS, R. Morrow R (1989) Anelasticity of the Iceland Plateau from surface wave analysis. J Geophys Res 94:5675-5688
Chiang Y-M, Birnie D III, Kingery WD (1997) Physical Ceramics: Principles for Ceramic Science and Engineering. Wiley, New York
Cho JH, Harmer MP, Chan HM, Rickman JM, Thompson AM (1997) Effect of yttrium and lanthanum on the tensile creep behavior of aluminum oxide. J Am Ceram Soc 80:1013-1017
Clarke DR (1987) On the equilibrium thickness of intergranular glass phases in ceramic materials. J Am Ceram Soc 70:15-22
Coble RL (1963) A model for boundary diffusion controlled creep in polycrystalline materials. J Appl Phys 34:1679-1682
Cook LS, Lakes RS (1995) Viscoelastic spectra of $Cd_{0.67}Mg_{0.33}$ in torsion and bending. Metall Mater Trans A 26A:2035-2039
Cooper RF (1990) Differential stress-induced melt migration: An experimental approach. J Geophys Res 95:6979-6992
Cooper RF, Kohlstedt DL (1982) Interfacial energies in the olivine-basalt system. Adv Earth Planet Sci 12:217-228
Cooper RF, Kohlstedt DL, Chyung K (1989) Solution-precipitation enhanced diffusional creep in solid-liquid aggregates which display a non-zero dihedral angle. Acta Metall 37:1759-1771
Courtney TH (1990) Mechanical Behavior of Materials. McGraw-Hill, New York
Covey-Crump SJ (1994) The application of Hart's state variable description of inelastic deformation to Carrara marble at $T < 450°C$. J Geophys Res 99:19793-19808
Crossman FW, Ashby MF (1975) The non-uniform flow of polycrystals by grain boundary sliding accommodated by power-law creep. Acta Metall 23:425-440
D'Anna G, Benoit W, Vinokur VM (1997) Internal friction and dislocation collective pinning in disordered quenched solid solutions. J Appl Phys 82:5983-5990
de Kloe R, Drury MR, van Roermund HLM (2000) Evidence for stable grain boundary melt films in experimentally deformed olivine-orthopyroxene rocks. Phys Chem Minerals 27:480-494
Dooris A, Lakes RS, Myers B, Stephens N (1999) High damping indium-tin alloys. Mech Time-Depend Mater 3:305-318

Findley WN, Lai JS, Onaran K (1976) Creep and Relaxation of Nonlinear Viscoelastic Materials. North-Holland, Amsterdam (corrected and republished: Dover Publications, New York, 1989)

Frost HJ, Ashby MF (1982) Deformation Mechanism Maps: The Plasticity and Creep of Metals and Ceramics. Pergamon Press, Oxford, UK

Glover G, Sellars CM (1973) Recovery and recrystallization during high-temperature deformation of α-iron. Metall Trans 4:765-775

Green DH, Cooper RF (1993) Dilatational anelasticity in partial melts: Viscosity, attenuation and velocity dispersion. J Geophys Res 98:19807-19817

Green DH, Cooper RF, Zhang S. (1990) Attenuation spectra of olivine/basalt partial melts: Transformation of Newtonian creep response. Geophys Res Lett 17:2097-2100

Gribb TT (1992) Low-frequency attenuation in microstructurally equilibrated silicate partial melts. MS thesis, University of Wisconsin–Madison, Madison, Wisconsin

Gribb TT, Cooper RF (1995) Anelastic behavior of silicate glass-ceramics and partial melts: Migration of the amorphous phase. In Plastic Deformation of Ceramics. Bradt RC, Brookes CA, Routbort JL (eds) Plenum Press, New York, p 87-97

Gribb TT, Cooper RF (1998) Low-frequency shear attenuation in polycrystalline olivine: Grain boundary diffusion and the physical significance of the Andrade model for viscoelastic rheology. J Geophys Res 103:27267-27279

Gribb TT, Cooper RF (2000) The effect of an equilibrated melt phase on the shear creep and attenuation behavior of polycrystalline olivine. Geophys Res Lett 27:2341-2344

Gribb TT, Zhang S, Cooper RF (1994) Melt migration and related attenuation in equilibrated partial melts. In Magmatic Systems. Ryan MP (ed) Academic Press, San Diego, p19-36

Griggs DT (1967) Hydrolytic weakening of quartz and other silicates. Geophys J Roy Astron Soc 14:19-31

Gueguen Y, Darot M, Mazot P, Woirgard J (1989) Q^{-1} of forsterite single crystals. Phys Earth Planet Inter 55:254-258

Hart EW (1970) A phenomenological theory for plastic deformation of polycrystalline metals. Acta Metall 21:295-307

Hart SR (1993) Equilibrium during mantle melting: A fractal tree model. Proc Natl Acad Sci USA 90:11914-11918

Herring C (1950) Diffusional viscosity of a polycrystalline solid. J Appl Phys 21:437-445

Hirth G, Kohlstedt DL (1995) Experimental constraints on the dynamics of the partially molten upper mantle: Deformation in the diffusion creep regime. J Geophys Res 100:1981-2001

Hirth G, Kohlstedt DL (1996) Water in the oceanic upper mantle: Implications for rheology, melt extraction and the evolution of the lithosphere. Earth Planet Sci Lett 144:93-108

Hobbs BE (1981) The influence of metamorphic environment upon the deformation of minerals. Tectonophysics 78:335-383

Holtzman B, Zimmerman ME, Kohlstedt DL, Phipps Morgan J (2001) Interactions of deformation and fluid migration I: Melt segregation in the viscous regime. EOS Trans, Am Geophys Union 82:F1107

Jackson I (1993) Progress in the experimental study of seismic wave attenuation. Ann Rev Earth Planet Sci 21:375-406

Jackson I (2000) Laboratory measurement of seismic wave dispersion and attenuation: Recent progress. In Earth's Deep Interior: Mineral Physics and Tomography from the Atomic to the Global Scale. Karato S-i, Forte AM, Liebermann RC, Masters G, Stixrude L (eds) American Geophysical Union, Washington, p 161-179

Jackson I, Faul UF, FitzGerald JD (2001) Laboratory measurements of seismic wave attenuation in upper-mantle materials: the effect of partial melting. EOS Trans, Am Geophys Union 82:F1163

Jackson I, FitzGerald JD, Faul UF, Tan BH (2002) Grain-size sensitive seismic-wave attenuation in polycrystalline olivine. J Geophys Res (in press)

Jackson I, Paterson MS, FitzGerald JD (1992) Seismic wave dispersion and attenuation in Åheim dunite: An experimental study. Geophys J Intl 108:517-534

Jamnik J, Raj R (1996) Space-charge-controlled diffusional creep: Volume diffusion case. J Am Ceram Soc 79:193-198

Jung H, Karato S-i (2001a) Effects of water on dynamically recrystallized grain size of olivine. J Struct Geol 23:1337-1344

Jung H, Karato S-i (2001b) Water-induced fabric transitions in olivine. Science 293:1460-1463

Karato S-i (1986) Does partial melting decrease the creep strength of the upper mantle? Nature 319:309-310

Karato S-i (1993) Importance of anelasticity in the interpretation of seismic tomography. Geophys Res Lett 20:1623-1626

Karato S-i (1995) Effect of water on seismic wave velocities in the upper mantle. Proc Japan Acad B 71:61-66

Karato S-i, Karki B (2001) Origin of lateral variation of seismic wave velocities and density in the deep mantle. J Geophys Res 106:21771-21783.

Karato S-i, Spetzler HA (1990) Defect microdynamics in minerals and solid-state mechanisms of seismic wave attenuation and velocity dispersion in the mantle. Rev Geophys 28:399-421

Karato S-i, Paterson MS, FitzGerald JD (1986) Rheology of synthetic olivine aggregates: Influence of grain size and water. J Geophys Res 91:8151-8176

Kohlstedt DL (1990) Chemical analysis of grain boundaries in an olivine-basalt aggregate using high resolution, analytical electron microscopy. *In* The Brittle Ductile Transition in Rocks: The Heard Volume. Duba AG, Durham WB, Handin JW, Wang HF (eds) American Geophysical Union, Washington, p 211-218

Kohlstedt DL, Weathers MS (1980) Deformation-induced microstructures, paleopiezometers, and differential stresses in deeply eroded fault zones. J Geophys Res 85:6269-6285

Kohlstedt DL, Keppler H, Rubie DC (1996) Solubility of water in the α, β and γ phases of $(Mg,Fe)_2SiO_4$. Contrib Mineral Petrol 123:345-357

Kê TS (1947) Experimental evidence of the viscous behavior of grain boundaries in metals. Phys Rev 71:533-546

Kê TS (1999) Fifty-year study of grain boundary relaxation. Metall Mater Trans A 30A:2267-2295

Lakes RS (1999) Viscoelastic Solids. CRC Press, Boca Raton, Florida

Lakes RS, Quackenbush J (1996) Viscoelastic behaviour in indium-tin alloys over a wide range of frequencies and times. Philos Mag Lett 74:227-232

Lerner I, Chiang S-W, Kohlstedt DL (1979) Load relaxation studies for four alkali halides. Acta Metall 27:1187-1196

Lifshits IM, Shikin VB (1965) The theory of diffusional viscous flow of polycrystalline solids. Soviet Physics-Solid State 6:2211-2218

Mackwell SJ, Kohlstedt DL (1990) Diffusion of hydrogen in olivine: Implications for water in the mantle. J Geophys Res 95:5079-5088

Mackwell SJ, Kohlstedt DL, Paterson MS (1985) The role of water in the deformation of olivine single crystals. J Geophys Res 90:1319-1333

Maier J (1994) Defect chemistry at interfaces. Solid State Ionics 70/71:43-51

Mavko GM (1980) Velocity and attenuation in partially molten rocks. J Geophys Res 85:5173-5189

Mavko GM, Nur A (1975) Melt squirt in the asthenosphere. J Geophys Res 80:1444-1448

McKenzie D (1984) The generation and compaction of partially molten rock. J Petrol 25:713-765

Mei S, Kohlstedt DL (2000) Influence of water on plastic deformation of olivine aggregates, 1. Diffusion creep regime. J Geophys Res 105:21457-21469

Molodenskiy SM, Zharkov VN (1982) Chandler wobble and frequency dependence of Q_μ of the Earth's mantle. Phys Solid Earth 18:245-254

Mosher DR, Raj R, Kossowsky (1976) Measurement of viscosity of the grain boundary phase in hot-pressed silicon nitride. J Mater Sci 11:49-53

Nabarro FRN (1948) Deformation of crystals by the motion of single ions. *In* Report of a Conference on the Strength of Solids. Physical Society of London, UK, p 75-90

Nowick AS, Berry BS (1972) Anelastic Relaxation in Crystalline Solids. Academic Press, San Diego

O'Connell RJ, Budiansky B (1978) Measures of dissipation in viscoelastic media. Geophys Res Lett 5:5-8.

Paterson MS (1976) Some current aspects of experimental rock mechanics. Philos Trans R Soc Lond A 283:163-172

Pezzotti G, Ota K, Kleebe H-J (1996) Grain boundary relaxation in high-purity silicon nitride. J Am Ceram Soc 79:2237-2246

Plookphol T (2001) Similarity and scaling properties in dislocation microstructures generated during high-temperature load relaxation and creep of rock salt and San Carlos olivine single crystals. PhD dissertation, University of Wisconsin–Madison, Madison, Wisconsin

Raj R (1975) Transient behavior of diffusion-induced creep and creep rupture. Metall Trans A 6A: 1499-1590

Raj R, Ashby MF (1971) On grain boundary sliding and diffusional creep. Metall Trans 2:1113-1127

Richter FM, McKenzie D (1984) Dynamical models for melt segregation from a deformable matrix. J Geol 92:729-740

Ricoult DL, Kohlstedt DL (1983) Structural width of low-angle grain boundaries in olivine. Phys Chem Minerals 9:133-138

Riley GN Jr, Kohlstedt DL (1991) Kinetics of melt migration in upper mantle-type rocks. Earth Planet Sci Lett 105:500-521

Romanowicz B, Durek JJ (2000) Seismological constraints on attenuation in the Earth: A review. *In* Earth's Deep Interior: Mineral Physics and Tomography from the Atomic to the Global Scale. Karato S-i,

Forte AM, Liebermann RC, Masters G, Stixrude L (eds) American Geophysical Union, Washington, p 161-179

Schmalzried H (1995) Chemical Kinetics of Solids. VCH, Weinheim, FRG

Schoeck G, Bisogni E, Shyne J (1964) The activation energy of high temperature internal friction. Acta Metall 12:1466-1468

Scott DR, Stevenson DJ (1986) Magma ascent by porous flow. J Geophys Res 91:9283-9296

Sickafus K, Sass SL (1985) Observation of a grain boundary phase transformation induced by solute segregation. J Vac Sci Tech 3:1525-1530

Simmons G, Wang HF (1971) Single Crystal Elastic Constants and Calculated Aggregate Properties: A Handbook (Second Edn). M.I.T. Press, Cambridge, Massachusetts

Smith CS (1952) Grain shapes and other metallurgical applications of topology. *In* Metal Interfaces. American Society for Metals, Cleveland, Ohio, p 65-108

Spiegelman M, Kenyon PM (1992) The requirements for chemical disequilibrium during magma migration. Earth Planet Sci Lett 109:611-620

Stevenson DJ (1986) On the role of surface tension in the migration of melts and fluids. Geophys Res Lett 13:1149-1152

Stone DS (1991) Scaling laws in dislocation creep. Acta Metall Mater 39:599-608

Sutton AP, Balluffi RW (1995) Interfaces in Crystalline Materials. Clarendon Press, Oxford, UK

Tan B, Jackson I, FitzGerald J (1997) Shear wave dispersion and attenuation in fine-grained synthetic olivine aggregates: Preliminary results. Geophys Res Lett 24:1055-1058

Tan BH, Jackson I, FitzGerald JD (2001) High-temperature viscoelasticity of fine-grained polycrystalline olivine. Phys Chem Minerals 28:641-664

Thigpen L, Hedstrom GW, Bonner BP (1983) Inversion of creep response for retardation spectra and dynamic viscoelastic functions. J Appl Mech 105:361-366

Toomey DR, Wilcock WSD, Solomon SC, Hammond WC, Orcutt JA (1998) Mantle seismic structure beneath the MELT region of the East Pacific Rise from P and S wave tomography. Science 280:1224-1227

Twiss RJ (1977) Theory and applicability of a recrystallized grain size paleopiezometer. Pure Appl Geophys 115:227-244

Vaughan PJ, Kohlstedt DL, Waff HS (1982) Distribution of the glass phase in hot-pressed, olivine-basalt aggregates: An electron microscopy study. Contrib Mineral Petrol 81:253-261

Waff HS (1980) Effects of the gravitational field on liquid distribution in partial melts within the upper mantle. J Geophys Res 85:1815-1825

Waff HS, Bulau JR (1979) Equilibrium fluid distribution in an ultramafic partial melt under hydrostatic stress conditions. J Geophys Res 84:6109-6114

Waff HS, Faul UH (1992) Effects of crystalline anisotropy on fluid distribution in ultra mafic partial melts. J Geophys Res 97:9003-9014

Walsh JB (1969) A new analysis of attenuation in partially melted rock. J Geophys Res 74:4333-4337

Webb S, Jackson I, FitzGerald J (1999) Viscoelasticity of the titanate perovskites $CaTiO_3$ and $SrTiO_3$ at high temperature. Phys Earth Planet Inter 115:259-291

Webb SC, Forsyth DW (1998) Structure of the upper mantle under the EPR from waveform inversion of regional events. Science 280:1227-1229

Xu Y, Zimmerman ME, Kohlstedt DL (2002) Deformation behavior of partially molten mantle rocks. *In* MARGINS Theoretical and Experimental Earth Science. Karner G (ed) Columbia University Press, New York (in press)

Yan SC, Kê TS (1987) Internal friction peaks associated with the polygonization boundaries in aluminum and dilute aluminum-copper alloys. Phys Stat Solidi 104:715-721

Zener C (1941) Theory of elasticity of polycrystals with viscous grain boundaries. Phys Rev 60:906-908

Zener C (1948) Elasticity and Anelasticity of Metals. University of Chicago Press, Chicago

10 Texture and Anisotropy

Hans-Rudolf Wenk

Department of Earth and Planetary Science
University of California
Berkeley, California 94720

INTRODUCTION

The study of preferred orientation of minerals in rocks dates back to Omalius d'Halloy (1833) who attributes a special significance to the alignment of crystals as an indicator of the formation process. Much later the influence of crystal alignment on physical properties was quantified (e.g., Weissenberg 1922, Voigt 1928, Reuss 1929). Only recently has this field emerged as a coherent part of earth science research linking such branches as mineralogy, petrology, structural geology, geodynamics and seismology. The reason for this was the emergence of quantitative methods to analyze preferred orientation, or "texture" as it was first called by Naumann (1850). These methods were largely developed in collaboration with materials science and mechanics. Quantitative measurements, detailed field studies, rigorous data analysis, theories to predict textures, and improvements in characterizing seismic anisotropy in the Earth are leading to a coherent picture that is now being refined.

Though seismologists have long accepted that there is a causal relationship between anisotropic propagation of seismic waves, the deformation field and crystal orientation, the prevailing view is still largely the mythological concept that seismic fast directions align with the flow direction. While this may be approximately the case for olivine deformed under certain conditions, it is certainly no universal law, as we will try to illustrate in this review. The "fast" direction of a crystal depends on the mineral species and its crystal structure. The alignment of crystals depends on microscopic, intra-crystalline deformation systems and the deformation history. Both relationships are complex and not intuitive, but there are well-established theories to compute single crystal physical properties as well as orientation patterns. Simulations can be compared with experimental data and then applied with some caution to the macroscopic Earth.

This review is intended to provide a brief introduction, highlighting some of the issues with examples, and refer new researchers in the field of texture and anisotropy to important publications. Since the classic book of Sander (1950), there have been newer books on texture analysis in earth science (e.g., Wenk 1985, Kocks et al. 2000), as well as published collections of research papers in conference proceedings (e.g., Bunge et al. 1994), special journal issues (Leiss et al. 2000) and the tri-annual *Proceedings of the International Conferences of Textures of Materials* (ICOTOM). These publications should be consulted for details.

MEASUREMENTS OF TEXTURES

Overview

Interpretation of textures has to rely on a quantitative description of orientation characteristics. Two types of preferred orientations are distinguished: The *lattice preferred orientation (LPO)* or 'texture' (also 'preferred crystallographic orientation') and the *shape preferred orientation* (or 'preferred morphological orientation'). Both can be correlated, such as in sheet silicates with a flaky morphology in schists, or fibers in fiber reinforced ceramics. In many cases they are not. In a rolled cubic metal the grain shape depends on the deformation rather than on the crystallography.

Many methods have been used to determine preferred orientation. Geologists have extensively applied the petrographic microscope equipped with a universal stage to measure the orientation of morphological and optical directions in individual grains (e.g., Phillips 1991, Wahlstrom 1979, Wenk 1985). Metallurgists have used a reflected light microscope to determine the orientation of cleavages and etch pits (e.g., Nauer-Gerhardt and Bunge 1986). With advances in image analysis, shape preferred orientation can be determined quantitatively and automatically with stereological techniques. Optical methods of LPO measurements of some minerals have also been automated (Heilbronner and Pauli 1993, Stöckhert and Duyster 1999)

Today diffraction techniques are most widely used to measure lattice preferred orientation (see e.g., Bunge 1986, Snyder et al. 1999, Wenk 2000). X-ray diffraction with a pole figure goniometer is a routine method. For some applications synchrotron X-rays provide unique opportunities. Neutron diffraction offers some distinct advantages, particularly for large bulk samples. Electron diffraction using the transmission (TEM) or scanning electron microscope (SEM) is gaining interest, because it permits one to correlate microstructures, neighbor relations and texture.

There are two distinct ways to measure orientations. One way is to average over a large volume of a polycrystalline aggregate. A pole figure collects signals from many crystals and spatial information is lost (for example, misorientations with neighbors), but also some orientation relations (such as how x, y, and z-axes of individual crystals correlate). The second method is to measure orientations of individual crystals. In that case orientations and the orientation distribution can be determined unambiguously. Also, if a map of the microstructure is available, the location of a grain can be determined and relationships with neighbors can be evaluated.

X-ray pole figure goniometer

X-ray diffraction was first employed by Wever (1924) to investigate preferred orientation in metals, but only with the introduction of the pole-figure goniometer and use of electronic detectors did it become a quantitative method (Schulz 1949). The principle is based on Bragg's law: in order to determine the orientation of a given lattice plane, *hkl*, of a single crystallite, the detector is first set to the proper Bragg angle, 2θ of the diffraction peak of interest, then the sample is rotated in the monochromatic beam until the lattice plane *hkl* is in the reflection condition (i.e., the normal to the lattice plane or diffraction vector is the bisectrix between incident and diffracted beam). The goniometer rotations are related to the angular coordinates that define a crystal orientation in the sample reference system. In the case of a polycrystalline sample, the intensity recorded at a certain sample orientation is proportional to the volume fraction of crystallites with their lattice planes in reflection geometry. Determination of texture can be done on a sample of large thickness and a plane surface on which X-rays are reflected (Fig. 1a, *reflection*) or on a thin slab which is penetrated by X-rays (Fig. 1b, *transmission*). Because X-rays are strongly absorbed by matter, the transmission method is generally only applicable to very thin foils or wires (<100 μm) and to materials with relatively low absorption. Because of defocusing effects, as the flat sample surface is inclined against the beam, variations in the irradiated volume, and absorption, intensity corrections are necessary, particularly in reflection geometry. In reflection geometry only incomplete pole figures can be measured, usually to a pole distance of 80°.

Synchrotron X-rays

Conventional X-ray tubes produce a broad beam of relatively low intensity. In a synchrotron, a very fine-focused, high-intensity beam with monochromatic or continuous wavelengths can be produced. The unique advantages of high intensity, small beam size (<5 μm) and free choice of wavelength opens a wide range of new possibilities (e.g., Heidelbach et al. 1999, Wcislak et al. 2002, Wenk and Grigull 2003).

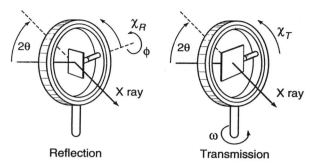

Figure 1. Geometry of an X-ray pole figure goniometer in (a) reflection and (b) transmission.

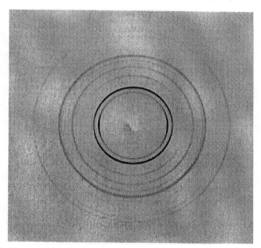

Figure 2. Synchrotron X-ray diffraction. Debye of a sheet of rolled titanium recorded with a CCD camera. Intensity variation along the rings immediately display the presence of texture.

Synchrotron diffraction images, recorded by CCD detectors almost instantaneously display the presence of texture expressed in systematic intensity variations along Debye rings, as illustrated for a rolled sheet of titanium (Fig. 2). While the presence of texture is immediately obvious, elaborate data processing is necessary to determine texture patterns quantitatively and interpret data in a satisfactory way. Figure 3a shows the geometry of a transmission diffraction experiment with incoming X-ray beam, sample and Debye cone with an opening angle 4θ, on which diffracted X-rays lie. If the sample is stationary only lattice planes *hkl* that are inclined by an angle (90°–θ) to the incoming beam diffract, and the corresponding reciprocal lattice vectors lie on a cone with an opening angle of (180°–2θ) which intersects the orientation sphere of the sample in a small circle (Fig. 3b). Intensity variations on the Debye ring are proportional (after corrections for absorption and sample volume) to pole density variations along the small circle. As is evident, coverage of the pole figure from a single image is minimal. The coverage can be improved by tilting the sample around an axis perpendicular to the incident X-ray and recording corresponding images. Each image contains information about pole densities on a rotated small circle in the pole figure (dotted circle in Fig. 3b). But with some data processing, information from a single synchrotron image can be sufficient to determine the orientation distribution and then to reconstruct complete pole figures. Often the use of high energy is advantageous because of good penetration and moderate absorption (Wcislak et al. 2002), as well as small 2θ angles.

Synchrotron analysis is useful for compounds with weak scattering (e.g., polymers and biological materials) and for investigating local texture variations (e.g., in extruded wires or bones). Other applications are *in situ* observations of texture changes during deformation at high pressure with diamond anvil cells (Merkel et al. 2002) and high temperature (e.g., recrystallization during annealing; Puig et al. 2001).

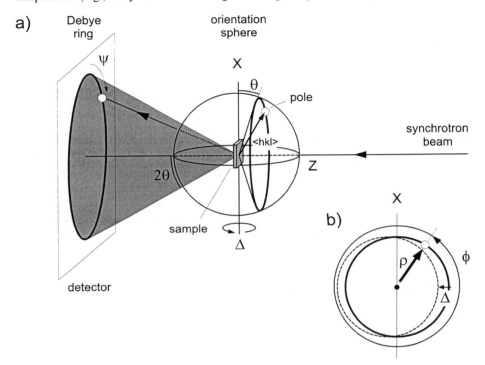

Figure 3. Geometry of a synchrotron X-ray diffraction experiment in transmission. (a) For a given reciprocal lattice vector *hkl* the accessible crystallite orientations lie on a cone which intersects the orientation sphere of the sample in a small circle. Diffracted X-rays lie on a cone (shaded) that intersects a planar detector along a circle. The diffraction spot of a single pole is indicated. (b) Pole figure representation of the small circle that is recorded and the location of a particular orientation, defined by azimuth and pole distance. The dashed circle represents the accessible orientations after tilting the sample by a small increment Δ about the vertical (x) axis (from Wenk and Grigull 2003).

Neutron diffraction

Neutron diffraction was first applied to textures by Brockhouse (1953). Neutron diffraction texture studies are done either at reactors with a constant flux of thermal neutrons, or with pulsed neutrons at spallation sources. The wavelength distribution of thermal neutrons is a broad spectrum with a peak at 1-2 Å. A disadvantage of neutrons is that the interaction of neutrons with matter is low, and long counting times are required. Weak interaction is also a great advantage because it provides high penetration and low absorption making neutrons suitable for bulk texture investigations of large sample volumes in transmission. Intensity corrections are generally unnecessary and complete pole figures can be determined in a single scan. Because of the low absorption, environmental stages (heating, cooling, straining) can be used for *in situ* observation of texture changes, e.g., during phase transformations.

A conventional neutron texture experiment uses monochromatic radiation produced with single crystal monochromators. As with an X-ray pole figure goniometer the detector is set at the Bragg angle for a selected lattice plane *hkl*. The pole densities of that lattice plane in different sample directions are scanned by rotating the sample around two axes to cover the entire orientation range.

It is also possible to use *position-sensitive detectors* that record intensities along a ring (1-D), or over an area (2-D), rather than at a point. A 2θ spectrum can be recorded simultaneously. Another method to measure spectra is with a single detector at a fixed position but with *polychromatic* neutrons and a detector system that can identify the energy of neutrons,, e.g., by measuring the time of flight (TOF) (Wenk 1994). The new TOF neutron diffractometer HIPPO at Los Alamos is dedicated to texture research with 50 detector panels that record simultaneously diffraction spectra from crystals that are in different orientations (Fig. 4) (Bennett et al 1999).

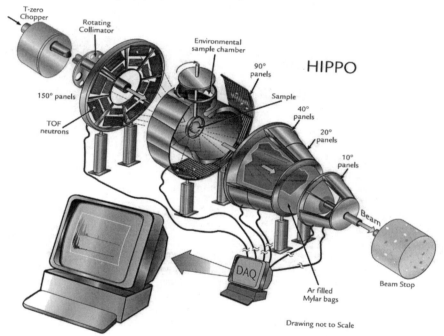

Figure 4. Schematic of the time-of-flight neutron diffractometer HIPPO at Los Alamos National Laboratory. Multiple detector banks are arranged on rings. Each ring (at different 2θ) records reflections of differently oriented lattice planes so that the pole figure is covered simultaneously (Bennett et al. 1999).

Transmission electron microscope (TEM and HVEM)

The transmission electron microscope (TEM) offers excellent opportunities to study textural details in fine-grained aggregates. Like light microscopy, the TEM not only provides information about orientation but also about grain shape and, more importantly, about dislocation microstructures indicative of active deformation mechanisms. There are many applications of Kikuchi patterns for orientation analysis (e.g., Humphreys 1988, Høier et al. 1994, Engler 1996). The procedure has recently been automated (Schwarzer and Sukkau 1998).

Figure 5. EBSP image of experimentally deformed halite.

Scanning electron microscope (SEM)

Local orientations can also be measured with the scanning electron microscope (SEM) and this technique is becoming very popular because it does not require much background in texture theory from the user and provides immediate results (e.g., Randle and Engler 2000). Unlike the TEM, the SEM it is not restricted to thin areas located along the edge of a hole in the specimen, but enables crystal orientations to be determined on surfaces of considerable extent. Interaction of the electron beam with the uppermost surface layer of the sample produces electron back-scatter diffraction patterns (EBSP or EBSD) that are analogous to Kikuchi patterns in the TEM. EBSPs are captured on a phosphor screen and recorded with a low intensity video camera or a CCD device. The procedure of indexing the patterns and scanning a specimen surface has been automated (Wright and Adams 1992). The sample is translated with a high precision mechanical stage or sample locations are reached by beam deflection in increments as small as 1 μm. At each position an EBSP is recorded. With a phosphor screen, back-scattered electrons are converted to light, this signal is transferred into a camera. The digital EBSP (Fig. 5 is an image of halite) is then entered into a computer and indexed, making use of the Hough transform (e.g., Kunze et al. 1993). Specimen coordinates, crystal orientation, a parameter describing the pattern quality and a parameter evaluating the pattern match are recorded. Then the sample is translated to the next position and the procedure is repeated. This sounds like an ideal technique. However, it is only applicable to crystals with fairly low dislocation densities, surface preparation is critical and the automatic indexing procedure is not always reliable. Failure to index and misindexing of patterns both are orientation-dependent and can produce texture artifacts. Furthermore there are statistical limitations that will be discussed below.

Comparison of methods

The actual choice of texture measurements depends on many variables, such as availability of equipment, material to be analyzed, and data requirements.

For routine metallurgical practice and many other applications in materials science and geology, *X-ray diffraction* in reflection geometry is generally adequate. It is fast, easily automated, and inexpensive both in acquisition and maintenance. Transmission geometry has been successfully used for texture analysis of sheet silicates in slates and shales (e.g., Oertel 1983, Ho et al. 1999). Conventional X-ray diffraction texture analysis is restricted to fine-grained materials (<1 mm) whose texture is homogeneous within the plane of the sample. Back-reflection provides incomplete pole figures, but this drawback can be overcome by adequate data-processing techniques. Pole figures can only be measured adequately if diffraction peaks are sufficiently separated. In geological samples and ceramics, X-ray diffraction is therefore generally limited to single-phase aggregates of orthorhombic or higher crystal symmetry. Synchrotron X-rays are used for *in situ* experiments at high pressure and temperature, generally of very fine-grained samples.

Neutron diffraction is advantageous because bulk samples rather than surfaces are measured, coarse-grained materials can be characterized, environmental cells (heating, cooling, straining) are available and angular resolution is generally better than for X-rays. It is possible to measure complex polyphase composites with many closely spaced diffraction peaks. Neutron diffraction requires the user to write proposals for a specific experiment and access is limited to a few days of beam time, generally only sufficient for a few selected examples.

Electron diffraction with a TEM is most time-consuming but provides, in addition to crystal orientation, valuable information about microstructures and, at least two-dimensionally, about interaction between neighbors and about heterogeneities within grains. These are important data to interpret deformation processes.

Recently *EBSPs*, measured with the *SEM* on polished surfaces, have become very popular. They allow for determination of local orientation correlations, which are important in the study of recrystallization, or strain concentrations where failure may occur. With the possibility of automation, this technique has become comparable in expense and effort to X-ray diffraction analysis, but information is different. There are limitations for samples with many planar and linear lattice defects.

DATA ANALYSIS

Orientation distributions

In quantitative texture analysis the coordinate systems of the sample and of the crystals need to be related. This requires three parameters, such as the classical Euler angles that relate two orthogonal coordinate systems (crystal: a,b,c and sample: x,y,z) through rotations (Fig. 6), or an axis-angle specification that brings the two coordinate systems to coincidence through a single rotation about a specific axis. (Both these representations of orientations were originally introduced by Leonhard Euler in 1775.) Other descriptions exist and have been discussed in detail in textbooks (e.g., Kocks et al. 2000). In the case of a polycrystal, an orientation becomes an orientation probability distribution of the three quantities and is described by an orientation distribution function (ODF). Knowledge about the ODF is required to determine physical tensor properties of polycrystals.

Experimental texture data have various forms. With electron microscopes and the Universal stage petrographic microscope single orientations are measured. This has the advantage that an ODF can be determined directly and unambiguously by entering orientations into the ODF and performing some smoothing and normalization. If continuous pole figures of certain lattice planes (hkl) are measured, then it is more difficult to retrieve the ODF.

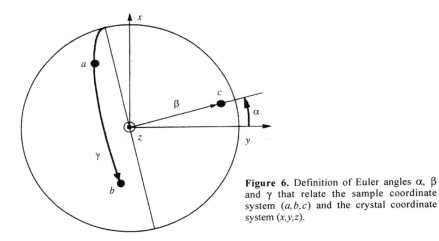

Figure 6. Definition of Euler angles α, β and γ that relate the sample coordinate system (a,b,c) and the crystal coordinate system (x,y,z).

From pole figures to ODF

A pole figure (also called a pole density distribution) is a two-dimensional distribution of a crystal direction (e.g., the pole to a lattice plane *hkl*) relative to sample coordinates. (A direction is specified by two angles.) Pole figures can be considered projections of the three-dimensional ODF. There are various methods to retrieve the ODF from measured pole figures (for a review see Kallend 2000). One set of methods works in direct space and uses basically algorithms of tomography. The WIMV (Williams-Imhof-Matthies-Vinel) method, introduced by Matthies and Vinel (1982), is most widely used. Variants are the vector method (Ruer and Baro 1977), ADC (Arbitrarily Defined Cells) (Pawlik et al. 1991), and the maximum entropy method (Liang et al. 1988, Schaeben 1988). Other methods work in Fourier space, most notably the harmonic method introduced by Bunge (1965) and Roe (1965). In this method pole densities are expanded with spherical harmonics. Harmonic coefficients from the pole figure expansion can then be used to determine harmonic coefficients of the ODF expansion. The expansion is carried to a finite order, usually between 22 and 32, providing an angular resolution of about 15-10°.

All methods yield similar results, at least for ideal test data. However, there is some ambiguity in continuous pole figures. This is most transparent for the harmonic method. Pole figures are, by their very nature, centrosymmetric, which means that odd coefficients vanish. The ODF in general is not centrosymmetric and therefore requires even and odd coefficients for a full representation, but the odd coefficients cannot be obtained from pole figures (Matthies 1979). Omissions of odd coefficients can introduce errors in pole densities in the ODF and add spurious maxima and minima, called ghosts (for a review see Wenk et al. 1988).

While the ODF is necessary to calculate pole figures and physical properties, the 3-D ODF is difficult to visualize (e.g., Wenk and Kocks 1987). In this review we will only use pole figures (representing the distribution of a crystal direction relative to sample coordinates) and inverse pole figures (representing the distribution of a sample direction relative to crystal coordinates) for texture representations. There are several software packages that calculate ODFs from pole figures and perform other operations to quantify textures in polycrystals (e.g., BEARTEX, Wenk et al. 1998; LaboTex, Pawlik et al. 1991; MulTex, Helming 1994; POPLA, Kallend et al. 1991, TexTools, Resmat Corp.). Details can be obtained from the Internet.

Use of whole diffraction spectra

Traditionally texture analysis has relied on pole figure measurements. This is efficient if only a few pole figures are required for the ODF analysis and if diffraction peaks are reasonably strong (relative to background) and well separated. The method becomes increasingly unsatisfactory for complex diffraction patterns of polyphase materials and low symmetry compounds with many closely spaced and partially or completely overlapped peaks. The amount of texture information is roughly proportional to the product of the number of pole figures (hkl) times the number of sample orientations. In conventional ODF analysis one relies on a few pole figures and many sample orientations. Another approach is to use many pole figures and few sample orientations. This is an obvious advantage for TOF neutron diffraction where many diffraction peaks are measured in a continuous spectrum.

Rietveld (1969) proposed a method to use powder patterns to obtain crystallographic information (e.g., Young 1993) and this method can be expanded to obtain texture information. In a powder with a random orientation of crystallites, the relative intensities are the same for all sample orientations and are due to the crystal structure. In a textured material, there are systematic intensity deviations from those observed in a powder as illustrated in Figure 7 for neutron diffraction data of eclogite. Intensities are linked to the crystal structure by means of the structure factor, they are also linked to the texture through the ODF.

As with the pole figure method described above, texture effects can be implemented in the Rietveld method either with Fourier or with direct methods. The finite number of harmonic ODF coefficients can be refined in a similar way as crystallographic parameters with a non-linear least-squares procedure (Popa 1992, Ferrari and Lutterotti 1994, Von

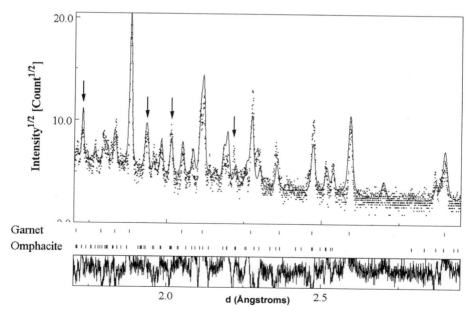

Figure 7. Rietveld fit of a time-of-flight neutron diffraction spectrum of eclogite from Dabie Shan, composed dominantly of omphacite and garnet. The solid line gives the fit, assuming a random aggregate; dots are experimental data. Deviations (some marked by arrows) are due to texture (Wenk et al. 2001).

Dreele 1997). With discrete methods ODF values are directly related to peak intensity values in the spectra. In this case it is more efficient to separate crystal structure and texture, and proceed in iterations. Intensity deviations can be extracted as arbitrary weights, which are then used to calculate the ODF.

In practice, first instrumental parameters (including a bulk scaling intensity, a peak width function, and a zero offset) are refined. Then spectral backgrounds are refined as polynomials. Next, lattice parameters as well as crystal structure parameters (atomic coordinates and temperature factors), are refined. The final step is to refine the texture. The Rietveld method has so far been applied to several mineral textures, including calcite (Lutterotti et al. 1997, Von Dreele 1997), plagioclase (Xie et al. 2002) and eclogite (Wenk et al. 2001). In all cases resolution was excellent. The method is implemented in software packages GSAS and MAUD that can both be downloaded from the internet.

Statistical considerations of single orientation measurements

With single orientation measurements that rely on surface coverage, the number of grains that can be measured is limited. This becomes apparent if we consider that a texture function (ODF) with 5° resolution has 181584 cells in the case of triclinic crystal symmetry. Even if we had that many grains and a random texture, some ODF cells would have 0 grains, most would have 1 grain and there would be cells with 2, 3, 4 or more grains, i.e., the ODF would range between 0 and 4 multiples of a random distribution (m.r.d.) For fewer grains the situation is worse. Either larger cells (and worse angular resolution) have to be used, or data have to be smoothed in a statistically correct way. In the case of a rock with a grain size of 0.5 mm, a 1 cm cube used for neutron diffraction contains 8000 grains, the surface of a 2 cm × 2 cm thin section contains at most 1600 and statistics are limited. For single orientations (rather than an averaged diffraction intensity) this is expressed in exaggerated pole densities and, particularly, a large texture index F_2. F_2 has been introduced by Bunge (1982) as a bulk measure of texture strength and is equal to the volume-averaged integral of squared orientation densities over the ODF. Therefore, F_2 is mainly influenced by sharp texture peaks (>>1 m.r.d.). The texture index is equal to 1.0 for a random texture. Matthies and Wagner (1996) have explored the relationship between number of measured grains N and F_2, and established a linear $1/N$ dependence of $F_2(N)$ that can be used to determine the smoothing function for a given sample.

This has been illustrated for a sample of anorthosite mylonite, with triclinic plagioclase measured both with neutron diffraction and EBSP (Xie et al. 2002). For neutron diffraction the texture index F_2 is 3.1. For EBSP data the texture index decreases as a function of the number of analyzed grains. Figure 8 shows the dependence of F_2 as a function of increasing number of orientations (from right to left), using constant $x = 1/N$-steps. As can be seen, F_2 decreases, first with stochastic oscillations, but for an increasing number of orientations more and more in a linear fashion. For 400 grains it is 600, for 1000 220, and for 5000 grains (the maximum measured in this EBSP experiment) it is still 50. If the linear behavior becomes stable and the resulting straight line is extrapolated for infinite N, the intersection of this line at $x = 0$ with the F_2-axis provides the "true" texture coefficient F_2 which is closer to the actual index determined by neutron diffraction (3.1). (For the physically meaningful N one has to take the number of the measured grains and not the number of measured EBSD-grid points!) The true texture strength can be obtained by smoothing discrete data, e.g., with a Gauss filter. The width of the filter function is not known a priori but depends on the texture type, the texture strength and the number of measured data points. In the case of the anorthosite illustrated in Figure 8, it is 18°.

Thus, EBSP data may provide good *qualitative* information on texture patterns, but neutron diffraction on bulk samples is often needed to obtain *quantitative* information on texture strength and thus anisotropy of physical properties. In most published EBSD texture analyses, arbitrary smoothing is applied (e.g., by fitting orientations with harmonic or Gauss functions) and results are, therefore, at best semi-quantitative.

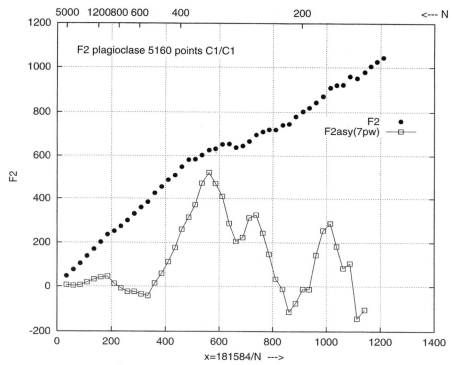

Figure 8. Dependence of the texture strength, described by the texture index F_2, on the number of orientations measured by EBSD, for the case of plagioclase. The "true" texture index, established by neutron diffraction measurements on the same sample is 3.2. Shown is the dependence of F_2 (●) and F2asy (□), which is the extrapolation for F_2 over the previous 7 data points (Xie et al. 2002).

From textures to elastic anisotropy

If we know the orientation distribution and single crystal physical properties, then we can calculate approximate polycrystal physical properties. In geophysics, the interest has been focused on elastic properties, described as a fourth rank tensor and related to the propagation of seismic waves. Single-crystal elastic constants for many materials under a variety of conditions have been compiled (e.g., Simmons and Wang 1971). Polycrystal elastic properties are obtained by a summation over all contributing single crystals, taking into account their orientation, the relationship with neighbors and microstructure, such as shape of crystallites, presence of pores and cracks. If there is preferred orientation (non-random ODF), a macroscopic anisotropy will result. In the case of polycrystal elastic properties, this summation ought to be done to both maintain continuity across grain boundaries when a stress is applied, and to minimize local stress concentrations. Recently practical analyses of this type have been carried out with finite element simulations (e.g., Dawson et al. 2001).

Mostly the local stress and strain distribution is neglected and the summation is done by simple averages. There are two extreme cases: The Voigt average assumes constant strain throughout the material and applies strictly to a microstructure composed of sheet-shaped laths with a tensile stress applied parallel to the sheets (Fig. 9, top). Values for elastic constants are a maximum. The Reuss average assumes that stress is constant and strictly applies to the case of a microstructure in which tensile stress is applied perpendicular to the sheets (Fig. 9, bottom). In this case aggregate elastic constants are a minimum. There are other averages that are intermediate between constant strain and constant stress. One example is the Hill average (1952), an arithmetic mean of Voigt and Reuss averages, or the geometric mean (Aleksandrov and Aisenberg 1966, revived by Matthies and Humbert 1993), or a self-consistent average proposed by Kröner (1961).

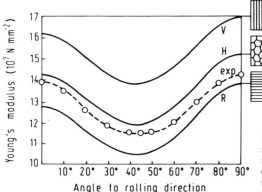

Figure 9. Young's modulus of a copper sheet as a function of the direction in the sheet plane (exp.) compared with the Reuss (R), Voigt (V) and Hill (H) approximations that are valid for different microstructures (Bunge 1985).

Detailed numerical procedures to calculate these averages from the ODF, either from discrete orientations or a continuous ODF, have been discussed in several reviews (e.g., Bunge 1985, Tomé 2000). The relationship between texture, single crystal properties and polycrystal properties is applied in various ways. In earth science, the seismic anisotropy of rocks or of model rocks with preferred orientation is determined to estimate the contribution of anisotropy from texture in the deep earth, compared to other factors such as compositional layering and melt; in engineering this relation is calculated to estimate performance of materials. Another application is to measure bulk elastic anisotropy and estimate from it texture. While such a determination is neither accurate nor complete, it can nevertheless be practical, e.g., to estimate crystal alignment and corresponding flow regimes in the lower mantle and inner core of the earth from seismic anisotropy measurements, or take advantage of destruction-free acoustic velocity measurements of large engineering components to ascertain specifications and conceivable damage. A field that is receiving a lot of attention is to determine single crystal elastic properties from diffraction measurements of elastic strain (changes in lattice spacings) on deformed textured polycrystals (e.g., Singh et al. 1998, Gnäupel-Herold et al. 1998, Matthies et al. 2001).

POLYCRYSTAL PLASTICITY SIMULATIONS
General comments

There are several reasons why it is desirable to simulate the development of texture and anisotropy during deformation. For engineering applications an important consideration is the cost of experiments versus simulations to obtain information on properties after specific treatments. In earth science it is often not possible to reproduce the complex strain paths that occur in nature. Most deformation experiments on rocks are done in compression geometry and, more recently, also in torsion. Both paths are very special cases that are rarely satisfied in actual geological conditions. A second reason is that conditions in nature can often not be reproduced in experiments, most significantly strain rates and grain size, e.g., at high pressure. If microscopic mechanisms are known that are active under a given set of conditions and if a good constitutive theory exists, then polycrystal behavior for any strain path can be simulated, including, for example, heterogeneous subduction of slabs to the bottom of the mantle.

Different mechanisms can produce or modify texture. Most important is dislocation glide, which we will discuss in some detail in the next section. Also significant is recrystallization, either dynamic or static, with nucleation of new domains and grain boundary mobility. If fluids are present, aspects of dissolution and growth in a stress field can have a profound influence on resulting orientation patterns (e.g., Spiers and Takeshita 1995, Bons and den Brok 2000).

Deformation

Deformation of a polycrystal is a very complicated heterogeneous process. When an external stress is applied to the polycrystal, it is transmitted to individual grains. Dislocations move on slip systems, dislocations interact and cause "hardening", grains change their shape and orientation, thereby interacting with neighbors and creating local stresses that need to be accommodated. To realistically model these processes is a formidable task and only recently have three-dimensional finite element formulations been developed to capture at least some aspects (Mika and Dawson 1999). The difficulty is that in real materials local stress equilibrium and local strain continuity are maintained, and this requires local heterogeneity at the microscopic level. Most of the polycrystal plasticity simulations have used highly simplistic approximations, e.g., that each grain is homogeneous, and yet arrived, at least for moderate strains, at useful results.

There are two extreme assumptions. Taylor (1938) suggested that in modeling plastic deformations, straining could be partitioned equally among all crystals. In Figure

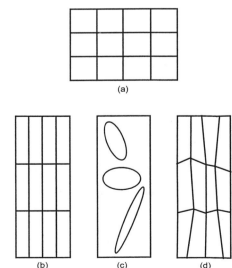

Figure 10. Two-dimensional cartoon comparing deformation of a square-shaped polycrystal (a) to a rectangle with different models. (b) In the Taylor model homogeneous deformation is imposed and all grains deform from a square to a rectangular shape independent of orientation and neighbors. (c) In the self-consistent model deformation of a grain as an inclusion in an anisotropic medium is assumed. Grains deform (and rotate) dependent on their orientation. (d) In the finite element approach deformation depends both on orientation and neighborhood, causing heterogeneous deformation but maintaining compatibility.

10 a microstructure of squares (a) is deformed to rectangles (b). This hypothesis has been used extensively for fcc and bcc metals (e.g., Van Houtte 1982, Kocks 2000). For this approach even to be viable, the individual crystals must each be able to accommodate an arbitrary deformation, requiring five independent slip systems. While the Taylor assumption is reasonable for materials comprised of crystals with many slip systems of comparable strength, using it in other situations can lead to prediction of excessively high stresses, incorrect texture components, or both. In the Taylor model, high stresses are required to activate slip systems, even in unfavorably oriented grains, and the model is therefore known as an *upper bound model*.

In contrast to the Taylor hypothesis, all crystals in a polycrystal can be required to exhibit identical stress, given that their behavior is rate-dependent at the slip system level. This is a variant of the original Sachs (1928) assumption for rate-independent behavior in which the stresses in the crystals throughout an aggregate share a common direction. The equal stress hypothesis is most effective for polycrystals comprised of crystals with fewer than five slip systems. It has also been used successfully in modeling the mechanical response of lower symmetry crystals that possess adequate numbers of independent slip systems, but whose slip systems display widely disparate strengths. The principal drawback is that deformation is often concentrated too highly in a small number of crystals, leading to inaccurate texture predictions. With the Sachs approach only the most favorable slip systems are activated and, therefore, stresses are low. This approach is known as a *lower bound model*.

Several other approaches have been developed for modeling the heterogeneous deformation of highly anisotropic polycrystals as is the case for many rocks. For example, Molinari et al. (1987) developed the viscoplastic self-consistent (VPSC) formulation for large strain deformation in which each grain is regarded as an inclusion embedded in a viscoplastic homogeneous equivalent medium whose properties coincide with the average properties of the polycrystal. An original sphere deforms to an ellipsoid. The deformation of individual grains depends on their orientation (Fig. 10c). The original VPSC formulation assumes the equivalent medium to be isotropic. A more general formulation with an anisotropic medium was introduced by Lebensohn and Tomé (1993). VPSC has been successfully applied to the prediction of plastic anisotropy and texture development of various metals (e.g., Tomé and Canova 2000) and geologic materials (e.g., Wenk 1999). In most cases, the VPSC simulations improve earlier predictions obtained with the Taylor model and provide better overall quantitative agreement with experiments. The VPSC method has been refined by adding adjustable parameters to either emphasize stress equilibrium or compatibility, thereby improving predictions of mechanical properties (Molinari and Toth 1994, Molinari 1999).

Another modeling approach is to employ finite element methodologies to compute deformations of an aggregate of crystals. In this case local heterogeneity can be taken into account. One example is the hybrid element polycrystal approach of Beaudoin et al. (1995) (Fig. 10d). Every grain is either discretized with a single element (Sarma and Dawson 1996) or with many elements (Mika and Dawson 1999). The boundary value problem resulting from the application of homogeneous macroscopic boundary conditions is solved to obtain the deformation of individual crystals. In this case the rotation of a grain and its deformation depends both on the orientation and the orientations of neighbors.

All polycrystal plasticity models are comprised of two basic parts: a set of crystal equations describing properties and orientations, and a set of equations that link individual crystals together into a polycrystal. The latter set provides the means to combine the single crystal quantities to define the polycrystal response on the basis of physically motivated assumptions regarding grain interactions. The single crystal equations are often approximated with a power law relation between the resolved shear stress on a slip (or twinning) system, and the rate of shearing on that system. Slip systems are defined by crystallographic slip plane and slip direction, a critical resolved shear stress and a stress exponent. Values are usually obtained from experiments. The slip system strengths may evolve with deformation. The evolution of the reference shear stress with strain is often described with a Voce type hardening law (Kocks 1976). As input for polycrystal plasticity simulations one also needs a set of initial orientations that may be random or display preferred orientations. Initial grains may have different grain shapes and different resolved shear stresses due to prior deformation history. Usually it is assumed that all initial grains are identical and that the grain shape is equiaxed but evolves during deformation.

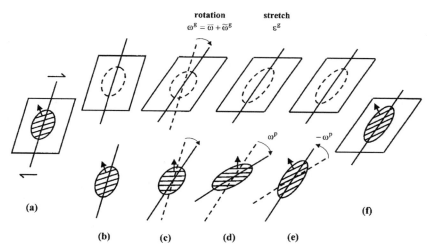

Figure 11. Idealized sequence illustrating the stretch and rotation of the grain during an incremental deformation step. The displacements of the grain (inclusion, bottom) and the cavity in which it fits (top) are described separately. (a) and (b) initial state; (c) rigid rotation of grain and cavity in simple shear without stretch; (d) stretch of grain and cavity (same) misaligns the main axes of the ellipsoid; (e) counter-rotation of the grain realigns the axes of grain and cavity; (f) final state with effective rotation (courtesy of C. Tomé).

As a crystal deforms by slip, it undergoes a lattice rotation. This rotation is illustrated in Figure 11 for the VPSC scheme by separating crystal (inclusion) (bottom) and the corresponding cavity in the medium (top). A macroscopic rigid-body rotation in simple shear deforms the sample to a different shape (Fig. 11b→c). Activity of slip systems deforms the crystal to a new shape (stretch) and misaligns the main axes of the ellipsoid (Fig. 11c→d). A counter-rotation of the grain (Fig. 11d→e) realigns the axes of grain and cavity and produces a final state with an effective crystal rotation (Fig. 11f) compared to the initial state (Fig. 11a). These lattice rotations are the cause of development of preferred orientations in aggregates with many component crystals. All polycrystal plasticity models deal with a highly non-linear system. A solution is obtained by working with a finite number of discrete grains (given as orientations) and deformation is applied in increments. The deformation is defined by a displacement gradient tensor that may be constant or may change with deformation. Rotations of all grains are calculated after each strain increment and the orientations, as well as their shape and slip system activities are then updated. There are other models for texture development, not based on polycrystal plasticity, but using continuum mechanics.

Recrystallization

The relationship between slip and crystal rotations is straightforward. Other processes such as climb, grain boundary sliding, diffusion in general may also affect orientation distributions. Of particular importance in geological situations is recrystallization.

For metamorphic petrologists, recrystallization means growth of new mineral assemblages to attain thermodynamic equilibrium in different temperature and pressure conditions. In deformation studies, recrystallization is the development of strain-free regions, either during deformation (dynamic recrystallization, Guillope and Poirier 1979) or after deformation (static recrystallization). It is agreed that also here energy considerations are responsible for the development of recrystallized domains. Some theories suggest that thermodynamic equilibrium in a non-hydrostatic stress field controls recrystallization (discussed in various forms by Kamb 1961, Paterson 1973, Green 1980, Shimizu 1992), but at least in metals it is clear that strain energies produced by accumulations of dislocations far exceed thermodynamic energies imposed by the stress field (Humphreys and Hatherly 1995). More highly deformed grains have a higher strain energy than less deformed grains (Haessner 1978). Those grains may be consumed through boundary migration by less deformed grains (growth). Alternatively, dislocation-free nuclei may form in highly deformed grains and then grow at the expense of others. If growth is controlling, "hard" grains with little deformation dominate the texture and highly deformed grains will disappear. If nucleation is prevalent, "soft" grains develop nuclei that will ultimately grow and dominate the fabric. In order for a boundary to become mobile, there is also a requirement for a significant misorientation. In geological materials recrystallization is often evident as grain-size reduction (e.g., Drury and Urai 1990, Karato 1987, 1988). This may be due to nucleation that often initiates along grain boundaries and twins, or it may be due to subgrain formation during recovery with large misorientations. The importance of changes in grain size distribution during recrystallization has been analyzed by Shimizu (1999).

The deformation state of a grain depends on its history and its orientation and can thus be predicted with polycrystal plasticity theory. The changes in texture and grain size that occur during annealing, and their dependence on microstructural mechanisms provides a logical link to develop detailed recrystallization models, which couple deformation models with probabilistic laws to simulate recovery and recrystallization. Among them, a model developed by Radhakrishnan et al. (1998) couples the finite element method (FEM) with the Monte Carlo technique so as to account for local effects in aggregates. Solas et al. (2001) introduced a self-consistent model in which each grain is divided into small domains.

A simpler approach which does not consider grain topology and misorientations across grain boundaries was proposed by Jessell (1990), evaluating grain deformation with the Taylor model. Heterogeneous orientation-dependent deformation can be more rigorously approached with the self-consistent theory, and this was used in a model to simulate texture development in static and dynamic recrystallization, based on a balance between grain growth and nucleation (Wenk et al. 1997).

Deformation simulations with the self-consistent model provide a population of grains with a variation in deformation and correspondingly in dislocation density. The microstructural hardening of slip systems during deformation provides an incremental strain energy to grains after each deformation step. Grains with a high stored energy are likely to be invaded by their neighbors with a lower stored energy. In the model the stored energy of each grain is compared with the average stored energy of the polycrystal (calculated from the increments in shear stresses). If the stored energy of a grain is lower than the average, it grows; if it is higher, it shrinks. The grain-boundary velocity is proportional to the difference in stored energy. A grain may disappear. After each simulation step grain sizes are renormalized so that the average over the whole aggregate remains constant.

If nucleation of strain-free domains accompanies boundary migration, a highly deformed parent grain divides upon reaching a threshold strain rate and produces an undeformed nucleus. The nucleus (which may be a subgrain or a bulge in a grain boundary) takes on the current orientation of the parent at the time of its formation, but its strain is reset to zero. This has an effect on the subsequent evolution, because these strain-free domains can grow much faster. In this case highly deformed grains dominate the final texture. Recrystallization conditions in the model are controlled by a few parameters such as a boundary migration velocity and a minimum grain volume below which a grain vanishes. Nucleation takes place if the strain increment exceeds a threshold value. With these parameters growth and nucleation can be balanced.

The model was successfully used to simulate static and dynamic recrystallization textures in geologic materials such as quartz (Takeshita et al. 1999), calcite (Lebensohn et al. 1998) and olivine (Wenk and Tomé 1999). A similar approach was applied by Thorsteinsson (2001, 2002) to ice, and Kaminski and Ribe (2001) applied a kinematic model to olivine. All these approaches are highly simplistic to the complex and still poorly understood process of recrystallization, but they provide methods to investigate possible changes in bulk anisotropy during recrystallization.

APPLICATIONS OF POLYCRYSTAL PLASTICITY

Introduction

Polycrystal plasticity simulations have been applied to many mineral systems, beginning with the early application of the Calnan and Clews (1950) model to calcite (Turner et al. 1956), the Taylor theory to halite (Siemes 1974) and quartz (Lister et al. 1978). Among the minerals that have been studied are quartz, calcite, dolomite, halite, periclase, olivine, enstatite, ε-iron and ice. There are several recent reviews of these geological applications of texture simulations (e.g., Schmid 1994, Skrotzki 1994, Wenk and Christie 1991, Wenk 1999). In this section we will just highlight a few case studies where polycrystal plasticity adds to a better understanding of plasticity in the crust, mantle and core of the earth. This discussion will then be followed by a section where, so far, simulations have failed to provide satisfactory answers.

Coaxial thinning versus non-coaxial shearing (calcite)

For structural geologists the aim is often to unravel the detailed strain history and textures are significant because they are generally sensitive to the path. For example, has a tectonic zone been subject to non-coaxial shearing in a shear zone (Fig. 12, bottom) or coaxial crustal thinning (Fig. 12, top). Both paths can lead to an identical finite strain. Geologists have used asymmetric microstructures to infer the presence of shearing and the sense of shear (e.g., Platt and Vissers 1980). Textures provide a means to quantify the amount of non-coaxial deformation. One of the universal principles of texture interpretation is symmetry and it has been widely applied in geological situations: The texture symmetry cannot be lower than the symmetry of the strain path (Paterson and Weiss 1961), if it started out uniform. For coaxial deformation one expects orthorhombic pole figures, whereas a non-coaxial path is likely to produce monoclinic pole figures.

Texture patterns of calcite in deformed marbles have been used to determine the partitioning of deformation into a coaxial deformation component and a non-coaxial (simple shear) component and to infer the deformation history. The philosophy is to first develop a deformation model and test it by comparing resulting texture patterns with experiments. Once mechanisms are established and the model adequately predicts the experiment, one can then apply the model to any arbitrary strain path, including those that cannot be approached experimentally. For calcite, calculated (0001) pole figures for plane strain document a symmetrical (orthorhombic) pure shear pole figure and an

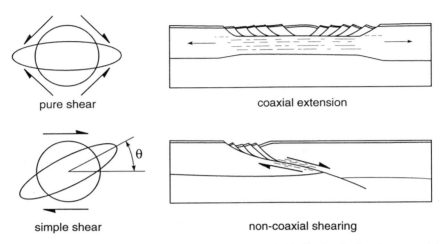

Figure 12. Cartoon comparing crustal deformation by coaxial thinning (top) and non-coaxial shearing (bottom). The finite strain (shape of ellipse) may be the same, but in simple shear the ellipse is inclined by an angle θ to the shear plane.

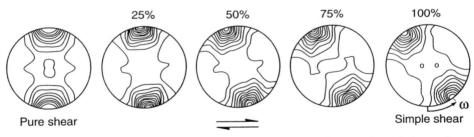

Figure 13. (0001) pole figures of calcite obtained with the Taylor model for 100% equivalent strain and using resolved shear stress ratios corresponding to low temperature deformation. Pure shear on left, mixed modes in the center, and simple shear on right. Note the increasing asymmetry of the maximum (inclined by ω to the shear plane normal). Sense of shear is indicated (from Wenk et al. 1987).

asymmetrical (monoclinic) simple shear pole figure (Fig. 13) and this agrees with experiments (Wagner et al. 1982, Kern and Wenk 1983, Schmid et al. 1987, Pieri et al. 2000). Simulations can provide intermediate states and the relative amount of simple shear can be quantified by measuring the angle of asymmetry ω between the (0001) maximum and the shear-plane normal. One can construct an empirical determinative diagram to assess the amount of simple shear from the asymmetry of the (0001) texture maximum (Wenk et al. 1987) (Fig. 14a).

In practice, geologists collect oriented rock samples in the field, then measure pole figures in the laboratory relative to geological coordinates, such as schistosity plane and lineation direction which define the shear plane and shear direction, respectively. From the asymmetry ω of the 0001 pole figure maximum relative to the shear plane the sense of shear can be inferred. From the angle of asymmetry and using the determinative diagram in Figure 14a, the strain partitioning can be estimated. Whereas many marbles in core complexes of the Western United States show almost symmetrical patterns of (0001) axes (Erskine et al. 1993; Fig. 14b) and presumably formed largely by coaxial crustal extension, limestones from the spreading nappes in the Alps have generally highly asymmetric texture patterns attributed to shearing on thrust planes (Dietrich and Song 1984, Ratschbacher et al. 1991; Fig. 14c).

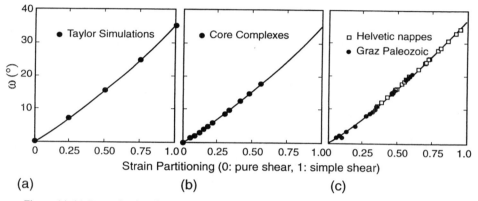

Figure 14. (a) Determinative diagram with angle of asymmetry ω versus strain partitioning factor as obtained from Taylor simulations (Fig. 13). Results from (b) marble mylonites from core complexes of the American Cordillera (Erskine et al. 1993), and (c) various limestones from Alpine spreading nappes (Ratschbacher et al. 1991).

Anisotropy in the upper mantle (olivine)

Anisotropy of physical properties in polycrystals can have various sources. It may be due to a layered arrangement of components or due to crystallographic preferred orientation. In geological materials the presence of oriented microfractures, presence of partial melt and, to a lesser extent, the grain shapes are also influential. Wave propagation through porous media and media with oriented fractures has been studied by Crampin (1981). Velocities through a rock with oriented fractures are much higher parallel to the fractures than across them. With increasing pressure microfractures close and their influence on anisotropy diminishes (e.g., Kern 1993). Above 500 MPa, microfractures are largely closed.

Seismic anisotropy in the upper mantle has received much attention from seismologists (e.g., Forsyth 1975, Silver 1996). The upper mantle is largely composed of olivine and anisotropy evolves as olivine rocks deform (e.g., Karato and Wu 1993). Olivine crystals exhibit about a 25% difference in longitudinal (P) velocities between the slowest [010] and the fastest crystal directions [100]. In an aggregate with preferred orientation of component crystals, one expects a directional dependence of seismic wave propagation. Savage and Silver (1993), based on seismic transverse (S) wave data, proposed a velocity model for the vicinity of the San Andreas fault in California with an upper layer in which the fast velocity direction is more or less parallel to the San Andreas fault (SE–NW) and a lower layer with a fast direction inclined about 60° to the San

Andreas fault (E–W) (Fig. 15). These results can be interpreted based on polycrystal plasticity modeling.

In the *upper layer* the crust of the western North American plate consists of many belts that are aligned in a SE–NW direction. These units are heterogeneous in composition. Many are rich in sheet and chain silicates with a similar SE–NW alignment, and these silicates have a very high intrinsic seismic anisotropy. Deformation has been predominantly brittle, as evidenced by numerous SE–NW trending faults indicated on Figure 15, and microscopic and macroscopic fractures are present that enhance anisotropy. One may expect, therefore, a seismic signal with a fast vibration direction parallel to the San Andreas fault on the basis of structural heterogeneity.

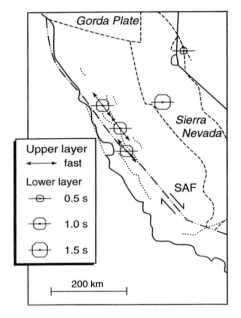

Figure 15. Seismic velocities along the San Andreas fault (SAF) in northern California (Savage and Silver 1993). Dashed lines separate tectonic units, dotted lines indicate structural trends. From interpretations of shear-wave splitting measurements they predict an upper layer with a fast direction parallel to the fault and a lower layer with a fast direction inclined to the fault.

The observed anisotropy in the *lower layer* is very different with a fast direction inclined to the San Andreas fault (E–W). Below 20 km depth, deformation was predominantly ductile and at the high pressure it is unlikely that open fractures could contribute to anisotropy. An imposed shear stress would not result in local fracture but in the development of a ductile shear zone with intercrystalline deformation of component crystals.

We can use polycrystal plasticity to predict the alignment of olivine in a vertical shear zone with a horizontal shear direction underneath the San Andreas fault. The orientation patterns for increasing simple shear (ε_{eq} = 50%, 100%, 150% and 200%) are shown as [100] (fast direction) pole figures in Figure 16 (center row). The patterns display weak preferred orientation at 50%, becoming distinct at 100%, increasing moderately to 200%. The [100] maximum is at about 25° to the shear plane, displaced against the sense of shear. The elastic tensor of these olivine simple shear textures is calculated by averaging and from the elastic tensors wave propagation surfaces are calculated (Fig. 16, bottom). The P-wave velocity maximum is inclined about 25° to the shear plane which is just about what is observed (Fig. 15). If the movements of the Pacific and the North American plates are driven by similar shear displacements in the upper mantle, then the observed type of anisotropy should develop. From the pattern of seismic anisotropy, and assuming that olivine is the major constituent of the shear zone, one can also interpret the correct right lateral sense of shear.

The geologic processes that have been discussed so far occur close to the surface and are accessible to direct sampling and detailed measurements. Indirect observations suggest that deformation occurs also in the deeper parts of the Earth and that anisotropy is produced during convection. In the mantle large cells of convection are induced by instabilities and driven by temperature gradients (e.g., Bunge et al. 1998). The strain distribution in a convection cell is very heterogeneous but can easily be approached with

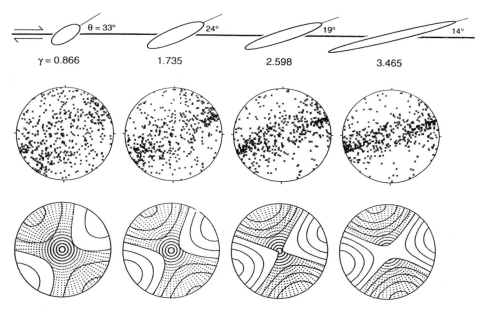

Figure 16. Deformation of olivine in simple shear with increasing strain γ, simulated with the self-consistent theory (Wenk 2000), [100] pole figures represented in equal area projection, shear plane is horizontal. At the bottom P-wave velocity surfaces. Contour interval is 0.1 km/s, dotted below 8.5 km/s. From left to right increasing shear strain ε_{eq} = 50%, 100%, 150% and 200% corresponding to $\gamma = \varepsilon_{eq} \sqrt{3} = 0.87, 1.74, 2.60$ and 3.47.

polycrystal plasticity. On a microscopic scale, olivine is deformed in the upper mantle by intracrystalline processes such as slip of dislocations, and accompanying dynamic recrystallization. Chastel et al. (1993) have used the finite element method that incorporates as a constitutive equation polycrystal plasticity to investigate the development of anisotropy during mantle convection (see also Blackman et al. 1996, Wenk et al. 1999). Figure 17 follows texture development of olivine in the upper mantle along a streamline in [100] pole figures for various model assumptions such as (from left to right) lower bound deformation, self-consistent deformation, growth-dominated dynamic recrystallization and nucleation-dominated recrystallization (Blackman et al. 2002). In all cases a strong texture develops very rapidly during upwelling. The preferred orientation stabilizes during spreading and attenuates during subduction. The pole figures are distinctly asymmetric due to the component of simple shear. While the finite strain along a streamline increases monotonically, the texture does not. There are minor differences in corresponding patterns from different models; for example simulations with the lower bound model produce consistently sharper textures than those with the self-consistent model.

Dawson and Wenk (2000) have mapped textures and seismic anisotropies in a hypothetical convection cell and documented great heterogeneity, even without considering compositional or temperature variations. It is clear that the deforming upper mantle shows structural as well as compositional and thermal variations; all have an influence on seismic velocities that are of similar magnitude. By including polycrystal plasticity, Geophysical deformation modeling is entering a new state of refinement and future convection modeling should include the complexities of anisotropy and mechanical properties in addition to compositional heterogeneity.

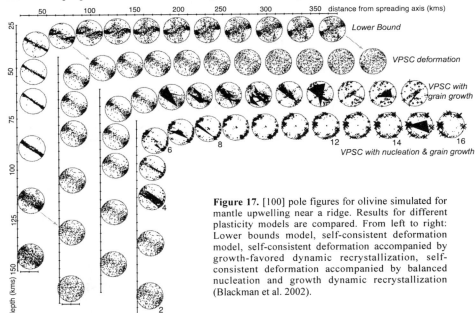

Figure 17. [100] pole figures for olivine simulated for mantle upwelling near a ridge. Results for different plasticity models are compared. From left to right: Lower bounds model, self-consistent deformation model, self-consistent deformation accompanied by growth-favored dynamic recrystallization, self-consistent deformation accompanied by balanced nucleation and growth dynamic recrystallization (Blackman et al. 2002).

Lower mantle

Much less is known about the deeper Earth, because pressures are beyond conditions reached by ordinary deformation devices such as the Griggs, Heard or Paterson apparatus (Tullis and Tullis 1986). The phases that are present in the lower mantle tend to have simpler crystal structures than minerals in the crust, as established by high-pressure experiments and theory (Fiquet 2001). Major phases in the lower mantle include $CaSiO_3$ perovskite, $(Mg,Fe)SiO_3$ silicate perovskite, $(Mg,Fe)O$ magnesiowüstite and possibly SiO_2 stishovite. Very little is known about the deformation mechanisms of these phases at the conditions relevant to the Earth's mantle.

With diamond anvil cells and radial geometry (Fig. 18) texture changes during deformation can be observed *in situ* over the whole pressure and temperature range of the lower mantle. Polycrystalline samples subjected to non-hydrostatic conditions in these experiments can develop preferred orientation. During such experiments a powder is first compacted and then plastically deformed by loading which has a directional stress component that produces elastic and plastic deformation. Merkel et al. (2002) performed experiments to study the shear strength, elastic moduli, and deformation mechanisms in MgO from ambient pressure to 47 GPa at room temperature. The axial stress component in the polycrystalline MgO sample is found to increase rapidly to ~8.5 GPa at a

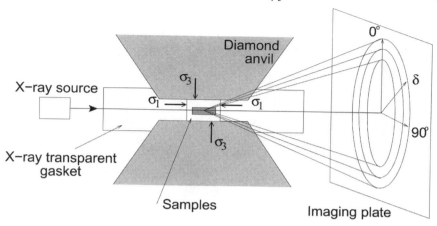

Figure 18. Diamond anvil cell, using radial geometry to display texture in Debye cones (Merkel et al. 2002).

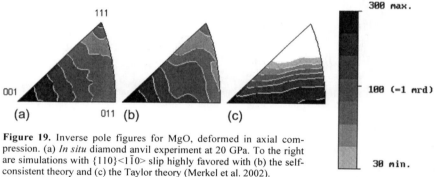

Figure 19. Inverse pole figures for MgO, deformed in axial compression. (a) *In situ* diamond anvil experiment at 20 GPa. To the right are simulations with {110}<1$\bar{1}$0> slip highly favored with (b) the self-consistent theory and (c) the Taylor theory (Merkel et al. 2002).

hydrostatic pressure of 10 GPa. Under axial compression a strong cube texture develops which was recorded *in situ*. It is likely that the preferred orientation of MgO is due to deformation by slip and a comparison between the experimental textures (Fig. 19a) and results from self-consistent polycrystal plasticity (Fig. 19b) suggest that {110}<1$\bar{1}$0> is the only significantly active slip system under very high confining pressure at room temperature, producing a maximum of (100) poles parallel to the compression direction. If other slip systems are equally active (in Taylor simulations they are activated to maintain compatibility), the texture type is very different with a maximum at (110) (Fig. 19c). The elastic moduli, obtained in these diamond anvil experiments from shifts in *d*-spacings with direction, are in agreement with Brillouin spectroscopy studies (Sinogeikin and Bass 2000).

Naturally, in the lower mantle deformation occurs also at high temperature and, in analogy to halite (Carter and Heard 1970), several slip systems may be active. Indeed, torsion experiments on magnesiowüstite by Stretton et al. (2001) (Fig. 20a) and shear experiments by Yamazaki and Karato (2002) can be only explained by simultaneous activity of {111}, {110} and {100} slip systems, all with the [110] slip direction (Fig. 20c). At high strains, with partial recrystallization, the texture changes (Fig. 20b) and those orientations that are easily deformed in {111} slip appear to dominate (Fig. 20d).

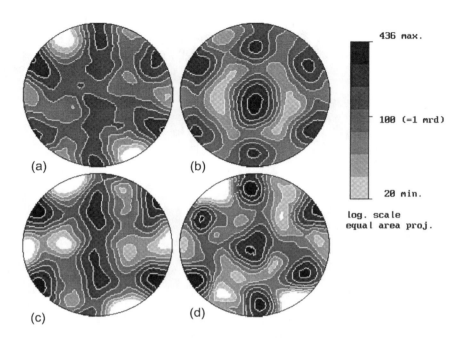

Figure 20. (a,b) Experimental (from Stretton et al. 2001) and (c,d) simulated 111 pole figures for magnesiowuestite deformed in torsion. (a,c) are for intermediate strain, (b,d) for high strain with pervasive recrystallization and grain size reduction. The shear plane is horizontal. The simulation in (d) collects grains that are most easily deformed in {111} slip.

During subduction of upper mantle slabs into the lower mantle, geodynamic modeling suggests heterogeneous deformation with complicated streamlines (Kellogg et al. 1999, McNamara et al. 2001; Fig. 21). Using slip systems that were active in the high temperature torsion experiments, we can then again predict texture evolution along a streamline in the subducting slab with increasing depth (Fig. 22). Textures are strong, and, since single crystal anisotropy of MgO is high at high pressure and temperature (Chen et al. 1998), significant seismic anisotropy is expected in the deep lower mantle (Karato 1998).

So far no convincing deformation experiments have been done with silicate perovskite. Experiments on the analog $CaTiO_3$ suggest that, while twinning may produce a texture at low temperature, at higher temperature superplasticity may be active with no preferred orientation (Karato et al. 1995). Naturally there is uncertainty with such conclusions from analogs and particularly superplastic behavior in the experiments may have been due to the small grain size. In the future experiments need to be done with large samples and compositions corresponding to those present in the lower mantle.

Core

It seems firmly established, both from body waves and free oscillation observations, that compressional P-waves travel through the solid inner core 3 to 4% faster along the (vertical) axis of the Earth, than in the equatorial plane (e.g., Creager 1992, Romanowicz et al. 1996, Shearer 1994, Song 1997, Su and Dziewonski 1995). The reason for seismic anisotropy is most likely an alignment of crystals which could have taken place from

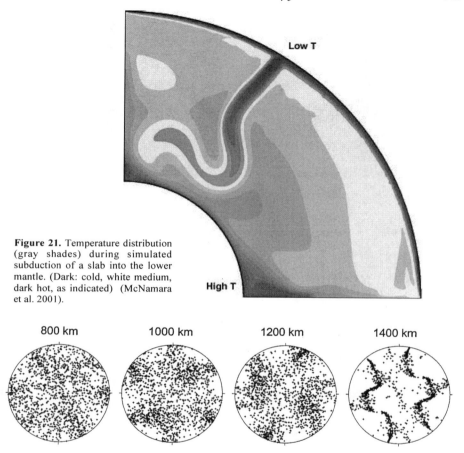

Figure 21. Temperature distribution (gray shades) during simulated subduction of a slab into the lower mantle. (Dark: cold, white medium, dark hot, as indicated) (McNamara et al. 2001).

Figure 22. Simulation of texture development of magnesiowüstite during slab subduction in the lower mantle along a streamline of the McNamara et al. (2001) model at four different depths. (100) pole figures. It is assumed that {110} and {111} slip are equally active.

deformation-induced texturing during convection (Jeanloz and Wenk 1988, Wenk et al. 2000), growth in a stress field (Yoshida et al. 1996), deformation in a magnetic field (Buffett and Wenk 2001, Karato 1999), or solidification texturing at the boundary with the liquid outer core (Bergman 1997, Karato 1993). The main component of the inner core is an iron-rich alloy, most likely with a hexagonal close-packed (ε-iron, hcp) structure (e.g., experiments by Mao et al.1998, and first principles calculations by Wasserman et al. 1996).

Buffett and Wenk (2001) explored the effects of deformation induced by magnetic stresses and used polycrystal plasticity to predict the development of preferred orientation and anisotropy. The largest electromagnetic (Maxwell) shear stresses in the Earth's geodynamo arise from the combined influence of the radial and azimuthal components of the magnetic field and are on the order of several Pa. Strain gradually accumulates, as the inner core grows by solidification to about 50% in 1 Myr. The azimuthal component of the Maxwell stress, which is about an order of magnitude larger than the radial component, imposes a strong simple shear deformation.

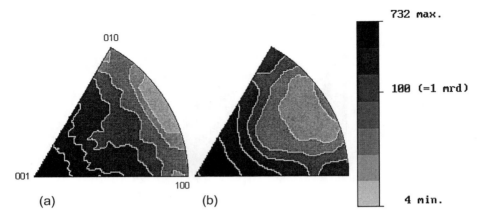

Figure 23. Inverse pole figures for ε-iron (hcp) deformed in compression. (a) *In situ* diamond anvil texture determination at 220 GPa. (b) Texture simulation with the self-consistent theory for conditions that favor basal slip, 50% strain (Wenk et al. 2000).

A prerequisite for texture simulations is the knowledge about slip systems that are active in ε-iron. This phase is not stable at ambient conditions and *in situ* deformation experiments and texture measurements need to be performed at high pressure. This can be done with diamond anvil cells described above, and a comparison of the strong texture patterns that were observed at pressures close to those in the inner core (220 GPa) (Fig. 23a) with polycrystal plasticity simulations (Fig. 23b) indicate that basal and prismatic slip are active (Wenk et al. 2000), consistent with *ab initio* predictions (Poirier and Price 1999).

With this information about intracrystalline mechanisms one can then apply the Maxwell stresses to the solid core and predict orientation patterns. Figure 24a-c shows c-axis pole figures for different locations in the core. Averaging the single crystal elastic properties over the simulated orientation distributions, an anisotropy pattern similar to that observed by seismologists is obtained with faster P-wave velocities parallel to the Earth's axis (Fig. 24d-f). Because of the many uncertainties such predictions are highly speculative and other processes may also contribute to anisotropy. Single crystal elastic constants at core pressure and temperature are not very well known. Recent *ab initio* calculations suggest that at low temperature the c-axis is the stiffest direction, while at high temperature the c-axis becomes the softest direction (Steinle-Neumann et al. 2001). There is no experimental evidence for this reversal, but Matthies et al. (2001) have shown that *in situ* diamond anvil diffraction experiments on textured polycrystals could be used to determine single crystal elastic properties from systematic peak shifts, if the accuracy of measurements could be improved by an order of magnitude.

PROBLEMS AND OPPORTUNITIES

The previous section illustrated how polycrystal plasticity has helped in interpreting natural and experimental textures, as well as observed seismic anisotropy. Convincing cases have been made in the literature, but research papers always have to be convincing and have to play down flaws, otherwise they would not get published. In a review one has more freedom to also address some cases that remain puzzling and illustrate with examples some unresolved issues where more work is needed. There is indeed a wide range of texture issues that are not understood.

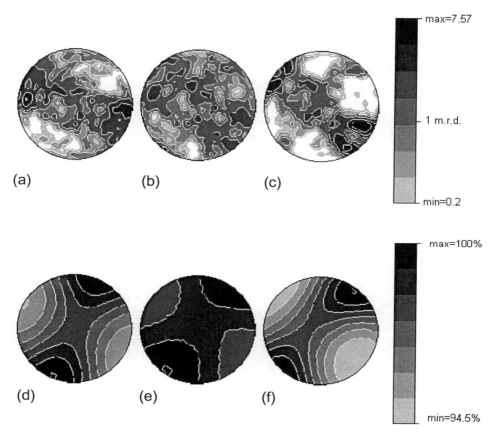

Figure 24. Simulated texture development of ε-iron in the inner core, induced by Maxwell stresses. (a-c) (0001) pole figures at three different locations are shown. The plots are oriented such that the north pole is in the center and the equator at the periphery. (d-f) Corresponding relative *P*-wave velocity surfaces of the textured aggregates (Buffett and Wenk 2001).

Large strain torsion experiments on calcite marble have produced a recrystallized microstructure with strong texture (Pieri et al. 2000). At low strain calcite produced a texture that can be well explained with conventional slip systems such as $r\{10\bar{1}4\}<20\bar{2}\bar{1}>$, $f\{01\bar{1}2\}<20\bar{2}\bar{1}>$ and c $(0001)<11\bar{2}0>$ slip (Fig. 25a,b). At larger strain ($\gamma > 2$) the coarse marble underwent grainsize reduction, with small grains first nucleating along grain boundaries and then replacing old grains. A new texture developed with a $(10\bar{1}4)$ plane parallel to the shear plane and a $[11\bar{2}0]$ direction parallel to the shear direction (Fig. 25c). This texture remained stable up to very large strains ($\gamma > 11$). It would seem logical that this end-orientation corresponds to an "easy-slip" orientation where the slip plane is parallel to the shear plane and the slip direction parallel to the shear direction. For such a geometry, and only in simple shear, there is no crystal rotation. Indeed polycrystal plasticity with the recrystallization model of Wenk et al. (1997) predicts exactly this texture if it is assumed that $\{10\bar{1}4\}<11\bar{2}0>$ is the dominant slip system (Fig. 25d). Based on this evidence it was suggested that this is a new slip system in calcite, active at high temperature. This did not seem unreasonable, because it

Next page >>>>

Figure 25. Calcite deformed in torsion (Pieri et al. 2000); (0001) and (10$\bar{1}$4) pole figures. (a) Experiment after moderate strain (γ 1), (b) Texture simulations using hypothetical {10$\bar{1}$4}<11$\bar{2}$0> slip as dominant system, (c) Experiment after large strain (γ = 5), (d) Simulation for dynamic recrystallization with hypothetical slip system.

~~~~~~

is an important slip system in dolomite (Barber et al. 1981). The problem was then followed up with single crystal experiments. Mechanical data and optical microscopy (deBresser, unpublished) and TEM investigations (Barber, unpublished) failed to produce any evidence for {10$\bar{1}$4}<11$\bar{2}$0> slip in calcite and thus the question on how the texture in the torsion experiments evolved remains unresolved.

Plane strain deformation experiments on calcite showed a texture transition from a texture with a *c*-axis maximum parallel to the compression direction at low temperature to a double maximum at higher temperature. This could be explained with polycrystal plasticity as an effect of slip system activity (Takeshita et al. 1987) with pervasive mechanical twinning at low temperature and slip-dominated deformation at higher temperature (Fig. 26a,b). Most naturally deformed calcite rocks (including those discussed above, Figs. 12-14) display the low temperature type (Fig. 26c). Yet there are a few examples of marbles from Carrara with a split *c*-maximum (Fig. 26d) that bear resemblance to the high temperature texture (Leiss and Weiss 2000). On closer inspection it became apparent that this resemblance is superficial, because (10$\bar{1}$4) pole figures in the high temperature simulations and the naturally deformed Carrara marble are distinctly different. While it has been possible to simulate the *c*-axis pole figure, assuming reasonable calcite slip systems, all attempts have failed so far to generate a texture pattern that satisfies both the *c*-axis and (10$\bar{1}$4) pole distributions, for different strain paths, different slip system activities and different recrystallization parameters. Perhaps the texture pattern represents a multistage deformation. The example illustrates the importance of considering the full orientation distribution, rather than single pole figures, in the interpretation of textures. At this point there is no explanation for the curious Carrara marble textures.

Natural olivine textures have been compiled by Ben Ismail and Mainprice (1998). By far the most common texture type has (010) poles nearly perpendicular to the foliation plane and [100] axes subparallel to the lineation direction. Most of these rocks are found predominantly in ophiolites and textures are attributed to deformation in the upper mantle. There are a few exceptions. Möckel (1969) and Buiskool-Toxopaeus (1977) described a texture type from metamorphic rocks at Alpe Arami in the Central Alps with (100) perpendicular to the foliation (Fig. 27a). A similar type has recently been described from a close-by locality of Cima di Gagnone (Freese and Trommsdorff 1999) and deformation seems to be related to subduction during Alpine orogenesis to a depth of 80-100 km, at relatively low temperature.

There have been many studies to determine slip systems in olivine, including Raleigh (1968), Nicolas et al. (1973), Bai et al. (1989). At high temperature (010)[100] is the preferred slip system and simulations produce the high temperature texture observed in mantle peridotites (Fig. 27b). At lower temperature Raleigh (1968) established (100)[001] slip and {hk0}[001] pencil glide, and indeed the Arami texture type can be simulated with those systems for plane strain pure shear deformation (Fig. 27c).

Jung and Karato (2001) investigated the influence of stress and water content on olivine texture development in simple shear at 1500 K and large strains. They could document a fabric transition from a (010) normal to the foliation type at low H_2O and low

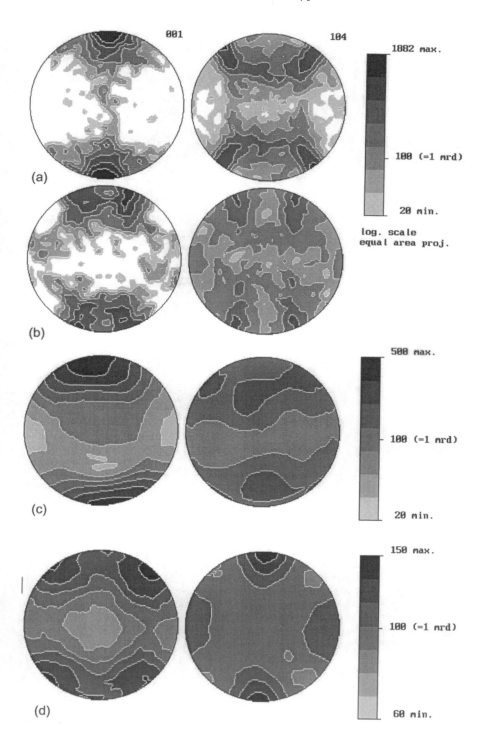

Next page >>>>

Figure 26. Texture types in deformed carbonate rocks represented in (0001) and (10$\bar{1}$4) pole figures. (a,b) simulated pole figures with the Taylor theory (Tomé et al. 1991); (a) conditions with prevailing *r*-slip and *e*-twinning, (b) conditions with *r*- and *f*-slip and minor *e*-twinning, 50% strain. (c,d) natural textures; (c) limestone from an Alpine spreading nappe in the Graz Paleozoic (Ratschbacher et al. 1991), (d) texture observed in the Alpe Appuane near Carrara (Leiss and Weiss 2000).

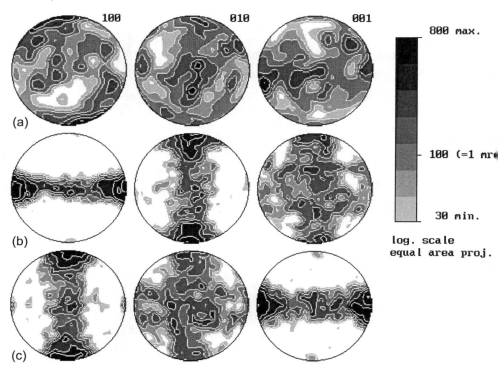

Figure 27. Texture types in deformed olivine. (a) is the natural metamorphic texture from the deformed peripheral zone of the Alpe Arami ultramafic body, Central Alps (measured by EBSP). (b) and (c) are self-consistent simulations for plane strain, pure shear, assuming slip system activities representative of high temperature (b) and low temperature (c).

stress, to a (100) normal to the foliation type at high stress and higher water content. The latter fabric is similar to the Arami texture but in the experiments the material underwent dynamic recrystallization. Also, microstructures at Alpe Arami and Cima di Gagnone do not show indications of simple shear and comparisons are not straightforward. The rare (100) texture type has also been observed in lherzolite mylonites of the Horoman complex in Japan (Furusho and Kanagawa 1999). Also here the texture is not related to mantle peridotites but to crustal deformation. Under unusual conditions metamorphic olivine rocks in the crust do develop preferred orientation. In most cases though, including many foliated peridotites in the Alps, there is no preferred orientation and these cases go unreported.

The understanding of plasticity in polyphase materials is still very rudimentary, for rocks as well as for metals and other composite materials (Wenk 1994). Different phases may have very different mechanical properties and corresponding aggregates will deform

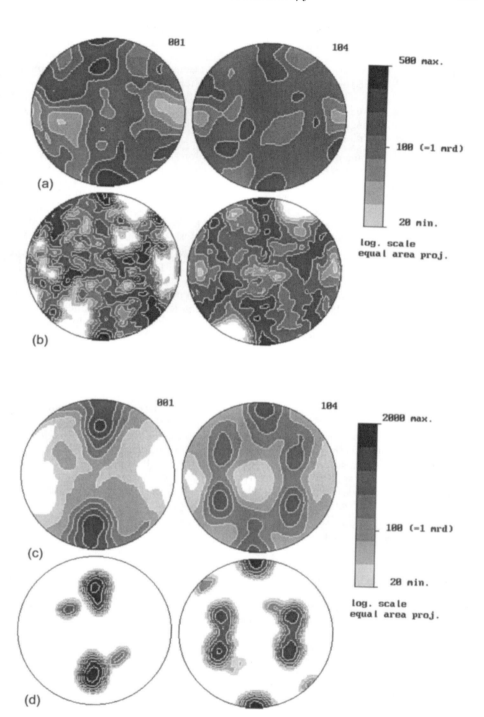

very heterogeneously, as is well known in mylonites with feldspar clasts and quartz flowing around them and undergoing recrystallization. The problem of percolation needs to be considered. If the soft phase dominates, it controls the deformation behavior; but if the hard phase is dominant and hard grains are in contact and form a frame the hard grains control deformation (Raj and Ghosh 1981). Compositional banding is commonly observed in deformed polymineralic rocks. Since topological relationships between neighboring grains are essential, polycrystal plasticity models that deal with neighbor interaction in a statistical way (self-consistent) or not at all (upper and lower bounds) are inadequate. If different phases are present, the recrystallization behavior is affected, since grain boundary migration is precluded.

Texture development in polymineralic metamorphic rocks cannot be explained purely with polycrystal plasticity. Models must include processes of dissolution and growth (Spiers and Takeshita 1995), interaction with aqueous solutions (Karato et al. 1986), chemical reactions and phase transformations. Apart from orientation state and dislocation microstructure, the chemical composition, grain size and shape are important parameters (Heilbronner and Bruhn 1998, Shimizu 1999, Wang 1994). If deformation is highly heterogeneous, grain-boundary sliding may occur.

Such considerations need not only to be introduced into models but experiments need to be designed that take these factors into account. Unresolved questions are the development of preferred orientation patterns in such common rocks as gneiss and amphibolite. Are mica and hornblende crystals aligned due to the anisotropic grain shape (Jeffery 1923, March 1932)? Or is this a growth fabric, perhaps controlled by stress (Shimizu 1992)? The field of polyphase deformation is wide open, with many challenges. After all, most rocks in the crust, upper mantle and lower mantle are polyphased.

CONCLUSIONS

The discussion has illustrated that texture evolution and anisotropy in the Earth are complex and heterogeneous. Methods do exist to simulate patterns for a range of conditions but much work remains to be done. Clearly there is no simple answer to anisotropy in the Earth such as "fast directions align with the flow direction". Preferred orientation of minerals in rocks is an important aspect of the plastic behavior and advanced computational methods help to link microscopic mechanisms and macroscopic properties, such as seismic anisotropy. From observed texture patterns we can infer deformation mechanisms. If mechanisms are known we can predict anisotropy that should evolve in geodynamic situations and, in turn, use seismic observations of anisotropy to interpret deformation processes in the deep Earth.

Texture, i.e., the alignment of crystals, is an integral part of rocks as much as mineralogic and chemical composition, grain size and shape distribution, and the topology of mineral constituents. Ever since the discovery of directionality in rocks by d'Halloys (1833), have structural geologists been fascinated by the alignment of crystals in rocks. The alignment may be due to growth or due to deformation and causes anisotropy of physical properties, such as seismic wave propagation in the Earth. There have been many books (including Sander's famous monograph) and a multitude of scientific papers published on various experimental and theoretical aspects of preferred orientation, yet many questions remain. It is appropriate to explore the future role of petrofabric analysis and define some important research projects. While much is known about monomineralic rocks such as quartzite, marble and olivinite, we still know practically nothing about polymineralic rocks and do not understand in any quantitative way such basic questions as to how mica crystals align in gneiss. While the theoretical framework to deal with such materials is still in its infancy (having to balance growth, deformation, chemical reactions and recrystallization), it is also difficult to characterize the orientations of crystals and orientation relationships between neighbors. New advanced techniques of texture analysis, such as EBSP, synchrotron X-rays and neutron diffraction will become important. It cannot be overemphasized how important experiments will be, with sufficient complexity to mimic accompanying chemical reactions and deformation during metamorphism. Clearly, there is a wealth of opportunities for future research to improve our understanding of rock-forming processes. One of the big attractions of this research has been the stimulating interaction between structural geologists, petrologists, geophysicists, mineralogists and materials scientists. A close collaboration will be essential in the future.

REFERENCES

Aleksandrov KS, Aisenberg LA (1966) Method of calculating the physical constants of polycrystalline materials. Soviet Phys–Doklady 11:323-325
Bai Q, Mackwell SJ, Kohlstedt DL (1991) High temperature creep of olivine single crystals. I. Mechanical results for buffered samples. J Geophys Res 96:2441-2463.
Barber DJ Heard HC, Wenk H-R (1981) Deformation of dolomite single crystals from 200-800°C. Phys Chem Minerals 7:271-286
Beaudoin AJ, Dawson PR, Mathur KK, Kocks UF (1995) A hybrid finite element formulation for polycrystal plasticity with consideration of macrostructural and microstructural linking. Intl J Plast 11:501-521
Ben Ismaïl W, Mainprice D (1998) An olivine fabric database: an overview of upper mantle fabrics and seismic anisotropy, Tectonophysics 296:145-157
Bennett K, Von Dreele RB, Wenk H-R (1999) "HIPPO", a new high intensity neutron diffractometer for characterization of bulk materials, Proc ICOTOM-12. NRC Research Press, p 129-134
Bergman MI (1997) Measurements of the elastic anisotropy due to solidification texturing and the implication for the Earth's inner core. Nature 389:60-63
Blackman DK, Kendall J-M, Dawson PR, Wenk H-R, Boyce D, Phipps Morgan J (1996) Teleseismic imaging of subaxial flow at mid-ocean ridges: travel-time effects of anisotropic mineral texture in the mantle. Geophys J Intl 127:415-426
Blackman DK, Wenk H-R, Kendall J-M (2002) Seismic anisotropy in the upper mantle: 1. Factors that affect mineral texture and effective elastic properties. Geochem Geophys Geosystems 3:10.1029
Bons PD, Den Brok B (2000) Crystallographic preferred orientation development by dissolution-precipitation creep. J Struct Geol 22:1713-1722
Brockhouse BN (1953) The initial magnetization of nickel under tension. Can J Phys 31:339-355.
Buffet B, Wenk H-R (2001) Texturing of the inner core by Maxwell stresses. Nature 413:60-63
Buiskool Toxopeus JMA (1977) Deformation and recrystallization of olivine during mono- and poly-phase deformation. A transmission electron microscope study. N Jahrb Mineral Abh 129:233-268
Bunge H-J (1965) Zur Darstellung allgemeiner Texturen. Z Metallk 56:872-874.
Bunge H-J (1982) Texture Analysis in Materials Science – Mathematical Methods. London: Butterworths, 593 p
Bunge H-J (1985) Physical properties of polycrystals. In Preferred Orientation in Deformed Metals and Rocks. H-R Wenk (ed) An Introduction to Modern Texture Analysis. Academic Press, p 507-525
Bunge H-J (ed) (1986) Experimental Techniques of Texture Analysis. DMG Informationsgemeinschaft, Oberursel, Germany, 442 p
Bunge H-J, Siegesmund S, Skrotzki W, Weber K (1994) Textures of Geological Materials. DMG Informationsgesellschaft, Oberursel, Germany, 399 p
Bunge H-P, Richards MA, Lithgow-Bertelloni C, Baumgardner JR, Grand SP, Romanowicz B (1998) Time scales and heterogeneous structure in geodynamic Earth models. Science 280:91-95
Calnan EA, Clews CJB (1950) Deformation textures of cubic face-centered metals. Phil Mag 41:1085-1100
Carter NL, Heard HC (1970) Temperature and rate-dependent deformation of halite. Am J Sci 269:193-249
Chastel YB, Dawson PR, Wenk H-R, Bennett K (1993) Anisotropic convection with implications for the upper mantle. J Geophys Res B98:17757-17771
Chen G, Liebermann RC, Weidner DJ (1998) Elasticity of single crystal MgO to 8 gigapascals and 1600 kelvin. Science 280:1913-1916

Crampin S (1981) A review of wave motion in anisotropic and cracked elastic media. Wave Motion 3: 242-391
Creager KC (1992) Anisotropy of the inner core from differential travel times of the phases PKP and PKIKP. Nature 356:309-314
D'Halloy OJJ (1833) Introduction à la Géologie. Levrault, Paris, 894 p
Dawson P, Boyce D, MacEwen S, Rogge R (2001). On the influence of crystal elastic moduli on computed lattice strains in AA-5182 following plastic straining. Mater Sci Eng A313:123-144
Dawson PR, Wenk H-R (2000) Texturing of the upper mantle during convection. Phil Mag A 80:573-598
Drury MR, Urai JL (1990) Deformation-related recrystallization processes, Tectonophysics 172:235-253
Engler O (1996) Nucleation and growth during recrystallisation of aluminium alloys investigated by local texture analysis. Mater Sci Technol 12:859-872
Erskine BG, Heidelbach F, Wenk H-R (1993) Lattice preferred orientations and microstructures of deformed Cordilleran marbles: correlation of shear indicators and determination of strain path. J Struct Geol 15:1189-1205
Euler L (1775) Formulae generales. Nov Comm Acad Sci Imp Petrop 20:189-207
Ferrari M, Lutterotti L (1994) Method of simultaneous determination of anisotropic residual stresses and texture by X-ray diffraction. J Appl Phys 76:7246-7255
Fiquet G (2001) Mineral phases of the earth's mantle. Z Kristallogr 216:248-271
Forsyth DW (1975) The early structural evolution and anisotropy of the oceanic upper mantle. Geophys J Royal Astron Soc 43:103-162
Frese K, Trommsdorff V, Kunze K (2001) Metamorphic [100] olivine lattice preferred orientations (LPO) of garnet lherzolites, Central Alps. 6th Intl Eclogite Conf, Ehime Prefectural Science Museum, Niihama, Japan, p 40-41 (abstr)
Furusho M, Kanagawa K (1999) Transformation-induced strain localization in a lherzolite mylonite from the Hidaka metamorphic belt of central Hokkaido, Japan. Tectonophysics 313:511-432
Gnäupel-Herold T, Brand, PC Prask HJ (1998) The calculation of single crystal elastic constants for cubic crystal symmetry from powder diffraction data. J Appl Crystallogr 31:929-935
Green HW (1980) On the thermodynamics of non-hydrostatically stressed solids. Phil Mag A 41: 637-647
Guillopé M, Poirier J-P (1979) Dynamic recrystallization during creep of single-crystalline halite: An experimental study. J Geophys Res 84:5557-5567
Haessner F (1978) Recrystallization of Metallic Materials. Stuttgart: Riederer
Heidelbach F, Riekel C, Wenk H-R (1999) Quantitative texture analysis of small domains with synchrotron X-rays. J Appl Crystallogr 32:841-849
Heilbronner RP, Bruhn D (1998) The influence of three-dimensional grain size on the rheology of polyphase rocks. J Struct Geol 20:695-705
Heilbronner RP, Pauli C (1993) Integrated spatial and orientation analyisis of quartz c-axes by computer aided microscopy. J Struct Geol 15:369-382
Helming K (1994) Some applications of the texture component method. Proc 10th Intl Conf on Texture of Materials. Clausthal, Materials Science Forum 157-162:363-368
Hill R (1952) The elastic behavior of a crystalline aggregate. Proc Phys Soc A65:349-354
Ho N-C, Peacor DR, van der Pluijm BA (1999). Preferred orientation of phyllosilicates in Gulf Coast mudstones and relation to the smectite-illite transition. Clays Clay Minerals 47:495-504
Høier R, Bentdal J, Daaland O, Nes E (1994) A high resolution transmission electron diffraction method for on-line texture analysis. In Textures of Materials ICOTOM-10. H-J Bunge (ed) Zürich: Trans Tech Publ 143-148
Humphreys FJ (1988) Experimental techniques for microtexture determination. In Eighth Intl Conf on Textures of Materials. JS Kallend, G Gottstein, (eds) Warrendale, Pennsylvania: The Metallurgical Society, p 171-182.
Humphreys FJ, Hatherly M (1995), Recrystallization and Related Annealing Phenomena. Oxford University Press, Oxford, UK, 497 p
Jeanloz R, Wenk H-R (1988) Convection and anisotropy of the inner core. Geophys Res Lett 15:72-75
Jeffery GB (1923) The motion of ellipsoidal particles immersed in a viscous fluid. Proc R Soc London A102:161-179.
Jessel MW (1988) Simulation of fabric development in recrystallizing aggregates – I. Description of the model. J Struct Geol 10:771-778.
Jung H, Karato S (2001) Water-induced fabric transitions in olivine. Science 293:1460-1463
Kallend J S (2000) Determination of the orientation distribution from pole figure data. In Texture and Anisotropy. Preferred Orientations in Polycrystals and Their Effect on Materials Properties, 2nd

Edition. Kocks UF, Tomé CN, Wenk H-R (eds) Cambridge, UK: Cambridge University Press, p 102-124

Kallend JS, Kocks UF, Rollett AD, Wenk H-R (1991) Operational texture analysis. Mater Sci Eng A132: 1-11

Kamb WB (1961) The thermodynamic theory of nonhydrostatically stressed solids. J Geophys Res 66: 259-271

Kaminski E, Ribe NM (2001) A kinematic model for recrystallization and texture development in olivine polycrystals. Earth Planet Sci Lett 189:253-267

Karato S-I (1987) Seismic anisotropy due to lattice preferred orientation of minerals: kinematic or dynamic? In High Pressure Research in Mineral Physics]MH Manghnani, Y Syono (eds) Terra Publ, Tokyo, p 317-333

Karato S-I (1988) The role of recrystallization in preferred orientation of olivine. Phys Planet Inter 51:107-122

Karato S (1993) Inner core anisotropy due to magnetic field-induced preferred orientation of iron. Science 262:1708-1711

Karato S-I (1998) Some remarks on the origin of seismic anisotropy in the D" layer. Earth Planet Space 50:1019-1028

Karato S-I (1999) Seismic anisotropy of the Earth's inner core resulting from flow induced by Maxwell stresses. Nature 402:871-873

Karato S-I, Wu P (1993) Rheology of the upper mantle— a synthesis. Science 260:771-778

Karato S-I, Toriumi M (1989) Rheology of Solids and of the Earth. Oxford, 440 p

Karato S-I, Paterson MS, FitzGerald JD (1986) Rheology of synthetic olivine aggregates: Influence of grain size and water. J Geophys Res 95:17631-17642

Karato S-I, Zhang S, Wenk H-R (1995) Superplasticity in the Earth's lower mantle. Evidence from seismic anisotropy and rock physics. Science 270:458-461

Kellogg LH, Hager BH, van der Hilst RD (1999) Compositional stratification in the deep mantle. Science 283:1881-1884

Kern H, Wenk H-R (1983) Calcite texture development in experimentally induced ductile shear zones. Contrib. Mineral Petrol. 83:231-236

Kern H (1993) P- and S-wave anisotropy and shear wave splitting at pressure and temperature in possible mantle rocks and their relation to the rock fabric. Phys Earth. Planet Inter 78:245-256

Kocks UF (1976) Laws for work-hardening and low-temperature creep. J Eng Mater Technol 98:76-85

Kocks UF (2000) Simulation of deformation texture development for cubic metals, Chapter 9 in Texture and Anisotropy. Preferred Orientations in Polycrystals and Their Effect on Materials Properties, 2nd Edition. Kocks UF, Tomé CN, Wenk H-R (eds) Cambridge, UK: Cambridge University Press, p 390-418

Kocks UF, Tomé CN, Wenk H-R (eds) (2000) Texture and Anisotropy. Preferred Orientations in Polycrystals and Their Effect on Materials Properties, 2nd Edition. Cambridge, UK: Cambridge University Press, 676 p

Kröner E (1961) Zur plastischen Verformung des Vielkristalls. Acta Metall 9:155-161

Kunze K, Wright SI, Adams BL, Dingley DJ (1993) Advances in automatic EBSP single orientation measurements. Textures Microstruc 20:41-54

Lebensohn RA, Wenk H-R, Tomé C (1988) Modelling deformation and recrystallization textures in calcite. Acta Mater 46:2683-2693

Lebensohn RA, Tomé, CN (1993) A self-consistent anisotropic approach for the simulation of plastic deformation and texture development of polycrystals—Application to zirconium alloys. Acta metall mater 41:2611-2624

Leiss B, Weiss T (2000) Fabric anisotropy and its influence on physical weathering of different types of Carrara marbles. J Struct Geol 22:1737-1745

Leiss B, Ullemeyer, K Weber K Edts. (2000) Textures and Physical Properties of Rocks. J Struct Geol 22:1527-1873

Liang Z, Wang F, Xu J (1988) Inverse pole figure determination according to the maximum entropy method. *In* 8th Intl Conf on Textures of Materials. JS Kallend, G Gottstein (eds) Warrendale, Pennsylvania: The Metallurgical Society, p 111-114

Lister GS, Paterson MS, Hobbs BE (1978) The simulation of fabric development during plastic deformation and its application to quartzite: The model. Tectonophysics 45:107-158

Lutterotti L, Matthies S, Wenk H-R, Schultz A J, Richardson J W (1997) Combined texture and structure analysis of deformed limestone from time-of-flight neutron diffraction spectra, J Appl Phys 81: 594-600

Mao HK, Wu Y, Chen LC, Shu JF (1990) Static compression of iron to 300 Gpa and $Fe_{0.8}Ni_{0.2}$ alloy to 260 Gpa—Implications for composition of the core. J Geophys Res 95:21737-21742

March A (1932) Mathematische Theorie der Regelung nach der Korngestalt bei affiner Deformation. Z Kristallogr 81:285-297.

Matthies S, Merkel S, Wenk H-R, Hemley RJ, Mao H-K (2001) Effects of texture on the determination of elasticity of polycrystalline ε-iron from diffraction measurements. Earth Planet Sci Lett 194:201-212

Matthies S, Wagner F (1996) On a 1/n law in texture related single orientation analysis. Phys Stat Solidi 196:K11-K15

Matthies S (1979) On the reproducibility of the orientation distribution function of texture samples from pole figures (ghost phenomena). Phys Stat Solidi (b) 92:K135-138

Matthies S, Humbert M (1993) The realization of the concept of the geometric mean for calculating physical constants of polycrystalline materials. Phys Stat Solidi (b) 177:K47-K50

Matthies S, Vinel GW (1982) On the reproduction of the orientation distribution function of textured samples from reduced pole figures using the concept of conditional ghost correction. Phys Stat Solidi (b) 112:K111-114

Matthies S, Priesmeyer HG, Daymond MR (2001) On the diffractive determination of single crystal constants using polycrystalline samples, J Appl Crystallogr 34:585-601

McNamara AK, Karato S-I, van Keken PE (2001) Localization of dislocation creep in the lower mantle: implications for the origin of seismic anisotropy. Earth Planet Sci Lett 191:85-99

Merkel S, Wenk H-R, Shu J Shen G, Gillet P, Mao H-K, Hemley RJ (2002) Deformation of MgO aggregates at pressures of the lower mantle. J Geophys Res (in press)

Mika DP, Dawson PR (1999) Polycrystal plasticity modeling of intracrystalline boundary textures. Acta Mater 47:1355-1369

Möckel JR (1969) Structural petrology of the garnet-peridotite of Alpe Arami (Ticino, Switzerlamd). Leidse Geol Med 42:61-130

Molinari A, Canova GR, Ahzi S (1987) A selfconsistent approach of the large deformation polycrystal viscoplasticity. Acta Metall 35:2983-2994

Molinari A, Toth LS (1994) Tuning a selfconsistent viscoplastic model by finite-element results. I. Modeling. Acta metall mater 42:2453-2458

Molinari A (1999) Extensions of the self-consistent tangent model. Modelling Simulation Mater Sci Eng 7:683-697

Nauer-Gerhardt CU, Bunge H-J (1986), Orientation determination by optical methods. *In* Experimental Techniques of Texture Analysis. H-J Bunge (ed) DMG Informationsgesellschaft, Oberursel, Deutsche Gesell Metallkunde, p 125-145

Naumann CF (1850) Lehrbuch der Geognosie. Leipzig: Engelmann, 1000 p

Nicolas A, Boudier F, Bouiller AM (1973) Mechanisms of flow in naturally and experimentally deformed peridotites. Am J Sci 273:853-876

Oertel G (1983) The relationship of strain and preferred orientation of phyllosilicate grains in rocks—review. Tectonophysics 100:413-447

Paterson MS, Weiss LE (1961) Symmetry concepts in the structural analysis of deformed rocks. Geol Soc Am Bull 72:841-882.

Paterson MS (1973) Nonhydrostatic thermodynamics and its geologic applications. Rev Geophys 11:355-389

Pawlik K, Pospiech J, Lücke K (1991) The ODF approximation from pole figures with the aid of the ADC method. Textures Microstruc 14-18:25-30

Phillips WR (1971). Mineral Optics, Principles and Techniques. San Francisco: Freeman, 249 p

Pieri M, Stretton I, Kunze K, Burlini L, Olgaard DL, Burg J-P, Wenk H-R (2000) Texture development in calcite through deformation and dynamic recrystallization at 1000 K during torsion to large strains. Tectonophysics 330:119-140

Platt JP, Vissers RLM (1980) Extensional structures in anisotropic rocks. J Struct Geol 2:397-410

Poirier J-P, Price GD (1999) Primary slip system of epsilon-iron and anisotropy of the Earth's inner core. Phys Earth Planet Inter 110:147-156

Popa NC (1992) Texture in Rietveld refinement. J Appl Crystallogr 25:611-616

Puig-Molina A, Gorges B Graafsma H (2001) A 1000°C furnace for *in situ* X-ray diffraction. J Appl Crystallogr 34:677-678

Radhakrishnan B, Sarma GB, Zacharia T (1998) Modeling the kinetics and microstructural evolution during static recrystallization—Monte Carlo simulation of recrystallization. Acta Mater 46:4415-4433

Raj R, Ghosh AK (1981) Micromechanical modeling of creep using distributed parameters. Acta Metall 29:283-292

Raleigh CB (1968) Mechanisms of plastic deformation in olivine. J Geophys Res 73:5391-5406

Randle V, Engler O (2000) Introduction to Texture Analysis: Macrotexture, Microtexture and Orientation Mapping. Gordon and Breach Science Publishers

Ratschbacher L, Wenk H-R, Sintubin M (1991) Calcite textures: examples from nappes with strain-path partitioning. J Struct Geol 13:369-384.

Reuss A (1929) Berechnung der Fliessgrenze von Mischkristallen auf Grund der Plastizitätsbedingung für Einkristalle. Z Angew Math Mech 9:49-58

Rietveld HM (1969) A profile refinement method for nuclear and magnetic structures, J Appl Crystallogr 2:65-71

Roe R-J (1965) Description of crystallite orientation in polycrystalline materials III, general solution to pole figure inversion. J Appl Phys 36:2024-2031.

Romanowicz B, Li XD, Durek J (1996) Anisotropy of the inner core: Could it be due to low-order convection? Science 274:963-966

Ruer D, Baro R (1977) A new method for the determination of the texture of materials of cubic structure from incomplete reflection pole figures. Adv X-Ray Analysis 20:187-200

Sachs G (1928) Zur Ableitung einer Fliessbedingung. Z Ver Dtsch Ing 72:734-736

Sander B (1950) Einführung in die Gefügekunde der Geologischen Körper, Vol. 2. Vienna: Springer-Verlag, 409 p

Sarma GB, Dawson PR (1996) Effects of interaction among crystals on the inhomogeneous deformations of polycrystals. Acta Mater 44:1937-1953

Savage MK, Silver PG (1993) Mantle deformation and tectonics: constraints from seismic anisotropy in the western United States. Phys Earth Planet Inter 78:207-227.

Schaeben H (1988) Entropy optimization in texture goniometry. Phys Stat Solidi (b) 148:63-72

Schmid SM (1994) Textures of geological materials: computer model predictions versus empirical interpretations based on rock deformation experiments and field studies. *In* Bunge et al. (eds) Textures of Geological Materials. DMG Informationsgesellschaft, Oberursel, p 279-301

Schmid SM, Panozzo R Bauer S (1987) Simple shear experiment on calcite rocks: rheology and microfabric. J Struct Geol 9:747-778

Schulz LG (1949) A direct method of determining preferred orientation of a flat transmission sample using a Geiger counter X-ray spectrometer. J Appl Phys 20:1030-1033

Schwarzer RA, Sukkau J (1998) Automated crystal orientation mapping (ACOM) with a computer-controlled TEM by interpreting transmission Kikuchi patterns. Mater Sci Forum 273-275:215-222

Shearer PM (1994). Constraints on inner core anisotropy from PKP(DF) travel times, J Geophys Res 99:19647-19659

Shimizu I (1992) Nonhydrostatic and nonequilibrium thermodynamics of deformable materials. J Geophys Res 97B:4587-4597

Shimizu I (1999) A stochastic model of grain size distribution during dynamic recrystallization. Phil Mag. A 79:1217-1231

Siemes H (1974) Anwendung der Taylor-Theorie auf die Regelung von kubischen Mineralen. Contrib Mineral Petrol 43:149-157

Silver PG (1996) Seismic anisotropy beneath the continents: Probing the depths of geology, Ann Rev Earth Planet Sci 24:385-432

Simmons G, Wang H (1971) Single-Crystal Elastic Constants and Calculated Aggregate Properties: A Handbook, 2nd Edition. Cambridge, Massachusetts: The M.I.T. Press, 370 p

Singh AK, Mao H-K, Shu J, Hemley RJ (1998) Estimation of single-crystal elastic moduli from polycrystalline X-ray diffraction at high pressure: application to FeO and iron. Phys Rev Lett 80:2157-2160

Sinogeikin SV, Bass JD (2000) Single crystal elasticity of pyrope and MgO to 20 GPa by Brillouin spectroscopy scattering in the diamond cell. Phys Earth Planet Inter 120:43--62

Skrotzki W (1994) Mechanisms of texture development in rocks. *In* Bunge et al. (eds) Textures of Geological Materials. DMG Informationsgesellschaft, Oberursel, p 167-186

Snyder RL, Fiala J, Bunge H-J (eds) (1999) Defect and Microstructure Analysis by Diffraction. Oxford Univ Press IUC Monograph on Crystallography 10, 785 p

Solas DE, Tomé CN, Engler O, Wenk, H-R (2001) Deformation and recrystallisation of hexagonal metals. Modeling and experimental results for zinc. Acta Mater 49:3791-3801

Song X (1997) Anisotropy of the Earth's inner core. Rev Geophys 35:297-313

Spiers CJ, Takeshita T (eds) (1995) Influence of Fluids on Pressure Solution in Rocks. Tectonophysics, Spec Issue, 245:117-297

Steinle-Neumann G, Stixrude L, Cohen RE, Gülseren O (2001) Elasticity of iron at the temperature of the Earth's inner core. Nature 413:57-60

Stöckhert B, Duyster J (1999) Discontiuous grain growth in recrystallized vein quartz—implications for grain boundary structure, grain boundary mobility, crystallographic preferred orientation, and stress history. J Struct Geol 21:1477-1490

Stretton I, Heidelbach F, Mackwell S, Langenhorst F (2001) Dislocation creep of magnesiowüstite ($Mg_{0.8}Fe_{0.20}O$). Earth Planet Sci Lett 194:229-240

Su W-J, Dziewonski AM (1995) Inner core anisotropy in three dimensions. J Geophys Res 100:9831-9852

Takeshita T, Wenk H-R, Lebensohn R (1999) Development of preferred orientation and microstructure in sheared quartzite: comparison of natural and simulated data. Tectonophysics 312:133-155

Takeshita T, Tomé CN, Wenk H-R, Kocks UF (1987) Single-crystal yield surface for trigonal lattices: Application to texture transitions in calcite polycrystals. J Geophys Res B92:12917-12920.

Taylor GI (1938) Plastic strain in metals. J Inst Metals 62:307-324

Thorsteinsson T (2001) An analytical approach to deformation of anisotropic ice-crystal aggregates. J Glaciol 47:507-516

Thorsteinsson T (2002) Fabric development with nearest-neighbor interaction and dynamic recrystallization. J Geophys Res 107:148-227

Tomé CN (2000) Tensor properties of textured polycrystals. *In* Texture and Anisotropy. Preferred Orientations in Polycrystals and Their Effect on Materials Properties, 2nd Paperback Edn. UF Kocks, CN Tomé, H-R Wenk (eds) Cambridge Univ Press, p 282-324

Tomé CN, Canova GR (2000) Self-consistent modeling of heterogeneous plasticity. *In* Texture and Anisotropy. Preferred Orientations in Polycrystals and Their Effect on Materials Properties, 2nd Edition. UF Kocks, CN Tomé, H-R Wenk (eds) Cambridge, UK: Cambridge Univ Press, p 466-510

Tomé CN, Wenk H-R, Canova GR, Kocks UF (1991b), Simulations of texture development in calcite: comparison of polycrystal plasticity theories. J Geophys Res 96:11865-11875

Tullis TE, Tullis J (1986) Experimental rock deformation techniques, in Mineral and Rock Deformations: Laboratories Studies. BE Hobbs, HC Heard (eds) Washington, DC: Am Geophys Union, p 297-324

Turner FJ, Griggs DT, Clark RH, Dixon RH (1956) Deformation of Yule marble, part VII: Development of oriented fabrics at 300°C-400°C. Geol Soc Am Bull 67:1259-1294

Van Houtte P (1982) On the equivalence of the relaxed Taylor theory and the Bishop-Hill theory for partially constrained plastic deformation of crystals. Mater Sci Eng 55:69-77

Voigt W (1928) Lehrbuch der Kristallphysik. Leipzig: Teubner, 978 p

Von Dreele RB (1997) Quantitative texture analysis by Rietveld refinement. J Appl Crystallogr 30:517-525

Wagner F, Wenk H-R, Kern H, Van Houtte P, Esling C (1982) Development of preferred orientation in plane strain deformed limestone: Experiment and theory. Contrib Mineral Petrol 80:132-139

Wahlstrom EE (1979) Optical Crystallography, 5th Edition. New York: John Wiley, 488 p

Wang JN (1994) The effect of grain size distribution on the rheological behavior of polycrystalline materials. J Struct Geol 16:961-970

Wasserman E, Stixrude L, Cohen RE (1996) Thermal properties of close-packed phases of iron at high pressures and temperatures. Phys Rev B53:8296-8309

Wcislak L, Klein H, Bunge H-J, Garbe U, Tschentscher T, Schneider JR (2002). Texture analysis with high-energy synchrotron radiation. J Appl Crystallogr 35:82-95

Weissenberg K (1922) Statistische Anisotropie in kristallinen Medien. Ann Phys 69:409-435

Wenk H-R (ed) (1985) Preferred Orientation in Deformed Metals and Rocks: An Introduction to Modern Texture Analysis. New York: Academic Press, 610 p

Wenk H-R (ed) (1994) NATO Advanced Research Workshop on Polyphase Polycrystal Plasticity. Mater Sci Eng A175:1-277

Wenk H-R (1994) Texture analysis with TOF neutrons. *In* Time-of-Flight Diffraction at Pulsed Neutron Sources. JD Joergensen, AJ Schultz (eds) Trans Am Crystallogr Assoc 29:95-108

Wenk H-R (1999) A voyage through the deformed Earth with the self-consistent model. Modeling Mater Sci Eng 7:699-722

Wenk H-R (2000) Pole figure measurements with diffraction techniques. *In* Kocks UF, Tomé CN, Wenk H-R (eds) Texture and Anisotropy. Preferred Orientations in Polycrystals and Their Effect on Materials Properties, 2nd Edition. Cambridge, UK: Cambridge University Press, p 127-177

Wenk H-R, Christie JM (1991) Review paper: Comments on the interpretation of deformation textures in rocks. J Struct Geol 13:1091-1110

Wenk H-R, Kocks UF (1987) The representation of orientation distributions. Metall Trans 18A:1083-1092

Wenk H-R, Grigull S (2003) Texture determination by synchrotron X-ray diffraction. J Appl Crystallogr (submitted)
Wenk H-R, Tomé CN (1999) Modeling dynamic recrystallization of olivine aggregates deformed in simple shear. J Geophys Res 104:25,513-25,527
Wenk H-R, Takeshita T, Bechler E, Erskine BG, Matthies S (1987) Pure shear and simple shear calcite textures. Comparison of experimental, theoretical and natural data. J Struct Geol 9:731-745
Wenk H-R, Canova G, Brechet Y, Flandin L (1997) A deformation-based model for recrystallization, Acta Mater 45:3283-3296
Wenk H-R, Matthies S, Donovan J, Chateigner D (1998) BEARTEX, a Windows-based program system for quantitative texture analysis. J Appl Crystallogr 31:262-269
Wenk H-R, Dawson P, Pelkie C, Chastel Y (1999) Texturing of Rocks in the Earth's Mantle. A Convection Model Based on Polycrystal Plasticity. *Video*. Am Geophys Union, Washington, DC
Wenk H-R, Matthies S, Hemley RJ, Mao H-K, Shu J (2000) The plastic deformation of iron at pressures of the Earth's inner core. Nature 405:1044-1047
Wenk H-R, Cont L, Xie Y, Lutterotti L, Ratschbacher L, Richardson J (2001) Rietveld texture analysis of Dabie Shan Eclogite from TOF neutron diffraction spectra. J Appl Crystallogr 34:442-453
Wever F (1924) Über die Walzstruktur kubisch kristallisierender Metalle. Z Phys 28:69-90
Wright SI, Adams BL (1992) Automatic analysis of electron backscatter diffraction patterns. Metall Trans A23:759-767
Xie Y, Wenk H-R, Matthies S (2002) Plagioclase preferred orientation by TOF neutron diffraction and SEM-EBSD. Tectonophysics (in press)
Yamazaki D, Karato S (2002) Shear deformation of (Mg,Fe)O: Implications for seismic anisotropy in Earth's lower mantle. Phys Earth Planet Inter (in press)
Yoshida S, Sumita I, Kumazawa M (1996) Growth model of the inner core coupled with the outer core dynamics and the resulting elastic anisotropy. J Geophys Res 101:28085-28103
Young RA (1993) The Rietveld Method. Oxford, UK: Oxford University Press, 298 p

11 Modeling Deformation of Polycrystalline Rocks

Paul R. Dawson

Sibley School of Mechanical and Aerospace Engineering
Cornell University
Ithaca, New York 14853

MODELING PRELIMINARIES

The finite element method is a powerful complement to polycrystal plasticity theory for modeling the non-uniform deformation of crystalline solids. Polycrystal plasticity provides a micro-mechanical model for slip-dominated plastic flow and serves as a constitutive theory for deformation simulations (Kocks et al. 1998). The finite element method offers a numerical means to solve partial differential equations, such as the field equations of elasticity or plasticity (Zienkiewicz et al. 1989). The two can be combined in different ways depending on the goals of a modeling effort.

Length scales

Finite elements and polycrystal plasticity may be applied to the detailed modeling of a collection of grains that represent a sample of the material. In this case, there are one or more finite elements discretizing each grain and balance laws for momentum and mass are applied at the level of individual crystals. A second combination of finite elements and polycrystal plasticity is to embed polycrystal theory within a finite element formulation for physical systems that are far larger than the dimension of a grain. Polycrystal plasticity serves as a constitutive theory in essentially the same way as continuum elastoplasticity models. Balance laws are applied at the larger continuum scale. We refer to these as small-scale and large-scale applications, respectively. Care must be exercised in assuring consistency between the macroscopic material element volume and the polycrystal dimensions.

With respect to characterizing these applications, it is useful to define a geometric parameter, ζ, as the relative sizes of a finite element and a crystal. Allowing h to be a characteristic dimension of an element and d to be the representative grain size the parameter ζ simply is h/d. Here, large ζ implies large numbers of crystals in each element; small ζ implies many elements within each crystal. For small-scale applications, there is a limit to how finely a crystal may be resolved and thus how small ζ may become. The size of an element must be much larger than the distance between slip planes for process of slip to be homogenized in a meaningful way. For large-scale applications, although the macroscopic deformation gradients may be steep over characteristic lengths of the body, they may only vary slowly across the dimension of an polycrystal. As such we may consider the polycrystal to be subjected to a uniform deformation locally. Locally here refers to a point on macroscopic scale, so that we permit only single (tensor) values of stress and velocity gradient. In this case, the dimension of a crystal must be small compared to the dimension over which the macroscopic velocity gradient changes appreciably. In turn, the polycrystal must contain a sufficient number of crystals and inherent appropriate symmetry relations such that the above arguments of homogenization are justified.

General comments

Regardless of the scale of the simulation, a number of attributes of the models and solution procedures will be the same for the methodologies discussed in this summary. The balance laws applied are balance of linear momentum, conservation of mass, and conservation of energy. Inertia is neglected, so that all motions are quasistatic. Elasticity is neglected in the examples presented here, although the formulations have been extended to include both elastic and plastic responses (Dawson and Marin 1998). Plastic deformation occurs solely by slip, with other mechanisms being neglected (including twinning, diffusion, and grain boundary sliding). The deformations are assumed to be isochoric (constant volume).

The governing equations are solved numerically using the finite element method. Depending on the formulation, the solution is based on weak forms of the differential equations. For conservation of mass, a consistent penalty approach is used. A nonlinear solution procedure must be invoked, as the equations governing slip are highly nonlinear. The solution is carried out incrementally, with the geometry and state evolution being closely coupled over the deformation history.

Because the deformations are assumed to occur on a restricted number of slip systems, each involving a particular crystallographic plane and direction, the orientation of the crystal lattice is of paramount concern. For polycrystalline materials, there is a distribution associated with the lattice orientations (Wenk 1985), referred to as the crystallographic texture (or just texture when the meaning is clear). The initialization of the texture is a central issue in the application of polycrystal-based models. This is true for both small-scale and large-scale applications. For small-scale applications, it is necessary to define the spatial relation of crystals. This may be done by randomly assigning orientations to elements; alternatively, with recently developed experimental capability, the elements of a finite element mesh can be associated with specific orientations measured within a material sample. Here, the lattice orientations of all crystals are assigned from an orientation distribution for the material. For large-scale applications, a representation of the orientation distribution must exist at every point where properties are evaluated. This may be accomplished with an orientation distribution function or with a discrete set of orientations drawn from the distribution.

Small-scale simulations

The manner in which deformations are partitioned among the crystals of a polycrystal affects the average stress exhibited by the polycrystal and the subsequent evolution of its crystallographic texture. For materials comprised of crystals that exhibit low anisotropy in the single crystal yield surface, the assumption of uniform straining put forward by Taylor (Taylor 1938) is adequate for predicting the stress response and gross features of the texture, but is deficient in describing various texture details. For materials having high anisotropy in the single crystal yield surface, the assumption of equal straining is less satisfactory, and in many instances is not plausible. For example, for crystals with fewer than five independent slip systems, equal straining is not possible because each grain lacks sufficient deformation modes to accommodate a general deformation. In such cases, deformation is not uniformly partitioned among the crystals and consideration of the topology of the aggregate is necessary to obtain accurate predictions The inclusion of topology, however, may be important in obtaining more accurate predictions due to the influence of each crystal's specific neighborhood on its response.

For small-scale applications, an aggregate of crystals is constructed with finite elements discretizing each grain. While the finite element model has considerable generality and flexibility, we use it here to explore the implications of aggregate topology for materials undergoing intra-crystal slip. In particular, we examine the effects of the variability of straining from crystal to crystal on texture evolution. The results are intended to help develop a more comprehensive understanding of polycrystal behavior and motivate better models for plastic response.

As applied to the analysis of aggregates, the finite element formulation has several features. Each crystal is discretized by one or more finite elements. Standard finite element methodology is utilized to impose compatibility both within crystals and across their boundaries. Equilibrium is enforced by requiring that, in a weighted residual sense, the interface tractions vanish. This is accomplished using a hybrid finite element formulation in which the full body is divided into physically identifiable domains and the traction constraint is applied between domains (Beaudoin et al. 1995). Within the context of a hybrid formulation, domains are defined so that finite elements correspond to individual crystals or parts of those crystals The single crystal slip relation is satisfied approximately over each crystal domain via a weighted residual using trial functions for the stress that *a priori* satisfy equilibrium within an element. In this way the aggregate is a polycrystalline body to which the balance laws and constitutive relations are applied.

Large-scale simulations

Large scale applications are those in which the body, whether an engineered component or a geologic formation, is very much larger than individual crystals. In these applications, the polycrystal is not a body itself, but rather is a representation of the microstructural state. To emphasize the difference between the roles assumed by crystals in the large-scale and small-scale models, we refer to polycrystal in terms of the orientation distribution or the set of orientations drawn from it.

As a state variable representation for plastic flow, polycrystal plasticity theory has several distinct advantages. Implicit with the use of state variable models is the ability to initialize the state. Orientation distributions are directly accessible via diffraction measurements and well-established methodologies for interpreting those measurements. The slip system strengths are not as direct, but for simplified assumptions regarding hardening (such as that all slip systems harden the same way), the strengths may be initialized from simple compression testing. Polycrystal theory also provides a direct means for updating the material state via integration of the evolution equations for the crystal lattice orientation and the slip system strength.

In applying polycrystal plasticity, it is assumed that the properties at any point in the body are determined from a collection of anisotropic crystals that underlies that point. An assumption, or rule, is employed to partition the macroscopic (average) strain over the set of orientations. A variety of rules are possible, some of which constitute bounds. By modeling the linkage between the microscopic (crystal) and macroscopic (continuum) scales with partitioning rules, the need to define each crystal's neighbors is circumvented. Instead, the set of orientations can be thought of as a set of co-existing orientations whose averaged responses define the macroscopic properties. The orientations play the role of state variables, together with the crystal strengths and the grain shape. As state variables, the orientations replace the need to remember the deformation path, but require initialization to begin an analysis.

Different assumptions have been used to link the crystal behavior to the macroscopic deformation and thus to provide a means to partition the deformations and the stresses among the set of orientations. For crystal with high symmetry, such as FCC or BCC crystals with many slip systems of comparable strength, and nearly initial equiaxed shape, it often is assumed that each crystal of the set experiences the same velocity gradient. This partitioning rule (Asaro and Needleman 1985) is an extension of the Taylor hypothesis (Taylor 1938). For crystals which have markedly anisotropic yield surfaces, say those with fewer than five independent slip systems, equal partitioning of the deformation over the set of orientations leads to physically unrealistic conditions. A crystal may be forced to deform in a mode that is not available to it from combinations of the existing slip systems or may exhibit the very high stress levels needed to activate unfavorable slip systems.

In these cases, assumptions must be invoked that permit unequal straining. For example an equilibrium assumption may be made (all crystals experience the same stress state), or crystals may be permitted to deform using only favorable slip systems using a constrained hybrid approach such as offered by Parks and Ahzi (Parks and Ahzi 1990).

As applied to the analysis of large-scale systems, the finite element formulation has features common to many viscous flow formulations, with modifications necessary to utilize polycrystal plasticity (Mathur and Dawson 1989). In particular, the constitutive relation between deviatioric stress and deformation rate is given by the orientational average of the single crystal relation according to the linking hypothesis discussed above. As before, standard finite element methodology is utilized to impose compatibility both within crystals and across their boundaries. Incompressibility is enforced using a consistent penalty method. In this case, equilibrium is enforced via a weighted residual of the local balance of linear momentum equation. The simulation results are useful in better understanding how the material microstructure (specifically the crystallographic texture) both influences and is influenced by the straining present during the deformation of geologic formations.

SINGLE CRYSTAL CONSTITUTIVE BEHAVIOR

Slip systems

The mechanisms that contribute to the plastic deformations of minerals are dependent on strain rate and temperature (Wenk 1985; Kocks et al. 1998). Depending on the regime of strain rate and temperature, some mechanisms will dominate over others and be principally responsible for shape changes occurring in the material. Here, we assume that the crystals are deforming solely by slip, rather than by other mechanisms or by combinations of mechanisms. Although slip occurs by the motion of dislocations through the lattice, we do not resolve behavior of individual dislocations. Rather, the net effect of dislocation motion is captured by the activity of slip systems, each of which contributes to a single shear mode of deformation.

A slip system can be described mathematically by vectors that describe the normal to the slip plane (m) and the slip direction (s) (Wenk 1985). Forming the tensor product of these gives the Schmid tensor

$$T = s \otimes m, \qquad (1)$$

which has symmetric and skew parts denoted by

$$P = \mathrm{sym}\,(T) \qquad (2)$$

and

$$Q = \mathrm{skw}\,(T). \qquad (3)$$

Minerals often exhibit only a few potentially active slip systems. However, it takes five independent slip systems to accommodate all possible components of deviatoric part of the deformation rate. With the slip systems defined, the plastic velocity gradient, L^p, in a crystal is written as a combination of the slip system responses

$$L^p = \sum_\alpha T^\alpha \dot\gamma^\alpha = \sum_\alpha P^\alpha \dot\gamma^\alpha + \sum_\alpha Q^\alpha \dot\gamma^\alpha, \qquad (4)$$

where $\dot\gamma^\alpha$ is the rate of shear on the α slip system. (A superscript (α) is used to designate one of the systems.) This relationship is a central part of the mathematical structure of polycrystal plasticity as it relates the net result of dislocation movement to crystal shape change and lattice rotation.

Crystal kinematics

A comprehensive treatment of the crystal kinematics begins with the mapping of coordinates of points within the crystal over time (Dawson and Marin 1998). This mapping is the motion of the crystal and is given by

$$x = x(X, t), \tag{5}$$

in which x are the current coordinates and X are reference coordinates. From the mapping the crystal deformation gradient, F^c, is determined as

$$F^c = \frac{\partial x}{\partial X}. \tag{6}$$

The full deformation gradient is decomposed into several parts

$$F^c = V^* \cdot R^* \cdot F^p, \tag{7}$$

where V^* is the elastic stretch, R^* is the lattice rotation, and F^p is the plastic deformation gradient. We restrict our attention to the purely inelastic response, assuming that the elastic strains are always small in relation to unity, such that $V^* \approx I$. The crystal deformation gradient then becomes

$$F^c = R^* F^p, \tag{8}$$

which is differentiated with respect to time and divided into symmetric and skew parts to give

$$D^c = D^p = \sum_\alpha \dot{\gamma}^\alpha P^\alpha \tag{9}$$

and

$$W^c = \dot{R}^* R^{*T} + \sum_\alpha \dot{\gamma}^\alpha Q^\alpha = \dot{R}^* R^{*T} + W^p. \tag{10}$$

where D^c is the crystal deformation rate, D^p is the plastic deformation rate, W^c is the crystal spin, and W^p is the plastic spin. Equation (9) relates the crystal shape change to the net motion of dislocations, while Equation (10) provides a relation between the crystal spin and the lattice reorientation. Together, D^p and W^p form the plastic velocity gradient given in Equation (4). Note that $D' = D$ for incompressible deformations. (Primed quantities are deviatoric throughout the paper).

Crystal compliance and stiffness

The above kinematic framework is not sufficient by itself to define the crystal stiffness; a relationship between the crystal deviatoric stress, σ'^c, and slip system shearing rates, $\dot{\gamma}^\alpha$, is also required. This is obtained from the kinetic relation for slip on a slip system and the geometric relation between the crystal stress and its component on the slip plane and in the slip direction. The kinetics of slip is assumed to be well represented with a power law relation

$$\dot{\gamma}^\alpha = \dot{\gamma}_0 \left| \frac{\tau^\alpha}{\hat{\tau}} \right|^{\frac{1}{m}} \text{sgn}(\tau^\alpha) = f(\tau^\alpha, \hat{\tau}), \tag{11}$$

where m is the strain rate sensitivity, $\dot{\gamma}_0$ is a model parameter, $\hat{\tau}$ is the temperature dependent slip system strength, and τ^α is the resolved shear stress for the αth system. The resolved shear stress is

$$\tau^\alpha = P^\alpha \cdot \sigma'^c. \tag{12}$$

Factoring a term that is linear in τ^α from Equation (11), and combining this with Equations (9) and (12) gives

$$D'^c = \mathcal{M}^c \cdot \sigma'^c, \tag{13}$$

where

$$\mathcal{M}^c = \sum_\alpha \left(\frac{f(\tau^\alpha, \hat{\tau})}{\tau^\alpha} \right) P^\alpha P^{\alpha T}. \tag{14}$$

Inverting Equation (13) gives

$$\sigma'^c = \mathcal{M}^{c^{-1}} \cdot D'^c = \mathcal{S}^c \cdot D'^c. \tag{15}$$

The compliance $\mathcal{M}^c$ or the stiffness $\mathcal{S}^c$ is used in constructing the macroscopic stiffness $\mathcal{S}$ in accordance with the assumptions that link the crystal and macroscopic scales, discussed in the following section.

POLYCRYSTAL CONSTITUTIVE EQUATIONS

Orientational averages

In the class of large-scale applications we present, every macroscopic point has associated with it a representation of the crystallographic texture of a small volume of material containing the point. We represent the texture by a probability density function, A, that prescribes an orientation distribution (Wenk 1985). The texture enters the simulations through the determination of mechanical properties at the macroscopic level. The macroscopic stiffness is obtained by averaging the crystal stiffness (or by inverting the average of the crystal compliance) over the orientation distribution. This requires that the crystallographic texture be available at each instant when the velocity field is computed. For this purpose, we present methods for evolving the texture for imposed deformation histories, histories that will be evaluated as part of the full simulation procedure.

The orientation distribution associated with an aggregate is given by the probability density, $A(\boldsymbol{r})$, with the property that the volume fraction of crystals v_f^* in a subset of orientation space Ω^* is given by

$$v_f^* = \int_{\Omega^*} A(\boldsymbol{r}) \, d\Omega, \tag{16}$$

where Ω is the domain of orientation space and $\boldsymbol{r}$ is the vector describing the crystal orientation using an angle-axis representation. $A(\boldsymbol{r})$ is normalized such that

$$\int_\Omega A(\boldsymbol{r}) \, d\Omega = 1. \tag{17}$$

The extent of the domain Ω depends on the specific parameterization of orientation space and the symmetries exhibited by the crystal.

The macroscopic behavior is the average of the behaviors of the individual crystals. We obtain the average by an integration of the crystal quantity, weighted by the probability density, over orientation space. For an arbitrary crystal quantity, Γ^c, the average value, $<\Gamma^c>$, is given by

$$\Gamma = <\Gamma^c> = \int_\Omega \Gamma^c(\boldsymbol{r}) A(\boldsymbol{r}) \, d\Omega. \tag{18}$$

In particular, we require that the deviatoric stress and deformation rate at the macroscopic level reflect the average of their counterparts at the crystal level so that

$$D' = <D'^c> = \int_\Omega D'^c(\boldsymbol{r}) A(\boldsymbol{r}) \, d\Omega \tag{19}$$

and

$$\sigma' = <\sigma'^c> = \int_\Omega \sigma'^c(\boldsymbol{r}) A(\boldsymbol{r}) \, d\Omega. \tag{20}$$

In instances in which the texture representation consists of a discrete sample of orientations taken from the orientation distribution, the macroscopic values are simply the weighted averages of the crystal quantities

$$D' = <D'^c> = \sum_c w^c D'^c \tag{21}$$

and

$$\sigma' = <\sigma'^c> = \sum_c w^c \sigma'^c, \tag{22}$$

where w^c is a set of weights assigned with one-to-one correspondence to the orientations (Kocks et al. 1998). The values of weights are chosen to emphasize those crystals corresponding to regions in the orientation domain where A is high.

Linking crystal responses to continuum scale motion

Recall from Equations (14) and (15) that the the relationship between the deformation rate and deviatoric stress in a single crystal is an invertible relationship of the form

$$D'^c = \mathcal{M}^c \cdot \sigma'^c \tag{23}$$

or

$$\sigma'^c = \mathcal{S}^c \cdot D'^c = (\mathcal{M}^c)^{-1} \cdot D'^c, \tag{24}$$

where $\mathcal{M}^c$ and $\mathcal{S}^c$ are the single crystal viscoplastic compliance and stiffness, respectively. If we proceed with averaging these relationships we obtain

$$<D'^c> = <\mathcal{M}^c \cdot \sigma'^c> \tag{25}$$

and

$$<\sigma'^c> = <\mathcal{S}^c \cdot D'^c>. \tag{26}$$

At the macroscopic scale, we assume the deviatoric stress and deformation rate are related in an analogous form

$$\sigma' = \mathcal{S} \cdot D'. \tag{27}$$

To relate Equations (25) and (26) with Equation (27), we must designate which hypothesis to invoke to link the microscopic (single crystal) and macroscopic behaviors. Two possibilities discussed earlier are: (1) the velocity gradient is identical from crystal to crystal over an aggregate (which can be thought of as an extension of the hypothesis of Taylor (Taylor 1938)) and (2) the stress is identical in each crystal (Prantil et al. 1995), (which can be thought of as an extension to the hypothesis of Sachs (Sachs 1928)). For the first case we note in Equation (26), the crystal deformation rate can be removed from the orientation average if it is identical for all orientations to give

$$\sigma' = <\sigma'^c> = <\mathcal{S}^c \cdot D'^c> = <\mathcal{S}^c> \cdot D' = \mathcal{S} \cdot D'. \tag{28}$$

Using this assumption renders equal straining in all crystals of an aggregate. For the second case, the identical stress may be taken outside the orientational average to give

$$D' = <D'^c> = <\mathcal{M}^c \cdot \sigma'^c> = <\mathcal{M}^c> \cdot \sigma'^c = <\mathcal{M}^c> \cdot \sigma'. \tag{29}$$

To obtain the form of Equation (27), $<\mathcal{M}^c>$ is inverted

$$\sigma' = <\mathcal{M}^c>^{-1} \cdot D' = \mathcal{S} \cdot D'. \tag{30}$$

Estimating $\mathcal{S}$ by these two approaches gives upper and lower bounds of the stiffness, respectively, for a specific texture. With the Taylor hypothesis, compatibility is enforced, but equilibrium may be violated. With the equal stress constraint, equilibrium is satisfied identically, but the crystal deformations may not be compatible.

SMALL-SCALE HYBRID ELEMENT FORMULATION

Small-scale formulations are performed using a hybrid finite element formulation. Central to hybrid formulations is the concept of domain decomposition, which here coincides with crystals of a polycrystal. The stress is represented in a manner that satisfies equilibrium within the domain *a priori* so that enforcement of equilibrium for the full polycrystal consists of matching tractions across the domain (crystal) interfaces. To this end, the traction equilibrium residual, after utilizing the divergence theorem, is given as

$$R_t = \sum_e \left[\int_{\Omega_e} (\boldsymbol{\sigma}'^c - p\boldsymbol{I}) : \nabla(\boldsymbol{\Phi}) \, d\Omega - \int_{\Gamma_e} \boldsymbol{\Phi} \cdot \boldsymbol{t} \, d\Gamma \right] = 0, \tag{31}$$

where $\boldsymbol{t}$ is the surface traction, p is the pressure, $\boldsymbol{I}$ is the identity tensor, $\boldsymbol{\Phi}$ are vector weights and Γ_e is the surface bounding Ω_e. Using Equation (13), which gives a linearized version of the relation that quantifies intracrystal slip, a residual is constructed over the each subdomain volume, Ω_e

$$R_\sigma = \int_{\Omega_e} \boldsymbol{\Psi} \cdot (\boldsymbol{\mathcal{M}}^c : \boldsymbol{\sigma}'^c - \boldsymbol{D}^c) \, d\Omega = 0, \tag{32}$$

where $\boldsymbol{\mathcal{M}}^c$ serves as a crystal viscoplastic matrix that relates the crystal deviatoric stress, $\boldsymbol{\sigma}'^c$, to the crystal deformation rate, $\boldsymbol{D}^c$, and $\boldsymbol{\Psi}$ are weighting functions. Consistent with the order of Equation (13), $\boldsymbol{\Psi}$ are second order quantities. Finally, allowing for slight compressibility a penalized residual on the incompressibility condition is constructed for each domain using scalar weights, Υ

$$R_v = \int_{\Omega_e} \Upsilon p \, d\Omega + \frac{1}{\epsilon} \int_{\Omega_e} \Upsilon \, \text{div} \boldsymbol{u} \, d\Omega = 0, \tag{33}$$

where ϵ is a penalty parameter.

Trial functions are introduced in the residuals for the interpolated field variables. In the hybrid formulation, both the velocity and the stress (deviatoric and spherical portions) are represented with trial functions

$$\boldsymbol{\sigma}'^c = \boldsymbol{N}^s S^h; \quad p = \boldsymbol{N}^p P^h; \quad \boldsymbol{u} = \boldsymbol{N}^u U^h. \tag{34}$$

Here, $\boldsymbol{N}^s$, $\boldsymbol{N}^p$, and $\boldsymbol{N}^u$ are interpolations functions and S^h, P^h, and U^h are nodal quantities for each variable being represented. For the deviatoric stress, piecewise discontinuous functions are specified which satisfy equilibrium within an element *a priori* (Bratianu and Atluri 1983). The weights are constructed according to a Galerkin methodology as

$$\boldsymbol{\Psi} = \boldsymbol{N}^s \Psi^h; \quad \Upsilon = \boldsymbol{N}^p \Upsilon^h; \quad \boldsymbol{\Phi} = \boldsymbol{N}^u \Psi^h. \tag{35}$$

Each residual is written in matrix form using the trial and weight functions. For the crystal slip relation, Equation (32), nodal stresses are related to the nodal velocities

$$\boldsymbol{H} \cdot S^h - \boldsymbol{R}^T \cdot U^h = 0, \tag{36}$$

where

$$\boldsymbol{H} = \int_{\Omega_e} \boldsymbol{N}^{sT} \cdot \boldsymbol{\mathcal{M}}^c \cdot \boldsymbol{N}^s \, d\Omega \tag{37}$$

and

$$\boldsymbol{R} = \int_{\Omega_e} \boldsymbol{B}^T \cdot \boldsymbol{N}^s \, d\Omega \tag{38}$$

where $\boldsymbol{B}$ contains spatial derivatives of $\boldsymbol{N}^u$ such that

$$\boldsymbol{D} = \boldsymbol{B} U^h \tag{39}$$

and $\boldsymbol{D}$ is a vector containing the components of the deformation rate. Solving for the nodal stress values gives

$$S^h = \boldsymbol{H}^{-1} \cdot \boldsymbol{R}^T \cdot U^h = 0. \tag{40}$$

In a similar way, the incompressibility constraint is written in matrix form as

$$\boldsymbol{M}^p \cdot \boldsymbol{P}^h = -\frac{1}{\epsilon} \boldsymbol{G}^T \boldsymbol{U}^h, \tag{41}$$

where

$$\boldsymbol{M}^p = \int_\Omega \boldsymbol{N}^{pT} \boldsymbol{N}^p d\Omega \tag{42}$$

and

$$\boldsymbol{G} = \int_\Omega \boldsymbol{B}^T \boldsymbol{h}^T \boldsymbol{N}^p d\Omega. \tag{43}$$

Here, $\boldsymbol{h}$ is the trace operator. It is possible to solve for P^h over each domain independently, giving

$$P^h = -\frac{1}{\epsilon} \boldsymbol{M}^{p-1} \cdot \boldsymbol{G}^T \boldsymbol{U}^h. \tag{44}$$

The traction residual becomes

$$\sum_e \left(\boldsymbol{R} \cdot S^h - \boldsymbol{G} \cdot P^h - \boldsymbol{f}\right) = 0. \tag{45}$$

Substituting Equations (40) and (44) into Equation (45) for S^h and P_h, repectively, give

$$\sum_e \left(\left[\boldsymbol{R} \cdot \boldsymbol{H}^{-1} \cdot \boldsymbol{R}^T - \boldsymbol{G} \cdot \boldsymbol{M}^{p-1} \cdot \boldsymbol{G}^T\right] \cdot U^h - \boldsymbol{f}\right) = 0, \tag{46}$$

where

$$\boldsymbol{f} = \int_{\Gamma_e} \boldsymbol{N}^{uT} \cdot \boldsymbol{t} \, d\Gamma. \tag{47}$$

This discretized traction residual is solved simultaneously for the velocity field, from which the stress may be recovered.

LARGE-SCALE VELOCITY-PRESSURE FORMULATION

For large-scale applications, the finite element formulation is similar to traditional developments for viscous flow (Mathur and Dawson 1989). The linear momentum and conservation of mass balance laws are used to form the residuals for the finite element approximations constructed to evaluate the velocity fields. Coupling to the solution of the temperature distribution (Zienkiewicz and Taylor 1989). is straightforward and will not be discussed here. A weighted residual for equilibrium (Johnson 1987) is written over the body volume, Ω, using vector weights, $\boldsymbol{\Phi}$ as

$$R_u = \int_\Omega \boldsymbol{\Phi} \cdot \left(\mathrm{div}\boldsymbol{\sigma}^T + \boldsymbol{b}\right) d\Omega = 0. \tag{48}$$

This is modified to obtain the weak form through integration by parts and application of the divergence theorem to give

$$R_u = -\int_\Omega \mathrm{tr}\left(\boldsymbol{\sigma}'^T \cdot \nabla \boldsymbol{\Phi}\right) d\Omega + \int_\Omega p \, \mathrm{div}\boldsymbol{\Phi} \, d\Omega + \int_\Gamma \boldsymbol{t} \cdot \boldsymbol{\Phi} \, d\Gamma + \int_\Omega \boldsymbol{b} \cdot \boldsymbol{\Phi} \, d\Omega = 0, \tag{49}$$

where the Cauchy stress has been divided into its deviatoric ($\boldsymbol{\sigma}'$) and spherical parts ($p\boldsymbol{I}$). The body force is $\boldsymbol{b}$ and $\boldsymbol{t}$ is the traction acting on the boundary Γ of Ω. For an incompressible medium

$$\mathrm{tr}\boldsymbol{D} = \mathrm{div}\boldsymbol{u} = 0, \tag{50}$$

which can be enforced using a consistent penalty approach (Engelman et al. 1982). First a residual is written using the incompressibility constraint (Equation (50))

$$R_v = \int_\Omega \Upsilon \, \mathrm{div}\boldsymbol{u} \, d\Omega = 0, \tag{51}$$

where Υ are scalar weights. Slight compressibility is admitted, giving the possibility to write a penalty constraint in a form consistent with the finite element discretization

$$R_v = \int_\Omega \Upsilon p d\Omega + \frac{1}{\epsilon} \int_\Omega \Upsilon \, \text{div} \boldsymbol{u} \, d\Omega = 0, \qquad (52)$$

where ϵ is a penalty parameter. Finite element interpolation is introduced as

$$\boldsymbol{u} = \boldsymbol{N}^u U^h \quad \text{and} \quad p = \boldsymbol{N}^p P^h \qquad (53)$$

for the trial functions and

$$\boldsymbol{\Psi} = \boldsymbol{N}^u \Psi^h \quad \text{and} \quad \Upsilon = \boldsymbol{N}^p \Phi^h \qquad (54)$$

for the weighting functions where $\boldsymbol{N}^u$ and $\boldsymbol{N}^p$ are functions of the position $\boldsymbol{x}$. The trial functions for $\boldsymbol{u}$ are continuous over element boundaries, but for p discontinuous interpolation is specified to match the interpolation for the gradient of the velocity. With discontinuous pressure interpolation, the penalized incompressibility constraint can be written element-by-element as

$$P_h = -\frac{1}{\epsilon} \boldsymbol{M}^{p-1} \boldsymbol{G}^T \boldsymbol{U}^h, \qquad (55)$$

where

$$\boldsymbol{M}^p = \int_\Omega \boldsymbol{N}^{pT} \boldsymbol{N}^p d\Omega \qquad (56)$$

and

$$\boldsymbol{G} = \int_\Omega \boldsymbol{B}^T \boldsymbol{h}^T \boldsymbol{N}^p d\Omega. \qquad (57)$$

Here $\boldsymbol{B}$ is defined in Equations (39) and $\boldsymbol{h}$ again is the trace operator.

We introduce the mechanical behavior from Equation (27) in matrix form as

$$\boldsymbol{\sigma}' = \boldsymbol{S}\boldsymbol{D}'. \qquad (58)$$

Substitution of this relationship into Equation (49) eliminates the deviatoric stress. Taken with Equation (55) for eliminating the pressure, the residual is written for arbitrary variations in the weights as

$$\left(\boldsymbol{K}_s + \frac{1}{\epsilon} \boldsymbol{G} \boldsymbol{M}^{p-1} \boldsymbol{G}^T \right) U^h = \boldsymbol{F}, \qquad (59)$$

where

$$\boldsymbol{K}_s = \int_\Omega \boldsymbol{B}^T \boldsymbol{S} \boldsymbol{B} d\Omega \qquad (60)$$

and

$$\boldsymbol{F} = \int_\Gamma \boldsymbol{N}^T t d\Gamma + \int_\Omega \boldsymbol{N}^T b d\Omega. \qquad (61)$$

From Equation (59), the velocity field is determined for a specific combination of geometry, loading, and state.

HALITE TEXTURE EVOLUTION: A SMALL-SCALE APPLICATION

Generalities

Modeling the plastic deformations and associated texture evolution in halite is challenging due to its high level of plastic anisotropy, at least at low homologous temperature. Small-scale finite element modeling provides an avenue to study how the anisotropy influences the degree of heterogeneity of deformation at the crystal scale and how this in

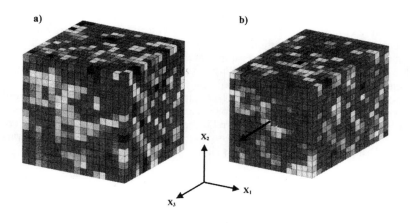

Figure 1. The finite element meshes of the inner 1000 crystals of an aggregate of halite crystals corresponding to the initial and deformed shapes (repaired). Gray shades show the longitudinal component of the deformation rate.

turn affects the texture evolution. Although halite possesses a cubic structure, its response under mechanical load is highly anisotropic because of the sparseness of available slip systems. At the level of a single crystal, some modes of deformation require considerably higher stress levels than others. As a consequence, there is a tendency in polycrystals for crystals that are more favorably oriented to deform more than others that are less favorably oriented. A consequence of this deformation heterogeneity is that assuming that all grains deform identically when modeling the mechanical response leads to poor texture predictions. As example of the use of the small-scale formulation we show results reported in (Lebensohn et al. 2003) in which predictions obtained using finite element simulations, self-consistent simulations and Taylor simulations were compared. The intent of the simulations is to learn more about how straining is partitioned among crystals in an aggregate and from there to better understand how texturing proceeds under continued deformation.

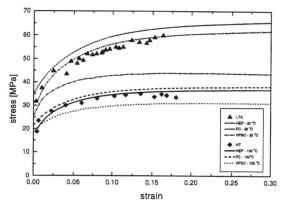

Figure 2. Measured and computed stress strain curves for halite deformed in extension at 20°C and 100°C computed using the finite element model, a Taylor assumption, and a viscoplastic model.

Simulation specifics

Simulations were performed on a sample of halite consisting of 4096 crystals. The starting texture corresponded to a uniform orientation distribution, with orientations randomly chosen from the distribution and assigned to the crystals comprising the sample. The initial shape of all crystals (and sample) was a cube. Boundary conditions were imposed to replicate conditions corresponding to an extension experiment with constant true strain rate. As the deformation preceeded, the crystals quickly became distorted and irregular due to the strain heterogeneity. To compensate for this, the crystals periodically were given a brick shape corresponding to the current exterior shape of the sample without altering the lattice orientations as shown in Figure 1. The slip system parameters were chosen so that the computed average stress, as determined by the integrated surface tractions, matched measured stress strain curves, as shown in Figure 2.

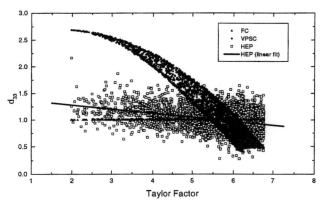

Figure 3. Tensile component of the deformation rate at high temperature (100°C) for the halite crystals under three modeling assumptions.

Simulation results

A goal of modeling is to accurately capture the heterogeneity of deformation among the crystals of an aggregate. A tradeoff exists between the complexity of the model and its ability to capture this heterogeneity. Modelers seek a balance such that essential features of a deformation are accurately computed with unnecessary details and accompanying computational burdens. To help understand where this balance lies, we compare the variation in deformation rate over the aggregate computed with the finite element model (HEP) to two other modeling assumptions: a Taylor assumption (FC, indicating full kinematic constraints) and a viscoplastic self-consistent formulation (VPSC). Figure 3 shows how the deformation rate varies as a function of crystal orientation as quantified with the Taylor factor. The FC model, by definition, has no variation. The VPSC model has a strong systematic dependence. The HEP model has weaker dependence with Taylor than the VPSC overall, but greater variation of deformation at fixed Taylor factor. This stems from the unique neighborhood that each finite element has, resulting in different deformation rates for two crystals of similar orientation but different surroundings. This difference in local behavior of the grains influences the evolution of properties with continuing deformation. The lattice orientations evolve with deformation according to the equations for crystal plasticity reviewed in the prior sections of the paper. Textures after various amounts of deformation are shown as inverse pole figures of the extension direction in Figure 4 for the three modeling assumptions. As is evident, certain component of the texture are elevated and other suppressed by the degree to which the deformations can vary over the polycrystal. Texture development is strongest for the Taylor (FC) model and weakest for the finite element (HEP) model. In all cases, a strong 111 component develops, but for the VPSC a 100 component also is significant.

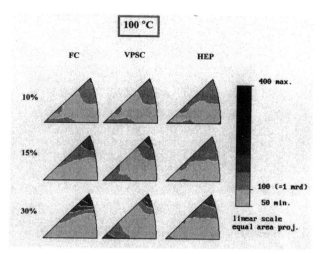

Figure 4. Computed inverse pole figures for texture evolution of halite at 100°C for increasing amount of extension. Directions in inverse pole figures are: lower left – 001; lower right – 011; upper right – 111.

MANTLE CONVECTION – A LARGE-SCALE APPLICATION

Generalities

Of particular importance for processes in the crust, such as volcanic activity, earthquakes, and mountain building, is the deformation of the mantle. Within the mantle, large cells of convection are induced by instabilities and driven by temperature and density gradients. The convective flow of the mantle substantially alters the internal structure, and the mechanical properties derived from it. The main constituent of the upper mantle is olivine, which exhibits about a 25% difference in acoustic speed between the slowest and the fastest crystal directions. Due to this difference, a polycrystal with preferred orientations of component crystals will display a directional dependence (anisotropy) of seismic wave propagation. Indeed, seismic waves do travel faster (about 5%) perpendicular to the oceanic ridges than parallel to them (Morris et al. 1969). As early as 1964, Hess interpreted this as a result of a preferential alignment of crystals with directional properties, and proposed that this alignment was attained during mantle convection (Hess 1964).

Chastel and coworkers presented model for predicting the development of anisotropy in a convection cell and discussed implications for convection in the Earth's mantle (Chastel et al. 1993). The results indicate that during convection, crystals reorient into characteristic textures that cause anisotropic physical properties. It was observed that these patterns are highly heterogeneous over the span of the convection cell and depend on the specific straining history along individual streamlines of the cell. The rotations that reorient crystal lattices are due to activation of slip systems in deforming single crystals as well as the macroscopic rotation of material progressing along a streamline. Both causes of lattice reorientation are most intense during upwelling at ridges and during subduction. Thus, texture development is particularly strong in these regions. Using the crystallographic texture throughout the convection cell and the physical properties of single crystals, seismic

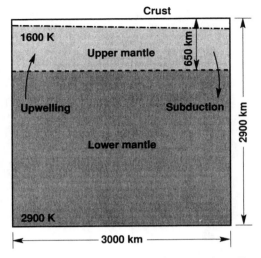

Figure 5. Schematic diagram of the convection cell showing the upper and lower regions of the mantle and the crust.

velocity averages over mantle sectors were computed and the directional dependence of wave propagation determined. The predicted azimuthal variation of p-wave anisotropy of 5-10% reported in that publication agrees well with actual observed values.

The convection cell used by Chastel et al. was highly idealized. In a later investigation reported by Dawson and Wenk (Dawson and Wenk 2000) a more realistic mantle was modeled, one with dimensions corresponding to those in the Earth and with material properties that are closer to those which are believed to exist. The convective cell was discretized with a greater number of finite elements to better resolve gradients of the flow and use a larger number of crystals to represent each polycrystal. Apart from that, the procedure was similar to Chastel's. The 'mantle' was still idealized: the model was two-dimensional; there was no interaction with the crust (e.g., during subduction); and the sole phase change (at 650 km) affected only the mechanical properties. Above 650 km, the mantle was assumed to be solely composed of olivine. Below 650 km, the mantle was assumed to be

Table 1. Parameters used in the simulations (Turcotte and Schubert 1990; Blankenbach 1989). If ranges are given, the first value is for the top of the mantle and the second for the bottom.

Quantity	Symbol	Units	Value
Density	ρ	kg/m^{-3}	$4(3.4\text{-}5.6)\times10^{3}$
Heat capacity	C_p	$J/(kgK)$	1.25×10^{3}
Thermal conductivity	k	$W/(mK)$	5
Volumetric thermal expansion	α	$1/K$	2.5×10^{-5}
Gravity	g	m/s^{-2}	$9.8(9.86\text{-}10.68)$
Youngs modulus	E	Pa	$1.6\text{-}7.6\times10^{11}$
Compressibility	β	$1/Pa$	$7.5(8.0\text{-}1.5)\times10^{-12}$
Kinematic viscosity	ν	m^2/s	$2.5\times10^{17\text{-}19}$
Dynamic viscosity ($\mu = \nu\rho$)	μ	$Pa\,s$	$1\times10^{21\text{-}23}$

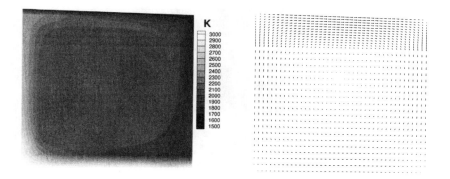

Figure 6. Left: temperature distribution during mantle convection showing the effects of upwelling (on the left) and subduction (on the right). Right: convective velocity field shown using scaled vectors directed tangent to the flow.

Table 2. Single crystal elastic moduli (GPa) (Simmons and Wang 1971). Voigt notation.

C_{11}	C_{12}	C_{13}	C_{22}	C_{23}	C_{33}	C_{44}	C_{55}	C_{66}
324	59	79	249	78	249	66.7	81	79.3

isotropic with a viscosity higher that that of the upper mantle. The model was intentionally kept simple to emphasize the effects of the convective flow on the internal structure of the mantle material and on the resulting anisotropy in its mechanical properties. Here we summarize some of those results.

Simulation specifics

The convection cell is 2900 km deep and 3000 km wide. Within this region are three zones: the lower mantle extends from 650 km to 2900 km depth across the width of the cell; the upper mantle extends from the bottom of the crust to the lower mantle, also across the full width of the cell; the crust extends from the surface to the top of the upper mantle, varying in thickness from 50 km (along the boundary with upwelling) to 125 km (along the boundary with subduction).

The cell geometry is shown schematically in Figure 5. Along the bottom of cell (lower extent of the lower mantle), no vertical velocity is allowed. Along both lateral surfaces, no horizontal velocity is allowed. The crust is assumed rigid, and at the boundary between the upper mantle and the crust the velocity of the upper mantle is required to be tangential to the crust/upper mantle boundary. It is not necessary to model the crust explicitly with these boundary conditions, and so only the upper and lower mantle zones have been discretized with finite elements. The top surface of the mantle is required to be 1600 K; the bottom surface of the lower mantle is fixed at 2900 K. The lateral boundaries are adiabatic.

A total of 240 higher order, isoparametric elements are used to discretize the full mantle, 120 in the upper mantle and 120 in the lower mantle. An Eulerian procedure is employed, so the elements are spatially fixed and mantle material flows though them. The driving force for the flow arises from the density gradients associated with the nonuniform temperature field. To accommodate this, an iterative procedure is used to couple the heat transfer solution with the solution for the velocity field.

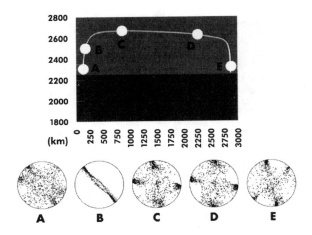

Figure 7. 100 pole figures along one streamline illustrating the evolution of texture for a rock traveling near the cell perimeter. 100 pole figures A-E correspond to positions A-E, respectively.

The macroscopic material stiffness given by Equation (27) is derived as the average of single crystal responses using either an upper or lower bound assumption. The slip system response for the single crystals is determined by matching laboratory data for compression of polycrystalline olivine, as discussed earlier. In the previous study (Chastel et al. 1993), the lower bound stiffness was used in the finite element equations and convective velocity fields were computed for two different sets of boundary conditions. Each set restricted the magnitude of the velocity to be consistent with observed convection rates. In the later study of Dawson and Wenk (Dawson and Wenk 2000), the anisotropic stiffness derived this way was replaced with an isotropic stiffness based on a single viscosity, to avoid having to place an explicit constraint on the magnitude of the velocity field. These isotropic properties were chosen in accordance with the geodynamics literature (Turcotte and Schubert 1990) and benchmark calculations (Blankenbach 1989). Corresponding values are listed in Table 1. Values used in the present simulations correspond to the center of this range. The reason for making this change was that the viscosity implied by the anisotropic stiffness as derived from experiment is orders of magnitude lower than that inferred from convective rates or glacial rebound. While anisotropy was neglected for the purposes of computing the velocity field, the simulations still based the texturing on the polycrystal model, including the single crystal behavior for olivine and the lower bound linking hypothesis. In this respect coupling between the macroscopic velocity field and the evolving microstructure was only from the velocity field to the microstructure, and not from the microstructure to the velocity field.

Simulation results

Figure 6 illustrates the temperature distribution and the velocity field when convection has reached a steady state, displaying patterns dominated by convective heat exchange.

The evolution of texture is best evaluated by tracing material flow along various streamlines as shown in Figure 7. In the lower mantle the material is assumed to be isotropic. When it enters the upper mantle and transforms to olivine, the texture evolution begins starting from an assumed uniform (untextured) orientation distribution as was assumed in (Chastel et al. 1993; Dawson and Wenk 2000). Strong preferred orientation (represented

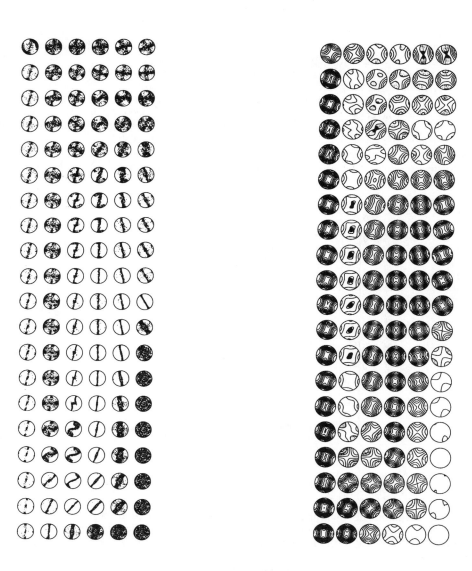

Figure 8. Left: 100 pole figures displayed at the centroids of the finite elements of the upper mantle (not to scale). Right: longitudinal *p*-wave propagation surfaces at the centroids of the finite elements of the upper mantle corresponding to the texture patterns. Linear contours with contour interval 100 m/s. Dot pattern below 7200 m/s. [Note: The plot was rotated by 90° counterclockwise on the page to facilitate display.] Horizontal Earth direction is vertical on the page with the surface of the Earth to the left. Upwelling occurs at the left side of the rotated diagram, spreading is toward the right, and downwelling is on the right side.

as 100 pole figures) develops during upwelling with high shear. The pattern changes as the flow lines change direction from vertical to horizontal. During horizontal spreading orientation changes are minimal. Finally, during subduction, a new pattern develops which is generally weaker but does not revert to randomness at the interface with the lower mantle. In our model the time for the convection path of an outer streamline in the upper mantle is about 120 million years, corresponding to a spreading velocity of 2 cm/y.

The texture patterns for the element centroids of the upper mantle are presented in Figure 8 as 100 pole figures. Great heterogeneity is observed, vertically and laterally, with regions having strong texture while others have almost random orientation distributions. The uppermost layer is strongly textured, but immediately underneath there is a fairly isotropic layer. The center of the cell is again strongly textured. All pole figures display a statistical diadic symmetry, consistent with the two-dimensional (plane strain) deformation. The pole figures have no mirror planes because the deformation has a simple shear component (texture calculations and representations were done with BEARTEX (Wenk et al. 1998).)

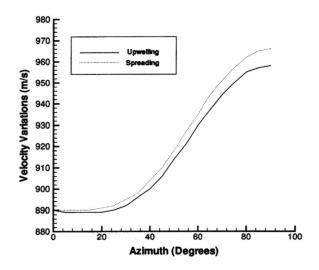

Figure 9. Seismic p-wave velocity variations in regions of upwelling and spreading.

If the orientation distribution and the tensor properties of the single crystal are known, then the physical properties of the textured polycrystal can be calculated by appropriate averaging. In the case of the mantle, elastic properties are of main interest. The single crystal elastic moduli at ambient conditions are given in Table 2 for olivine. Using these values, averaging over all 1000 crystallites in each cell was performed with the self consistent method assuming spherical grain shape (Kocks et al. 1998). From the averaged elastic constants and the density (4.5 g/cm for the central region of the mantle) propagation surfaces for longitudinal (p) and transverse (s) waves were calculated, with those for p-waves shown for all element centroids in Figure 8. The velocity maps are shown in equal area projection. We are aware that both elastic constants and density change with temperature and pressure, but in our representation we used constant values to emphasize the effects of anisotropy. All deviations from the value 7280 m/s for an isotropic polycrystal are due to texture. Similar to pole figures in Figure 8, there are large deviations from isotropy and also a heterogeneous distribution of anisotropy over the convection cell.

Since the flow pattern is two-dimensional, the deformation corresponds to plane strain, and textures and velocity surfaces have statistically a symmetry plane and a two-fold rotation axis perpendicular to it. Overall, there is a correspondence between fast velocities and 100 concentrations, since 100 is the fast single crystal direction. If earthquake travel times were recorded from waves which pass through the upper mantle at high angles, zones with high anisotropy would mainly appear as low velocity zones, since fast 100 directions are dominantly horizontal.

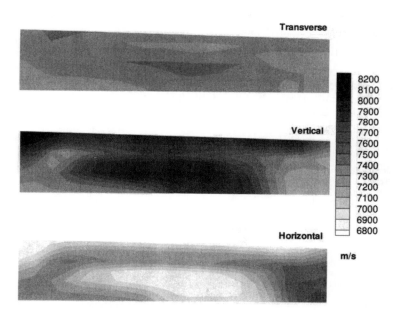

Figure 10. Seismic p-wave velocity distributions over the upper mantle corresponding to horizontal, vertical and normal (transverse) propagation directions. Note that in the original report of Dawson and Wenk (2000) the gray shades were in error (inverted) for two directions.

In addition to the local averages, averages have been computed over larger scales, particularly for regions of upwelling and spreading (see Fig. 9). This figure shows the azimuthal variation in p-wave velocity, as would be observed with surface waves (horizontal). The velocity variation is 5-10%, in good correspondence with observed data from Hawaii (Morris et al. 1969), the Mendocino ridge (Raitt et al. 1969) and the Southern Pacific (Nishimura and Forsyth 1989).

Finally contour maps of vertical, horizontal and normal velocities (again, assuming constant density and single crystal properties) are presented for the full convective cell. This represents the local velocities if corresponding seismic waves were propagating in the corresponding directions (see Fig. 10). Due to the two-dimensional approach deviations from average in the normal direction are minimal. However, in the other two maps there are again variations of 5-10% with patterns resembling seismic tomography maps (Anderson and Djewonski 1984).

SUMMARY

Anisotropy at the level of individual crystals strongly influences the mechanical response of minerals. Polycrystal plasticity provides the theoretical basis for modeling this anisotropy. Using polycrystal plasticity it is possible to compute the dependence of the mechanical properties on the crystallographic texture and to evolve the texture with deformation. The finite element method provides a powerful tool that can be used in conjunction with polycrystal plasticity to solve complex boundary value problems. These can be at a macroscopic scale, as in the case of mantle convection, or crystal scale, as in the detailed simulation of crystal aggregates. The potential for a more fundamental understanding of mineral deformation is open through coordinated efforts of experiments and simulations using these tools.

ACKNOWLEDGMENTS

This work has been supported in part by the Cornell Theory Center. The author thanks Chris Pelkie of Cornell Theory Center for assistance in the graphic presentation of the simulation results, Donald Boyce of Cornell University for help with the texture evolution computations, and Carlos Tomé of Los Alamos National Laboratory for assistance with the seismic velocity computations using the self-consistent approach.

REFERENCES

Anderson DL, Djewonski AM (1984) Seismic tomography. Sci Am 251:60-68
Asaro R, Needleman A(1985) Texture development and strain hardening in rate-dependent polycrystals. Acta Metall 23:923-953
Beaudoin AJ, Dawson PR, Mathur KK, Kocks UF (1995) A hybrid finite element formulation for polycrystal plasticity with consideration of macrostructural and microstructural linking. Intl J Plasticity 11:501-521
Blankenbach B (1989) A benchmark comparison for mantle convection codes. Geophys J Intl 98:23-38
Bratianu C, Atlui S (1983) A hybrid finite element method for stokes flow: Part I—formulation and numerical studies. Computer Methods Appl Mech Eng 36:23-30
Chastel YB, Dawson PR, Wenk H-R, Bennett K (1993) Anisotropic convection with implications for the upper mantle. J Geophys Res 98:17,757-17,771
Dawson PR, Marin EB (1998) Computational mechanics for metal deformation processes using polycrystal plasticity. *In* van der Giessen E et al. (eds) Advances in Applied Mechanics 34:78-169
Dawson PR, Wenk H-R (2000) Texturing the upper mantle during convection. Philos Mag A 80:573-598
Engelman M, Sani R, Gresho P, Bercovier M (1982) Consistent vs. reduced integration penalty methods for incompressible media using several old and new elements. Intl J Numer Methods Fluids 2:25
Hess HH (1964) Seismic anisotropy of the uppermost mantle under oceans. Nature 203:629-640
Johnson C (1987) Numerical Solution of Partial Differential Equations by the Finite Element Method. Cambridge University Press, Cambridge, UK
Kocks UF, Tome CN, Wenk H-R (1998) Texture and Anisotropy. Cambridge University Press, Cambridge, UK
Lebensohn R, Dawson PR, Kern H, Wenk H-R (in press) Heterogeneous deformations and texture development in halite polycrystals: comparison of different modeling approaches and experimental data. Tectonophysics
Mathur KK, Dawson PR (1989) On modeling the development of crystallographic texture in bulk forming processes. Intl J Plasticity 5:67-94
Morris GB, Raitt RW, Shor GG (1969) Velocity anisotropy and delay time maps of the mantle near Hawaii. J Geophys Res 74:4300-4316
Nishimura C, Forsyth D (1989) The anisotropic structure of the upper mantle in the pacific. Geophys J 96:203-229
Parks DM, Ahzi S (1990) Polycrystalline plastic deformation and texture evolution for crystals lacking five independent slip systems. J Mech Phys Solids 38:701-724
Prantil V, Dawson PR, Chastel YB (1995) Comparison of equilibrium-based plasticity models and a Taylor-like hybrid formulation fo deformation of constrained crystal systems. Modeling Simul Mater Sci Eng 3:215-234
Raitt RW, Shor GG, Francis TJG, Morris GB (1969) J Geophys Res 74:3095-3109
Sachs G (1928) Zur Ableitung einerFliessbedingung. Z Verein Deut Ing 72:734-736

Simmons G, Wang H (1971) Single Crystal Elastic Constants and Calculated Aggregate Properties. M I T Press, Cambridge, Massachusetts
Taylor G (1938) Plastic strains in metals. J Inst Metals 62:307-324
Turcotte DL, Schubert G (1990) Geodynamics Applications of Continuum Physics to Geological Problems. Wiley, New York
Wenk H-R (1985) Preferred Oreintations of Deformed Metals and Rocks: An Introduction to Modern Texture Analysis. Academic Press
Wenk H-R, Mathies S, Donovan J, Chateigner D (1998) Beartex: A Windows-based program system for quantitative texture analysis. J Appl Crystallogr 31:262-272
Zienkiewicz O, Taylor R (1989) The Finite Element Method. McGraw-Hill, London

12   Seismic Anisotropy and Global Geodynamics

Jean-Paul Montagner[1,2,3] and Laurent Guillot[1]

[1] *Seismological Laboratory, CNRS URA 195*
Institut de Physique du Globe, Paris, France

[2] On leave at: *Seismological Laboratory*
California Institute of Technology, 252-21
Pasadena, California 91125

[3] On leave at: *Jet Propulsion Laboratory*
California Institute of Technology
4800 Oak Grove Drive, Pasadena, California 91109

INTRODUCTION

For many years, seismic anisotropy was often neglected, mostly because of the inherent heavy mathematical and computational tools needed to describe and model its effects on seismic waves. The usual basic knowledge about propagation in isotropic media cannot easily apply to anisotropic media, where new phenomena come up, such as birefringence (or shear-wave splitting), or difference between directions of propagation of phase velocity and of group velocity. Consequently, geophysicists often claimed that it was a second-order effect, and considered the Earth as isotropic.

This hypothesis was assumed to be a good approximation, because of the random orientation of crystals in most parts of the Earth, and of the random sampling of anisotropic regions by seismic rays. This assumption furthermore made easier the description of wave propagation, as well as the parameterization of media in inverse problems. An isotropic elastic medium can be described by two independent elastic parameters (λ and μ Lamé parameters), but the simplest anisotropic medium (transverse isotropy with a vertical symmetry axis) requires 5 independent parameters (Love 1927, Anderson 1961). To date, seismic observations have been explained in terms of isotropic (and often thermal) lateral heterogeneities, ignoring manifestations of anisotropy. However, since the 1960s, it was recognized that most parts of the Earth are not only laterally heterogeneous but also anisotropic and that seismic anisotropy provides a simple explanation of different observational data:

- azimuthal variation of Pn-velocities below oceans (discovered in the 1960s; Hess (1964) explained it by anisotropy),
- Rayleigh-Love wave discrepancy: It is impossible to simultaneously explain Rayleigh and Love wave dispersion by an isotropic model (Anderson 1961, Aki and Kaminuma 1963, Mc Evilly 1964),
- Shear-wave splitting (or birefringence), the most unambiguous observation of anisotropy, particularly for SKS waves (Vinnik et al. 1984).

Seismic anisotropy, contrary to isotropy, is reflecting some inherent *organization* of the matter. Different geophysical fields are involved in the investigation of the manifestations of anisotropy of Earth materials: mineral physics and geology for the study of the microscopic scale, seismology and geodynamics for scales larger than typically one kilometer. The origin of seismic anisotropy is non-unique. In the crust, the crack distribution seems to play a major role (Crampin and Booth 1985). In the upper mantle, it is usually explained by the lattice-preferred orientation of α-olivine (Nicolas and Christensen 1987, Zhang and Karato 1995) and is related to plate-tectonic processes. More generally the intrinsic anisotropy of minerals (olivine and to a less extent orthopyroxene and clinopyroxene) associated with their lattice-preferred orientation may induce large-scale observable and unambiguous effects, either on body waves (S-wave splitting observed on SKS (Vinnik et al. 1984), P-wave anisotropy (Babuska et al. 1984) or surface waves through the azimuthal anisotropy (Forsyth 1975) and the radial (improperly named "polarization") anisotropy (Schlue and Knopoff 1977). These different measurements of anisotropy performed at different spatial scales were difficult to reconcile.

Investigating deeper lower mantle anisotropy is a formidable challenge, because the anisotropy signal is small, often masked by upper mantle anisotropy, and its physical explanation is controversial. Most of the lower mantle seems to be isotropic (Meade et al. 1995), except the D"-layer (Vinnik et al. 1989b, Maupin 1994) where anisotropy could result from the layering of old subducted slabs and/or melted materials. Even deeper, anisotropy has also been found in the inner core from free oscillations (Woodhouse et al. 1986) and from the P-wave travel times reported in the ISC bulletins (Morelli et al 1986). The origin, amplitude and mechanisms creating the anisotropy in the core are still subject of controversy (e.g., see Singh et al. 2000). Since these early observations of seismic anisotropy, a large and rapidly growing number of studies have confirmed its existence in the different depth ranges of the Earth. A complete and exhaustive review of these studies is beyond the scope of this paper, and, we will only underline their geodynamic implications.

From the global geodynamics point of view, seismic anisotropy has many applications, although it is still in its infancy. We will show how it makes it possible to define the root of continents and to investigate the coupling between the lithosphere and the rest of the mantle (Montagner and Tanimoto 1991, Silver 1996), and more generally to gain insight into mantle convection (Anderson and Regan 1984, Montagner 1994). Mantle convection characterized by a high Rayleigh number is highly chaotic and numerical modeling demonstrates that most of the deformation takes place in boundary layers. Conversely, since seismic anisotropy is closely related to large-scale deformation (Nicolas and Christensen 1987, Karato 1989), boundary layers can be detected by the existence of seismic anisotropy (Montagner 1998, Karato 1998).

We present in this paper the different depth ranges in the mantle, where seismic anisotropy was detected, i.e., uppermost mantle, transition zone and D"-layer. In the top boundary layer of the mantle, seismic anisotropy can be directly compared to geological observations. The robust features of these different investigations in different depth ranges are presented and the enormous scientific potential of seismic anisotropy is emphasized.

CAUSES OF SEISMIC ANISOTROPY
FROM MICROSCOPIC TO LARGE SCALE

Many processes can give rise to seismic anisotropy, and we must be very careful when interpreting it. They can be related either to anisotropic structural settings in rocks, or to the intrinsic anisotropy of minerals in the Earth.

Shape Preferred Orientation (S.P.O.)

Anisotropic spatial organization of matter (even though isotropic) can induce seismic anisotropy:
- a fine layering of heterogeneous rocks (even though isotropic) is "seen" as transversely isotropic by a wave whose wavelength is larger than the typical thickness of layers (Backus 1962). Some authors proposed that this kind of structure, observed in ophiolites for instance, could result from the progressive mixing of rocks with heterogeneous mechanical properties and could explain the seismic anisotropy in the Earth's upper mantle (e.g., the "marble cake" of Allègre and Turcotte 1986);
- distribution of cracks (Crampin and Booth 1985) and/or fluid inclusions can induce

anisotropy (probably an important effect in the crust and in the inner core).

Lattice Preferred Orientation (L.P.O.)

Some major mineral phases of the Earth's mantle are anisotropic in elasticity. Under some conditions, plastic deformation of these minerals can result in a preferential orientation of their lattices. This phenomenon is often considered as the origin of the large-scale seismic anisotropy in the upper mantle. We present below in a very simplified manner the conditions required to develop such a seismic anisotropy. A complete discussion of these different mechanisms at different scales can be found in recent extensive and well-documented review papers (Kendall 2000, Mainprice et al. 2000, Savage 1999).

Intrinsic anisotropy of minerals. Some minerals present in the upper mantle are strongly anisotropic (see Mainprice et al. 2000 and references therein for an exhaustive review of the minerals and their anisotropic properties). The difference of P-wave velocity between the fast axis and the slow axis is larger than 20% for olivine, the main constituent of the upper mantle. Other important constituents such as orthopyroxene or clinopyroxene are anisotropic as well (>10%) (see Anderson 1989, or Babuska and Cara 1991). Some other constituents such as garnet display a very small intrinsic anisotropy. With increasing the depth, most of minerals undergo a series of phase transformations. There is some tendency (though not systematic), that with increasing pressure (whose effect is predominant in the mantle), the crystallographic structure evolves towards a more closely packed, more isotropic structure, such as cubic structure. For example, olivine transforms into β-spinel and then γ-spinel in the upper transition zone (410 to 660 km depth) and into perovskite and magnesiowüstite in the lower mantle; but perovskite $(Mg,Fe)SiO_3$ and the pure end-member of magnesiowüstite MgO are still anisotropic. That could explain the observed anisotropy in some parts of the lower mantle (Karato 1998, or MacNamara et al. 2002).

Efficient mechanisms of orientation of crystals. In order to observe seismic anisotropy, the crystals must be sensitive to the strain field, and a lattice preferred orientation must develop either from dislocation creep (activation of slip systems) or from dynamic recrystallization (see Karato (1989) and Poirier (1985), for a phenomenological description, and Ribe (1989) and Kaminski and Ribe (2001), e.g., for numerical modeling). Through the mechanisms of lattice-preferred orientation, it is found that the anisotropy of an aggregate of many minerals can be very large (Nicolas and Christensen 1987). In the lower mantle super-plasticity may be the predominant deformation mechanism, and that may cause the absence of large scale seismic anisotropy in this part of the Earth (Karato 1998).

Anisotropy of assemblages of minerals. Mantle rocks are assemblages of different minerals which are more or less anisotropic. The amount of anisotropy is largely dependent on the composition of the aggregates. The relative orientations of crystallographic axes in the different minerals must not counteract in destroying the intrinsic anisotropy of each mineral. For example, the anisotropy of peridotites, mainly composed of olivine and orthopyroxene, is affected by the relative orientation of their crystallographic axes (Christensen and Lundquist 1982). According to their observations, the fast axis of olivine is parallel to the intermediate axis of orthopyroxene in the shear plane and parallel to the flow direction, but the fast axis of orthopyroxene and the slow axis of olivine are orthogonal to the shear plane (Fig. 1). The resulting anisotropy for P-waves, however, is larger than 10% (Peselnick and Nicolas 1978). Such a large anisotropy is consistent on intermediate scale, for instance in massifs of ophiolites, which present such an anisotropy over several tens of kilometers (Nicolas 1993, Vauchez and Nicolas 1991).

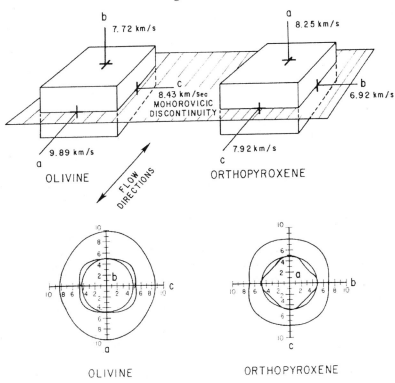

Figure 1. Figure of Christensen and Lundquist (see Anderson 1989) for olivine and orthopyroxene orientation). This observation is the base of the interpretation of seismic anisotropy in terms of convective flow.

Coherent strain field. At large scale, the deformation due to mantle convection must be coherent over large distances in order to preserve long wavelength anisotropy. From Pn studies and models of formation of the oceanic lithosphere, it is possible to infer that anisotropy remains uniform on horizontal length-scales in excess of 1000 km.

Under those assumptions it is possible to estimate the effect of LPO-induced anisotropy on body-wave speeds, and compare it with thermal effects (Montagner and Guillot 2000). The magnitude of seismic anisotropy can be estimated in a simplified geodynamical context such as an oceanic convective cell, for a simplified mineralogical composition (60% olivine, 40% orthopyroxene). The relative orientation of the crystalline populations in the kinematic field is that discussed in the previous section, and we neglect the dispersion of the orientations around this perfect or ideal orientation. The general method followed for the calculation of anisotropic parameters is summarized in Figure 2. Body-wave velocities are calculated along 4 vertical profiles, from the surface down to the 410-km-discontinuity (Fig. 3): at the mid-ocean ridge and along a slab, which are characteristic of hot or cold vertical currents; for a 7-Myr-old lithosphere and for a 30-Myr-old lithosphere from the ridge, where horizontal currents are dominant. Some other profiles are reported in Montagner and Guillot (2000) for a pure olivine upper mantle. The amplitude of the temperature effect (ΔT) can be roughly estimated by comparing the ridge and the slab profiles displaying the same orientations of minerals. It is found around 3 to 4%. The differences in PH-waves velocities (horizontally-polarized P-waves) between

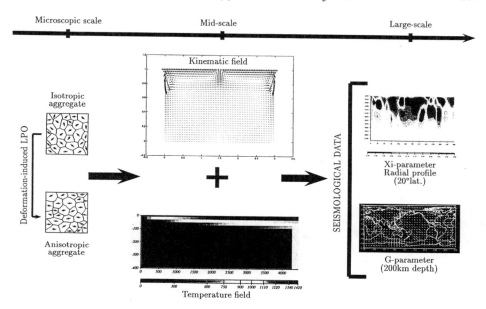

Figure 2. Different steps from mineralogical anisotropy at microscopic scale to observable seismic anisotropy at large scale. For a more accurate, color version of this figure, see Errata for Volume 51 at http://www.minsocam.org/MSA/RIM/ .

vertical and horizontal currents are mostly due to the orientation of crystals (effect symbolized by $\Delta\alpha$ on Fig. 3). When comparing these velocities with those obtained for an equivalent isotropic system (Voigt average, see Appendix B), the magnitude of anisotropic effects on PH-wave (resp. SV-wave) velocities (Fig. 3) is about ±3%, which is slightly higher than thermal effects in vertical currents (except at the ridge). Even though anisotropy in the upper mantle should be weaker than in those experiments (natural dispersion of the crystallites), its effect on seismic properties might be as high as thermal effects; that underlines the importance of including anisotropic effects in tomographic studies.

EFFECT OF ANISOTROPY ON SEISMIC WAVES

The effect of seismic anisotropy on seismic waves has been extensively investigated since Love (1927). The reader is referred to the classical textbooks of Fedorov (1968) and Helbig (1994) and their application to Earth sciences (Anderson 1989, Babuska and Cara 1991). The basic theory of seismic wave propagation in anisotropic media is briefly presented in the Appendix A. The complexity of the effect of elastic anisotropy results from the general linear relationship between the stress tensor components σ_{ij} and the strain tensor components ε_{kl} which involve a fourth-order elastic tensor C_{ijkl} This relationship is named the generalized Hooke's law:

$$\sigma_{ij} = C_{ijkl}\,\varepsilon_{kl} \qquad (1)$$

The elastic tensor, C, in the most general case, presents 21 independent parameters. How to deal with this tensor is briefly described in the Appendix B. There are two ways to manage such complexity of C: (1) by considering either that the medium possesses symmetry properties (in order to decrease the number of independent elastic moduli) and

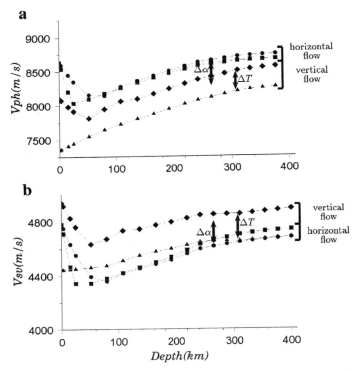

Figure 3. V_{PH} and V_{SV} velocity profiles for a 60-40 olivine-orthopyroxene upper mantle within the convecting cell displayed in Figure 1 (from Montagner and Guillot 2000). Diamond: slab; △ : ridge; □ : 7-Myr-old plate; ○ : 30-Myr-old plate. The arrows represent rough approximations of the effects of orientation ($\Delta\alpha$) and temperature (ΔT) on body wave velocities.

ultimately that it is isotropic, (2) or that the seismic anisotropy is a first-order perturbation, which permits the application of classical perturbation theories (Backus 1965). Except an isotropic medium defined by 2 elastic moduli (λ and μ), the simplest case of anisotropic medium is the transversely isotropic (TI) medium (hexagonal symmetry), characterized by 5 independent parameters A, C, F, L, N (Love 1927) when the symmetry axis is vertical. This parameterization with a vertical symmetry axis was used in deriving PREM (preliminary reference Earth model) by Dziewonski and Anderson (1981). It corresponds to the most general case for a spherically symmetric model. A slightly more complex parameterization is the TI case with a tilted symmetry axis (hexagonal symmetry), which adds 2 additional angular parameters defining the orientation of the axis. The perturbation theory is well suited for investigating the anisotropic structure of the deep Earth. As demonstrated in the first section, seismic anisotropy is not a second-order effect but, due to several averaging processes, it is usually smaller than 10%, enabling the application of perturbation theories. Seismologists are working in this framework for body waves, surface waves and normal modes.

Body waves

From a theoretical point of view, a first application of perturbation theory to body waves was performed by Backus (1965) and then extensively applied (Crampin 1984, for a first review paper). In the case of a weak anisotropy, we can consider that the polarization of the three solutions of the Christoffel equation (see Appendix A) is very close to P, SH and SV waves. Let us call V_P, V_{S1} and V_{S2} the velocities of quasi-P, quasi-SV and quasi-SH waves. We draw the attention that since the sagittal plane (plane defined by the ray trajectory) is not necessarily a symmetry plane, the SV and SH components of the incoming wave may be different from the S_1 and S_2 components of quasi SH and SV waves. For a wave propagating horizontally in the plane (1,2) with azimuth Ψ and where 3- is the vertical direction, the different velocities are given by the following expressions in the symmetry planes:

$$\rho V_P^2 = 1/8\ (C_{1111}+2(C_{1122}+C_{1212})+3C_{2222}) + 1/2\ (C_{1111}-C_{2222})\cos2\psi + (C_{2111}+C_{1222})\sin2\psi$$
$$+ 1/8\ (C_{1111}-2(C_{1122}+C_{1212})+C_{2222})\cos4\psi + 1/2\ (C_{2111}-C_{1222})\sin4\psi \tag{2a}$$

$$\rho V_{S1}^2 = 1/8\ (C_{1313}+C_{2323}) + 1/2\ (C_{1313}-C_{2323})\cos2\psi + C_{2313}\sin2\psi \tag{2b}$$

$$\rho V_{S2}^2 = 1/8\ (C_{1111}-2(C_{1122}-C_{1212})+C_{2222}) - 1/8\ (C_{1111}-2(C_{1122}+C_{1212})+C_{2222})\cos4\psi$$
$$- 1/2\ (C_{2111}-C_{1222})\sin4\psi \tag{2c}$$

These expressions are not general, but are extensively used when assuming a TI medium with any symmetry axis. This case is a very good approximation for the upper mantle pyrolitic model (Estey and Douglas 1986).

For body waves, the evidence of anisotropy primarily results from the investigation of the splitting in teleseismic shear waves (Fig. 4a) such as SKS (Vinnik et al. 1984, 1989a,b, 1991, 1992; Silver and Chan 1988, 1991; Ansel and Nataf 1989), ScS (Ando 1984, Fukao 1984) and S (Ando et al. 1983; Bowman and Ando 1987; Fischer et al. 1996; Gaherty and Jordan 1995), which is corroborated by the evidence of P-wave anisotropy (Babuska et al. 1984, 1993). Among these different observations, the splitting information derived from SKS is probably the less ambiguous and has been extensively used in teleseismic anisotropy investigations. See Silver (1996) and Kendall (2000) for reviews. The drawbacks of SKS observations are, that it is almost impossible to precisely locate the depth of anisotropic regions, nor to take account for a dipping symmetry axis, and that only continental areas can be extensively investigated. The rapid variation of directions of fast velocity which can be observed in some continental regions on a short spatial scale (Vinnik et al. 1989a, Hirn et al. 1995), cannot be explained by a very deep anisotropy and the origin of anisotropy is confined to the first 410 km, either in the lithosphere or in the top of the asthenosphere. However, there are some first attempts to introduce more complex radial models of anisotropy. A close examination of SKS data makes it necessary, in several areas, to introduce at least two layers of anisotropy with different amplitudes and directions (Silver and Savage 1994, Farra and Vinnik 1994, Girardin and Farra 1998, Savage 1998, Wolfe and Silver 1998). A map summarizing the different observations of SKS splitting is presented in Figure 4b, which is an upgrade by Savage (1999) of the compilation of Silver (1996).

Surface waves

Surface waves are also well suited for investigating upper mantle anisotropy. Contrary to body waves, surface waves enable us to locate anisotropy at depth but, so far, its lateral resolution (several thousands of kilometers) is very poor. Two kinds of observable anisotropy have been considered. The first one results from the well-known discrepancy between Love and Rayleigh waves (Anderson 1961, McEvilly 1964), often referred as the "polarization" anisotropy (Schlue and Knopoff 1977) or the radial anisotropy. In order to remove this discrepancy, it is sufficient to consider a TI medium with a vertical symmetry axis, characterized by five elastic parameters plus density. On a global scale, Nataf et al. (1984, 1986), by the simultaneous inversion of Rayleigh and Love wave dispersion, derived the geographical distributions of S-wave radial anisotropy

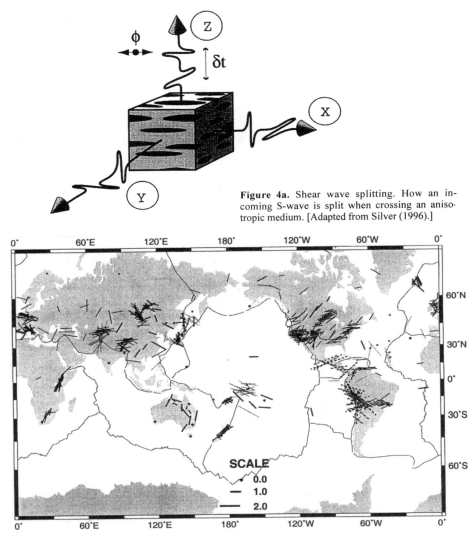

Figure 4a. Shear wave splitting. How an incoming S-wave is split when crossing an anisotropic medium. [Adapted from Silver (1996).]

Figure 4b. Shear wave splitting. Compilation of worldwide shear wave splitting measurements. [Adapted from Savage (1999).]. This dataset includes primarily SKS data but also S, ScS, SS, and PS data.

at different depths. Lévêque and Cara (1985), Cara and Lévêque (1988) used higher mode data to display radial anisotropy under the Pacific Ocean and North America down to at least 300 km. The second kind of observable anisotropy is the azimuthal anisotropy, directly derived from the azimuthal variation of phase velocity of surface waves. It was observed for the first time on Rayleigh waves by Forsyth (1975) in Nazca plate. Since these pioneering studies, global and regional models have been derived for both kinds of anisotropy (Mitchell and Yu 1980, Montagner 1985). Tanimoto and Anderson (1985) obtained a global distribution of the Rayleigh wave azimuthal anisotropy at different periods. On a regional scale, several tomographic investigations reported in the eighties the existence of azimuthal anisotropy in the Indian Ocean (Montagner 1986a), Pacific Ocean (Suetsugu and Nakanishi 1987, Nishimura and Forsyth 1989) and in Africa (Hadiouche et al. 1989).

The radial anisotropy (or "polarization" anisotropy) and the azimuthal anisotropy are two different manifestations of a same phenomenon, the anisotropy of the upper mantle. Montagner and Nataf (1986) derived a technique that makes it possible to simultaneously explain these two forms of seismically observable anisotropy. The principles of this technique will be only briefly described, for the most general case of anisotropy (provided that it is small). A complete description of the whole procedure can be found in Montagner (1996, 1998). Most of these measurements are based on the phase of waves. But it is now possible to use amplitude data by looking at polarization anomalies (Yu and Park 1994, Larson et al. 1998, Laske and Masters 1998, Pettersen and Maupin 2002).

From a theoretical point of view, a general slight elastic anisotropy in a plane-layered medium gives rise to an azimuthal dependence of the local phase or group velocities of Love and Rayleigh waves at point (θ, ϕ) along the ray of the form (Smith and Dahlen 1973, 1975):

$$V(\omega, \theta, \phi, \Psi) - V_0(\omega, \Psi) = \alpha_0(\omega, \theta, \phi) + \alpha_1(\omega, \theta, \phi)\cos 2\Psi + \alpha_2(\omega, \theta, \phi)\sin 2\Psi + \alpha_3(\omega, \theta, \phi)\cos 4\Psi + \alpha_4(\omega, \theta, \phi)\sin 4\Psi \quad (3)$$

where ω is the frequency of the wave, $V_0(\omega, \Psi)$ the reference velocity of the unperturbed medium, and Ψ is the azimuth along the path. The local phase velocity $V(\omega, \theta, \phi, \Psi)$ is used to calculate the travel time between the epicenter E and the receiver R, which is easily related to the measurement of the along-path averaged phase velocity $V_d(\omega)$:

$$t_{E \to R} = \Delta/V_d(\omega) = \int_E^R \frac{ds}{V(\omega, \theta, \phi, \Psi)} \quad (4)$$

Montagner and Nataf (1986), following the same approach as Smith and Dahlen (1973), displayed that simple linear combinations of the elastic tensor components C_{ij} are sufficient to describe the two seismically observable effects of anisotropy on surface waves. For the Voigt notation of the elastic constants (two indices instead of four), see Appendix B. The 0-Ψ term corresponds to the average over all azimuths and involves 5 independent parameters, A, C, F, L, N, which express the equivalent transversely isotropic medium with vertical symmetry axis. The other azimuthal terms (2-Ψ and 4-Ψ) depend on 4 groups of 2 parameters, B, G, H, respectively describing the 2-Ψ azimuthal variation of A, L, F, and E describing the 4-Ψ azimuthal variation of A and N. Therefore, the different azimuthal terms α_0, α_1, α_2, α_3, α_4, depend on 13 3-dimensional parameters, which are assumed independent:

- Constant term (0-Ψ azimuthal term: α_0)

$$A = \rho V_{PH}^2 = 3/8(C_{11}+C_{22}) + 1/4 C_{12} + 1/2 C_{66}$$
$$C = \rho V_{PV}^2 = C_{33}$$
$$F = 1/2(C_{13}+C_{23}) \quad (5a)$$
$$L = \rho V_{SV}^2 = 1/2(C_{44}+C_{55})$$
$$N = \rho V_{SH}^2 = 1/8(C_{11}+C_{22}) - 1/4 C_{12} + 1/2 C_{66}$$

- 2-Ψ azimuthal term:

$$B_C = 1/2(C_{11}-C_{22})$$
$$G_C = 1/2(C_{55}-C_{44})$$
$$H_C = 1/2(C_{13}-C_{23})$$

$$B_S = C_{16} + C_{26}$$
$$G_S = C_{54} \quad (5b)$$
$$H_S = C_{36}$$

- 4-Ψ azimuthal term:
$$E_C = 1/8(C_{11}+C_{22}) - 1/4 C_{12} - 1/2 C_{66} \quad (5c)$$
$$E_S = 1/2(C_{16}-C_{26}),$$

where indices 1 and 2 refer to horizontal coordinates (1: North; 2: East) and index 3 refers to vertical coordinate. ρ is the density, V_{PH}, V_{PV} are respectively horizontal and vertical *propagating* P-wave velocities, V_{SH}, V_{SV} horizontal and vertical *polarized* S-wave velocities. So, the different parameters present in the different azimuthal terms are simply related to elastic moduli C_{ij}. The corresponding kernels are detailed and some of their variations at depth are plotted in Montagner and Nataf (1986). The complete description of anisotropic effects on normal modes in the spherical case can be found Mochizuki (1986) and Tanimoto (1986). In the most general case, 13 parameters are necessary to explain surface wave data (Rayleigh and Love waves) for small anisotropy, but only 4 parameters are well resolved (Montagner and Jobert 1988): the azimuthally averaged S-wave velocity V_S, the radial anisotropy expressed through the ξ parameter [$\xi = (V_{SH}/V_{SV})^2$] where V_{SH} (respectively V_{SV}) is the velocity of S-wave propagating horizontally with horizontal transverse polarization (respectively with vertical polarization), and the G (G_C, G_S) parameters expressing the horizontal azimuthal variation of V_{SV}. ξ was introduced in the reference Earth model PREM (Dziewonski and Anderson 1981) down to 220 km in order to explain a large dataset of free oscillation eigenfrequencies and body wave travel times. The other elastic parameters can be derived by using constraints from petrology in order to reduce the parameter space (Montagner and Anderson 1989a). Two extreme petrological models were used to derive the necessary correlations between anisotropic parameters, the pyrolite model (Ringwood 1975) and the piclogite model (Anderson and Bass 1984, 1986; Bass and Anderson 1984). In the depth inversion process, the smallest correlations between parameters of both models were kept and included in the *a priori* correlation matrix on parameters. This approach was followed by Montagner and Anderson (1989b) to derive an average reference earth model, and by Montagner and Tanimoto (1991) for the first global 3-D anisotropic model.

Figure 5 shows what is expected for the observable parameters V_S, ξ, G, ψ_G in the case of a simple convective cell with LPO. In terms of convective flow, radial anisotropy ξ expresses its vertical ($\xi < 1$) or horizontal character ($\xi > 1$), and the azimuthal anisotropy G, can be related to the horizontal flow direction. Conversely, the three maps of V_S, ξ, G, can be interpreted in terms of convective flow. These three pieces of information are necessary to correctly interpret the data. For example, upwellings or downwellings are both characterized by a weak or negative ξ parameter, but a correlative positive or negative δV_S discriminates between these possibilities. By simultaneously inverting at depth for the different azimuthal terms of Rayleigh and Love waves, it is therefore possible to separate the lateral variations in temperature from those induced by the orientation of minerals. Such an interpretation should however be erroneous in water-rich mantle regions where LPO of minerals such as olivine is no simply related to the strain field (Jung and Karato 2001). Also note that this interpretation of anisotropy in terms of convective flow, using S-waves speeds only, could be erroneous in a olivine-poor mantle; see Mainprice et al. (2000), e.g., for the relations between lattice preferred orientations and V_P and V_S in the whole orientation space.

The complete tomographic technique (regionalization + inversion at depth) has been applied for investigating either regional structures of the Indian Ocean (Montagner and

Jobert 1988, Debayle and Lévêque 1997), of the Atlantic Ocean (Mocquet et al. 1989, Silveira et al. 1998), of Africa (Hadiouche et al. 1989, Debayle et al. 2001), of Pacific Ocean (Nishimura and Forsyth 1989, Bussy et al. 1993), of Antarctica (Roult et al. 1994), Australia (Debayle and Kennett 2000, Simmons et al. 2002) and Central Asia (Griot et al. 1998a,b) or global structure (Montagner and Tanimoto 1990, 1991; Montagner 2002).

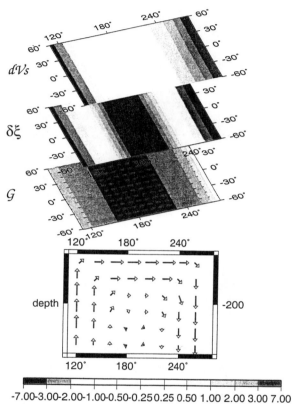

Figure 5. The seismic observable parameters V_S, ξ, G, ψ_G associated with a simple convecting cell in the upper mantle, assuming lattice-preferred orientation of anisotropic minerals such as olivine. A vertical flow is characterized by a negative ξ radial anisotropy (ratio between V_{SH} and V_{SV}) and a small azimuthal anisotropy ($G \approx 0$). An upwelling (resp. downwelling) is characterized by a large positive (resp. negative) temperature anomaly inducing $\delta V_S < 0$ (resp. $\delta V_S > 0$). A predominant large-scale horizontal flow will be translated into a significant amplitude of the G azimuthal anisotropy and its orientation will reflect the direction of flow (with a 180° ambiguity). For a more accurate, color version of this figure, see Errata for Volume 51 at http://www.minsocam.org/MSA/RIM/ .

Comparison between surface wave anisotropy and SKS splitting data

It can be noted that the anisotropic parameters, linear combinations of elastic moduli C_{ij}, derived from surface waves, also come up when you consider the propagation of body waves in symmetry planes for a slightly anisotropic medium (see section entitled *Body Waves*). A global investigation of anisotropy inferred from SKS body wave splitting measurements (delay times and directions of maximum velocities) has been undertaken by different authors (Vinnik et al. 1992, Silver 1996, Savage 1999). Unfortunately, most of SKS measurements have been done in continental parts of the Earth, and very few in oceans. It turns out that a direct comparison of body wave and surface wave datasets is now possible (Montagner et al. 2000). If the anisotropic medium is assumed to be characterized by a horizontal symmetry axis with any orientation (that is a very strong assumption), a synthetic dataset of SKS delay times and azimuths can be calculated from the global distribution of anisotropy derived from surface waves, by using the following equation:

$$\delta t_{SKS} = \int_0^h dz \sqrt{\frac{\rho}{L}} \left[\frac{Gc(z)}{L(z)} \cos 2\Psi(z) + \frac{Gs(z)}{L(z)} \sin 2\Psi(z) \right] \quad (6)$$

where δt_{SKS} is the integrated travel time for the depth range 0-h, where the anisotropic parameters $G_c(z), G_s(z)$ and $L(z)$ are the anisotropic parameters retrieved from surface waves at different depths. It is remarkable to realize that only the G-parameter (expressing the SV-wave azimuthal variation) is present in this equation. From Equation (6), we can infer the maximum value of delay time δt_{SKS}^{max} and the corresponding azimuth Ψ_{SKS}:

$$\delta t_{SKS}^{max} = \sqrt{\left\{ \int_0^h dz \sqrt{\frac{\rho}{L}} \frac{Gc(z)}{L(z)} \right\}^2 + \left\{ \int_0^h dz \sqrt{\frac{\rho}{L}} \frac{Gs(z)}{L(z)} \right\}^2} \quad (7)$$

$$\tan 2\Psi_{SKS} = \frac{\int_0^h dz \sqrt{\frac{\rho}{L}} \frac{Gs(z)}{L(z)}}{\int_0^h dz \sqrt{\frac{\rho}{L}} \frac{Gc(z)}{L(z)}} \quad (8)$$

However, Equation (6) is approximate and only valid when the wavelength is much larger than the thickness of layers. It is possible to make more precise calculations by using the technique derived for 2 layers by Silver and Savage (1994) or by using the general expressions given in Rumpker and Silver (1998) and Montagner et al. (2000). With Equations (7) and (8), a synthetic map of the maximum value of delay time δt_{SKS}^{max} can be obtained by using a 3-D anisotropic surface wave model. A detailed comparison between synthetic SKS derived from AUM (Montagner and Tanimoto 1991) and observed SKS (Silver 1996) was presented in Montagner et al. (2000). Figure 6 shows such a map for the Pacific hemisphere, by using a new anisotropic surface wave model (Montagner 2002) derived from the data of Ekström and Dziewonski (1998). First of all, the comparison shows that both datasets are compatible in magnitude but not necessarily in directions. Some contradictions between measurements derived from surface waves and from body waves have been noted. The agreement of directions is correct in tectonically active areas but not in old cratonic zones. The discrepancy in these areas results from the rapid lateral change of directions of anisotropy at a small scale. These changes stem from the complex history of these areas, which have been built by successive collages of continental pieces. It might also result from the hypothesis of horizontal symmetry axis, which was shown to be invalid in many areas (Plomerova et al. 1996). The positive consequence of this discrepancy is that a small scale mapping of anisotropy in such areas might provide clues for understanding the processes of growth of continents and mountain building.

Contrary to surface waves, SKS-waves have a good lateral resolution, and are sensitive to the short wavelength anisotropy just below the stations. But their drawback is that they have a poor vertical resolution. On the other hand, global anisotropy tomography derived from surface waves only provides long wavelength anisotropy (poor lateral resolution) but enables the location of anisotropy at depth.

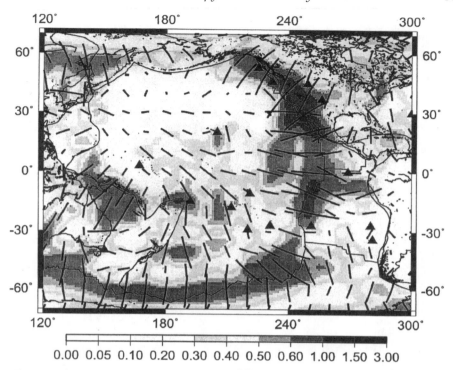

Figure 6. Distributions of synthetic delay time δt_{SKS}^{max} and azimuth Ψ_{SKS} at the surface of the Earth, such as derived from the anisotropic tomographic model of Montagner (2002), derived from data of Ekström and Dziewonski 1998). The crustal part of the 3SMAC -model (Ricard et al. 1996) has been removed. The synthetic map of SKS is calculated by using the method of Montagner et al. (2000), from the G-distribution of V_{SV} azimuthal anisotropy. The length of lines is proportional to δt_{SKS}. For a more accurate, color version of this figure, see Errata for Volume 51 at http://www.minsocam.org/MSA/RIM/ .

The long wavelength anisotropy derived from surface waves will display the same direction as the short wavelength anisotropy inferred from body waves only when large-scale vertical coherent processes are predominant. As demonstrated by Montagner et al. (2000), the best agreement between observed and synthetic SKS can be found when only layers in the uppermost 200 km of the mantle are taken into account. In some continental areas, short scale anisotropy, the result of a complex history, might be important and even might mask the large-scale anisotropy more related to present convective processes. This last statement settles on the following observation: when making a comparison with plate velocities directions, it is found a good statistical agreement between both directions (Fig. 7); this result shows that to first order, seismic anisotropy is reflecting the large scale plate tectonic motions. The differences between anisotropy and tectonic plate directions are related to more complex processes, as we will see below in the section *Oceanic Plates*.

ANISOTROPY IN THE DIFFERENT LAYERS OF THE EARTH AND THEIR GEODYNAMIC APPLICATIONS

Before considering measurements of anisotropy at global, regional or local scales, let us consider the laterally averaged Earth or equivalently, the spherically symmetric reference Earth models.

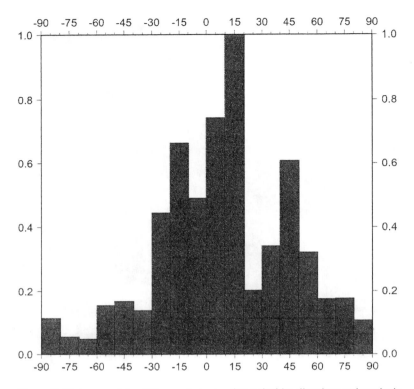

Figure 7. Histogram of the difference between plate velocities directions and synthetic SKS anisotropy azimuths in the Pacific plate.

Reference 1-D Earth models

Because tomographic models are derived from a reference model by linearized inversion schemes, there is a need for good starting radial reference models. The quality of the reference model will strongly condition the outcome of inversions for 3-D models. The most general case of anisotropy for a spherically symmetric earth is the transverse isotropy with vertical symmetry axis (also termed radial anisotropy). As discussed in the previous section, such a medium can be described using six functions of radius r, the density ρ, the wave velocities, V_{PH}, V_{SV}, and 3 anisotropic parameters ξ, ϕ and η readily related to A, C, F, L, N defined in the previous section.

PREM (Dziewonski and Anderson 1981) first introduced the radial anisotropy in the uppermost 220 km of the mantle. However, some aspects of the normal mode data are not well explained by PREM (Montagner and Anderson 1989a), for example fundamental toroidal mode and body wave travel times. There were several attempts to reconcile normal mode data and body wave data (Widmer 1991, Montagner and Kennett 1996).

Several robust features have been found regarding the radial anisotropy, from such inversions based on the different body-wave models [IASP91 (Kennett and Engdahl 1991); SP6 (Morelli and Dziewonski 1993); AK135 and AK303 (Kennett et al. 1996)]. First of all, anisotropy is significant in the whole upper mantle with a minimum value in the depth range 300-500 km. It is very small in the whole lower mantle except in a zone we call the "lower transition zone" (between the 660-km-discontinuity and 900-km depth) and in the D″-layer. An interesting feature of these models is the existence of a complex radial anisotropy in the transition zone (410-660 km) and at the uppermost lower mantle (660-900 km), each zone presenting a different ξ $(= (V_{SH}/V_{SV})^2 - 1)$ -value pattern. The existence of anisotropy in the D″-layer seems robust as well, but is difficult to interpret, with a small S-wave anisotropy but large P-wave and η anisotropies. Therefore, these new reference Earth models provide some indication of the existence of anisotropy in three depth ranges, the first one in D″-layer at the core-mantle boundary, the second one around the 660-km-discontinuity. Independent seismological studies tend to corroborate these findings in the upper mantle, the transition zone, and the D″-layer.

Evidence of anisotropy in the upper 410 km of the mantle

The uppermost mantle down to 410 km is the range where the existence of seismic anisotropy is now widely recognized and well documented. Azimuthal variations have been found for body waves and surface waves in different areas of the world. During the last years, the shear wave splitting, primarily on SKS waves, was extensively used for studying continental deformation, but very few studies using body waves are devoted to oceanic areas. Conversely, global anisotropic tomographic models have been derived during the last 10 years from surface waves, but they are the most reliable below oceanic areas. Therefore the comparison of body wave and surface wave data is still in its infancy. However, as shown by Montagner et al. (2000) and Vinnik et al. (2002), such a comparison is providing encouraging results.

The seismic anisotropy in the Earth can therefore be retrieved by different methods from different datasets. We will only present some examples of interesting applications of anisotropy in large scale geodynamics and tectonics. The application of seismic anisotropy to geodynamics in the upper mantle is straightforward, if we assume that fast-polarization axis of mineralogical assemblages is in the flow plane parallel to the direction of flow (Fig. 1). Seismic anisotropy in the mantle is therefore reflecting the strain field prevailing in past (frozen-in anisotropy) for shallow layers or present convective processes in deeper layers. Therefore, it makes it possible to map convection in the mantle. It must be noted that, when only the radial anisotropy is retrieved, its interpretation is non-unique. A fine layering of the mantle can also generate such a kind of anisotropy, and neglecting the azimuthal anisotropy can bias the amplitude of radial anisotropy.

As mentioned in the previous section (Fig. 5), a complete interpretation of anisotropic tomography maps makes it necessary to simultaneously consider the maps of V_S, ξ and G parameters in order to separate the effects of temperature and orientation of flow. A discussion for the Earth can be found in Montagner and Guillot (2000). We will only focus on geodynamic consequences for oceanic plates and continents.

We will tentatively suppose that the direction of the flow in the mantle is directly related to the direction of azimuthal anisotropy or the sign of the radial anisotropy. This is approximately true in a rich-olivine mantle. A complete modeling of the relations between mantle flow and preferential orientations doesn't exist now, but this hypothesis seems quite reasonable, especially in regions where the strain field could have been quasi-stationary for millions of years.

Oceanic plates

Figure 8 shows 3 maps at depths of 100 km and 200 km displaying V_{SV} velocity anomalies (Fig. 8a) and the 2 kinds of anisotropy, which can be retrieved by simultaneous inversion of Rayleigh and Love waves constant 0-Ψ and azimuthal terms of Equation (5) from the model of Montagner (2002). On Figure 8b, the equivalent radial anisotropy of the medium, for S-wave expressed through the ξ parameter, is displayed. The maps of

Figure 8c are the distributions of the G-parameter related to the azimuthal variation of SV-wave velocity. The maximum amplitude of G is around 5% and is rapidly decreasing as depth is increasing. The distribution of anisotropy has completely different patterns and amplitudes at these 2 depths (100 and 200 km).

It shows that, to the first order, the agreement of directions of maximum velocity with plate tectonics is correct in the depth range 100 to 300 km (Montagner 1994). However, the azimuth of G-parameter can largely vary as a function of depth (Montagner and Tanimoto 1991). For instance, at shallow depths (down to 60 km), the maximum velocity can be parallel to mountain belts or plate boundaries (Vinnik et al. 1991, Silver 1996, Babuska et al. 1998), and orthogonal at large depth. This means that, at a given place, the orientation of fast axis is a function of depth, making difficult the interpretation of SKS splitting.

Since convective flow below oceans is dominated by large-scale plate motions, the long wavelength anisotropy found in oceanic lithospheric plates, should be similar to the smaller-scale anisotropy which should be measured from body waves. By the way, one of the first evidences of anisotropy was found in the Pacific Ocean by Hess (1964) for Pn-waves. Since that time, there were many measurements of the subcrustal anisotropy (see Babuska and Cara 1991, for a review). So far, there are very few measurements of anisotropy by SKS splitting in the oceans. Due to the lack of seismic stations on the seafloor (with the exception of H2O halfway between Hawaii and California), the only measurements available for SKS were performed in stations located on ocean islands (Ansel and Nataf 1989, Kuo and Forsyth 1992, Russo and Okal 1999, Wolfe and Silver 1998), which are by nature anomalous objects, such as volcanic hotspots where the strain field is perturbed by the ascending material and not necessarily representative of the main mantle flow field. SKS splitting was measured during the temporary MELT experiment on the East-Pacific Rise (Wolfe and Solomon 1998) but the orientation of the splitting is in disagreement with the predictions of Blackman et al. (1996). Walker et al. (2001) presented a first measurement of SKS splitting at H2O, but it is in disagreement with independent SKS splitting measurement at the same station by Vinnik et al. (2002) and with surface wave anisotropy (Montagner 2002).

There were other attempts of determining anisotropy for other kinds of body waves such as ScS and multiple ScS (Ando 1984, Farra and Vinnik 1994, Gaherty et al. 1995), or differential times of sS-S, or SS-S waves (Kuo et al. 1987, Sheehan and Solomon 1991, Fischer and Yang 1994, Gaherty et al. 1995, Fouch and Fischer 1996, Yang and Fischer 1994), however, only surface wave inversion can provide a 3-D map of anisotropy below oceans. Radial cross-sections show that the ξ parameter is usually negative and small (in any case smaller than average), where flow is primarily radial (mid-ocean ridges and subducting zones). Between plate boundaries, oceans display very large areas with a large positive radial anisotropy such as in the Pacific Ocean (Ekström and Dziewonski 1997), characteristic of an overall horizontal flow field. Oceanic plates are zones where the comparison between directions of plate velocities (Minster and Jordan 1978) or NUVEL-1 (DeMets et al. 1990) and directions of G-parameter is the most successful

Figure 8 (next page). Result of the simultaneous inversion of Rayleigh and Love waves dispersion and their azimuthal variations at 100 km depth (left) and 200 km depth (right). (adapted from Montagner 2002). (Top): Distribution of the V_{SV} parameter in %. (Middle): ξ distributions in % with respect to ACY400 (Montagner and Anderson 1989b). Be aware that ξ anomalies are plotted with respect to a reference value different from 0. (Bottom): Anisotropy map of the G-parameter (V_{SV} azimuthal anisotropy). For a more accurate, color version of this figure, see Errata for Volume 51 at http://www.minsocam.org/MSA/RIM/ .

Seismic Anisotropy and Global Geodynamics

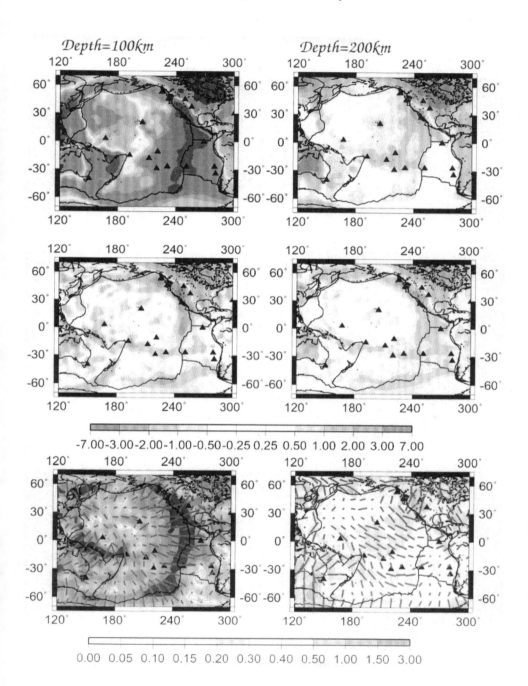

in the asthenosphere down to 250-300 km (Montagner 1994). Conversely, such a comparison is more controversial below plates bearing a large proportion of continents, such as the European-Asian plate, characterized by a very small absolute motion in the hotspot coordinate system.

Oceans are the areas where Plate tectonics applies almost perfectly. The large-scale azimuthal anisotropy within and below lithosphere is closely related to plate motions (Montagner 1994) and modeled in this framework (Tommasi et al. 1996). The map of the G-parameter at 100 km shows that the azimuthal anisotropy is very large along spreading ridges with a large asymmetry for the East Pacific rise. The direction of anisotropy is in very good agreement with plate motion. The anisotropy is large as well in the middle of the Pacific plate, but it can be observed that there is a line of very small azimuthal anisotropy almost parallel to the EPR. This linear area of small anisotropy was coined Low Anisotropy Channel by Montagner (2002). They are presumably related to cracking within the Pacific plate and/or to secondary convection within and below the rigid lithosphere, predicted by numerical and analog experiments. These new features provide strong constraints on the decoupling between the plate and asthenosphere. The existence and location of these LACs might be related to the current active volcanoes and hotspots (possibly plumes) in Central Pacific. LACs, which are dividing the Pacific Plate into smaller units, might indicate a future reorganization of plates with ridge migrations in the Pacific Ocean. They call for more thorough numerical modelling.

Continents

Seismic anisotropy can provide fundamental information on the structure of continents, their root and the geodynamic processes involved in mountain building and collision between continents (Vinnik et al. 1992, Silver 1996) such as in Central Asia (Griot et al. 1998a,b). ξ is usually very heterogeneous below continents in the first 150-200 km of depth with positive or negative areas according to geology. But it seems to display a systematic tendency of being positive at larger depth (down to 300 km), whereas it is very large in the oceanic lithosphere in the depth range 50 to 200 km and decreases rapidly at larger depths (Montagner 1994). Conversely, radial anisotropy is displaying a maximum (though smaller than in oceanic lithosphere) below very old continents (such as Siberian and Canadian Shield) in the depth range 200 to 400 km. Seismic anisotropy below continents, sometimes confined to the upper 220 km (Gaherty and Jordan 1995) can still be significant below. A more quantitative comparison of radial anisotropy between different continental provinces is presented in Babuska et al. (1998), and demonstrates systematic differences according to the tectonic context. The existence of positive large-scale radial anisotropy below continents at depth might be a good indicator of the continental root which was largely debated since the presentation of the model of tectosphere by Jordan (1978, 1981). If we assume that this maximum of anisotropy is related to an intense strain field in this depth range, it might be characteristic of the boundary between continental lithosphere and "normal" upper mantle material. And our results show that the root of continents as defined by anisotropy is located between 200 and 300 km. Such a result was recently confirmed for the Australian continent by Debayle and Kennett (2000) and Simmons et al. (2002).

It must be emphasized that the anisotropy near the surface is probably different from the deep one. Part of the observed anisotropy might be related to the fossil strain field prevailing during the setting of materials and the deeper part is related to the present strain field. If we bear in mind that the anisotropy displayed from surface waves is the long wavelength filtered anisotropy (approximately 1500 km), it can be easily understood that the average anisotropy displayed from surface waves in the first 200 km might be very different from the one found from body waves. The tectonic structure of continents is the result of a long and complex history. The characteristic length scale under continents is related to the size of blocks successively accreted to existing initial cores and probably smaller than 1500 km. This statement is supported by different studies of SKS body waves that demonstrate that the direction of maximum velocity can change on horizontal scales smaller than 100 km (Vinnik et al. 1991, Silver 1996). The fact that we do not observe a systematic behaviour in the first hundreds of kilometers for similar continental geological zones, does not mean that anisotropy is not present but only that its characteristic scale is different from one region to another one. Due to the low pass filtering effect of surface wave tomographic technique, the long wavelength signature of anisotropy is diluted.

The measurement of seismic anisotropy can provide fundamental information for the understanding of tectonic processes. Silver (1996) and Savage (1999) present reviews of the information provided by shear-wave splitting beneath continents. The poor lateral resolution of global scale anisotropic tomography can be considered as a strong limitation in continental areas. This technique can only be efficiently applied in areas where large-scale coherent processes are present. The best candidate where this condition is fulfilled in continents, is the collision zone between India and Asia, where the applicability of plate tectonics can be questioned.

Griot et al. (1998a) undertook such an investigation in Central Asia. The primary goal of this study was the discrimination between two competing extreme models of deformations, the heterogenous model of Avouac and Tapponnier (1993) and the homogeneous model of England and Houseman (1986). It was necessary to use shorter wavelength surface waves (40-200 s) in order to obtain a lateral resolution of 350 km. Synthetic models of seismic anisotropy can be inferred from the heterogeneous and homogeneous models. In order to perform correct and quantitative comparisons between observed seismic anisotropy and the deformation models, the short wavelengths of the synthetic models (spatial scale smaller than 350 km) were filtered out. The statistical comparison between observed and synthetic azimuthal anisotropies for both models enables to determine in different depth ranges, which deformation model dominates. Griot et al. (1998b) show that the heterogeneous model is in better agreement with observations in the first 200 km, whereas the homogeneous model better fits the deep anisotropy below 200 km. We must be aware that such a comparison is only valid from a statistical point of view and that a comparison at a more local scale (the scale of body wave measurements) might display some differences with the observed SKS anisotropy. This kind of investigation only underlines large scale ongoing and prevailing active processes and is not devoted to a precise measurement of anisotropy at any specific place. It demonstrates that in the uppermost 200 km, the tectonics is well described by the relative movement of continental blocks and that plate tectonics can apply to these plate-like areas.

Seismic anisotropy can now be used for making quantitative measurements. By using simultaneously splitting data and geodetic data, Silver and Holt (2002) were able to calculate the relative motion of lithospheric plate and asthenosphere. They show that the mantle trails the hotspot motion of the plate. This results show a weak coupling between lithosphere and asthenosphere, in contrast with the Pacific plate, where the coupling (reflected by plate direction) is the first-order effect in the uppermost 200 km, whereas LACs (Low Anisotropy channels) suggesting some decoupling, is a second-order effect.

The fossil shallow anisotropy (first 100 kilometers) that reflects the past strain field, should be very useful for understanding the processes involved in surficial tectonics (see review in Savage 1999). If we were able to determine the age of establishment of this shallow anisotropy, the measurement of this kind of anisotropy should open wide a new field in Earth sciences: the Paleo-Seismology which might provide fundamental information to structural seology.

Anisotropy in the transition zone

The transition zone plays a key role in mantle dynamics, particularly the 660-km-discontinuity which might inhibit the passage of matter between the upper and the lower mantle. Its seismic investigation is made difficult on global scale by the poor sensitivity of fundamental surface waves in this depth range and by the fact that teleseismic body waves recorded at continental stations from earthquakes primarily occurring along plate boundaries have their turning point below the transition zone. However, Montagner and Kennett (1996), by using eigen-frequency data, display some evidence of radial anisotropy in the upper (410-660 km) and lower (660-900 km) transition zones. Another important feature of the Transition zone is that, contrarily to the rest of the upper mantle, the upper transition zone is characterized by a large degree 2 pattern (Masters et al. 1982), and to a less extent, the degree 6. Montagner and Romanowicz (1993) explained this degree 2 pattern by the predominance of a simple large-scale flow pattern characterized by two upwellings in central Pacific Ocean and Eastern Africa and two downwellings in the Western and Eastern Pacific Ocean. This scheme was corroborated by the existence in the upper transition zone of a slight but significant radial anisotropy displayed by Montagner and Tanimoto (1991) and Roult et al. (1990). In the Transition zone, the pattern of radial anisotropy is dominated by degree 4 in agreement with the prediction of this model. Therefore, the observations of the geographical distributions of degrees 2, 4, 6 in the transition zone are coherent and spatially dependent. Montagner (1994) compared these different degrees to the corresponding degrees of the hotspot and slab distribution. In this simple framework, the distribution of plumes and slabs are merely a consequence of the large-scale simple flow in the transition zone.

The existence of anisotropy close to the 660-km-discontinuity was also found by Vinnik and Montagner (1996) below Germany and Vinnik et al. (1998) in central Africa. By studying P-to-S converted waves at the GRF network and at GEOSCOPE station BNG in central Africa, they observed that part of the initial P-wave is converted into SH-wave. This signal can be observed on the transverse component of seismograms. The amplitude of this SH-wave cannot be explained by a dipping 660-km-discontinuity and it constitutes a good evidence for the existence of anisotropy just above this discontinuity. However, there is some evidence of lateral variation of anisotropy in the transition zone as found by the investigation of several subduction zones (Fischer and Yang 1994, Fischer and Wiens 1996). Fouch and Fischer (1996) present a synthesis of these different studies and show that some subduction zones such as Sakhalin Islands requires deep anisotropy in the transition zone, whereas others such as Tonga do not need any anisotropy. They conclude that their data might be reconciled by considering the upper transition zone (410-520 km) intermittently anisotropic, and the rest of the transition zone might be isotropic.

The evidence of anisotropy in the transition zone was confirmed recently by 2 independent studies, using different datasets. Wookey et al. (2002) present evidence of very large S-wave splitting (up to 7 s) in the vicinity of the 660-km-discontinuity between Tonga-Kermadec subduction zone and Australia. On a global scale, Trampert and van Heijst (2002) show a robust long-wavelength azimuthal anisotropic structure in the transition zone. The rms amplitude of lateral variations of G is about 1%. The interpretation of these new exciting results is not obvious but they confirm that the transition zone might be a mid-mantle boundary layer.

Anisotropy in the D''-layer

During the last years, the structure of the mysterious D''-layer has been extensively investigated. It is not the goal of this paper to review these different studies which make use of different kinds of body waves but only to draw attention to recent results which make evident the presence of anisotropy in the D''-layer (Vinnik et al. 1989, Lay and Young 1991). The search for anisotropic structures in the D''-layer might provide fundamental information on boundary layer shear flow, partial melt or slab remnants at core-mantle boundary. There seems to be a general consensus that, except the D''-layer, the lower mantle does not contribute significantly to shear wave splitting observation. However, we must be aware that it might not be valid everywhere, and part of the anisotropy attributed to D''-layer might originate in the lowermost mantle. For example, differential splitting between phases such as $P_{660}S$ and SKS phases, or between SKS and SKKS, indicate that at least some regions of the lower mantle contain significant anisotropy (Iidaka and Niu 1998, Barruol and Hoffman 1999, Vinnik et al. 1997). Actually, the main drawback to investigate anisotropy in the lower mantle results from the limited crossing ray coverage. In addition, it is necessary to get rid of the influence of the anisotropy in the upper mantle. Therefore, it is not yet clear if lower mantle anisotropy is only localized in the bottom boundary layer (D'' layer) or if there are some areas within the lower mantle with significant anisotropy.

So far, there is clear evidence of anisotropy in the D''-layer. By studying Sd-waves (S diffracted waves at the Core-Mantle boundary), Vinnik et al. (1995) found that SVd-waves is delayed relative to SHd-waves by 3s. Their observations are characteristic of a transversely isotropic medium with vertical symmetry axis with $V_{SH} > V_{SV}$. Other seismic observations such as anomalous diffraction of body waves (Maupin 1994) confirm this kind of anisotropy and its rapid lateral variation is made evident in many areas around the world (see review in Lay et al. (1998)) such as in Caribbean Sea (Kendall and Silver 1996), North America (Matzel et al. 1996, Garnero and Lay 1996), Indian Ocean (Ritsema 2000). The origin of anisotropy observed in D''-layer is still debated (Kendall and Silver 1996, Kendall 2000) and might be due to horizontal layering in connection with paleo-subduction and/or aligned inclusions inducing different velocities for SV and SH (SPO anisotropy). On the same line, McNamara et al. (2001) showed that near cold downwellings, LPO may dominate, while near warm upwellings, SPO might be a more likely candidate. Numerical calculations of mantle convection (McNamara 2002) show that the slab deformation in the deep mantle, inducing large-strain deformation at high stresses can explain the presence of strong anisotropy in D''-layer.

NUMERICAL MODELING AND BOUNDARY LAYERS

In the previous section, we have highlighted the presence of seismic anisotropy in different parts of the Earth, using both body and surface wave data. This review of the presence of anisotropy in different layers of the Earth demonstrated that the anisotropy is a very general feature. However, it is not present in all depth ranges nor at all scales. As discussed in the first section, the observation of seismic anisotropy due to LPO at large scales requires several strong conditions, starting with the presence of anisotropic crystals up to the existence of an efficient large scale present or past strain field. Theoretical studies suggest that, when anisotropic minerals such as olivine are deformed, their crystallographic axes develop a systematic relationship to the principal axes of finite strain, and the amplitudes of the anisotropy is grossly proportional to the amplitude of the finite strain until it reaches a steady state (McKenzie 1979, Ribe 1989, Ribe and Yu 1991, Kaminski and Ribe 2001). Many numerical modelings of the convective mantle show that in a convective system, the strain field is not spatially uniform (Chastel et al. 1993). Streamlines are much more concentrated in boundary layers than in the middle of the cells. The consequence is that the amplitude of the strain field and of the seismic anisotropy resulting from the lattice preferred orientation (Tommasi et al. 2000), is very heterogeneous and the largest in boundary layers.

For filling the gap between grain scale modeling and large scale anisotropy measurements, there is now a real need for making more quantitative comparisons between seismic anisotropy and numerical modeling. Gaboret et al. (2002) calculated the convective circulation in the mantle by converting perturbations of S-wave velocity into density perturbations. Figure 9 shows 2 cross-sections through the Pacific hemisphere and the associated flow lines derived from the tomographic model of Ekstrom and Dziewonski (1998). This kind of modeling enables to calculate the strain tensor and to test different hypotheses for the prevailing mechanisms of alignment, by comparison with seismic data. We present in Figure 10 a first attempt to compare azimuthal anisotropy of the global anisotropic model of Montagner (2002) and the maximum horizontal stretching rate. The cosine of the difference between the fast axis of V_{SV} and the axis of horizontal extensional rate, is plotted. The best agreement is found for young ages in the Pacific ocean. More generally, the comparison is correct for oceanic plates but not so good for continental plates, where the resulting anisotropy displays a more complex distribution.

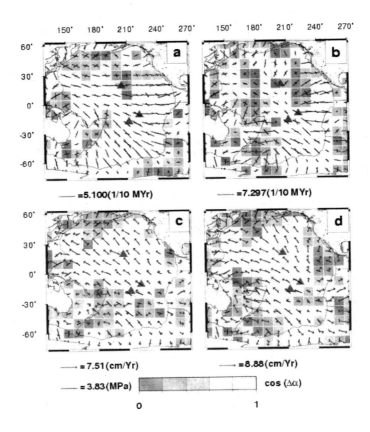

Figure 9. Two whole mantle cross-sections through the Ekstrom and Dziewonski tomographic model (1998) along great circles. Also shown in the inset maps are triangles which represent the locations of 3 Pacific hotspots. The superimposed black arrows in the cross-sections represent the mantle flow velocities predicted on the basis of the buoyancy forces derived from shear velocity anomalies. (adapted from Gaboret et al. 2002). For a more accurate, color version of this figure, see Errata for Volume 51 at http://www.minsocam.org/MSA/RIM/ .

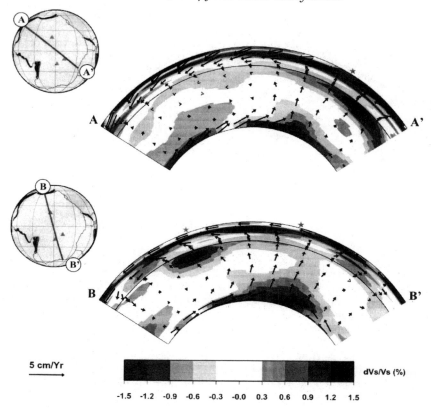

Figure 10. Comparison of azimuthal seismic anisotropy and the direction of maximum stretching predicted by the flow model presented in Figure 9. [(Adapted from Gaboret et al. (2002).] For a more accurate, color version of this figure, see Errata for Volume 51 at http://www.minsocam.org/MSA/RIM/.

Conversely, we can assume that the observation of mantle seismic anisotropy is the indication of a strong present-day strain field (at the exception of crust, topmost oceanic, and continental lithosphere where fossil anisotropy may be present). This strain field can be associated with boundary layers. In the previous sections, we saw that there are good evidences of the presence of seismic anisotropy in the D″-layer, in the transition zone and in the uppermost mantle. These findings are summarized in Figure 11.

The D″-layer and the uppermost mantle have been related to boundary layers of the mantle convective system for a long time. The D″-layer above the core-mantle boundary is characterized by a large degree of seismic heterogeneities and anisotropy with V_{SH} larger than V_{SV}. It might be at the same time, the graveyard of subducted slabs and the source of megaplumes. D″- anisotropy can be related either to horizontal layering (lattice- and shape- preferred orientations) of cold material or the presence of aligned inclusions owing to the presence of melt (Kendall and Silver 1996, McNamara et al. 2002). For the uppermost oceanic mantle, seismic anisotropy is present in both lithosphere and asthenosphere, and for oceans, some finite-element models are able to quantitatively relate lithospheric and asthenospheric strain to anisotropy (Tommasi et al. 1996). But the

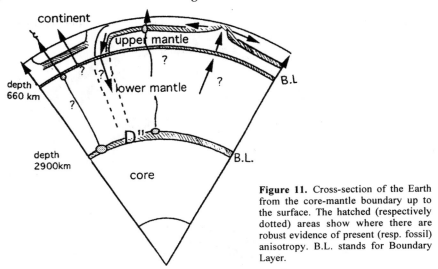

Figure 11. Cross-section of the Earth from the core-mantle boundary up to the surface. The hatched (respectively dotted) areas show where there are robust evidence of present (resp. fossil) anisotropy. B.L. stands for Boundary Layer.

presence of anisotropy in the transition zone (Montagner and Kennett 1996, Vinnik and Montagner 1996, Wookey et al. 2002, Trampert and van Heijst 2002) is fundamental and problematic, because it provides a new clue that the transition zone could act as a secondary boundary layer. The first evidence of anisotropy in the transition zone tends to favor the predominance of horizontal flow over vertical flow. The major consequence of this finding is that the transition zone (down to about 900-1000 km) is dividing, on average, the mantle into two convective systems, the upper mantle and the lower mantle. This general statement does not rule out the possibility that flow circulation between the upper and the lower mantle is occuring. But, it means that the exchange of matter between the upper and the lower mantle is difficult. It is too early to assess the amount of matter going through, from measurements of seismic anisotropy, but the complete mapping of seismic anisotropy in the transition zone (Trampert and van Heijst 2002) associated with numerical modeling might enable to put constraints on the flow between the upper and lower mantles.

CONCLUSIONS : FROM SEISMIC ANISOTROPY TO ANISOTROPIC SEISMOLOGY

We have presented in this paper different observations of seismic anisotropy and their applications in geology and geodynamics. Seismic anisotropy is able to define continental roots, to investigate the coupling between different layers, to discriminate different geodynamic processes and competing convective models. Three boundary layers were so far detected by seismic anisotropy: the uppermost mantle, the transition zone (though new work is necessary to understand its role) and the D″-layer. Other applications of seismic anisotropy can be easily found. For example, Montagner and Anderson (1989a) show that the different anisotropic parameters might be used for discriminating competing petrological models such as pyrolite or piclogite. Some seismologists claim that the temporal variation of anisotropy in the crust might be an efficient tool for investigating and monitoring the earthquake cycle (Crampin and Booth 1985, Crampin and Volti 1999). The temporal changes in shear-wave polarization were successfully used to show the variation of the stress regime before and after a volcanic eruption (Miller and Savage 2001). Seismic anisotropy starts to be used in seismic exploration for monitoring oil reservoirs. Therefore, to study time-dependent phenomena, seismic anisotropy turns out to be very efficient. The scientific potential of seismic anisotropy is enormous and largely unexploited. In conclusion, the seismic anisotropy provides a new dimension in the investigation of processes of our dynamic Earth.

ACKNOWLEDGMENTS

We are grateful to Don Anderson, Adam Dziewonski, Andrea Morelli, Lev Vinnik, Mike Gurnis, Jeroen Ritsema, Jeroen Tromp, Barbara Romanowicz, Anne Davaille, Alessandro Forte, Goran Ekström, for fruitful discussions.

This is I.P.G.P. contribution # 1878.

APPENDIX A. BASIC THEORY OF WAVE PROPAGATION IN ANISOTROPIC MEDIA

In any elastic medium, and in neglecting all additional terms due to prestress, gravity, inertial terms, the equation of motion is:

$$\rho \frac{\partial^2 u_i}{\partial t^2} = \frac{\partial}{\partial x_j} \sigma_{ij}$$

where ρ is the density, u_i the components of the displacement, σ_{ij} the components of the stress tensor. There is a linear relationship between the stress tensor and the strain tensor

$$\varepsilon_{kl} = \frac{1}{2}\left(u_{k,l} + u_{l,k}\right),$$

such that:

$$\sigma_{ij} = C_{ijkl}\, \varepsilon_{kl}$$

Let us consider ω the angular frequency of the wave, and k its propagation vector, while $\mathbf{v} = k/k$ is a unit vector in the direction of k. The equations of motion simplify to:

$$\rho \omega^2 u_i = k_j k_k C_{ijkl}\, u_l$$

We look for a solution on the form of a plane wave:

$$u_i = a_i \exp\left(i\omega\left(t - \frac{v_q x_q}{V}\right)\right)$$

where V is the phase velocity, a_i the components of the polarization vector $\mathbf{a}$, v_k the unit vector of the wave vector $\mathbf{v}$. For a wave propagating in direction e_1:

$$\rho V^2 a_1 = C_{1111}\, a_1 + C_{1112}\, a_2 + C_{1113}\, a_3$$
$$\rho V^2 a_2 = C_{2111}\, a_1 + C_{2112}\, a_2 + C_{2113}\, a_3$$
$$\rho V^2 a_3 = C_{3111}\, a_1 + C_{3112}\, a_2 + C_{3113}\, a_3$$

These three equations can be written in a condensed way:

$$(\mathbf{T} - \rho V^2 \mathbf{I})\, \mathbf{a} = 0$$

where

$$T = \begin{pmatrix} C_{1111} & C_{1112} & C_{1113} \\ C_{2111} & C_{2112} & C_{2113} \\ C_{3111} & C_{3112} & C_{3113} \end{pmatrix}$$

is the Christoffel matrix, I the identity matrix. That is an eigenvalue problem which has 3 real positive roots for ρV^2 with orthogonal eigenvectors. Therefore, these equations demonstrate that there are 3 body waves in every direction of phase propagation with orthogonal particle motion and with velocities which in general are different and vary with direction. In case of isotropic medium the elastic tensor is invariant with rotation, and T is a diagonal matrix:

$$T = \begin{pmatrix} \lambda + 2\mu & 0 & 0 \\ 0 & \mu & 0 \\ 0 & 0 & \mu \end{pmatrix}$$

The well-known isotropic velocities can be written immediately:

$$V = V_P = \sqrt{\frac{\lambda + 2\mu}{\rho}}, \quad \text{and} \quad V_P = \sqrt{\frac{\mu}{\rho}} \quad \text{(repeated root)}.$$

In case of weak anisotropy, we can define quasi P-waves and quasi S-waves, but in case of very strong anisotropy, the eigenvectors can have any orientation which can be very different from quasi P- and quasi S-waves. An important consequence is that the propagation of energy of a plane wave is no longer in the direction of the propagation vector.

APPENDIX B. TENSORS AND MATRICES MANIPULATIONS

The effect of intrinsic anisotropy has been extensively studied in laboratory and in the field. Most crystals present in the Earth are strongly anisotropic (olivine, pyroxenes) except garnet.

A medium is perfectly elastic when a linear relationship (Hooke's law) relates the stress tensor σ_{ij} and the strain tensor ε_{kl}. A general elastic medium is anisotropic since its physical properties are dependent on its geometrical orientation. The anisotropic properties of any elastic medium are usually described by its fourth-order elastic tensor C_{ijkl} which linearly relates the stress tensor σ and the strain tensor ε, as seen in the previous section.

Contraction of indices. Instead of keeping 4 indices, it is common, for sake of simplicity to write the elastic tensor C_{ijkl} as a matrix (6 ×6) by using the following rules (the reader must be aware that for avoiding mathematical errors, it is easier to perform calculations by using the fourth-order elastic tensor C_{ijkl}):

c_{ijkl} C_{pq}
(3×3×3×3)   (6×6)

$p = i = j$ if $i = j$; $q = k = l$ if $k = l$
$p = 9-i-j$ if $i \neq j$; $q = 9-k-l$ if $k \neq l$

This is the Voigt notation. It allows to obtain simplified matrix representations for elastic tensors:

- Isotropic elastic tensor (2 independent parameters):

$$\begin{pmatrix} \lambda+2\mu & \lambda & \lambda & 0 & 0 & 0 \\ 0 & \lambda+2\mu & \lambda & 0 & 0 & 0 \\ 0 & 0 & \lambda+2\mu & 0 & 0 & 0 \\ 0 & 0 & 0 & \mu & 0 & 0 \\ 0 & 0 & 0 & 0 & \mu & 0 \\ 0 & 0 & 0 & 0 & 0 & \mu \end{pmatrix}$$

- Elastic tensor of a transversely isotropic medium with a vertical symmetry axis (5 independent parameters):

$$\begin{pmatrix} A & A-2N & F & 0 & 0 & 0 \\ A-2N & A & F & 0 & 0 & 0 \\ F & F & C & 0 & 0 & 0 \\ 0 & 0 & 0 & L & 0 & 0 \\ 0 & 0 & 0 & 0 & L & 0 \\ 0 & 0 & 0 & 0 & 0 & N \end{pmatrix}$$

- Elastic tensor of a general orthorhombic medium with three mutually perpendicular symmetry planes (9 independent parameters):

$$\begin{pmatrix} a & b & c & 0 & 0 & 0 \\ b & d & e & 0 & 0 & 0 \\ c & e & f & 0 & 0 & 0 \\ 0 & 0 & 0 & g & 0 & 0 \\ 0 & 0 & 0 & 0 & h & 0 \\ 0 & 0 & 0 & 0 & 0 & i \end{pmatrix}$$

These expressions are valid, for a given symmetry, only in a specific co-ordinate system, the crystal reference frame, where the reference axes are simply related to the symmetry elements of the medium (see Nye 1957, e.g.). In the general case, the components of the tensor are expressed in another reference frame, which can be related to the crystal reference frame by a 3-D rotation. This rotation is a linear transformation of the basis vectors of the crystal reference frame, and as any linear transformation it can be represented by a matrix A, whose components are written $A_i{}^j$. The components C'_{mnpq} of the fourth-order elastic tensor in the new reference frame, after rotation, are related to the crystal frame components C_{ijkl} as:

$$C'_{mnpq} = A_m^i A_n^j A_p^k A_q^l C_{ijkl}$$

Note that this expression is only valid for the covariant components of the elastic tensor, in the general case. For the contravariant and mixed components, e.g., see Hladik (1995). In this paper, the elastic tensor is expressed in the usual 3-D (real) Cartesian reference frame, A represents a rotation and is then an orthogonal real matrix, so the expression above is valid for all types of components.

Elastic matrix: scale of aggregates. For polycrystals, problems of averaging arise. There are different ways to make averages. Let n_m be the number of different crystalline populations, and α_q their proportion in the aggregate. Each mineral of a given population has its elasticity described by the elastic tensor C, with components C_{ijkl}^q. These quantities are called stiffnesses. We also introduce the compliance tensor S, whose components S_{ijkl} are defined by: $\varepsilon_{ij} = S_{ijkl} \sigma_{kl}$.

Different averages among the most commonly used are (Watt et al. 1976):

- Voigt average: Hypothesis of constant deformation. We obtain for the stiffnesses:

$$C_{ijkl}^V = \sum_{q=1}^{nm} \alpha_q C_{ijkl}^q$$

- Reuss average: Hypothesis of constant stress. We obtain for the compliances:

$$S_{ijkl}^V = \sum_{q=1}^{nm} \alpha_q S_{ijkl}^q$$

After some manipulations (matrix inversion in the most general case), the associated stiffnesses C_{ijkl}^R can be calculated (Nye 1957).

- Voigt-Reuss-Hill average: It is only the arithmetic mean of the Voigt and Reuss averages, and has no theoretical justification.
- Hashin-Shtrikman: Minimization of the elastic energy (no analytical solution in the most general case).

A more thorough description of these averages as well as others can be found in Mainprice et al. (2000).

REFERENCES

Aki K, Kaminuma K (1963) Phase velocity of Love waves in Japan (part 1): Love waves from the Aleutian shock of March 1957. Bull Earthquake Res Inst 41:243-259
Anderson DL (1961) Elastic wave propagation in layered anisotropic media. J Geophys Res 66:2953-2963
Anderson DL (1989) Theory of the Earth. Blackwell Scientific Publications, Oxford, UK
Anderson DL, Bass JD (1984) Mineralogy and composition of the upper mantle. Geophys Res Lett 1: 637-640
Anderson DL, Bass JD (1986) Transition region of the Earth's upper mantle. Nature 320:321-328
Anderson DL, Regan J (1983) Upper mantle anisotropy and the oceanic lithosphere. Geophys Res Lett 10:841-844
Ando M (1984) ScS polarization anisotropy around the Pacific Ocean. J Phys Earth 32:179-196
Ando M, Ishikawa Y, Yamazaki F (1983) Shear wave polarization anisotropy in the upper mantle beneath Honshu, Japan. J Geophys Res 88:5850-5864
Ansel V, Nataf HC (1989) Anisotropy beneath 9 stations of the Geoscope broadband network as deduced from shear wave splitting. Geophys Res Lett 16:409-412
Avouac JP, Tapponnier P (1993) Kinematic model of active deformation in central Asia. Geophys Res Lett 20:895-898
Babuska V, Cara M (1991) Seismic Anisotropy in the Earth. Kluwer Academic Press, Dordrecht, The Netherlands
Babuska V, Montagner JP, Plomerova J, Girardin N (1998) Age-dependent large-scale fabric of the mantle lithosphere as derived from surface-wave velocity anisotropy. Pure Appl Geophys 151:257-280
Backus GE (1962) Long wave elastic anisotropy produced by horizontal layering. J Geophys Res 67: 4427-4440
Backus GE (1965) Possible forms of seismic anisotropy of the upper mantle under oceans. J Geophys Res 70:3249-3439
Barruol G, Hoffman R (1999) Upper mantle anisotropy beneath the GEOSCOPE stations. J Geophys Res 104:10757-10774
Blackman DK, Kendall JM, Dawson PR, Wenk HR, Boyce D, Morgan JP (1996) Teleseismic imaging of subaxial flow at mid-ocean ridges: Travel-time effects of anisotropic mineral texture in the mantle. Geophys J Intl 127:415-426

Bostock MG (1997) Anisotropic upper-mantle stratigraphy and architecture of the Slave craton. Nature 390:392-395
Bowman JR, Ando M (1987) Shear-wave splitting in the upper mantle wedge above the Tonga subduction zone. Geophys J R Astron Soc 88:25-41
Cara M, Leveque JJ (1988) Anisotropy of the asthenosphere: The higher mode data of the Pacific revisited. Geophys Res Lett 15:205-208
Chastel YB, Dawson PR, Wenk HR, Bennett K (1993) Anisotropic convection with implications for the upper mantle. J Geophys Res 98:17757-17771
Christensen NI, Lundquist S (1982) Pyroxene orientation within the upper mantle. Bull Geol Soc Am 93:279-288
Crampin S (1984) An introduction to wave propagation in anisotropic media. Geophys J R Astron Soc 76:17-28
Crampin S, Booth DC (1985) Shear-wave polarizations near the North Anatolian fault, II: Interpretation in terms of crack-induced anisotropy. Geophys J R Astron Soc 83:75-92
Crampin S, Volti T, Stefansson R (1999) A successfully stress-forecast earthquake. Geophys J Intl 138: F1-F5
Debayle E (1999) SV-wave azimuthal anisotropy in the Australian upper mantle: preliminary results from automated Rayleigh waveform inversion. Geophys J Intl 137:747-754
Debayle E, Lévêque JJ (1997) Upper mantle heterogeneities in the Indian Ocean from waveform inversion. Geophys Res Lett 24:245-248
Debayle E, Kennett BLN (2000) Anisotropy in the Australasian upper mantle from Love and Rayleigh waveform inversion. Earth Planet Sci Lett 184:339-351
Dziewonski AM, Anderson DL (1981) Preliminary Reference Earth Model. Phys Earth Planet Inter 25: 297-356
Ekström G, Dziewonski AM (1998) The unique anisotropy of the Pacific upper mantle. Nature 394:168-172
England P, Houseman G (1986) Finite strain calculations of continental deformation, 2. Comparison with the India-Asia collision zone. J Geophys Res 91:3664-3676
Estey LH, Douglas BJ (1986) Upper mantle anisotropy: A preliminary model. J Geophys Res 91:11393-11406
Farra V, Vinnik LP (1994) Shear-wave splitting in the mantle of the Pacific. Geophys J Intl 119:195-218
Fedorov F (1968) Theory of Elastic waves in crystals. Plenum Press, New York
Fischer KM, Wiens DA (1996) The depth distribution of mantle anisotropy beneath the Tonga subduction zone. Earth Planet Sci Lett 142:253-260
Fischer KM, Yang X (1994) Anisotropy in Kuril-Kamtchatka subduction zone structure. Geophys Res Lett 21:5-8
Forsyth DW (1975) The early structural evolution and anisotropy of the oceanic upper mantle. Geophys J R Astron Soc 43:103-162
Forsyth DW, Webb SC, Dorman LM, Shen Y (1998) Phase velocities of Rayleigh waves in MELT experiment on the East Pacific Rise. Science 280:1235-1238
Fouch MJ, Fischer KM (1996) Mantle anisotropy beneath northwest Pacific subduction zones. J Geophys Res 101:15987-16002
Fouch MJ, Fischer KM, Parmentier EM, Wysession ME, Clarke TJ (2000) Shear wave splitting, continental keels, patterns of mantle flow. J Geophys Res 105:6255-6275
Fouch MJ, Fischer KM, Wysession ME (2001) Lowermost mantle anisotropy beneath the Pacific: Imaging the source of the Hawaiian plume. Earth Planet Sci Lett 190:167-180
Fukao Y (1984) Evidence from core- reflected shear waves for anisotropy in the Earth's mantle. Nature 309:695-698
Gaboret C, Forte A, Montagner JP (2002) The unique dynamics of the Pacific hemisphere mantle and its signature on seismic anisotropy. Earth Planet Sci Lett (submitted)
Gaherty JB, Jordan TH (1995) Lehmann discontinuity as the base of the anisotropic layer beneath continents. Science 268:1468-1471
Garnero EJ, Lay T (1996) Lateral variation in lowermost mantle shear wave anisotropy beneath the North Pacific and Alaska. J Geophys Res 102:8121-8135
Griot DA, Montagner JP, Tapponnier P (1998a) Surface wave phase velocity and azimuthal anisotropy in Central Asia. J Geophys Res 103:21215-21232
Griot DA, Montagner JP, Tapponnier P (1998b) Heterogeneous versus homogeneous strain in Central Asia. Geophys Res Lett 25:1447-1450
Hadiouche O, Jobert N, Montagner JP (1989) Anisotropy of the African continent inferred from surface waves. Phys Earth Planet Inter 58:61-81
Hager B, O'Connell R (1979) Kinematic models of large-scale flow in the Earth's mantle. J Geophys Res 84:1031-1048

Helbig K (1994) Foundations of Anisotropy for Exploration Seismics. Pergamon/Elsevier Science, Oxford, UK

Hess H (1964) Seismic anisotropy of the uppermost mantle under the oceans. Nature 203:629-631

Hladik J (1995) Le calcul tensoriel en physique. Paris, Masson, France.

Hirn A, Jiang M, Sapin M, Diaz J, Nercessian A, Lu QT, Lépine JC, Shi DN, Sachpazi M, Pandy MR, Ma K, Gallart J (1995) Seismic anisotropy as an indicator of mantle flow beneath the Himalayas and Tibet. Nature 375:571-574

Iidaka T, Niu F (1998) Evidence for an anisotropic lower mantle beneath eastern Asia: comparison of shear wave splitting data of SKS and P660s. Geophys Res Lett 25:675-678

Jordan TH (1978) Composition and development of the continental tectosphere. Nature 274:544-548

Jordan TH (1981) Continents as a chemical boundary layer. Phil Trans R Soc London Ser A 301:359-373

Jung HY, Karato SI (2001) Water-induced fabric transitions in olivine. Science 293:1460-1462

Kaminski E, Ribe NM (2001) A kinematic model for recrystallization and texture development in olivine polycrystals. Earth Planet Sci Lett 189:253-267

Karato S-i (1989) Seismic anisotropy: mechanisms and tectonic implications. *In* Rheology of Solids and of the Earth. S-i Karato, M Toriumi (eds) Oxford University Press, Oxford, p 393-42

Karato SI, Li P (1993) Diffusive creep in perovskite: Implications for the rheology of the lower mantle. Science 255:771-778

Kendall JM (2000) Seismic anisotropy in the boundary layers of the mantle. *In* Earth's Deep Interior: Mineral physics and tomography from the atomic to the global scale. Karato S-i, Forte A, Liebermann RC, Masters G, Stixrude L (eds) Am Geophys Union Monograph, Washington, DC

Kendall JM, Silver PG (1996) Constraints from seismic anisotropy on the nature of the lowermost mantle. Nature 381:409-412

Kennett BLN, Engdahl ER (1991) Traveltimes for global earthquake location and phase identification. Geophys J Intl 105:429-465

Larson EWF, Tromp J, Ekstrom G (1998) Effects of slight anisotropy on surface waves. Geophys J Intl 132:654-666

Laske G, Masters G (1998) Surface-wave polarization data and global anisotropic structure. Geophys J Intl 132:508-520

Lavé J, Avouac JP, Lacassin R, Tapponnier P, Montagner JP (1996) Seismic anisotropy beneath Tibet: Evidence for eastward extrusion of the Tibetan lithosphere? J Geophys Res 140:83-96

Lay T, Williams Q, Garnero EJ, Kellog L, Wysession ME (1998) Seismic wave anisotropy in the D" region and its implications. *In* Core-Mantle Boundary Region. Gurnis M, Wysession ME, Knittle E, Buffett B (eds) Geodynamic Series 28:229-318

Lévêque JJ, Cara M (1985) Inversion of multimode surface wave data: evidence for sub--lithospheric anisotropy. Geophys J R Astron Soc 83:753-773

Lévêque JJ, Cara M, Rouland D (1991) Waveform inversion of surface-wave data: a new tool for systematic investigation of upper mantle structures. Geophys J Intl 104:565-581

Levshin A, Ratnikova L (1984) Apparent anisotropy in inhomogeneous media. Geophys J R Astron Soc 76:65-69

Love AEH (1927) A Treatise on the Theory of Elasticity, 4th Edition. Cambridge University Press, Cambridge, UK

Mainprice DG, Barruol G, Ben Ismail W (2000) The seismic anisotropy of the Earth's. mantle: From single crystal to Polycrystal. *In* Earth's Deep Interior: Mineral Physics and Tomography From the Atomic Scale to the Global Scale. Geophys Monogr 117:237-264

Masters G, Jordan TH, Silver PG, Gilbert F (1982) Aspherical Earth structure from fundamental spheroidal-mode data. Nature 298:609-613

Matzel E, Sen MK, Grand SP (1996) Evidence for anisotropy in the deep mantle beneath Alaska. Geophys Res Lett 23:2417-2420

Maupin V (1994) On the possibility of anisotropy in the D" layer as inferred from the polarization of diffracted S-waves. Phys Earth Planet Inter 87:1-32

McEvilly TV (1964) Central U.S. crust-upper mantle structure from Love and Rayleigh wave phase velocity inversion. Bull Seism Soc Am 54:1997-2015

McKenzie D (1979) Finite deformation during fluid flow. Geophys J R Astron Soc 58:687-715

McNamara AK, Karato SI, van Keken PE (2001) Localization of dislocation creep in the lower mantle; implications for the origin of seismic anisotropy. Earth Planet Sci Lett 191:85-99

McNamara AK, van Keken PE, Karato SI (2002) Development of anisotropic structure in the Earth's lower mantle by solid-state convection. Nature 416:310-314

Meade C, Silver PG, Kaneshima S (1995) Laboratory and seismological observations of lower mantle anisotropy. Geophys Res Lett 22:1293-1296

Miller V, Savage S (2001) Changes in seismic anisotropy after volcanic eruptions: Evidence from Mount Ruapehu. Science 293:2231-2233

Minster JB, Jordan TH (1978) Present-day plate motions. J Geophys Res 83:5331-5354
Mitchell BJ, Yu GK (1980) Surface wave dispersion, regionalized velocity models and anisotropy of the Pacific crust and upper mantle. Geophys J R Astron Soc 63:497-514
Mochizuki E (1986) The free oscillations of an anisotropic and heterogeneous Earth. Geophys J R Astron Soc 86:167-176
Montagner JP (1985) Seismic anisotropy of the Pacific Ocean inferred from long-period surface wave dispersion. Phys Earth Planet Inter 38:28-50
Montagner JP (1986a) First results on the three dimensional structure of the Indian Ocean inferred from long period surface waves. Geophys Res Lett 13:315-318
Montagner JP (1986b) Regional three-dimensional structures using long-period surface waves. Ann Geophys 4:B3:283-294
Montagner JP (1994) What can seismology tell us about mantle convection? Rev Geophys 32:2:115-137
Montagner JP (1998) Where can seismic anisotopy be detected in the Earth's mantle? In boundary layers. Pure Appl Geophys 151:223-256
Montagner JP (2002) Upper mantle low anisotropy channels below the Pacific Plate. Earth Planet Sci Lett (in press)
Montagner JP, Anderson DL (1989a) Constraints on elastic combinations inferred from petrological models. Phys Earth Planet Inter 54:82-105
Montagner JP, Anderson DL (1989b) Constrained reference mantle model. Phys Earth Planet Inter 58:205-227
Montagner JP, Griot DA, Lavé J (2000) How to relate body wave and surface wave anisotropies? J Geophys Res 105:19015-19027
Montagner JP, Guillot L (2000) Seismic anisotropy tomography. In Problems in Geophysics for the Next Millennium. E Boschi, G Ekström, A Morelli (eds) Editrice Compositori, Bologna, Italy, p 217-254
Montagner JP, Jobert N (1988) Vectorial Tomography. II: Application to the Indian Ocean. Geophys J R Astron Soc 94:309-344
Montagner JP, Kennett BLN (1996) How to reconcile body-wave and normal-mode reference Earth models? Geophys J Intl 125:229-248
Montagner JP, Lognonné P, Beauduin R, Roult G, Karczewski JF, Stutzmann E (1998) Towards multiscale and multiparameter networks for the next century: The French efforts. Phys Earth Planet Inter 108:155-174
Montagner JP, Nataf HC (1986) On the inversion of the azimuthal anisotropy of surface waves. J Geophys Res 91:511-520
Montagner JP, Nataf HC (1988) Vectorial Tomography. I: Theory. Geophys J R Astron Soc 94:295-307
Montagner JP, Tanimoto T (1990) Global anisotropy in the upper mantle inferred from the regionalization of phase velocities. J Geophys Res 95:4797-4819
Montagner JP, Tanimoto T (1991) Global upper mantle tomography of seismic velocities and anisotropies. J Geophys Res 96:20337-20351
Montagner JP, Romanowicz B, Karczewski JF (1994) A first step towards an Oceanic Geophysical Observatory. EOS, Trans Am Geophys Union 75:150-154
Montagner JP, Romanowicz B (1993) Degrees 2, 4, 6 inferred from seismic tomography. Geophys Res Lett 20:631-634
Morelli A, Dziewonski AM, Woodhouse JH (1986) Anisotropy of the inner core inferred PKIKP travel times. Geophys Res Lett 13:1545-1548
Morelli A, Dziewonski AM (1993) Body wave traveltimes and a spherically symmetric P- and S-wave velocity model. Geophys J Intl 112:178-194
Nataf HC, Nakanishi I, Anderson DL (1984) Anisotropy and shear velocity heterogeneities in the upper mantle. Geophys Res Lett 11:109-112
Nataf HC, Nakanishi I, Anderson DL (1986) Measurement of mantle wave velocities and inversion for lateral heterogeneity and anisotropy, III. Inversion. J Geophys Res 91:7261-7307
Nicolas A (1993) Why fast polarization directions of SKS seismic waves are parallel to mountain belts? Phys Earth Planet Inter 78:337-342
Nicolas A, Boudier F, Boullier AM (1973) Mechanisms of flow in naturally and experimentally deformed peridototes. Am J Sci 273:853-876
Nicolas A, Christensen NI (1987) Formation of anisotropy in upper mantle peridotites: A review. In Composition, Structure and Dynamics of the Lithosphere/Asthenosphere System. Fuchs K, Froidevaux C (eds) American Geophysical Union, Washington, DC, p 111-123
Nishimura CE, Forsyth DW (1989) The anisotropic structure of the upper mantle in the Pacific. Geophys J 96:203-229
Nye JF (1985) Physical Properties of Crystals: Their Representation by Tensors and Matrices. Reprint (with corrections). Clarendon Press, Oxford, UK
Park J, Levin V (2002) Seismic anisotropy: tracing plate dynamics in the mantle. Science 296:485-489

Peselnick L, Nicolas A, Stevenson PR (1974) Velocity anisotropy in a mantle peridotite from Ivrea zone: Application to uper mantle anisotropy. J Geophys Res 79:1175-1182

Peselnick L, Nicolas A (1978) Seismic anisotropy in an ophiolite peridotite: Application to oceanic upper mantle. J Geophys Res 83:1227-1235

Pettersen O, Maupin V (2002) Lithospheric anisotropy on the Kerguelen hotspot track inferred from Rayleigh wave polarisation anomalies. Geophys J Intl 149:225-246

Ribe NM (1989) Seismic anisotropy and mantle flow. J Geophys Res 94:4213-4223

Ribe NM, Yu Y (1991) A theory for plastic deformation and textural evolution of olivine polycrystals. J Geophys Res 96:8325-8335

Ricard Y, Nataf HC, Montagner JP (1996) The 3S-Mac model: confrontation with seismic data. J Geophys Res 101:8457-8472

Ringwood AE (1975) Composition and Petrology of the Earth's Mantle. McGraw-Hill, New York, 618 p

Ritsema J (2000) Evidence for shear wave anisotropy in the lowermost mantle beneath the Indian Ocean. Geophys Res Lett 27:1041-1044

Roult G, Rouland D, Montagner JP (1994) Antarctica II: Upper mantle structure from velocity and anisotropy. Phys Earth Planet Inter 84:33-57

Rümpker G, Silver PG (1998) Apparent shear-wave splitting in the presence of vertically varying anisotropy. Geophys J Intl 135:790-800

Russo R, Silver PG (1994) Trench-parallel flow beneath the Nazca plate from seismic anisotropy. Science 263:1105-1111

Russo RM, Okal EA (1999) Shear wave splitting and upper mantle deformation in French Polynesia: Evidence for small-scale heterogeneity related to the Society hotspot. J Geophys Res 103:15089-15107

Savage MK (1999) Seismic anisotropy and mantle deformation: What have we learned from shear wave splitting? Rev Geophys 37:65-106

Schlue JW, Knopoff L (1977) Shear--wave polarization anisotropy in the Pacific Ocean. Geophys J R Astron Soc 49:145-165

Silver PG (1996) Seismic anisotropy beneath the continents: Probing the depths of geology. Ann Rev Earth Planet Sci 24:385-432

Silver PG, Chan WW (1988) Implications for continental structure and evolution from seismic anisotropy. Nature 335:34-39

Silver PG, Chan WW (1991) Shear wave splitting and subcontinental mantle deformation. J Geophys Res 96:16429-16454

Silver PG, Holt WE (2002) The mantle flow field beneath North America. Science 295:1054-1057

Silveira G, Stutzmann E, Montagner JP (1998) Mendes-Victor L. Anisotropic tomography of the Atlantic Ocean from Rayleigh surface waves. Phys Earth Planet Inter 106:259-275

Simmons FJ, van der Hilst R, Montagner JP, Zielhuis A (2002) Multimode Rayleigh wave inversion for shear wave speed heterogeneity and azimuthal anisotropy of the Australian upper mantle. Geophys J Intl (in press)

Singh S, Taylor M, Montagner JP (2002) On the presence of liquid in Earth's inner core. Science 287:2471-2474

Smith ML, Dahlen FA (1973) The azimuthal dependence of Love and Rayleigh wave propagation in a slightly anisotropic medium. J Geophys Res 78:3321-3333

Smith ML, Dahlen FA (1975) Correction to 'The azimuthal dependence of Love and Rayleigh wave propagation in a slightly anisotropic medium'. J Geophys Res 80:1923

Suetsugu D, Nakanishi I (1987) Regional and azimuthal dependence of phase velocities of mantle Rayleigh waves in the Pacific Ocean. Phys Earth Planet Inter 47:230-245

Tanimoto T (1986) Free oscillations in a slightly anisotropic Earth. Geophys J R Astron Soc 87:493-517

Tanimoto T, Anderson DL (1985) Lateral heterogeneity and azimuthal anisotropy of the upper mantle: Love and Rayleigh waves 100-250s. J Geophys Res 90:1842-1858

Tommasi A, Mainprice D, Canova G, Chastel Y (2000) Viscoplastic sel-consistent and equilibrium-based modeling of oilivine preferred orientations: Implications for the upper mantle anisotropy. J Geophys Res 105:7893-7908

Tommasi A, Vauchez A, Russo R (1996) Seismic anisotropy in ocean basins: Resistive drag of the sublithospheric mantle? Geophys Res Lett 23:2991-2994

Trampert J, Woodhouse JH (1996) Global phase velocity maps of Love and Rayleigh waves between 40 and 150 seconds. Geophys J Intl 212:675-690

Trampert J, van Heijst HJ (2002) Global anisotropy azimythal anisotropy in the transition zone. Science 296:1297-1299

van der Hilst RD, Karason H (1999) Compositional heterogeneity in the bottom 1000 km of the Earth's mantle: Toward a hybrid convection model. Science 283:1885-1888

Vauchez A, Nicolas A (1991) Mountain building: strike-parallel motion and mantle anisotropy. Tectonophysics 185:183-191

Vinnik LP, Chevrot S, Montagner JP (1998) Seismic evidence of flow at the base of the upper mantle. Geophys Res Lett 25:1995-1998

Vinnik LP, Farra V, Romanowicz B (1989b) Azimuthal anisotropy in the earth from observations of SKS at GEOSCOPE and NARS broadband stations. Bull Seism Soc Am 79:1542-1558

Vinnik LP, Kosarev GL, Makeyeva LI (1984) Anisotropiya litosfery po nablyudeniyam voln SKS and SKKS. Dokl Akad Nauk USSR 278:1335-1339

Vinnik LP, Kind R, Kosarev GL, Makeyeva LI (1989a) Azimuthal Anisotropy in the lithosphere from observations of long-period S-waves. Geophys J Intl 99:549-559

Vinnik LP, Makayeva LI, Milev A, Usenko AY (1992) Global patterns of azimuthal anisotropy and deformations in the continental mantle. Geophys J Intl 111:433-447

Vinnik LP, Montagner JP (1996) Shear wave splitting in the mantle from Ps phases. Geophys Res Lett 23:2449-2452

Vinnik LP, Romanowicz B, Le Stunff Y, Makayeva L (1995) Seismic anisotropy in D"-layer. Geophys Res Lett 22:1657-1660

Vinnik LP, Montagner JP, Girardin N, Dricker I (2002) Saul, Shear wave splitting at H2O: A comment. Geophys Res Lett (submitted)

Walker KT, Bokelmann GH, Klemperer SL (2001) Shear-wave splitting to test mantle Deformation models around Hawaii. Geophys Res Lett 28:4319-4322

Watt JP, Davies GF, O'Connell (1976) The elastic properties of composite materials. Rev Geophys Space Phys 14:541-565

Widmer R, Masters G, Gilbert F (1993) Spherically symmetric attenuation within the Earth from normal mode data. Geoph J Intl 104:541-553

Wolfe JW, Silver PG (1998) Seismic anisotropy of oceanic upper mantle: Shear wave splitting methodologies and observations. J Geophys Res 103:749-771

Woodhouse JH, Giardini D, Li XD (1986) Evidence for inner core anisotropy from free oscillations. Geophys Res Lett 13:1549-1552

Wookey J, Kendall JM, Barruol G (2002) Mid-mantle deformation inferred from seismic anisotropy. Nature 415:777-780

Yu Y, Park J (1993) Anisotropy and coupled long-period surface waves. Geophys J Intl 114:473-489

Zhang S, Karato S-i (1995) Lattice preferred orientation of olivines aggregates deformed in simple shear. Nature 375:774-777

13 Theoretical Analysis of Shear Localization in the Lithosphere

David Bercovici and Shun-ichiro Karato

Department of Geology and Geophysics
Yale University
New Haven, Connecticut 06511

INTRODUCTION

Deformation in the Earth is rarely homogeneous and often occurs in narrow regions of concentrated strain referred to as zones of shear localization. Mylonites found in the continental crust are the manifestation of shear localization (White et al. 1980) and, at the largest scale, shear localization is proposed to be crucial for the generation of tectonic plates from a convecting mantle (Bercovici 1993, 1995b,a, 1996, 1998; Bercovici et al. 2000; Trompert and Hansen 1998; Tackley 1998, 2000a,b,c,d). Shear localization is apparent in many field of physics, including, for example, metallurgy (Lemonds and Needleman 1986), rock mechanics (Poirier 1980; Jin et al. 1998), granular dynamics (Scott 1996; Géminard et al. 1999, e.g.) and glaciology (Yuen and Schubert 1979). However, basic solid-state rheologies, such as elasticity, visco-elasticity, viscous flow and even steady-state non-Newtonian viscous flow, are insufficient by themselves to generate shear-localization; this is because in all such rheologies an increase in deformation or rate of deformation results in greater resistance (i.e., stress) instead of self-weakening and loss of strength. Shear-localization tends to require dynamic feedback mechanisms wherein self-weakening is controlled by a macroscopic variable (such as temperature) or microscopic structure (grain size or microcrack density) whose evolution and concentration are themselves determined by deformation; in this way deformation can induce weakening, which subsequently causes deformation to concentrate on the weak zone (being most easily deformed), causing further weakening, and thus more focusing of deformation, and so on.

One of the most fundamental manifestations of such feedback mechanisms arises from the coupling of viscous heating and temperature-dependent viscosity wherein the zone of dissipative heating weakens and thus focuses deformation, leading to further heating and weakening; this mechanism is thought applicable to problems in metal weakening, glacial surges, and lithosphere dynamics (Schubert and Turcotte 1972; Yuen and Schubert 1979; Poirier 1980; Balachandar et al. 1995; Bercovici 1996; Thatcher and England 1998). In granular media, localization is due to dilation of the medium leading to effectively weaker rarifed zones that concentrate deformation, which in turn agitates the medium causing further dilation (Géminard et al. 1999); this phenomenon is of potential importance in earthquake dynamics (Scott 1996; Marone 1998; Segall and Rice 1995; Sleep 1995, 1997; Mora and Place 1998). Another shear localization mechanism arises from the reduction of grain size under stress in solid-state creep mechanisms which have grainsize-dependent viscosities (i.e., grain reduction leads to zones of weakness which concentrate deformation, increasing stress and thus further reducing grain size and enhancing weakening); this mechanism has been proposed to be the cause for mylonitic shear zones and is thought to be a basic ingredient of crustal and lithospheric deformation (Kameyama et al. 1997; Jin et al. 1998; Braun et al. 1999). Finally, material that undergoes brittle or combined brittle-ductile deformation can, with large strain, experience concentration of microcrack and void populations developing weak bands on which deformation focuses, leading to further cracking, weakening, and invariably shear localization. This phenomenon occurs in the process of dilatant plasticity in metals as well as crustal rocks (Lemonds and Needleman 1986; Ashby and Sammis 1990; Lemaitre 1992; Hansen and Schreyer 1992; Lockner 1995; Mathur et al. 1996; Krajcinovic 1996; Lyakhovsky et al. 1997; Regenauer-Lieb 1999). The inference that much of the lithosphere undergoes combined brittle-ductile behavior (Kohlstedt et al. 1995; Evans and Kohlstedt 1995), combined with the longevity of plate boundaries (Gurnis et al. 2000), have motivated some (Bercovici 1998; Tackley 1998; Bercovici et al. 2001a,b; Bercovici and Ricard 2003; Auth et al. 2002) to consider that the primary weakening mechanism that leads to plate boundary formation is due to damage, that is, semi-ductile void and microcrack formation.

In this paper we examine in some mathematical detail the fundamental physics of shear-localizing feedback mechanisms by using the basic model of a simple-shear layer. Such a layer could represent anything from a strike-slip shear zone to a subducting slab to a glacial gravity flow, etc. Simple one-dimensional shear-flows have been considered many times before, primarily for viscous-heating and thermal-runaway effects (e.g., Yuen and Schubert 1977; Schubert and Yuen 1978; Fleitout and Froidevaux 1980). Here, we will use this model to examine three basic feedback mechanisms thought relevant to the mantle and lithosphere, i.e., (1) viscous heating with temperature-dependent viscosity; (2) a simple damage formulation to model semiductile-semibrittle behavior; and (3) grainsize-dependent rheology and recrystallization. (While we focus on only three basic mechanisms, a more complete survey of many shear-localizing mechanisms, but with less mathematical detail, is provided by Regenauer-Lieb and Yuen (2002).) An important aspect of any feedback mechanism is not only how a rheology-controlling macroscopic or microscopic property is enhanced by deformation, but also how that property is restored, lost, equilibrated or "healed"; thus we will also consider some basic healing mechanisms. After our foray into simply theory, we will discuss recent progress on some of the more complicated aspects of these feedback mechanisms, and will close with a brief summary.

THEORETICAL PRELIMINARIES

To illustrate the essential physics of shear localization with various rheological mechanisms, we consider a model of unidirectional simple-shear flow in an infinitely long channel. The channel is infinite in the x and z directions and is $2L$ wide, with boundaries at $y = \pm L$ that move in opposite directions along the x direction. Since the channel is uniform and infinite in x and z, none of the properties of the medium (stress, velocity, temperature, etc.) are functions of x and z, but only of y and time t. In this case, the general force balance equation (i.e., assuming slow creeping flow and thus negligible acceleration terms) is

$$0 = -\nabla P + \nabla \cdot \underline{\sigma} + \rho \mathbf{g} \tag{1}$$

where P is pressure or isotropic stress, $\underline{\sigma}$ the deviatoric stress tensor, ρ the density and $\mathbf{g}$ the gravitational acceleration vector. The component of this equation in the x direction, given that there is no dependence on x and z (assuming that $\mathbf{g}$ is only in the z direction) becomes

$$0 = \frac{\partial \sigma_{yx}}{\partial y} \quad (2)$$

where σ_{yx} is the shear stress that is acting on surfaces facing in the y direction but pulling on them in the x direction. The above equation implies that $\sigma_{yx} = \sigma$ which is a constant in y.

For viscous flows, shear localization is measured in terms of strain-rate,

$$\underline{\dot{\varepsilon}} = \tfrac{1}{2}(\boldsymbol{\nabla} \boldsymbol{v} + [\boldsymbol{\nabla} \boldsymbol{v}]^t) \quad (3)$$

where $\underline{\dot{\varepsilon}}$ is the strain-rate tensor and $\boldsymbol{v}$ the velocity vector. In our example $\boldsymbol{v} = u(y)\hat{\boldsymbol{x}}$ where u is the velocity in the x direction and $\hat{\boldsymbol{x}}$ the unit vector in the x direction; thus the strain-rate tensor has only one unique component $\dot{\varepsilon}_{yx} = \tfrac{1}{2}\frac{\partial u}{\partial y}$.

We make the simplifying assumption that stress is related to strain-rate plus only one other variable which we'll denote for now by Θ which is either temperature, void fraction or grainsize. The constitutive relation between stress and strain rate is thus

$$\underline{\sigma} = 2\mu(\Theta, \dot{\varepsilon})\underline{\dot{\varepsilon}} + \eta \boldsymbol{\nabla} \cdot \boldsymbol{v} \underline{I} \quad (4)$$

where μ is viscosity, $\dot{\varepsilon}^2 = \tfrac{1}{2}\underline{\dot{\varepsilon}} : \underline{\dot{\varepsilon}}$ (which is related to the 2nd strain-rate invariant), and η is a material property which includes the bulk viscosity. In our example, since $\boldsymbol{v} = u(y)\hat{\boldsymbol{x}}$ there is ony one stress and one strain-rate and our constitutive law becomes

$$\sigma = 2\mu(\Theta, \dot{\varepsilon})\dot{\varepsilon} \quad (5)$$

where $\dot{\varepsilon} = \tfrac{1}{2}\frac{\partial u}{\partial y}$.

If the relation between stress and strain-rate is monotonic (i.e., an increase in one always causes an increase in the other), and there is no feedback between strain-rate and the state variable Θ, then there can be no localization in our example. In particular, say we employ a power-law rheology, then $\mu = \tfrac{1}{2}A\dot{\varepsilon}^{\frac{1-n}{n}}$ where A is a constant and n is the power-law index. For pseudo-plastics (like many food products such as mustard and ketchup, and silicates in the dislocation-creep regime) $n > 1$ so that viscosity decreases with increasing strain-rate. However, stress and thus resistance to motion, or strength, $\sigma = A\dot{\varepsilon}^{1/n}$ always increases with strain-rate, even if the viscosity decreases. Moreover, in our example, $\dot{\varepsilon} = (\sigma/A)^n$ is a constant across the layer because σ and A are constants, and thus without any heterogeneity in strain-rate, there is no shear-localization anywhere in the layer.

For the model system to develop shear-localization, it must allow non-constant strain-rate across the layer (the localization is a region of high strain-rate surrounded by low-strain rate). This can be accomplished in our model if the effective stress–strain-rate relation is non-monotonic (i.e., there is a maximum or minimum stress), such that one stress can yield at least two-possible strain-rates. As we show below, such an effective rheology generally arises from a feedback between the state variable Θ and the stress or strain-rate.

SHEAR LOCALIZING FEEDBACK MECHANISMS

One requirement of a feedback mechanism that permits shear-localization is that the evolution of Θ must depend on the deformation, i.e., strain or strain-rate of the system; in our examples we will focus on completely irrecoverable deformation and thus work with strain-rate (i.e., we will take a viscous-flow approach). Thus, for a feedback mechanism to exist, an increase in strain-rate must lead to a change in Θ that subsequently reduces the viscosity, causing (through Eqn. 5) yet larger strain-rate, and so on. We will examine our three basic feedback mechanisms (thermal, simple-damage and grainsize controlled) for simple decay-type loss mechanisms that facilitate the analysis. The decay-loss mechanism, however, is not always realistic in the case of thermally-controlled, and possibly even damage-controlled feedback mechanisms, wherein diffusive loss is as important; therefore, we will analyze these feedbacks for diffusive loss as well, which leads to different mathematical solutions but the interpretation is related to the simpler decay-loss cases.

Thermal feedback with decay-loss healing

In considering the feedback between viscous heating and temperature-dependent viscosity, we define Θ to be temperature T. The evolution equation for T involving temporal change, advective transport, decay loss and a deformation-related source term is

$$\rho c_p \left(\frac{\partial T}{\partial t} + \boldsymbol{v} \cdot \boldsymbol{\nabla} T \right) = -K(T - T_0) + 2\mu(T, \dot{\varepsilon})\dot{\varepsilon}^2 \tag{6}$$

where ρ is density, c_p is heat capacity, K is a heat-transfer coefficient, and T_0 is the value of T to which the unforced system equilibrates (e.g., it could be the temperature of heat reservoirs in contact with the boundaries of the channel). A heat-transfer coefficient can be used instead of diffusive loss if the layer thickness L is very small (in which case $K \sim k/L^2$ where k is thermal conductivity). The last term on the right represents the deformational work, which for simple single-phase systems goes into dissipative heating and thus an increase in internal/thermal energy (i.e., via an increase in entropy). As discussed in the next subsection, in multi-phase systems, part of this work can also go into creation of surface energy by, say, creating new microcracks, voids and/or defects.

Given the nature of our simple one-dimensional system, Equation (6) becomes

$$\rho c_p \frac{\partial T}{\partial t} = -K(T - T_0) + 2\mu(T, \dot{\varepsilon})\dot{\varepsilon}^2 \tag{7}$$

Let us first consider steady-state solutions such that Equation (7) becomes

$$K(T - T_0) = 2\mu(T, \dot{\varepsilon})\dot{\varepsilon}^2 \tag{8}$$

To solve this system we must specify the viscosity law for $\mu(T, \dot{\varepsilon})$. Materials that undergo solid-state creep (diffusion or dislocation creep) have a temperature-dependence of viscosity given by an Arrhenius factor; for our 1-D system the viscosity law can be written in as

$$\mu = \tfrac{1}{2} A e^{\frac{H}{nRT}} \dot{\varepsilon}^{\frac{1-n}{n}} \tag{9}$$

where A is constant, H is the activation enthalpy, R the gas constant and n the power-law index; for Newtonian diffusion creep $n = 1$, while typical values for non-Newtonian dislocation creep in rocks is $n = 3$ to 5. We can reduce the number of free parameters by defining variables in terms of the natural scales of the system. We define $T = T_0 + T^*\theta$ and $\dot{\varepsilon} = \dot{\varepsilon}^*\dot{\varepsilon}'$ where T^* and $\dot{\varepsilon}^*$ are as-yet-unknown scales, and θ and $\dot{\varepsilon}'$ are dimensionless temperature and strain-rate, respectively. Substituting these expressions into Equation (8) while using Equation (9) leads to (dropping primes for simplicity)

$$\left[\frac{KT^*}{A(\dot{\varepsilon}^*)^{\frac{1+n}{n}}}\right]\theta = e^{\left[\frac{H}{RT^*}\right]\frac{1}{n(\theta_0+\theta)}}\dot{\varepsilon}^{\frac{1+n}{n}} \qquad (10)$$

where $\theta_0 = T_0/T^*$. Likewise, the constitutive law Equation (5)–along with Equation (9)– becomes

$$\left[\frac{\sigma}{A(\dot{\varepsilon}^*)^{1/n}}\right] = e^{\left[\frac{H}{RT^*}\right]\frac{1}{n(\theta_0+\theta)}}\dot{\varepsilon}^{\frac{1}{n}} \qquad (11)$$

The expressions in square brackets are dimensionless groups, and we can choose T^* and $\dot{\varepsilon}^*$ to make two of these go to unity. In particular, we define

$$T^* = \frac{H}{R} \quad \text{and} \quad \dot{\varepsilon}^* = \left(\frac{KH}{RA}\right)^{\frac{n}{n+1}} \qquad (12)$$

such that Equation (10) and Equation (11) become our steady-state governing equations:

$$\theta = \left(e^{\frac{1}{\theta_0+\theta}}\dot{\varepsilon}^{n+1}\right)^{\frac{1}{n}} \qquad (13)$$

and

$$\tau = \left(e^{\frac{1}{\theta_0+\theta}}\dot{\varepsilon}\right)^{\frac{1}{n}} \qquad (14)$$

respectively, where

$$\tau = \sigma\left(\frac{R}{A^n KH}\right)^{\frac{1}{n+1}} \qquad (15)$$

Although we have reduced the number of free parameters, our system still depends on the dimensionless boundary or background temperature θ_0 and the power-law index n. Note that since we have chosen $H/R \approx 6 \times 10^4$ K as our temperature scale (i.e., $H \approx 500$ kJ/mol and and $R \approx 8$ J/K/mol), then our typical backround lithospheric nondimensional temperature θ_0 will be of order 10^{-2}. We will, however, also examine higher background temperatures for the sake of comparison.

Equation (13) and Equation (14) lead to a parametric equation for stress and strain-rate (i.e., τ and $\dot{\varepsilon}$ are functions of each other through the "parameter" θ)

$$\tau = \left(\theta e^{\frac{1}{\theta_0+\theta}}\right)^{\frac{1}{n+1}}, \quad \dot{\varepsilon} = \left(\theta^n e^{-\frac{1}{\theta_0+\theta}}\right)^{\frac{1}{n+1}} \qquad (16)$$

which is thus an effective constitutive law. Curves for the effective constitutive relations for τ versus $\dot{\varepsilon}$ for various n and θ_0 are shown in Figure 1. In this case a variety of behavior is available. For large θ_0 one can obtain a simple power-law like behavior wherein stress monotonically increases with strain-rate and thus no shear-localization is possible. For relatively low θ_0 a non-monotonic effective stress–strain-rate constitutive relation occurs which therefore permits multiple strain-rates across the layer and the possibility of shear

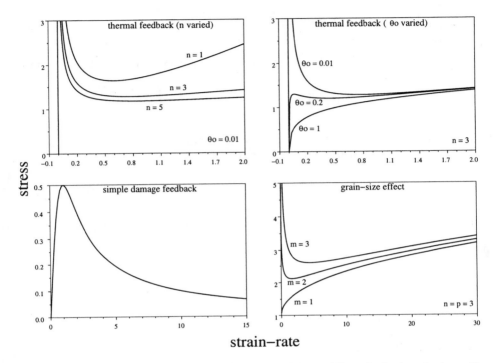

Figure 1. Effective "stress versus strain-rate" constitutive laws for three different feedback mechanisms with decay-type loss, as discussed in the text. The top two panels show the thermal feedback, i.e., where the sheared layer undergoes viscous heating that then influences the viscosity because of its temperature dependence. In these cases viscosity is possibly also a function of strain-rate through a power-law dependence that is controlled by the power-law index n (i.e., if $n = 1$ there is no dependence on strain-rate, and for $n > 1$ viscosity decreases with increasing strain-rate). The top left panel considers the thermal feedback with the background or boundary temperature θ_0 fixed at the value indicated, and for three different power-law indices n. The top right panel shows the same feedback for different background temperatures and one power-law index. For low enough background temperatures there is a region at low strain-rates where stress rapidly increases and then decreases with increasing strain-rate, which implies a resistance build-up and strength loss akin to stick-slip behavior. For yet higher strain-rates stress begins to increase again indicating a strength recovery; however, the rate of recovery with increasing strain-rate is much less for larger power-law indices n. For high background temperatures, there is no stick-slip-type region. The bottom two panels show feedback mechanisms for simple damage (deformation increases void/microcrack density which in turn reduces viscosity) and grainsize effects (stress reduces grain size which reduces viscosity). The damage mechanism gives a very simple stick-slip-type behavior with no strength-recovery. The grainsize effect is dependent on power-law index n, and indices that control the grainsize dependence of viscosity m (if $m = 0$ viscosity is grainsize independent) and the grainsize dependence of grain growth p (if $p = 1$ grain-growth is grain-size independent). Stick-slip behavior in the grainsize feedback mechanism occurs with no finite resistance build-up; i.e., resistance is effectively infinite when strain-rate is infinitesimally small. For increasing strain-rate there is strength loss and recovery also. However, the stick-slip type behavior only exists for sufficient grainsize dependence of viscosity, i.e., $m > 1$.

localization. Indeed, in these low-θ_0 cases, as many as three possible strain-rates can occur for a given stress (i.e., the constitutive curves have both a local maximum and a minimum).

The constitutive curves that involve multiple strain-rates for a given stress are also considered to have a pseudo-stick-slip behavior. In particular, at low strain-rates the material resistance is large and climbs rapidly with increased deformation-rate, what would be analogous to "stick" behavior. At a peak stress, however, stress and resistance give way to the "slip" behavior and loss of resistance or strength with increased strain-rate (the region of the constitutive curve with negative slope). In the thermal feedback, stress eventually begins to climb again at high strain-rates, albeit very gradually, as is characteristic of deformation in a hot, low-viscosity medium (which of course, at these high strain-rates and dissipative heating, it is). Such strength loss and stick-slip behavior is often considered closely associated with shear-localization.

The multiplicity of states (which then leads to mulitple strain-rates) in this system is illustrated by considering the possible solutions to temperature θ in the steady-state system; these are of course the roots to Equation (13) and Equation (14), which can be combined into one transcendental equation

$$\theta e^{\frac{1}{\theta_0+\theta}} = \tau^{n+1} \tag{17}$$

These roots as functions of stress τ are shown in Figure 2. Three roots are possible, and these represent a cold, highly viscous one (near $\theta = 0$), an intermediate-temperature one and a hot, low-viscosity one.

There are, however, several constraints on the existence of these three roots. For a background temperature $\theta_0 < 1/4$, the function $\theta e^{\frac{1}{\theta_0+\theta}}$ has two local extrema (a local maximum and minimum), and for $\theta_0 > 1/4$ it has no local extrema (only a monotonic increase with θ). Thus, for $\theta_0 < 1/4$ there are three possible solutions to Equation (17), but for larger θ_0 there is only one solution. Nevertheless, the temperature scale H/R is so large that essentially all realistic lithospheric temperatures will satisfy the constraint of $\theta_0 < 1/4$.

There is a further constraint on the possibility of shear localization in that the three possible roots only exist for τ^{n+1} beneath a maximum, as shown in Figure 2, which is itself a function of θ_0. For plausible lithospheric values of $\theta_0 \sim O(10^{-2})$, the maximum stress is given by $\tau^{n+1} \approx 10^{38}$ (see Fig. 2). The scale factor by which we redimensionalize τ is given by Equation (15); we can estimate this factor using typical silicate values for the given constants, namely $n = 3$, $A^n = 2.5 \times 10^{-6}$ MPa^3 s, (e.g., see Turcotte and Schubert 1982), $H/R \approx 6 \times 10^4$ K (see above), and $K = k/L^2 \approx 4 \times 10^{-14}$ MPa K^{-1} s^{-1} assuming $L \leq 10$ km. With the resulting scale factor, which is of the order 10^{-4} MPa, and the critical value of $\tau \approx 3 \times 10^9$ (having used $n = 3$) we find the maximum σ is of the order of 10^6 MPa. Thus essentially all tectonic stresses will fall below this maximum stress and thus permit multiple solutions and thus shear localization.

However, because steady-state solutions to these equations exist does not mean they are stable against perturbations. Although we could perform a linear stability analysis of these solutions, or even calculate a full nonlinear time-dependent solution, it is instructive to inspect the evolution equation. In particular, if we nondimensionalize time by $\rho c_p/K$, then

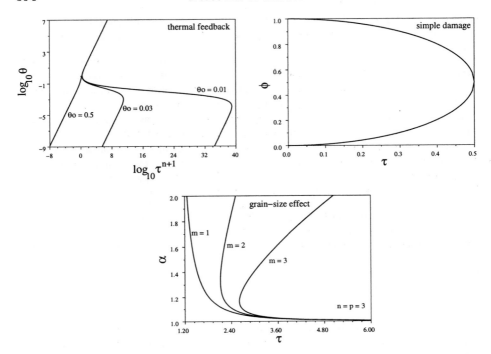

Figure 2. Solutions of the state variables (temperature θ, void density ϕ and grainsize α) as functions of stress τ for the same three feedback mechanisms of Figure 1. For each mechanism at least two possible solutions exist for a single stress τ. Since multiple possible states exist for one stress across the sheared layer, then mulitple viscosities and strain-rates can exist within the layer also, which is a condition for shear-localization (since it implies nonhomogenous shear). For the thermal feedback mechanism, three possible solutions exist for stresses beneath a maximum τ which is itself a function of background temperature θ_0 and power-law index n (note n is absorbed into the abscissa of the plot). The damage mechanism allows two possible solutions for a given stress beneath a maximum stress $\tau = 1/2$. For the grainsize mechanism, however, two possible solutions exist if the grainsize dependent of viscosity is sufficient (i.e., $m > 1$) and such multiple solutions exist only for a stress above a minimum value, instead of beneath a maximum as with the other two feedback mechanisms.

along with the other nondimensionalization steps discussed above, Equation (7) becomes

$$\frac{\partial \theta}{\partial t} = -\theta + \tau^{n+1} e^{-\frac{1}{\theta_0 + \theta}} \tag{18}$$

If we plot $\frac{\partial \theta}{\partial t}$ versus θ (Fig. 3) – a plot often referred to in dynamical systems as a phase diagram– we can see that the steady solutions occur where the curves intersect the $\frac{\partial \theta}{\partial t} = 0$ line. The stability of either steady solution is given by the the slope of the curve at the relevant interesection. If the slope is negative at the steady solution then it is stable; i.e., solutions on the curve to either side of the steady one will evolve to the steady solution (in particular, if the solution is to the left of the steady case with a smaller θ then its growth rate is positive and it will grow toward the steady value; if it is to the right with a larger θ, the growth rate is negative and the solution will decay to the steady value). In contrast if the slope is positive at the steady solution then that solution is unstable (solutions to either side of the steady one will diverge from it). In the case of Equation (18) (Fig. 3) we see

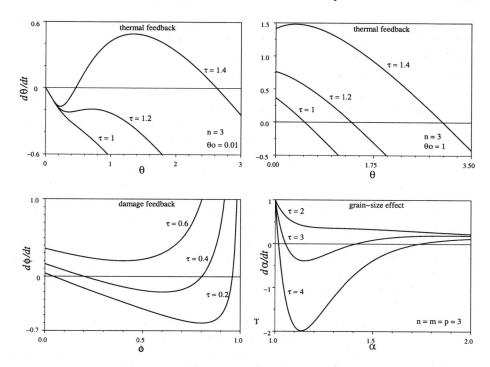

Figure 3. Phase diagrams for the same three feedback mechanisms of Figures 1 and 2, which show time-rate of change of the state variables versus the state variables themselves (i.e., temperature θ, void density ϕ and grainsize α). All panels show phase trajectories for different values of stress τ as indicated; the top two panels show the thermal feedback with two different background temperatures θ_0. These phase diagrams illustrate the temporal behavior of the feedback mechanisms and the stability of the steady solutions illustrated in Figure 2. Steady solutions occur where the phase trajectory crosses the line where the time-rate of change is zero. The trajectory slope at these intersections indicates the stability of the steady solution; i.e., the solution is unstable if the slope is positive, and stable if negative (see text for further discussion). Trajectories that do not cross the zero-rate line have no steady solution. See text for further discussion.

that if the background temperature θ_0 is sufficiently small (such that there are three possible steady solutions) then both the cold/viscous and the hot/low-viscosity cases are stable ($d\theta/dt$ vs. θ has a negative slope at that root), while the intermediate-temperature case is unstable. Therefore unsteady solutions in the vicinity of the intermediate-temperature root will evolve toward one or the other stable root; temperatures hotter than the intermediate case will grow to the hot steady state, and colder temperatures will go to the cold case. Therefore, since the layer will in the end have two different temperatures and two strain-rates, then weak zones and localization can be expected to occur (as long as the initial temperature is in the vicinity of the intermediate root). However, when a localized weak zone develops, it cannot weaken and localize without bound but must stop at the hot/low-viscosity solution. Thus a fault-like shear localization is impossible for this mechanism. However, it should be noted that given the temperature scale $H/R \approx 6 \times 10^4$ K both the intermediate and hot/low-viscosity roots are at temperatures near or well in excess of the melting temperature. Therefore, this mechanism would in fact give way to a melting instability well before these roots were reached. With such melting and extreme viscosity and strength drop, an extremely sharp localization is possible, although that is beyond the simple physics we are exploring in this model.

Simple-damage feedback with decay-loss healing

The large region of the lithosphere that is thought to undergo combined brittle-ductile behavior (Kohlstedt et al. 1995) is suggestive of a damage-induced shear localization whereby microcracks forming in zones of ductile deformation focus into weak zones that thereby concentrate deformation and accelerate their own nucleation and growth, in the end creating distinct shear bands that appear like faults. Damage-induced shear localization is also appealing given the various roles it offers for water, which is thought to be a facilitator of plate tectonics, and the cause for plate tectonics on Earth but not the other terrestrial planets (Tozer 1985). For example, volatiles such as water can either facilitate damage through hydro-fracturing, pore-pressure-reduction of friction and simple lubrication of slip zones; and/or the damaged material (with finite pore space and open grain boundaries) can facilitate hydrolytic weakening of minerals and/or pressure-solution creep (Guéguen and Palciauskas 1994; Ranalli 1995). Although the role of water has generally been thought to be confined to only the upper few tens of kilometers of the lithosphere, there is growing evidence that it is present and influencing Earth's rheology and evolution at hundreds of kilometers depth (Hirth and Kohlstedt 1996; Karato 2002).

A simplified viscous damage theory termed the "void-volatile" mechanism (Bercovici, 1998; see also Regenauer-Lieb, 1999) involves modelling the evolution of microcrack or void density with a scalar transport law in which deformational work provides the source for void creation (similar to the heat-transport equation with frictional heating included). In this case, the weakening variable Θ is set equivalent to void volume fraction (i.e., porosity) Φ, and we write a simple transport law which in our 1-D example is

$$\rho c \frac{\partial \Phi}{\partial t} = -B(\Phi - \Phi_0) + 2f\mu(\Phi, \dot{\varepsilon})\dot{\varepsilon}^2 \tag{19}$$

where c is the proportionality factor between Φ and energy per unit mass, B is analogous to K, and Φ_0 is the background value of Φ to which the unforced system equilibrates. In this equation we have assumed that some fraction f of the deformational work goes into creation of surface energy by creating new microcracks, voids and/or defects; this is often referred to as stored-work, or, as discussed in the 1920s by G.I. Taylor, through his experiments on torsional deformation of metals, as latent-heat of cold work (Farren and Taylor 1925; Taylor and Quinney 1934). Finally, we assume a linear dependence of viscosity on Φ, which is relevant for the sort of two-phase mixtures assumed here (Bercovici 1998; Ricard et al. 2001). In particular, if the mixture is of a solid matrix with viscosity μ_m and a void-filling fluid with viscosity μ_f, then simple mixture theory will yield an effective viscosity of $\mu_f \Phi + \mu_m(1 - \Phi)$, which can be recast as

$$\mu = \mu_r(1 - \lambda\Phi) \tag{20}$$

where $\mu_r = \mu_m$ is the reference viscosity and $\lambda = (\mu_m - \mu_f)/\mu_m$ controls viscosity variability.

Again we first consider steady solutions to Equation (19), i.e., solutions of

$$B(\Phi - \Phi_0) = 2f\mu_r(1 - \lambda\Phi)\dot{\varepsilon}^2 \tag{21}$$

To reduce the number of free parameters in the system, we again use the natural scales of the system by first defining

$$\Phi = \Phi_0 + \Phi^*\phi \tag{22}$$

where Φ^* is an as yet unknown porosity scale and ϕ is a rescaled porosity perturbation. In this case we rewrite viscosity as

$$\mu = \mu_0\left(1 - \frac{\lambda\Phi^*}{1-\lambda\Phi_0}\phi\right) \tag{23}$$

where $\mu_0 = \mu_r(1 - \lambda\Phi_0)$. At this point we can define the porosity scale as

$$\Phi^* = (1 - \lambda\Phi_0)/\lambda \tag{24}$$

such that our viscosity law becomes

$$\mu = \mu_0(1 - \phi) \tag{25}$$

We define strain-rate also in terms of a dimensionless quantity, namely $\dot{\varepsilon} = \dot{\varepsilon}^*\dot{\varepsilon}'$ where $\dot{\varepsilon}^*$ is a dimensional scale and $\dot{\varepsilon}'$ is a dimensionless variable. In this case, the constitutive law (Eqn. 5 using $\Theta = \Phi$) leads to

$$\left[\frac{\sigma}{2\mu_0\dot{\varepsilon}^*}\right] = (1-\phi)\dot{\varepsilon}' \tag{26}$$

and Equation (21) becomes

$$\left[\frac{B(1-\lambda\Phi_0)}{2f\lambda\mu_0(\dot{\varepsilon}^*)^2}\right]\phi = (1-\phi)\dot{\varepsilon}'^2 \tag{27}$$

where the terms in square brackets are dimensionless groups. We choose our strain-rate scale to set one of these groups to unity:

$$\dot{\varepsilon}^* = \sqrt{\frac{B(1-\lambda\Phi_0)}{2f\lambda\mu_0}} = \sqrt{\frac{B}{2f\lambda\mu_r}} \tag{28}$$

Thus, our two governing equations, Equation (26) and Equation (27) become, after dropping the primes,

$$\tau = (1-\phi)\dot{\varepsilon} \tag{29}$$

and

$$\phi = (1-\phi)\dot{\varepsilon}^2 \tag{30}$$

where

$$\tau = \sigma\sqrt{\frac{f\lambda}{2\mu_0 B(1-\lambda\Phi_0)}} \tag{31}$$

represents a dimensionless imposed stress. With these equations we can eliminate ϕ to arrive at an effective constitutive law

$$\tau = \frac{\dot{\varepsilon}}{1+\dot{\varepsilon}^2} \qquad (32)$$

which is shown in Figure 1. As noted in the previous section, because stress increases sharply below a certain strain-rate (in this case at $\dot{\varepsilon} < 1$) and then decreases for higher strain-rates, this effective constitutive law has been referred to explicitly as a stick-slip rheology (Whitehead and Gans 1974; Bercovici 1993). However, in contrast to the thermal feedback mechanism, the material continues to lose strength (stress decreases) as strain-rate goes to infinity; this implies that effectively infinite strain rates can be achieved, and the ongoing strength loss is indicative of a runaway effect, which will become more evident momentarily.

Apart from the effective constitutive law, the actual solutions for ϕ and $\dot{\varepsilon}$ in terms of τ are instructive for inferring shear-localization in the layer. The steady-state solutions require

$$\phi = \tfrac{1}{2}(1 \pm \sqrt{1-4\tau^2}) \quad \text{and} \quad \dot{\varepsilon} = \frac{1}{2\tau}(1 \pm \sqrt{1-4\tau^2}) \qquad (33)$$

and thus for a given τ there are two possible porosities and strain-rates, a high-porosity/fast (i.e., rapidly deforming) one associated with the $+$ root, and a low-porosity/slow case associated with the $-$ root (Fig. 2). Shear localization requires inhomogeneous strain-rate, i.e., for a given stress across the layer localization requires at least two possible strain-rates to co-exist, which this feedback mechanism also provides. Again, shear localization would be manifest if the layer contained some regions with the slow solution and some with the fast solution (the localizations would then be the fast zones).

Equation (33) shows that there are no steady-state solutions if $\tau > \tfrac{1}{2}$, in which case there are only time-dependent solutions. We can infer both the nature of time-dependent solutions and the stability of steady-state solutions by inspecting the evolution equation Equation (19). In particular, if we nondimensionalize time by $\rho c / B$, then along with the other nondimensionalization steps discussed above, Equation (19) becomes

$$\frac{\partial \phi}{\partial t} = -\phi + \frac{\tau^2}{1-\phi} \qquad (34)$$

If we plot $\frac{\partial \phi}{\partial t}$ versus ϕ (Fig. 3) we again see that the steady solutions occur where the curves intersect the $\frac{\partial \phi}{\partial t} = 0$ line and the stability of steady solutions is given by the the slope of the curve at the relevant interesection. In this case, the low-porosity/slow steady solution is stable and the high-porosity/fast one is unstable. However, this does not mean that the system will everywhere evolve to the slow solution; indeed, values of ϕ even slightly greater than that for the fast solution will grow toward even larger porosities. Although the maximum ϕ is 1 (lest we have a negative viscosity), solutions with porosities larger than that for the high-porosity/fast solution will undergo a run-away to $\phi = 1$ and hence zero viscosity. Indeed, by inspection of Equation (34), an urealistic solution of ϕ infinitesimally larger than 1 leads to an infinitely large negative growth rate which forces such unrealisic solutions to decay back to $\phi = 1$. In essence, $\phi = 1$ is a stable singularity. In those regions with $\phi = 1$

the strain-rate (which goes as $\tau/(1-\phi)$) will be infinite. However, the size of any zero-viscosity zone is of finite width since it is entirely dependent on the initial porosity field; it will be as large as any region whose initial porosity exceeded that for the high-porosity/fast steady solution and thus grew to $\phi = 1$. Thus although the strain-rate might be heterogeneous, the shear zone does not necessarily self-localize, although the infinite strain-rates require the velocity drop across a zero-viscosity zone to be infinite. Finally, if $\tau > 1/2$, there are no possible steady solutions (see Eqn. 33 and Fig. 1) and the entire system goes to $\phi = 1$, i.e., the entire layer undergoes a runaway and has an infinite velocity drop across it.

To put some of these dimensionless results in a physical perspective, we can try to redimensionalize the critical stress $\tau = \frac{1}{2}$ (below which there is shear-localization since two strain-rate solutions are permitted). The simple damage theory is idealized but we can relate some of the constants in the scaling equation Equation (31) to physical parameters. First, the partitioning fraction f is likely to be of order 0.1 (Ricard and Bercovici 2002), and λ is likely to be near unity. We can assume a very high lithospheric viscosity of $\mu_0 \approx 10^{25}$ Pa s (Beaumont 1976; Watts et al. 1982), and that background porosity $\Phi_0 \ll 1$. The quantitiy B is more difficult to quantify; it has units that represent the loss of energy density (J/m^3) per second. The energy density is likely given by surface energy on cracks, which in terms of volumetric energy density will be surface energy Γ–typically of order 1-100 J/m^2 for crack surfaces (Jaeger and Cook 1979; Cooper and Kohlstedt 1982; Atkinson 1987; Atkinson and Meredith 1987)– times void or crack curvature c (the inverse of void or crack size which is typically 100-1000 m^{-1}). The loss rate of this energy is unclear but we will assume that is controlled by loss of void-filling material, namely water, through Darcy-type flow; in this case the flux rate (Darcy velocity) of fluid is of the order $V = k_\phi \rho_w g/\mu_w$, where k_ϕ is permeability (of order 10^{-14} m^2 for porous rocks), ρ_w is water density (1000 kg/m^3), g is gravity, and μ_w is water viscosity (10^{-3} Pa s). The time-scale for energy loss will be typically L/V where L is the length-scale of the shear-zone (1-10 km), and thus B will scale as $\Gamma c V/L$. Given these values, the critical dimensional stress σ from Equation (31) is of the order several 100 to 1000 MPa, which is indeed typical of or larger than tectonic stresses. Thus we can speculate that tectonic stresses are sufficient to induce localization with this damage mechanism.

Grainsize feedback

Grainsize reduction is an important cause for shear localization. The association of shear localization and grainsize reduction is evident in mylonite shear zones (White et al. 1980; Jin et al. 1998). However, the link between grainsize reduction and shear localization is not clearly understood. The way in which grainsize is controlled during deformation is more complicated than other variables such as temperature. In fact, some recent analyses underscored the difficulties in explaining shear localization through grainsize reduction (Bresser et al. 2001; Montési and Zuber 2002), suggesting a need for more complete analysis of this problem. This section reviews some key concepts in grainsize sensitivity of deformation with the emphasis on the processes of dynamic recrystallization.

Grainsize controls the rheology through two contrasting ways. Under low temperature conditions where dislocation recovery and diffusion do not occur, grainboundaries act as a barrier for dislocation motion and the presence of many grainboundaries causes hardening (the Hall-Petch effect; see Cottrell 1953). In contrast, when diffusion and resultant dislocation recovery become effective (at high temperature T, i.e., exceeding roughly 50% of the melting temperature), then a smaller grainsize leads to weakening because deformation by diffusive mass transport occurs more efficiently for smaller grain sizes. Under most tectonic conditions except the shallow lithosphere, diffusive mass transfer is likely to be important, and we consider the effects of diffusion creep here. For simplicity, we assume (as in the previous sections) that viscosity depends only on grainsize a and strain-rate, i.e.,

$$\mu = \mu(a, \dot{\varepsilon}) = \mu_r \left(\frac{a}{a_0}\right)^m \left(\frac{\dot{\varepsilon}}{\dot{\varepsilon}_0}\right)^{\frac{1-n}{n}} \tag{35}$$

where $m = 2 - 3$ and $n = 1$ for diffusion creep, and $m \approx 0$ and $n = 3 - 5$ for dislocation creep; a_0 is a reference grainsize and $\dot{\varepsilon}_0$ is the reference strain-rate, both of which are defined below. The rheological law Equation (35) prescribes dependence of viscosity on grainsize and strain-rate, although strictly speaking these effects do not occur simultaneously since grainsize dependence occurs in diffusion creep, and strain-rate dependence occurs in dislocation creep. Moreover, one of the dominant processes leading to grainsize reduction, namely dynamic recrystallization, occurs only in the dislocation-creep regime, and thus the feedback between grainsize reduction (while in the dislocation-creep regime) and grainsize-weakening (while in the diffusion-creep regime) only occurs if the material essentially experiences both creep regimes. This would seem to constrain the feedback mechanism to only be effective near the transition between creep regimes. However, the actual rheology of the material is complicated because it is essentially composed of two phases, i.e., coarse-grained strong matrix (which deforms by dislocation creep) and fine-grained weak regions (deforming by diffusion creep). The flow law parameters such as n and m in such a two-phase material are not well known but are likely to be between those for dislocation and diffusion creep. Therefore when deformation reduces the grainsize, there is a feedback mechanism that could causes self-weakening and shear localization.

There are several processes by which grainsize evolves. The first is dynamic recrystallization wherein gradients in dislocation density drive formation of new grainboundaries and/or mobilization of existing boundaries leading to the modification of grainsize. Evidence of dynamic recrystallization is ubiquitous in naturally deformed rocks (Urai et al. 1986). The second process involves chemical reactions, including phase transformations (Furusho and Kanagawa 1999; Karato et al. 2001; Rubie 1983, 1984; Stünitz and Tullis 2001). In both cases, the kinetics of nucleation and growth controls the grainsize during deformation or chemical reactions, whereas after the completion of these processes, grain growth, through the reduction of grain-boundary surface energy, plays an important role when the grainsize is small enough. Whether grainsize modification causes shear localization depends on the competition between the weakening effect due to grainsize reduction and the hardening effect due to graingrowth, the details of which are specific to the particular processes involved. Here we consider the effects of grainsize evolution during dynamic recrystallization. In this case, and in the context of our simple-shear flow model, grainsize evolution may be described by

$$\frac{\partial a}{\partial t} = -\frac{\dot{\varepsilon}}{\varepsilon_r}(a - a_0) + \frac{K}{pa^{p-1}} \tag{36}$$

(Karato 1996; Kameyama et al. 1997) where the first term represents the effect of dynamic recrystallization on grainsize and the second term the effect of graingrowth; $\varepsilon_r \approx 1$ is the strain needed to complete dynamic recrystallization (Karato et al. 1980); K is a rate constant for graingrowth; $p = 2 - 3$ such that grain growth is fastest for smaller grains when grain-boundary surface energy is large and diminishes as grains increase in size; and a_0 is the equilibrium value of a if there were no grain-growth and is in effect the grainsize at which dynamic recrystallization is completed. This grainsize a_0 may be determined by the nucleation-growth and impingement of growing subgrains on one another (Derby and Ashby 1987), or by subgrain rotation and dislodgement under deformation (Shimizu 1998). The implicit assumptions behind Equation (36) are, first, that grainsize reduction and graingrowth occurs simultaneously, and, second, that the time scale to obtain a short-term equilibrium grainsize a_0 is significantly smaller than the time scale for graingrowth (therefore a_0 is assumed to be a constant).

As in previous sections, we nondimensionalize Equation (36) by defining $a = a_0 \alpha$, $t = \frac{p a_0^p}{K} t'$, and $\dot{\varepsilon} = \frac{K \varepsilon_r}{p a_0^p} \dot{\varepsilon}'$ and thus obtain (dropping the primes for simplicity)

$$\frac{\partial \alpha}{\partial t} = -\dot{\varepsilon}(\alpha - 1) + \alpha^{1-p} \tag{37}$$

Similarly, the constitutive law $\sigma = 2\mu\dot{\varepsilon}$ becomes, using Equation (35),

$$\tau = \alpha^m \dot{\varepsilon}^{1/n} \tag{38}$$

where

$$\tau = \frac{\sigma p a_0^p}{2\mu_r \varepsilon_r K} \tag{39}$$

In steady state, Equation (37) leads to

$$\dot{\varepsilon} = \frac{\alpha^{1-p}}{\alpha - 1} \tag{40}$$

in which case we can re-write Equation (38) as

$$\tau = \left(\frac{\alpha^{nm+1-p}}{\alpha - 1}\right)^{1/n} \tag{41}$$

We can therefore display an effective constitutive law, as we did previously, by writing τ as a function of $\dot{\varepsilon}$ parametrically through α, as shown in Figure 1. There is no steadystate for $\alpha < 1$ where grainsize increases both by dynamic recrystallization and by graingrowth (assuming that both the imposed τ and the resulting $\dot{\varepsilon}$ are > 0). We will thus consider only the case $\alpha > 1$, that is when an initially large grainsize is reduced by deformation.

For sufficiently large m (i.e., sufficient grainsize dependence), the effective constitutive law is similar to that for the thermal-feedback case in which there is a region at low strain-rates whereby stress decreases with strain-rate (the "slip" regions), and a region at higher-strain rates whereby stress increases gradually. The stress-loss region is, as discussed before, an apparent feature of shear-localizing mechanisms. In contrast to the other feedback mechanisms, however, there is no finite region at low-strain-rates where stress increases sharply with strain-rate. Instead the peak strength occurs at zero strain-rate, which is similar to true stick-slip behavior whereby the "stick" response involves no motion until a peak or yield stress is achieved, after which there is progressive loss of strength (i.e., failure ensues).

The stick-slip region only exists when there is a minimum stress in the effective constitutive curve (see Fig. 1) which means that somewere on the curve

$$\frac{d\tau}{d\dot{\varepsilon}} = \frac{d\alpha}{d\dot{\varepsilon}}\frac{d\tau}{d\alpha} = \frac{1}{n}\left(\frac{\alpha^{nm+1-p}}{\alpha-1}\right)^{\frac{1-n}{n}} \alpha^{nm}\frac{1-(nm-p)(\alpha-1)}{1+p(\alpha-1)} = 0 \qquad (42)$$

For $m = 0$ (i.e., no grainsize dependence) there can be no stress minimum and the constitutive curve has the appearance of a simple power-law relationship. For $m > 0$, the minimum occurs at $\alpha = 1 + \frac{1}{nm-p}$. Since we require $\alpha > 1$ then $nm > p$; therefore if, for example, $n = 3$ and $p = 3$ then even linear grainsize dependence $m = 1$ is not sufficient to give a stick-slip effect and the constitutive curve again looks like the case for $m = 0$ (see Fig. 1). For parameter combinations in which $nm > p$, stick-slip behavior can occur.

One should note, however, that in cases where stick-slip behavior occurs the stress spuriously goes to infinity as strain-rate goes to zero (see Fig. 1). This is an artifact of the model in that as grainsize α goes to infinity (associated with $\dot{\varepsilon} \to 0$; see Eqn. 40) the grain-size dependence of stress is assumed to persist (leading to $\tau \to \infty$; see Eqn. 41) even though it should not (i.e., dislocation creep must dominate). Although beyond the scope of this paper, a correction to this artifact, such as a grainsize-dependence index m that is itself grainsize-dependent (i.e., vanishes at some finite grainsize), would likely mitigate this effect and result in proper stick-slip curves similar to those shown for the other mechanisms of Figure 1.

The solutions for α in the steady-state system are the roots of Equation (41), which are shown in Figure 2. There are two solutions if the imposed stress is greater than the minimum stress

$$\tau > \tau_{min} = \left(\frac{(nm-p+1)^{nm-p+1}}{(nm-p)^{nm-p}}\right)^{1/n} \qquad (43)$$

The minimum stress τ_{min} is only guaranteed to exist for $nm > p$. For reasonable values of n, m and p, this minimum stress is $2 < \tau_{min} < 3$ which corresponds to 2-300 MPa for $\mu_r = 10^{18}$ Pa s and $\dot{\varepsilon}_0 = 10^{-10} - 10^{-12}$ s^{-1}. One of the solutions is very near $\alpha = 1$ (the equilibrium grainsize without graingrowth) and the other at large grainsize (Fig. 2). As stress increases and one moves further away from the minimum stress τ_{min}, the small grainsize solution decreases whereas the latter increases. For stresses less than τ_{min}, there are no steady-state solutions.

The existence of at least two steady solutions, which leads to two possible strain-rates as shown in Figure 1, is a minimum requirement for localization to occur in this, as well as other previously discussed decay-loss systems. Without two possible solutions, then only one strain-rate exists and then the entire shear-layer has uniform shear. If two strain-rates are possible, then the layer can have low-strain-rate regions surrounding high-strain-rate regions which would represent the zones of shear-localization.

However, an important difference between the grainsize feedback mechanism and the other two we have discussed (thermal and damage) is that the shear-localization regime exists for stresses greater than a minimum stress, i.e., it is a high-stress phenomenon; the other two mechanisms occur for stresses less than a maximum, i.e., they are low-stress phenomena.

The stability of these steady solutions is determined by combining Equation (37) and Equation (38) to obtain

$$\frac{\partial \alpha}{\partial t} = \tau^n \frac{1-\alpha}{\alpha^{nm}} + \alpha^{1-p} \tag{44}$$

whose phase-plot (i.e., $d\alpha/dt$ versus α) is displayed in Figure 3 for various values of stress τ. For a stress higher than the τ_{min} (such that two steady solutions are realized), the steady solution for the small grainsize is stable (the slope of the phase-plot curve is negative at the steady solution, i.e., where it crosses the line $d\alpha/dt = 0$) whereas the solution for large grainsize is unstable. For an unstable grainsize, shear localization occurs through bifurcation: grains larger than the unstable equilibrium size grow to make materials stronger (i.e., yield smaller strain-rate) and grains smaller than that size make materials softer hence yield higher strain-rates. For stresses smaller than τ_{min} there are no steady solutions and the grainsize everywhere grows indefinitely; in this case, stresses are too small to provide significant dynamic recrystallization and thus grain growth dominates. Therefore a necessary condition for shear localization, in this grainsize model, is given by Equation (43). However, certain values of n, m and p (in particular that $nm > p$) are also necessary to guarantee the existence of a minimum stress.

The degree to which shear localization develops depends not only on the stress but also on the viscosity contrast between the two regions which depends, in turn, on the magnitude of stress as well as temperature. The degree of localization will be higher for higher stresses and lower temperatures because of the smaller grainsize at higher stress, and because of the smaller activation energy for diffusion creep than that for dislocation creep in many minerals (Jin et al. 1998).

Note that if the presence of short-term equilibrium is ignored (i.e., $a_0 = 0$), then Equations (37) and (38) are reduced to

$$\frac{\partial \alpha}{\partial t} = -\dot{\varepsilon}\alpha + \alpha^{1-p} \tag{45}$$

and

$$\tau = \alpha^{\frac{nm+1-p}{n}} \tag{46}$$

respectively. Equation (46) has only one solution and hence no bifurcation and resultant shear localization will occur. We conclude therefore that the presence of a short-term, quasi-equilibrium state is crucial for the development of instability and localization associated with grainsize evolution due to dynamic recrystallization. Under these conditions, a single stress can allow two different grainsizes and thus heterogeneous strain-rate leading to localization; in essence, the material is composed of two "phases", one with a small grainsize (a weak phase) and the other with large grainsize (a strong phase), and therefore the presence of twophases is essential for the shear localization. This conclusion is consistent with the field and laboratory observations where the transient stage of dynamic recrystallization is characterized by a bimodal distribution of grainsize (Jin et al. 1998; Lee et al. 2002).

Various simplifying assumptions have been made that need to be discussed here. First, we assumed the presence of two time scales to control the grainsize. They correspond to two physical processes; one is the smallscale balance between growing grainboundaries and impingement (Derby and Ashby 1987) or local balance caused by subgrainrotation (Shimizu 1998), and another is the largerscale balance in regions of complete recrystallization. When the stress is low and diffusion creep dominates in these regions, then graingrowth driven by grain-boundary surface energy will occur, which would be balanced with grainsize reduction by dynamic recrystallization. The latter, longer-scale balance model, is similar to that by Bresser et al. (1998), but is different from their model. The De Bresser model assumes that the strain-rates due to diffusion creep and dislocation creep balance. There is no reason for the strain-rates of these processes to balance because they are independent processes (Frost and Ashby 1982; Poirier 1985), although grain-growth and grainsize reduction may balance under some limited conditions. Second, we assumed that grainsize reduction (by dynamic recrystallization) and graingrowth occur simultaneously. Strictly speaking, this is not correct because, for a given grainboundary, either dislocation density contrast or grainboundary curvature imbalance contributes to grainboundary migration. However, Equation (36) captures some physics to the extent that smaller grains tend to grow by grainboundary energy driven migration and larger grains tend to reduce their size through grainboundary migration driven by dislocation density heterogeneity. Third, the rheology of a material during the transient stage of dynamic recrystallization is likely to be sensitive to the geometry of a weaker (finegrained) material. When the weaker material assumes a connected network, then a significant reduction in strength occurs, which is likely to lead to unstable localized deformation (Handy 1994). This is analogous to the brittle fracture where the formation of interconnected cracks through crack-crack interaction is the key to the instability (e.g., Paterson 1978). Such an aspect is not included in the present analysis, but is starting to be approached by various damage theories, which is discussed below. Further refinement of the model is needed to treat these details to better understand the role of dynamic recrystallization in shear localization.

The degree of localization by this mechanism may be limited and dependent on the temperature. For example, for olivine-rich regions, when the initial grainsize is approximately 3 mm and a stress pulse of approximately 100 MPa is applied at temperatures around 1100 K, then the viscosity contrast between the coarse and finegrained portions (at the same stress) is $10^2 - 10^3$ but virtually no weakening would occur at $T > 1500$ K. Therefore shear localization due to this mechanism can occur only in the relatively shallow portions of the lithosphere. However, even at relatively low temperatures, extreme localization would be difficult with this mechanism alone. A combination of grainsize reduction with other mechanisms may be needed for extreme localization. Jin et al. (1998) reported evidence that localization caused by grainsize reduction led to shear melting that resulted in extreme localization (faulting).

The analyses by Braun et al. (1999) and Frederiksen and Braun (2001) also include the effects of transient behavior associated with grainsize reduction. In particular, the treatment by Frederiksen and Braun (2001) is equivalent to a two-phase behavior but the microscopic basis for the two-phase behavior was not examined in their paper.

Finally, other processes such as phase transformations (or chemical reactions) also lead to grainsize reduction and resultant rheological weakening and shear localization (Furusho and Kanagawa 1999; Stünitz and Tullis 2001) (see also Green and Marone, this volume).

The key concept in such a case is the degree of grainsize reduction associated with a phase transformation or chemical reaction and the enhancement of transformation by deformation or vice versa. Although, a quantitative model has been developed for the grainsize reduction due to a phase transformation (Riedel and Karato 1997), the understanding of deformation-transformation (or deformation-chemical reaction) interaction remains qualitative (Green and Burnley 1989; Kirby 1987).

Thermal and simple-damage feedbacks with diffusive-loss healing

Here we consider the nature of shear-localization when the loss mechanism is diffusive, i.e., dependent on gradients of gradients (or "curvature") in the state variable Θ. This loss mechanism is only relevant for the thermal feedback mechanism and perhaps the damage mechanism if the void-filling fluid undergoes chemical diffusion or hydrological conduction (i.e., spreads out to mitigate fluid pressure gradients). The mechanism is not relevant for the grainsize feedback mechanism. In this case, the governing evolution equations for temperature and porosity are similar enough that we can write them with one equation:

$$\rho c \frac{\partial \Theta}{\partial t} = k \frac{\partial^2 \Theta}{\partial y^2} + 2f\mu(\Theta, \dot{\varepsilon})\dot{\varepsilon}^2 \tag{47}$$

where Θ is either temperature or porosity, k is a conductivity (thermal, chemical or hydrological), and $f = 1$ for the thermal mechanism, or $0 < f < 1$ for the damage mechanism.

We can solve for the steady-state structure of Θ (in the 0-D examples, this structure was determined by the initial conditions) and thus the velocity field $u(y)$ as well. As with the previous examples using our simple-shear-layer thought-experiment, while shear stress is a constant σ across the layer, Θ is not necessarily so, and thus strain-rate is possibly variable. In terms of the shear-rate $\partial u/\partial y$, the constitutive law leads to

$$\frac{\partial u}{\partial y} = \frac{\sigma}{\mu(\Theta, \partial u/\partial y)} \tag{48}$$

We consider only the steady limit of Equation (47) which is

$$0 = k\frac{\partial^2 \Theta}{\partial y^2} + \tfrac{1}{2}f\mu(\Theta, \partial u/\partial y)\left(\frac{\partial u}{\partial y}\right)^2 \tag{49}$$

In addition to assuming that the channel's top and bottom boundaries (at $y = \pm L$) have equal and opposite motion, we also prescribe them to be held at a constant temperature or porosity Θ_0. Given the symmetry of the problem we assume that at $y = 0$ the velocity u is zero, and Θ is at its maximum (i.e., $\partial \Theta/\partial y = 0$ at $y = 0$). Given these boundary conditions, we can solve Equation (48) and Equation (49) to infer the structure of shear localization across the channel.

We again consider two different viscosity laws, i.e., Arrhenius for the thermal feedback, and linear for the damage feedback, each of which requires a slightly different scaling analysis, and significantly different analytic solutions.

Thermal feedback. Here we examine our basic feedback mechanism using the thermally activated Arrhenius power-law viscosity. Given the constant stress σ across the layer, the stress-strain-rate relation leads to (using Eqn. 9 and Eqn. 48)

$$\frac{\partial u}{\partial y} = 2\left(\frac{\sigma}{A}\right)^n e^{-\frac{H}{R\Theta}} \tag{50}$$

and a steady heat equation (again using Eqn. 9 and Eqn. 49)

$$0 = k\frac{\partial^2 \Theta}{\partial y^2} + A\left(\frac{1}{2}\frac{\partial u}{\partial y}\right)^{\frac{n+1}{n}} e^{\frac{H}{nR\Theta}} \tag{51}$$

where we assume that all deformational work goes into viscous heating and thus $f = 1$. Again using the nondimensionalizing scheme described previously, where now the temperature and velocity scales are $\Theta^* = H/R$ and $U^* = 2L\left(\frac{kH}{2L^2RA}\right)^{\frac{n}{n+1}}$, Equation (50) and Equation (51) become

$$\frac{\partial u}{\partial y} = \tau^n e^{-\frac{1}{\theta_0+\theta}} \tag{52}$$

$$0 = \frac{\partial^2 \theta}{\partial y^2} + \frac{1}{2}\tau^{n+1} e^{-\frac{1}{\theta_0+\theta}} \tag{53}$$

where now

$$\tau = \sigma\left(\frac{2L^2 R}{A^n k H}\right)^{\frac{1}{n+1}} \tag{54}$$

and $\theta_0 = R\Theta_0/H$ is the dimensionless boundary temperature. Again, τ is the dimensionless stress, and as with the 0-D decay-loss cases, both n and θ_0 appear in the dimensionless equations. (Although θ_0 could be absorbed into the differential equation readily by defining a variable $\vartheta = \theta_0 + \theta$, it would simply reappear in the boundary conditions.)

Multiplying both sides of Equation (53) by $2\partial\theta/\partial y$ and integrating we arrive at

$$\frac{\partial \theta}{\partial y} = \pm \tau^{\frac{n+1}{2}} \sqrt{\gamma^2 - F(\theta_0 + \theta)} \tag{55}$$

where

$$F(\psi) = \int e^{-1/\psi} d\psi = \psi e^{-1/\psi} - \text{Ei}(1, 1/\psi) \tag{56}$$

in which ψ is a dummy variable and

$$\text{Ei}(m, x) = \int_1^\infty t^{-m} e^{-xt} dt \tag{57}$$

is the error integral. We can find solutions to (55) that are parametric in the peak "temperature" θ_{max} which occurs at $y = 0$. First, since $\partial\theta/\partial y = 0$ at $y = 0$ (where $\theta = \theta_{max}$) then we obtain a relation for the integration constant γ:

$$\gamma^2 = F(\theta_0 + \theta_{max}) \tag{58}$$

which can be solved numerically for γ given a θ_0 and θ_{max}. Next, Equation (55) can be integrated from $y = -1$ to an arbitrary $y \leq 0$, and/or from $y = 1$ to any $y \geq 0$ to obtain

$$\int_0^\theta \frac{d\theta'}{\sqrt{\gamma^2 - F(\theta_0 + \theta')}} = \tau^{\frac{n+1}{2}} \left\{ \begin{array}{l} 1 + y, \text{ for } y \leq 0 \\ 1 - y, \text{ for } y \geq 0 \end{array} \right\} = \tau^{\frac{n+1}{2}} (1 - |y|) \tag{59}$$

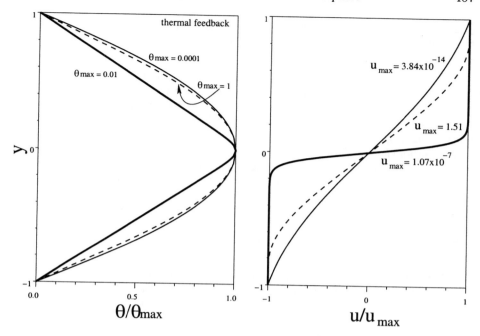

Figure 4. Temperature (θ) and velocity (u) versus distance across the sheared layer y, for the thermal feedback mechanism with diffusive loss and three different maximum temperature anomalies θ_{max} (indicated) and a boundary temperature $\theta_0 = 0.01$. Only intermediate temperatures ($\theta_{max} \approx \theta_0$, the thick solid curve in each frame) generate localization while cold ($\theta_{max} \ll \theta_0$, the thin solid curves) and hot ($\theta_{max} \gg \theta_0$, dashed curves) temperatures generate more distributed and nearly constant shear. Curves are normalized by their maxima (values indicated).

where we have used the plus sign in Equation (55) for $y < 0$ and the minus for $y > 0$ because we assume the temperature increases from the boundary at $y = -1$ to its maximum at $y = 0$ and then decreases again toward the boundary at $y = 1$. At $y = 0$ equation Equation (59) yields a relation for τ as a function of θ_{max}:

$$\tau^{\frac{n+1}{2}} = \int_0^{\theta_{max}} \frac{d\theta'}{\sqrt{\gamma^2 - F(\theta_0 + \theta')}} \tag{60}$$

Lastly, Equation (52) can be integrated to obtain

$$\begin{aligned} u &= u_0 + \tau^n \int e^{-\frac{1}{\theta_0+\theta}} dy \\ &= u_0 - \text{sgn}(y)\tau^{\frac{n-1}{2}} \int \frac{e^{-\frac{1}{\theta_0+\theta}}}{\sqrt{\gamma^2 - F(\theta_0 + \theta)}} d\theta \\ &= \text{sgn}(y) 2\tau^{\frac{n-1}{2}} \sqrt{\gamma^2 - F(\theta_0 + \theta)} \end{aligned} \tag{61}$$

where we have used the fact that $u = 0$ at $y = 0$ to eliminate u_0.

Both temperature and velocity fields for this case are shown in Figure 4 for various temperature-anomaly maxima θ_{max}. Localization only occurs for relatively cold boundaries (in the cases shown, $\theta_0 = 0.01$); with much hotter boundaries, there is never any significant localization since the entire viscosity field is near the minimum value given by the limit $\Theta \to \infty$ and so thermal perturbations have little effect.

Even with cold boundaries, localization only occurs for a finite range of peak "temperature" θ_{max}. If θ_{max} is too small (in particular $\ll \theta_0$) then the temperature is nearly constant across the layer and thus the viscosity field is approximately uniform and very large; this then leads to effectively Newtonian, constant-viscosity behavior with uniform shear. If θ_{max} is too large (i.e., $\gg \theta_0$) then most of the viscosity field (instead of just a narrow portion) approaches the low-viscosity limit at $\Theta \to \infty$ and thus viscosity and shear are mostly uniform. For $\theta_{max} \sim \theta_0$, the thermal anomalies are neither insignificant nor so large as to dominate the entire viscosity field, and thus one can obtain significant viscosity variations and shear localization across the layer. This result somewhat corresponds to the curve shown in Figure 1 wherein the "stick-slip" behavior for the Arrhenius mechanism occurs only over a finite parameter range (in those cases, over a finite range of strain-rates) outside of which behavior is mostly like that of simple viscous fluids.

Simple-damage feedback. For the simplified damage feedback we again use the simplified linearly temperature-dependent viscosity law $\mu = \mu_r(1 - \lambda\Theta)$, use $f < 1$ and rewrite our dependent and independent variables in terms of natural scales, i.e., $y = Ly'$, $\Theta = \Theta_0 + \Theta^*\theta$, and $u = U^*u'$ where y', θ and u' are dimensionless quantities, and Θ^* and U^* are unknown scales. In this regard, Equation (48) and Equation (49) become

$$\frac{\partial u'}{\partial y} = \left[\frac{\sigma L}{\mu_r(1 - \lambda\Theta_0)U^*}\right]\left(1 - \frac{\lambda\Theta^*}{1 - \lambda\Theta_0}\theta\right)^{-1} \tag{62}$$

$$0 = \frac{\partial^2\theta}{\partial y^2} + \frac{1}{2}\left[\frac{\mu_r(1 - \lambda\Theta_0)(U^*)^2}{k\Theta^*}\right]\left(1 - \frac{\lambda\Theta^*}{1 - \lambda\Theta_0}\theta\right)^{-1}\left(\frac{\partial u'}{\partial y}\right)^2 \tag{63}$$

Choosing $\Theta^* = (1 - \lambda\Theta_0)/\lambda$, $U^* = \sqrt{\frac{k}{\mu_r\lambda}}$, then Equation (62) and Equation (63) become (after dropping the primes)

$$\frac{\partial u}{\partial y} = \frac{\tau}{1 - \theta} \tag{64}$$

$$0 = \frac{\partial^2\theta}{\partial y^2} + \frac{\tau^2}{2}\frac{1}{1 - \theta} \tag{65}$$

where we have used Equation (64) to eliminate $\partial u/\partial y$ from Equation (65), and

$$\tau = \sigma\sqrt{\frac{\lambda L^2}{\mu_r(1 - \lambda\Theta_0)^2 k}} \tag{66}$$

is our only free parameter and, again, represents a dimensionless imposed stress. To solve Equation (65) we multiply both sides by $2\partial\theta/\partial y$ to arrive at

$$0 = \frac{\partial}{\partial y}\left(\frac{\partial\theta}{\partial y}\right)^2 = \tau^2\frac{\partial}{\partial y}\ln(1 - \theta) \tag{67}$$

which leads to

$$\int \frac{d\theta}{\sqrt{\gamma^2 + \ln(1-\theta)}} = \pm\tau(y - y_0) \tag{68}$$

where γ^2 and y_0 are integration constants; moreover, the plus sign on the right side corresponds to $y < 0$, and the minus sign to $y > 0$ since we expect θ to increase from 0 at $y = -1$ to a maximum at $y = 0$ and and then decrease to 0 again at $y = 1$. The above integral can be solved by making the substitution $\psi = \sqrt{\gamma^2 + \ln(1-\theta)}$ which eventually leads to the the implicit solution (i.e., y as a function of θ)

$$e^{-\gamma^2}\text{erfi}(\sqrt{\gamma^2 + \ln(1-\theta)}) = \pm\frac{\tau}{\sqrt{\pi}}(y_0 - y) \tag{69}$$

where erfi is the imaginary error function $\text{erfi}(x) = -i\text{erf}(ix)$ (and by definition $\text{erf}(x) = \frac{2}{\sqrt{\pi}}\int_{-\infty}^{x} e^{-x'^2}dx'$). At $y = -1$ the above equation yields $e^{-\gamma^2}\text{erfi}(\gamma) = \frac{\tau}{\sqrt{\pi}}(y_0 + 1)$ and at $y = +1$ it becomes $e^{-\gamma^2}\text{erfi}(\gamma) = \frac{\tau}{\sqrt{\pi}}(1 - y_0)$; both relations can only be true if $y_0 = 0$. This leads to a transcendental equation for γ

$$\text{erfi}(\gamma) = \frac{\tau}{\sqrt{\pi}}e^{\gamma^2} \tag{70}$$

which generally has two roots, but only for a finite range of τ; there is no solution for γ for $\tau > \tau_c$ where at τ_c the curves (i.e., functions of γ) on each side of Equation (70) just separate, i.e., intersect at one tangent point; this leads to $\text{erfi}(1/\tau_c) = \frac{\tau_c}{\sqrt{\pi}}e^{1/\tau_c^2}$, the solution for which is $\tau_c \approx 1.082$. Again, as shown in the earlier section on damage with decay-loss, this critical stress when redimensionalized will be of the order of 1kbar.

To obtain the velocity field we integrate Equation (64) to obtain

$$u = \tau e^{\gamma^2} \int \exp(-[\text{erfi}^{-1}(e^{\gamma^2}\tau y/\sqrt{\pi})]^2)dy = 2\text{erfi}^{-1}(e^{\gamma^2}\tau y/\sqrt{\pi}) \tag{71}$$

where erfi^{-1} is the inverse of the imaginary error function and we have used the condition that $u = 0$ at $y = 0$ to eliminate the integration constant. Evaluating Equation (71) at $y = \pm 1$, and using Equation (70), we find that the velocity at the boundaries is $\pm 2\gamma$; therefore γ represents the boundary or maximum velocity (within a factor or 2). By inspection of the argument to the erfi^{-1} function in Equation (71), it is clear that the width of the shear localization zone is

$$\delta_{sl} = \frac{\sqrt{\pi}}{\tau}e^{-\gamma^2} = \frac{1}{\text{erfi}\gamma} \tag{72}$$

which is an extremely rapidly decreasing function of γ; thus for large γ, δ_{sl} will be extremely small.

That there are two possible values of γ for each $\tau < \tau_c$ means there are two possible porosity and velocity fields, θ and u, for each imposed stress τ. These two solutions have small and large γ which correspond to the low-porosity/slow case (where the viscosity is large but strain-rate low) and a high-porosity/fast case (low viscosity but high strain-rate). That there are no steady-state solutions for $\tau > \tau_c$ is also evident in Figure 1 by the fact that effective constitutive law does not exist for stresses above the stress maximum.

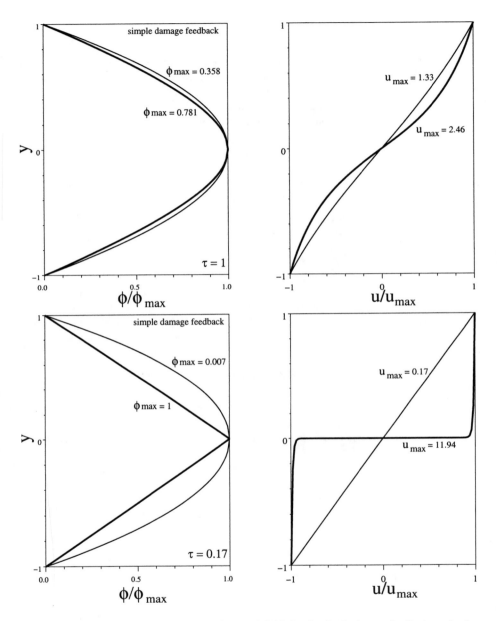

Figure 5. Void fraction or porosity (ϕ) and velocity (u) fields for the simple damage feedback mechanism with diffusive loss and for two different stresses τ (values indicated). Each stress is associated with two possible solutions, the low-porosity/slow case (thin solid line) and the high-porosity/fast (thick solidline) one. Curves are normalized by their maxima (values indicated). See text for discussion.

Figure 5 shows sample solutions for θ and u for various τ; At $\tau \approx \tau_c$ the low-porosity (slow) and high-porosity (fast) solutions coincide and involve relatively weak shear localization. As $\tau \to 0$, the slow solutions approach the constant-viscosity situation in which there is no shear-localization (γ is small and thus δ_{sl} is large), and the fast solutions involve increasingly larger boundary velocities and shear-localization at the midplane $y = 0$ that approaches a nearly fault-like quality (γ is large and thus δ_{sl} is extremely small). It is important to note again that in this case shear localization is associated with decreasing – not increasing – stress τ; i.e., the occurence of shear-localization is associated with a loss of strength whereby the layer is so weakened that even a very high strain-rate involves a small stress.

OTHER CONSIDERATIONS

Two-dimensional examples

Some of the feedback mechanisms discussed herein have been examined in various ways by many authors with regard to lithosphere and mantle dynamics and general rock mechanics (Schubert and Turcotte 1972; Schubert and Yuen 1978; Yuen and Schubert 1977, 1979; Fleitout and Froidevaux 1980; Poirier 1980; Balachandar et al. 1995; Bercovici 1996, 1998) and incorporated in various ways into mantle-convection calculations (Tackley 2000c,d,a; Auth et al. 2002). These applications have been discussed in other reviews in some detail (Bercovici et al. 2000; Tackley 2000b; Bercovici 2003). However, with immediate application to our analysis so far, Bercovici (1998) examined some simple two-dimensional flow calculations using some of the feedback mechanisms discussed above. In those calculations, a thin 2-D horizontal fluid layer (representing the lithosphere) is driven by a source and sink field (representing a mid-ocean ridge and a subduction zone, respectively). The fluid layer experiences the basic feedback mechanisms with either an exponential (e.g., $\mu \sim e^{-\lambda\Theta}$) or a linear viscosity law ($\mu \sim 1 - \lambda\Theta$). The exponential law represents a temperature-dependent viscosity (in fact, it is a semi-linearization of the Arrhenius law, i.e., the argument to the exponential is linearized), while the linear one represents a porosity-dependent law (with a damage feedback in which deformational work generates voids). The goal of the calculation was to see if the various feedback mechanisms could cause the source-sink driven flow to appear plate-like. A perfectly plate like flow would involve localized zones of strike-slip shear, or vertical vorticity, connecting the ends of the sources to the ends of the sinks (Bercovici 1993). In the calculations shown in Figure 6, the temperature-dependent viscosity law generates very little shear localization and flow that is only weakly plate-like (although, as noted previously, this neglects the likelihood of melting at very high dissipative heating rates), while the linear viscosity law generates a very narrow and intense localization. These more complicated two-dimensional results, therefore, are in keeping with the basic physics discussed with our one-dimensional analysis.

Incorporation of the void-volatile damage theory into two-dimensional convection calculations shows a regime of plate-like behavior involving discrete block-like portions of the cold upper thermal boundary layer and the development of passive spreading centers (Ogawa 2002) and dual low-angle fault-like zones above convergent and divergent zones (Auth et al. 2002). Yet, in fully three-dimensional convection calculations with a visco-plastic lithosphere both the void-volatile effect and stick-slip self-lubrication approach appears to lead to too much damage, causing plate-like regions to go unstable and disintegrate

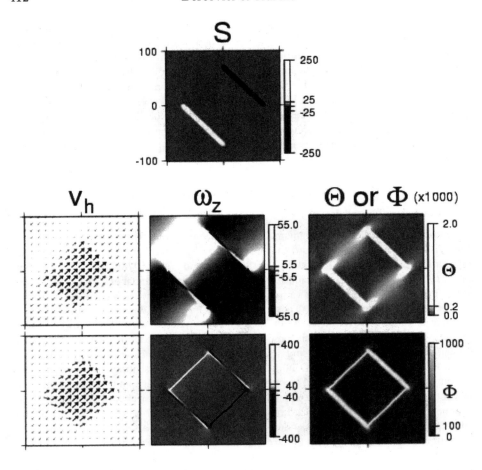

Figure 6. A simple source-sink flow model illustrating two different shear-localization mechanisms, i.e., (1) temperature-dependent viscosity and shear heating, and (2) void-volatile weakening (simple damage). The top frame, labeled S, shows the source-sink field used to drive horizontal flow in a two-dimensional shallow layer; white represents the source and black the sink. The three underlying columns are horizontal velocity v_h (in which gray shading represents velocity magnitude, black being the maximum magnitude); vertical vorticity or rate of strike-slip shear ω_z (white is positive vorticity or left-lateral shear, black is negative vorticity or right-lateral shear), and the scalar field that determines weakening, either temperature Θ or void/volatile volume fraction Φ (white is for large values, and black for small values; the values in the gray scales are multiplied by 1000, as indicated). Without any variable viscosity, the velocity field v_h would look more or less like a dipole field, and there would be no vertical vorticity ω_z. With variable viscosity, the velocity field looks more plate-like and there is signficant vorticity, or toroidal motion, generated. The first row of the 6 lower frames corresponds to the case with temperature-dependent viscosity and shear heating. Softening due to shear heating generates weak zones of strike-slip shear that connect the ends of the source and sink. Although the velocity superficially appears plate-like, in fact significant deformation of the plate-like region is occuring, as depicted in the diffuse zones of vorticity. The temperature field Θ also shows only a weak hot anomaly over the strike-slip zones, and thus a viscosity distribution that is not very plate-like either. With weakening due to creation of voids by damage (bottom row), the velocity field is more plate-like and the vorticity field is similar to what one expects for discontinuous strike-slip faults, i.e., very narrow, intense zones of deformation. The void volume fraction Φ is also very high over the strike slip zones, yielding a contiguous weak boundary surrounding a uniformly strong plate-like area. Modified after Bercovici (1998).

into smaller regions (Tackley 2000d); however, the use of both visco-plasticity and damage schemes is likely redundant since each effect separately represents lithospheric failure mechanisms.

More sophisticated damage theories

The simple void-volatile damage law used by several researchers (Bercovici 1998; Tackley 2000d; Ogawa 2002; Auth et al. 2002) and discussed in this paper, is idealized and is rather unspecific about the intrinsic physics controlling damage and shear localization. In short, while its simplicity is appealing, it is nevertheless ad hoc. A more rigorous damage approach however is difficult to define physically. A signficant body of literature exists on elasto-dynamics damage with applications from metallurgy to earthquakes (Ashby and Sammis 1990; Lemaitre 1992; Lyakhovsky et al. 1997; Krajcinovic 1996, 2000). These models introduce a new thermodynamic field variable (analagous to temperature) called the damage parameter, and given all the new thermodynamic coefficients thus required the models tend toward considerable complexity involving multiple free parameters. Viscous damage theories appropriate to geologic or convective time-scales that are compatible with convection models are rarer. Recently, a first-principles damage approach has been proposed (Bercovici et al. 2001a; Ricard et al. 2001; Bercovici et al. 2001b; Bercovici and Ricard 2003) that combines the essence of fracture mechanics with viscous continuum mechanics. This model is called a "two-phase damage" theory whereby, rather than invoke a new thermodynamic "damage" variable as in the elastodynamics theories, it treats damage through interface and surface thermodynamics. The two-phase damage theory assumes that the existence of a microcrack in a medium entails at least two phases or constituents, i.e., the host phase (the rock) and the void-filling phase (perhaps water). The energy necessary to create the microcrack is–as proposed in Griffiths crack theory (Griffith 1921)–the surface energy on the crack surface, or in the two-phase theory the energy on the interface between phases. Deformational work creates (or is stored as) this interfacial energy by generating more voids, thus inducing weak zones on which deformation concentrates, leading to more damage, void creation, localization and so on.

Preliminary simple calculations show that this mechanism can lead to a spectrum of shear localizing behavior, from more diffuse to very sharp or intense localization (Fig. 7), in addition to broadly distributed damage when the entire system essentially shatters (Bercovici et al. 2001b; Bercovici and Ricard 2003); the theory has also recently been used to predict shear-enhanced compaction as well as shear-localization (Ricard and Bercovici 2002). However, as simple as the two-phase theory purports to be, it is still rather complicated (as are all two-phase problems) and is thus a work in progress.

SUMMARY AND CONCLUSIONS

That a large amount of deformation on Earth occurs in narrow bands is evidence of the importance of shear-localization in the mantle-lithosphere system. In this review we have discussed some physical and theoretical aspects of various feedback mechanisms that lead to shear-localization. In particular, we examined in some mathematical detail the localization that occurs in a simple shear layer with three different rheological feedback mechanisms. These feedback mechanisms are perhaps the most classical ones considered in lithospheric deformation and include (1) a thermal feedback wherein viscous heating causes softening of the material through a temperature-dependent viscosity; (2) a simple

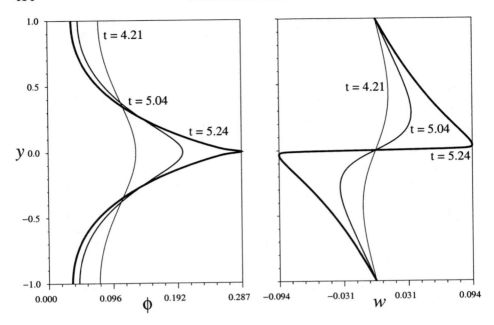

Figure 7. Simple one-dimensional shear-flow calculation demonstrating the two-phase damage theory. Damage occurs as a result of deformational work creating surface energy on void walls (here treated as the interface between two phases, e.g., rock and water); shear localization occurs because the voids cause weak zones on which deformation and further damage concentrates, leading to more voids, etc. The result here is for an infinitely long layer subjected to simple shear by an imposed shear stress (say, by a top boundary at $y = 1$ moving right and the bottom boundary at $y = -1$ moving left). The left panel shows the distribution of void volume density or porosity ϕ across the layer (in the y direction) which indicates the location of the generated weak zone at $y = 0$. The right panel shows the velocity w in the y direction of the matrix (rock) phase; upward motion ($w > 0$) above and downward motion ($w < 0$) below the centerline $y = 0$ indicates the matrix is dilating. The different thickness curves show different times (as indicated) in the calculation. The porosity ϕ evolves to a sharp cusp-like distribution indicating development of a nearly singular (fault-like) weak zone, and the cross-layer velocity w becomes nearly discontinuous at $y = 0$ indicating fault-like dilation. Modified after Bercovici and Ricard (2003).

damage feedback wherein deformational work creates voids that in turn weaken the material; and (3) a grainsize feedback wherein the material grainsize is reduced by dynamic recrystallization (related to propagation of dislocations) which then causes the viscosity to decrease. Each feedback mechanism is associated with a different controlling macroscopic or microscopic variable, which we refer to in general as the state variable; these are, for each feedback, temperature, void fraction (or porosity), and grainsize. We also considered two essential mathematical representations of loss or healing mechanisms, i.e., for each feedback mechanism, cooling, healing of voids, or grain growth and coalescence, respectively. The mathematical representation of loss accounts for either simple decay-type loss (i.e., loss depends only on the local value of the state variable) or diffusive loss (which depends on gradients in the state variable).

In general, the governing equations describe the constitutive relation between stress and strain rate (their ratio in essence being the viscosity), the dependence of viscosity on state variables (as well as strain-rate, in some cases); the force balance for the material at all points in the shear layer; and the evolution equation for the state variables. These equations in the end provide solutions for how the along-layer velocity (parallel to the boundaries of the shear layer) and the state variables are functions of both time t and the across-layer dimension y; e.g., strong variations in y of velocity and strain-rate as well as strong concentrations of temperature, void-fraction or grainsize, are indications of shear-localization. For our simple shear-layer, stress is constant in y; thus shear-localization, i.e., heterogeneity in strain-rate, only occurs if viscosity is nonuniform in y, which in the end requires nonuniform state variables.

For a decay-type loss mechanism, however, the governing equations make no explicit specifications about change in y; they are zero-dimensional ordinary differential equations with derivatives only in time t. In this case, the spatial structure of shear and the state variables (i.e., y dependence) will be governed by their initial conditions and how their values at each point in the layer evolve independently of any neighboring points (i.e., if there were y derivatives, the evolution of a variable would depend on neighboring values). Localization in this case thus requires the coexistence of at least two independently evolving states: a slow deformation state (involving either cold temperatures, low void-densities or large grainsize and thus high viscosity and low strain rates) and a fast deformation state (low viscosity and high strain rates). A deforming layer that permits both states will, even if the stress across the layer is constant, have hetergenous deformation with the fast deforming zone manifesting itself as the weak shear-localization. If only one deformation state is allowed for a given stress, then deformation is necessarily uniform and there is no localization. Even so, with the decay-type loss mechanism, the width and structure of the localization is not determined by the governing equations but by the initial conditions (which determine which state certain points in the layer evolve toward) and thus these localizations are not truly self-focussing.

In the cases using a decay-type loss, all the different feedback mechanisms generate at least two coexisting states, and thus permit some sort of shear-localization. Since a single stress can be associated with two or more strain-rates, the effective stress–strain-rate constitutive curve is (by definition) non-monotonic; this means that stress does not merely increase with strain-rate as with simple rheological laws but will, over certain regions of strain-rate, actually decrease with increased rate of deformation. The existence of stress-loss regions is associated with stick-slip-type behavior, wherein the "slip" refers to stress loss, and the "stick" to the high stress or resistance region that typically exists at smaller strain-rates.

With the thermal feedback mechanism, three coexisting states are possible if the background temperature (on top of which viscous heating contributes temperature anomalies) is small enough, on the order of (or smaller than) $H/(4R)$ (which corresponds to the condition $\theta_0 < 1/4$) where H is the activation enthalpy and R the gas constant; this is in fact sufficient for any lithospheric temperatures of the order 1000 K. Moreover, the three states only exist if the imposed stress is less than a critical value; at lithospheric temperatures, however, this critical stress is so large that essentially any realistic lithospheric stress will suffice to generate three solutions. These three solutions represent a cold viscous state (associated with "stick" behavior); an intermediate-temperature (cool) state (associated with "slip" behavior); and lastly a hot low-viscosity state involving strength recovery (i.e., stress once again increases with increasing strain-rate, albeit more gradually given the low viscosity). Both the cold/high-viscosity and hot/low-viscosity states are stable, while the intermediate-temperature "slip" state is unstable. Thus any initial state wherin there are temperatures both above and below the intermediate-temperature state will evolve toward both the cold and hot states and thus lead to inhomogeneous viscosities and strain-rates, and hence shear-localization. Since the system will evolve to one or more steady states, there is no runaway instability leading to intense or nearly singular plate-boundary-like localization. However, the hot/fast state typically has temperatures well in excess of the melting temperatures of silicates and thus the subsolidus creeping flow model we have used to analyze the thermal feedback mechanism is not valid at these temperatures; with melting, however, shear-localization is due to be more pronounced.

The idealized damage feedback mechanism is mathematically perhaps the simplest of all the feedbacks we considered here. In the case where loss is of decay-type, there are two possible steady deformation states if stress is less than a maximum that is given by a single dimensionless stress $\tau_{max} = 1/2$ (which we estimate to be dimensionally of the order 100 MPa). One of these states correlates to a low-porosity/slow-deformation (stick) state, and the other to a high-porosity/fast-deformation (slip) state. The low-porosity state is stable, while the high-porosity one is unstable. An initial configuration in which there are porosities both above and below the high-porosity state will lead to the porosities below decaying to the low-porosity state, and porosities above undergoing a runaway toward the maximum allowable porosity of 1. This suggests that with this simple feedback mechanism a localization will involve a high porosity band that grows toward a completely voided opening or macrocrack. In the case where the applied stress is greater than τ_{max} then the entire system will undergo a runaway toward a porosity of 1; i.e., the entire layer becomes a highly damaged zone.

Lastly, in the grainsize feedback mechanism there are also two possible deformation states as long as the grainssize dependence of viscosity is nonlinear (i.e., the parameter $m > 1$). However, in contrast to the other two feedback mechanisms, these multiple states only exist if the imposed stress is larger than a minimum value, which itself depends on several other parameters of the mathematical model but is dimensionally of the order 100 MPa or less. These two states involve small-grainsize/fast-deformation and a large-grainsize/slow-deformation states, respectively. Unlike the other feedback mechanisms, the fast-deformation state is stable and the slow one unstable. Thus any configuration that starts off in the vicinity of the slow-deformation state will develop stable localized regions at the fast-deformation state surrounded by slow-deformation regions that become increasingly slower and larger in grainsize. If stress is less than the critical value to obtain multiple states, then the entire system is undergoes grain-growth and progressive stiffening.

The three feedback mechanisms display similarities in that they allow coexisting multiple deformation states. However they differ significantly in other aspects. In particular, for the thermal and damage feedback cases, multiple states and thus shear-localization occurs for stresses beneath a critical value. Above that critical stress, the thermal feedback only allows a cold-viscous state, while the damage feedback allows no steady state and undergoes a runaway to a totally damaged zone. Below the critical stress, the coexisting states become more disparate as stress decreases, thus leading to more pronounced heterogeneity in strain-rate, and thus greater intensity of localization. For the grainsize effect, multiple states are allowed for stress above a critical value; for smaller stresses no steady state is allowed and the system evolves to an ever stiffer and larger-grainsized configuration, while for higher stresses localization becomes more intense. Therefore, both the thermal and damage feedbacks are in effect low stresses phenomena, while the grainsize feedback is a high-stress phenomenon.

In contrast to the decay-type loss mechanism, a diffusive-type healing mechanism leads to governing equations that depend explicitly on y; the solutions to the equations thus prescribe structure in y and thus permit self-focussing and localization without arbitrary dependence on initial conditions. As with the decay-type loss, multiple solutions and thus deformation states are allowed for a given imposed stress; however, these states do not co-exist in the deforming layer since each state is prescribed by the governing equations to exist over the entire width of the layer. For the thermal feedback mechanism, there are three possible states wherein the coldest and hottest have more or less uniform viscosity and very weak localization, and an intermediate state that involves moderate localization. The existence of these states also depends on the background or boundary temperature; if this temperature is too high then there is no intermediate localizing state, and thus in this case it is clear that localization is a lower-temperature phenomenon (although the background temperature that permits localization is typical of lithospheric temperature of the order 1000 K). For the damage feedback mechanism, there are two possible states: a low-porosity/slow-deformation state with little or no shear localization, and one with a high-porosity and rapidly deforming zone and thus very intense shear localization. Again with both cases, localization only occurs beneath a maximum stress, and the smaller the imposed stress the more intense the localization for the fast solutions, and the more uniform the strain-rate for the slow solutions. Lastly, it appears that the damage feedback mechanism permits much more intense localization than the thermal feedback, which also corresponds to the two-dimensional calculations discussed briefly above.

The diffusive loss for the grainsize mechanism was not studied since it is unlikely to be relevant for that mechanism. Moreover, the stability of cases with diffusive loss was not studied here because it is not analytically tractable, and it would require many numerical solutions to properly examine. However, the diffusive loss mechanism is likely to stabilize states whose analogs in the decay-loss mechanism cases were unstable; this is because the sharpening of a thermal anomaly during localization enhances diffusive loss while having no effect in the decay-loss scenario. However, diffusion does not completely stabilize these cases since it is known to still permit thermal instabilities and runaways to large-strain-rate solutions (Yuen and Schubert 1977; Schubert and Yuen 1978; Fleitout and Froidevaux 1980). Nevertheless, it should be noted that instabilities and runaway effects can be supressed if we impose a constant velocity across the shear layer, instead of a constant stress. This is because shear-localization instability and runaway involve changes in boundary velocities as well as strain-rates (e.g., Eqn. 71 to Eqn. 72 suggest that the boundary velocity and the sharpness of the shear-localization zone both increase as localization intensifies); however, if the boundary velocity is fixed then such instability is simply prohibited.

Finally, we should note that since the thermal feedback mechanism does not permit a runaway instability (at least not without melting) it is perhaps the weakest of the localization mechanisms (this is also evident in the structure of the steady-state solutions with diffusive loss, at least relative to the damage feedback mechanism; i.e. compare the localized solutions for Figs. 4 and 5). In contrast, the damage and grainsize feedback mechanisms do allow runaway effects and/or intense localization and thus are perhaps the more likely candidates to generate plate-boundary-like localizations in the lithosphere. The physics of these processes, however, are still not completely well understood and considerable research remains to elucidate their nature. However, it is also important to note the possible relation between the grainsize and damage mechanisms: both involve creation of surface energy, through the creation of more grain boundaries by propagating dislocations in one, and the creation of fracture surfaces through microcracking in the other, which are more or less manifestations of the same physics. In the end, shear-localization in a geological and geophysical context is a difficult but rich topic with considerable room for further progress and discovery from theoretical, experimental and observational aspects.

REFERENCES

Ashby M, Sammis C (1990) The damage mechanics of brittle solids in compression. Pure Appl Geophys 133:489–521
Atkinson B (1987) Introduction to fracture mechanics and its geophysical applications. *In* Atkinson B (ed) Fracture Mechanics of Rock. Academic, San Diego, CA, 1–26
Atkinson B, Meredith P (1987) Experimental fracture mechanics data for rocks and minerals. *In* Atkinson B (ed) Fracture Mechanics of Rock. Academic, San Diego, CA, 427–525
Auth C, Bercovici D, Christensen U (2002) Two-dimensional convection with a self-lubricating, simple-damage rheology. Geophys J Int (in press)
Balachandar S, Yuen D, Reuteler D (1995) Localization of toroidal motion and shear heating in 3-D high Rayleigh number convection with temperature-dependent viscosity. Geophys Res Lett 22:477–480
Beaumont C (1976) The evolution of sedimentary basins on a viscoelastic lithosphere. Geophys J R Astron Soc 55:471–497
Bercovici D (1993) A simple model of plate generation from mantle flow. Geophys J Int 114:635–650
Bercovici D (1995a) On the purpose of toroidal flow in a convecting mantle. Geophys Res Lett 22:3107–3110
Bercovici D (1995b) A source-sink model of the generation of plate tectonics from non-Newtonian mantle flow. J Geophys Res 100:2013–2030
Bercovici D (1996) Plate generation in a simple model of lithosphere-mantle flow with dynamic self-lubrication. Earth Planet Sci Lett 144:41–51
Bercovici D (1998) Generation of plate tectonics from lithosphere-mantle flow and void-volatile self-lubrication. Earth Planet Sci Lett 154:139–151
Bercovici D (2003) The Generation of plate tectonics from mantle convection. Earth Planet Sci Lett 205:107–121
Bercovici D, Ricard Y (2003) Energetics of a two-phase model of lithospheric damage, shear localization and plate-boundary formation. Geophys J Intl 152:1–16
Bercovici D, Ricard Y, Richards M (2000) The relation between mantle dynamics and plate tectonics: A primer. *In* Richards MA, Gordon R, van der Hilst R (eds) History and Dynamics of Global Plate Motions, Geophys Monogr Ser 121:5–46
Bercovici D, Ricard Y, Schubert G (2001a) A two-phase model of compaction and damage, 1. General theory. J Geophys Res 106:8887–8906
Bercovici D, Ricard Y, Schubert G (2001b) A two-phase model of compaction and damage, 3. Applications to shear localization and plate boundary formation. J Geophys Res 106:8925–8940
Braun J, Chery J, Poliakov A, Mainprice D, Vauchez A, Tomassi A, Daignieres M (1999) A simple parameterization of strain localization in the ductile regime due to grain size reduction: A case study for olivine. J Geophys Res 104:25,167–25,181
Bresser JD, Peach C, Reijs J, Spiers C (1998) On dynamic recrystallization during solid state flow: Effects of stress and temperature. Geophys Res Lett 25:3457–3460
Bresser JD, ter Heege J, Spiers C (2001) Grain size reduction by dynamic recrystallization: can it result in major rheological weakening? Intl J Earth Sci 90:28–45
Cooper R, Kohlstedt D (1982) Interfacial energies in the olivine-basalt system. *In* Akimoto S, Manghnani M (eds) High Pressure Research in Geophysics, Adv Earth Planet Sci 12:217–228
Cottrell A (1953) Dislocations and Plastic Flow in Crystals. Clarendon Press, Oxford, UK
Derby B, Ashby M (1987) On dynamic recrystallization. Scripta Metall 21:879–884
Evans B, Kohlstedt D (1995) Rheology of rocks. *In* Ahrens TJ (ed) Rock Physics and Phase Relations: A Handbook of Physical Constants. AGU Ref Shelf, vol 3. Am Geophys Union, Washington, DC, 148–165
Farren W, Taylor G (1925) The heat developed during plastic extension of metals. Proc R Soc London Ser A 107:422–451
Fleitout L, Froidevaux C (1980) Thermal and mechanical evolution of shear zones. J Struct Geol 2:159–164
Frederiksen S, Braun J (2001) Numerical modelling of strain localisation during extension of the continental lithosphere. Earth Planet Sci Lett 188:241–251
Frost H, Ashby M (1982) Deformation Mechanism Maps. Pergamon Press, Oxford, UK

Furusho M, Kanagawa K (1999) Reaction induced strain localization in a lherzolite mylonite from the Hidaka metamorphic belt of central Hokkaido, Japan. Tectonophysics 313:411–432

Géminard JC, Losert W, Gollub J (1999) Frictional mechanics of wet granular material. Phys Rev E 59:5881–5890

Green H, Burnley P (1989) A new self-organizing mechanism for deep focus earthquakes. Nature 341:733–737

Griffith A (1921) The phenomenon of rupture and flow in solids. Philos Trans R Soc London, Ser A 221:163–198

Guéguen Y, Palciauskas V (1994) Introduction to the Physics of Rocks. Princeton University Press, Princeton, New Jersey

Gurnis M, Zhong S, Toth J (2000) On the competing roles of fault reactivation and brittle failure in generating plate tectonics from mantle convection. In Richards M, Gordon R, van der Hilst R (eds) History and Dynamics of Global Plate Motions, Geophys Monogr Ser 121:73–94

Handy M (1994) Flow laws for rocks containing two nonlinear viscous phases: a phenomenological approach. J Struct Geol 16:287–301

Hansen N, Schreyer H (1992) Thermodynamically consistent theories for elastoplasticity coupled with damage. In Ju J, Valanis K (eds) Damage Mechanics and Localization. Am Soc Mech Eng, New York, 53–67

Hirth G, Kohlstedt D (1996) Water in the oceanic upper mantle: implications for rheology, melt extraction and the evolution of the lithosphere. Earth Planet Sci Lett 144:93–108

Jaeger J, Cook N (1979) Fundamentals of Rock Mechanics. 3rd edn. Chapman and Hall, New York

Jin D, Karato S, Obata M (1998) Mechanisms of shear localization in the continental lithosphere: Inference from the deformation microstructures of peridotites from the Ivrea zone, northwestern Italy. J Struct Geol 20:195–209

Kameyama M, Yuen D, Fujimoto H (1997) The interaction of viscous heating with grain-size dependent rheology in the formation of localized slip zones. Geophys Res Lett 24:2523–2526

Karato S (1996) Plastic flow in rocks. In Matsui T (ed) Continuum Physics, Earth and Planetary Sciences Series, vol 6. Iwani Shoten, Tokyo, 239–291

Karato S (2002) Mapping water content in the upper mantle. In Eiler J, Abers J (eds) Subduction Factory, Monograph, Am Geophys Union, Washington, DC, (in press)

Karato S, Riedel M, Yuen D (2001) Rheological structure and deformation of subducted slabs in the mantle transition zone: implications for mantle circulation and deep earthquakes. Phys Earth Planet Int 127:83–108

Karato S, Toriumi M, Fujii T (1980) Dynamic recrystallization of olivine single crystals during high temperature creep. Geophys Res Lett 7:649–652

Kirby S (1987) Localized polymorphic phase transformations in high-pressure faults and applications to the physical mechanisms of deep earthquakes. J Geophys Res 92:13,789–13,800

Kohlstedt D, Evans B, Mackwell S (1995) Strength of the lithosphere: Constraints imposed by laboratory experiments. J Geophys Res 100:17,587–17,602

Krajcinovic D (1996) Damage Mechanics. North-Holland, Amsterdam

Krajcinovic D (2000) Damage mechanics: accomplishments, trends and needs. Intl J Solids Struct 37:267–277

Lee K, Jiang Z, Karato S (2002) A scanning electron microscope study of effects of dynamic recrystallization on the lattice preferred orientation in olivine. Tectonophysics 351:331–341

Lemaitre J (1992) A Course on Damage Mechanics. Springer-Verlag, New York

Lemonds J, Needleman A (1986) Finite element analyses of shear localization in rate and temperature dependent solids. Mech Mater 5:339–361

Lockner D (1995) Rock failure. In Ahrens TJ (ed) Rock Physics and Phase Relations: A Handbook of Physical Constants, AGU Ref Shelf, vol. 3. Am Geophys Union, Washington, DC, 127–147

Lyakhovsky V, Ben-Zion Y, Agnon A (1997) Distributed damage, faulting, and friction. J Geophys Res 102:27,635–27,649

Marone C (1998) Laboratory-derived friction laws and their application to seismic faulting. Ann Rev Earth Planet Sci 26:643–696

Mathur K, Needleman A, Tvergaard V (1996) Three dimensional analysis of dynamic ductile crack growth in a thin plate. J Mech Phys Solids 44:439–464

Montési L, Zuber M (2002) A unified description of localization for application to largescale tectonics. J Geophys Res 107:10.1029/2001JB000465

Mora P, Place D (1998) Numerical simulation of earthquake faults with gouge: Toward a comprehensive explanation for the heat flow paradox. J Geophys Res 103:21,067–21,089

Ogawa M (2002) The plate-like regime of a numerically modeled thermal convection in a fluid with temperature-, pressure-, and stress-history-dependent viscosity. J Geophys Res (in press)

Paterson M (1978) Experimental Rock Deformation. Springer-Verlag, Berlin

Poirier J (1980) Shear localization and shear instability in materials in the ductile field. J Struct Geol 2:135–142

Poirier J (1985) Creep of Crystals. Cambridge University Press, Cambridge, UK

Ranalli G (1995) Rheology of the Earth. Chapman and Hall Publishers, London
Regenauer-Lieb K (1999) Dilatant plasticity applied to Alpine collision: Ductile void-growth in the intraplate area beneath the Eifel volcanic field. J Geodyn 27:1–21
Regenauer-Lieb K, Yuen D (2002) Modeling shear zones in geological and planetary sciences: solid- and fluid- thermal-mechanical approaches. Earth Sci Rev (in press)
Ricard Y, Bercovici D (2002) The void-matrix variation of a two-phase damage theory; applications to shear localization and shear-enhanced compaction. Geophys J Intl (submitted)
Ricard Y, Bercovici D, Schubert G (2001) A two-phase model of compaction and damage, 2, Applications to compaction, deformation, and the role of interfacial surface tension. J Geophys, Res 106:8907–8924
Riedel M, Karato S (1997) Grain-size evolution in subducted oceanic lithosphere associated with the olivine-spinel transformation and its effects on rheology. Earth Planet Sci Lett 148:27–43
Rubie D (1983) Reaction-enhanced ductility: The role of solid-solid univariant reactions in deformation of the crust and mantle. Tectonophysics 96:331–352
Rubie D (1984) The olivine-spinel transformation and the rheology of subducting lithosphere. Nature 308:505–508
Schubert G, Turcotte D (1972) One-dimensional model of shallow mantle convection. J Geophys Res 77:945–951
Schubert G, Yuen D (1978) Shear heating instability in Earth's upper mantle. Tectonophysics 50:197–205
Scott D (1996) Seismicity and stress rotation in a granular model of the brittle crust. Nature 381:592–595
Segall P, Rice JR (1995) Dilatancy, compaction, and slip instability of a fluid-infiltrated fault. J Geophys Res 100:22,155–22,171
Shimizu I (1998) Stress and temperature dependence of recrystallized grain size: A subgrain misorientation model. Geophys Res Lett 25:4237–4240
Sleep N (1995) Ductile creep, compaction, and rate and state dependent friction within major faults. J Geophys Res 100:13,065–13,080
Sleep N (1997) Application of a unified rate and state friction theory to the mechanics of fault zones with strain localization. J Geophys Res 102:2875–2895
Stünitz H, Tullis J (2001) Weakening and strain localization produced by syndeformational reaction of plagioclase. Intl J Earth Sci 90:136–148
Tackley P (1998) Self-consistent generation of tectonic plates in three-dimensional mantle convection. Earth Planet Sci Lett 157:9–22
Tackley P (2000a) Mantle convection and plate tectonics: Toward and integrated physical and chemical theory. Science 288:2002–2007
Tackley P (2000b) The quest for self-consistent generation of plate tectonics in mantle convection models. In Richards MA, Gordon R, van der Hilst R (eds) History and Dynamics of Global Plate Motions, Geophys Monogr Ser 121:47–72
Tackley P (2000c) Self-consistent generation of tectonic plates in time-dependent, three-dimensional mantle convection simulations, 1. Pseudoplastic yielding. Geochem Geophys Geosystems (G^3) 1:2000GC000,036
Tackley P (2000d) Self-consistent generation of tectonic plates in time-dependent, three-dimensional mantle convection simulations, 2. Strain weakening and asthenosphere. Geochem Geophys Geosystems (G^3) 1:2000GC000,043
Taylor G, Quinney H (1934) The latent energy remaining in metal after cold working. Proc R Soc London, Ser A 143:307–326
Thatcher W, England P (1998) Ductile shear zones beneath strike-slip faults: Implications for the thermomechanics of the San Andreas fault zone. J Geophys Res 103:891–905
Tozer D (1985) Heat transfer and planetary evolution. Geophys Surv 7:213–246
Trompert R, Hansen U (1998) Mantle convection simulations with rheologies that generate plate-like behavior. Nature 395:686–689
Turcotte D, Schubert G (1982) Geodynamics. John Wiley & Sons, New York
Urai J, Means W, Lister G (1986) Dynamic recrystallization in minerals. In Hobbs B, Heard H (eds) Mineral and Rock Deformation: Laboratory Studies. American Geophysical Union, Washington DC, 166–199
Watts A, Karner G, Steckler M (1982) Lithosphere flexure and the evolution of sedimentary basins. Philos Trans R Soc London, Ser A 305:249–281
White S, Burrows S, Carreras J, Shaw N, Humphreys F (1980) On mylonites in ductile shear zones. J Struc. Geol. 2:175–187
Whitehead J, Gans R (1974) A new, theoretically tractable earthquake model. Geophys J R Astron Soc 39:11–28
Yuen D, Schubert G (1977) Asthenospheric shear flow: thermally stable or unstable? Geophys Res Lett 4:503–506
Yuen D, Schubert G (1979) The role of shear heating in the dynamics of large ice masses. J Glaciol 24:195–212